Interactive LISREL: User's Guide

Mathilda du Toit
Stephen du Toit

with a contribution by Douglas Hawkins

Interactive LISREL: User's guide.

Cover based on an architectural detail from Frank Lloyd Wright's Robie House.

1 2 3 4 5 6 7 8 9 0 04 03 02 01

Published by:

Scientific Software International. Inc.
7383 North Lincoln Avenue, Suite 100
Lincolnwood, IL 60712—1704
Tel: + 1.847.675.0720
Fax: +1.847.675.2140
URL: http://www.ssicentral.com

ISBN: 0-89498-044-0

Preface

The LISREL methodology development started in 1970, when Karl Jöreskog presented a first LISREL model at a conference. The first generally available LISREL program (version 3) was published in 1975. The name LISREL is an acronym for "linear structural relations." The qualifier "linear" is too restrictive for the current version of the LISREL program, but the name LISREL has become synonymous with "structural equation modeling" or SEM.

In 1986, the first version of PRELIS was published. Its main function is the computation of the appropriate summary statistics for analysis with the LISREL program. Over time, its functionality has expanded to include a variety of exploratory tools to give users a better understanding of their data before they attempt a LISREL model. Its name stands for "preprocessor for LISREL."

SIMPLIS was publicly introduced in 1993 as an alternative syntax for the LISREL model specification. As the name implied, it meant a tremendous simplification over the original LISREL syntax, which required some understanding of matrix algebra as well as the memorization of several Greek characters. Another added capability was the drawing of path diagrams, with some first attempts to allow the user to interact with the program. Up to that point, the user communicated with the program in so-called batch mode. Some kind of script, also called "syntax file" or "command file," had to be written with a text editor, containing instructions for the program. Instructions about where the data file was to be found, how many variables it contained, the names of the variables, etc. This command file was then submitted to the actual program for execution.

In 1998, the first truly interactive version of LISREL (version 8.20, for MS Windows) introduced a dialog-box interface to facilitate the writing of those command files. As much as possible, typing of syntax was replaced with point-and-click actions in a series of dialog boxes. From these actions, the program then produced the equivalent command file, in PRELIS syntax for processing of raw data, or in SIMPLIS or LISREL syntax for the actual SEM analysis.

The path diagram feature became much more flexible, giving maximum control to the user, not only to specify or change a model, but also to produce a graphic representation of the model with publication quality that could be exported as a separate file for subsequent import in a document. Parallel with these developments regarding the way that the user interacts with the program, numerous statistical features were added, including a multilevel analysis module, exploratory factor analysis and principal component analysis, imputation for missing values, bootstrap and Monte Carlo procedures, etc.

The LISREL program makes all these additions available to the user through a choice of menus and submenus, toolbars, and command buttons. As a result, the LISREL user interface has grown to the point that a separate manual became necessary. This manual is specifically written to guide the user through the *interactive* part of the LISREL program. It does so mostly by example.

There are four other user's guides available with the LISREL program. These user's guides are platform independent by concentrating exclusively on the "batch mode" way of operation in the illustrations: the user writes a syntax (or command) file using a text editor and submits this file to the program for execution.

The four user's guides introduce the user to the methodology of structural equation modeling, both theoretically and through a variety of applications. Specifically, the *PRELIS 2: User's Reference Guide* focuses on the intricacies of multivariate data screening and preparing the data for subsequent model fitting. The *LISREL 8: User's Reference Guide* deals with the fitting and testing of structural equation models with the use of the LISREL syntax, while *LISREL 8: Structural Equation Modeling with the SIMPLIS Command Language* does the same, but using the SIMPLIS syntax. *LISREL 8: New Statistical Features* introduces multilevel modeling and various powerful statistical additions to LISREL 8, namely, two-stage least-squares estimation, exploratory factor analysis, principal components, normal scores, and latent variable scores.

In an undertaking of this nature, various people usually play an indispensable role. First, we would like to thank Professor Doug Hawkins from the University of Minnesota for allowing us to add his Formal Inference-Based Recursive Modeling (FIRM) program to the options available to a LISREL user. Professor Hawkins kindly provided us, not only with the program, but also the documentation on which the FIRM-related sections of this guide are based. We however accept full responsibility for the final versions of these sections. We would also like to thank Gerhard Mels and Bola King for their suggestions and unflagging enthusiasm for proofreading. Their contributions led to many improvements in the final form of this guide. Finally, we would like to thank Professors Karl Jöreskog and Dag Sörbom for allowing us to base numerous examples on material in the remaining four LISREL guides.

Mathilda du Toit

Stephen du Toit

Table of Contents

1 Introduction

This User's Guide is designed for use with the Windows version of LISREL 8 for Windows 95, 98, NT, 2000 and Millenium. It focuses on the fully interactive mode of working with LISREL that is only available on the Windows platform. This Guide details how the graphical user interface results in a highly improved work environment.

All features from LISREL 7 are still there and the user can run existing syntax files without change. Important differences from the previous version are the new SIMPLIS syntax, path diagrams, and the list output.

Not only is the quality of the path diagram now of publication-ready standard, all characteristics of the display are under the user's control, namely, the position, size, and shape of the variable objects, the paths among them, and the colors, fonts, and patterns. Regardless of changes made to the characteristics of the display, the program maintains the underlying model specification and a changed model can be re-estimated at any time. New with LISREL 8 is the option to start with the path diagram and draw a model specification from scratch. Conceptual path diagrams, *i.e.,* path diagrams without parameter estimates, may also be requested. Each path diagram may be exported as a Windows metafile (WMF format) or as a Graphics Interchange file (GIF format).

A series of dialog boxes guides the user through the different stages of the problem and model specification. When done, the program builds a multilevel modeling, SIMPLIS, or LISREL input file from the user's responses. This is the mode that Windows users have come to expect.

One of the new features of LISREL 8 is that one may switch from one mode to another: from dialog box mode to SIMPLIS syntax, from SIMPLIS syntax to LISREL syntax, from LISREL syntax to path diagram mode. Users who have become familiar with the program via the SIMPLIS command language can now type in existing LISREL command files (as found in the literature, for example) and let the program translate them in the more user-friendly SIMPLIS syntax or display them as path diagrams. The dialog box mode may also be used selectively to set up the problem specification, after which model specification may be continued in the path-diagram mode. This flexibility not only allows users to develop their own specific ways of working with the program, but it also greatly increases the applicability of the program in teaching situations.

PRELIS, the part of the program used to screen and summarize data before specifying the actual model using LISREL, is now much more integrated from the user's perspective. The program can read a variety of data formats, including data from all the major statistical packages, and present the data in spreadsheet format. Graphing capabilities allow the user to view the distributions of variables separately or to inspect univariate, bivariate or multivariate plots. The characteristics of the graphics can be modified by the user and exported as Windows metafiles (WMF format).

Pull-down menus and dialog boxes as an alternative to constructing a syntax file is a much easier way of running PRELIS. However, the batch mode is still available for those who prefer to write a syntax file or who wish to use an existing syntax file.

It is this wealth of features in LISREL and the need to walk the first-time user through it step-by-step, using a variety of applications, that gave rise to this User's Guide.

The remainder of this guide is organized in four major sections. Chapter 2 provides an overview of the graphical user interface while Chapters 3 to 6 contain examples. Chapter 7 describes the new statistical features which are as yet undocumented elsewhere. Finally, Chapter 8 gives a summary of the syntax which has been added to LISREL since the release of LISREL 8.30.

2 Overview of the Interface

2.1 Introduction

The easiest way to start using LISREL 8.50 for Windows is to create a shortcut on your desktop or to click ***Start, Programs***, ***LISREL 8.50 for Windows***. LISREL opens with the basic menu bar shown below.

Other toolbars automatically appear when specific types of files are opened. These toolbars are designed to assist the user in performing specific tasks. The three most frequently used toolbars are shown below.

2.1.1 Text Editor Toolbar (text files)

2.1.2 PRELIS System File Toolbar (*.psf files)

2.1.3 Path Diagram Toolbar (*.pth files)

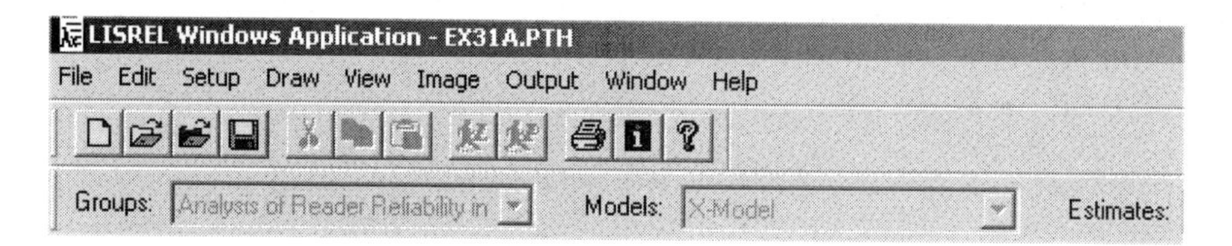

This chapter gives an overview of each element of the LISREL interface. If you are unsure what LISREL's button icons mean, position the mouse pointer over the icon to display the button's name.

We start with a discussion of the basic toolbar and menu options.

2.1.4 File Menu

The ***File*** menu is the starting point for running LISREL. When the ***File*** menu is selected, the following options are displayed.

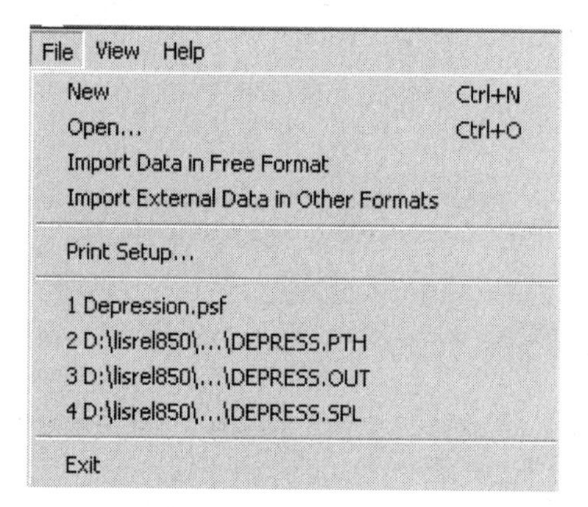

A short discussion of the ***File*** menu options follows.

New Option

The ***New*** option on the ***File*** menu is associated with the ***New*** dialog box shown below. This dialog box contains five file types that may be opened as new files.

These types are:

- Syntax Only (*.**pr2**, *.**ls8**, etc.)
- PRELIS Data (see Chapter 7) (*.**psf**)
- SIMPLIS Project (*.**spj**)
- LISREL Project, and (*.**lpj**)
- Path Diagram (*.**pth**)

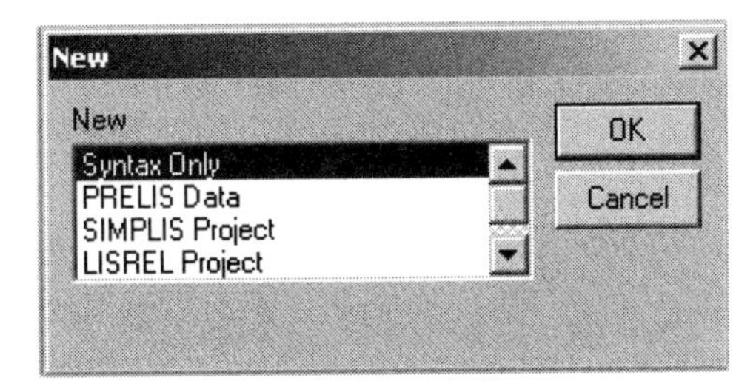

Syntax Only Option

LISREL opens with a new blank document ready for editing. The user can type in new LISREL, SIMPLIS or PRELIS syntax, copy text from a clipboard, etc.

Note:

Be sure to save the new text by assigning a filename to it.

PRELIS Data Option

When the ***PRELIS Data*** option is selected, LISREL opens a blank data spreadsheet (see Section 3.10 for a description of how to insert variable names and cases). PRELIS system files are saved as *.**psf** files.

To change variable names, make graphical displays of the data, calculate polychoric correlation coefficients, do regression analysis, etc. see the various examples in Chapter 3.

SIMPLIS Project Option

A SIMPLIS project has a *.**spj** file extension, and once opened is used to build or modify SIMPLIS syntax interactively. This is described in detail in Chapter 4, Section 4.6.

LISREL Project Option

A LISREL project has a *.**lpj** file extension. For more information on how to build or modify LISREL syntax interactively, consult Chapter 4, Section 4.7.

Open Option

The ***Open*** option on the ***File*** menu allows you to open a file that has been created previously. It is associated with the ***Open*** dialog box shown below. You may select any of the following file types:

- Syntax Only
- PRELIS Data (*.**psf**)
- SIMPLIS Project (*.**spj**)
- LISREL Project (*.**lpj**)
- Path Diagram (*.**pth**)
- All Files (*.*)

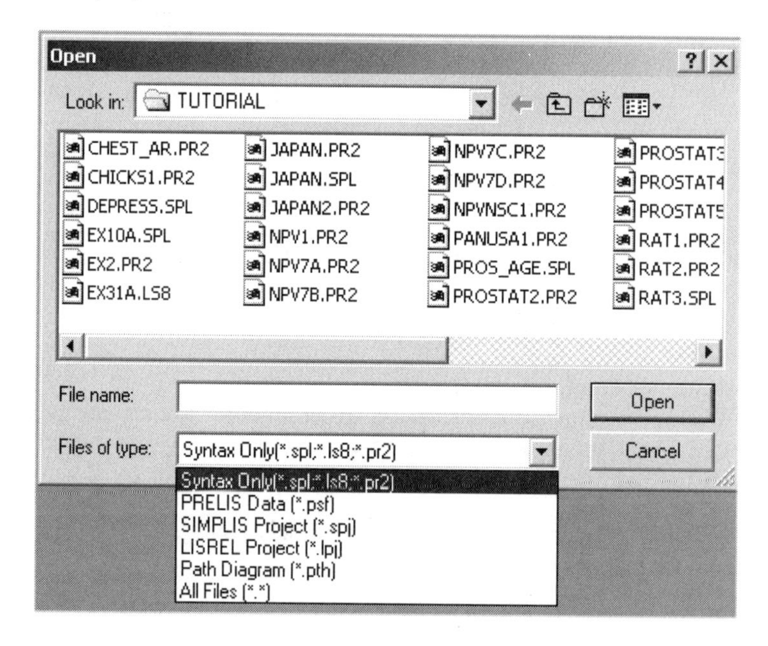

Syntax Only Option

In the examples that accompany the program, the following filename extensions are used to distinguish between PRELIS, SIMPLIS and LISREL syntax files.

- **pr2** : PRELIS syntax file
- **spl** : SIMPLIS syntax file
- **ls8** : LISREL syntax file

PRELIS Data Option

If a file of type *.**psf** is opened, it is displayed in the form of a spreadsheet. For example, if **holz.psf** is selected from the **tutorial** folder, the spreadsheet shown below is obtained.

HOLZ.PSF

	GENDER	AGEYEAR	BIRTHMON	VISPERC	CUBES	LOZENGES
1	0.00	13.00	2.00	20.00	31.00	3.00
2	1.00	13.00	8.00	32.00	21.00	17.00
3	1.00	13.00	2.00	27.00	21.00	15.00
4	0.00	13.00	3.00	32.00	31.00	24.00
5	1.00	12.00	3.00	29.00	19.00	7.00
6	1.00	14.00	2.00	32.00	20.00	18.00
7	0.00	12.00	2.00	17.00	24.00	8.00

Note:

PRELIS system files are associated with the `SY` command in PRELIS syntax, and the `RA` and `RAW Data from` LISREL and SIMPLIS commands respectively.

SIMPLIS Project Option

A *.**spj** file is obtained by creating SIMPLIS syntax interactively (see Section 4.6) or by building syntax from a path diagram (see Sections 4.3 to 4.5).

When a *.**spj** file is opened, the SIMPLIS syntax and ***Relationships*** keypad are displayed. The keypad may be used to add new syntax or to delete or modify existing syntax. For more information, see Section 4.6.

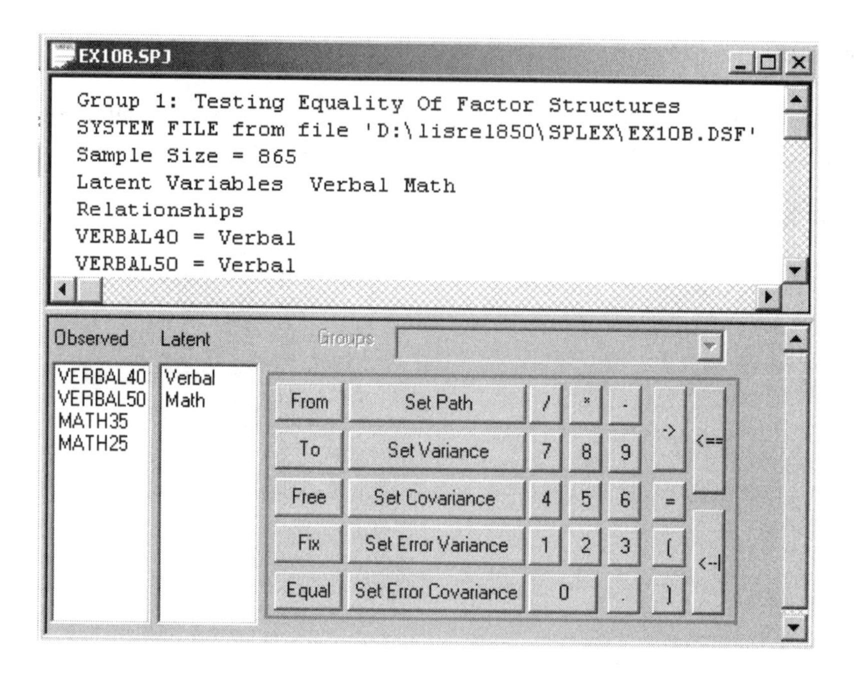

LISREL Project Option

A *.**lpj** file is obtained by creating LISREL syntax interactively (see Section 4.7) or by building syntax from a path diagram (see Sections 4.3 to 4.5 for examples).

One may use the ***Setup***, ***Model*** or ***Output*** options on the main menu bar to add, modify or delete existing text.

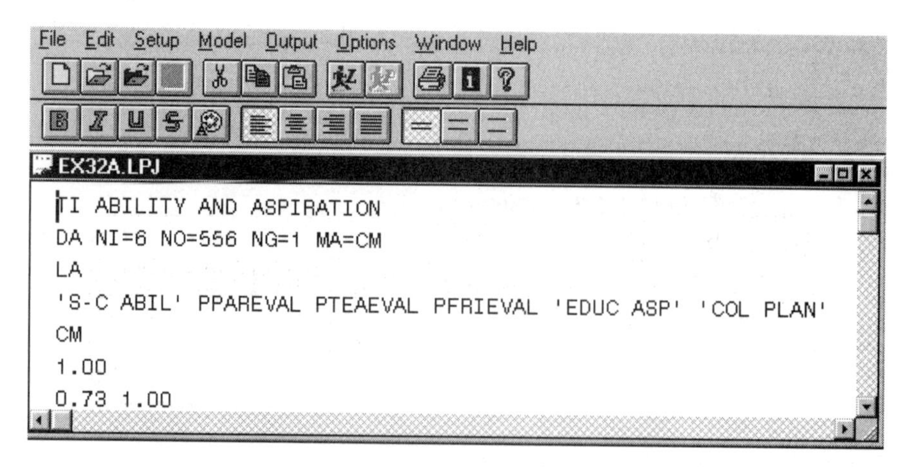

Path Diagram Option

If the ***Path Diagram*** option is selected, all the files of a given folder with file extension *.**pth** are displayed. LISREL stores path diagrams as *.**pth** files.

A *.**pth** file is displayed below.

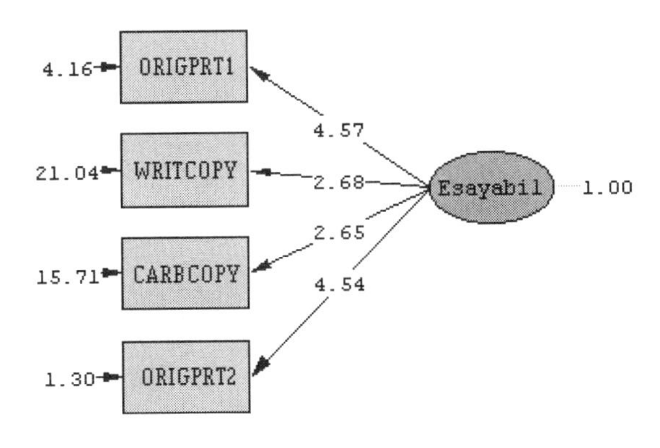

All Files Option

If the ***All Files*** option is selected, all the files of a given folder are displayed. Below is an illustration showing the selection of the file **boys.mea**. Note that LISREL, SIMPLIS and PRELIS syntax files have the default extensions of *.**ls8**, *.**spl** and *.**pr2** respectively. The user may choose any other extension when creating a syntax file.

Once ***Open*** is clicked, the file **boys.mea** is displayed. This file may be modified by using the cut, copy, paste etc. functions of the LISREL text editor.

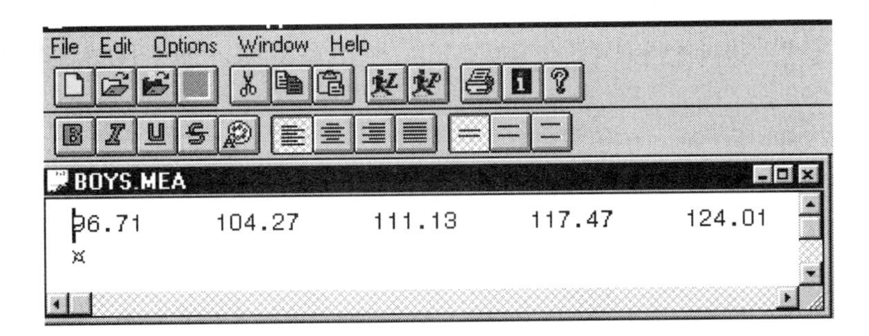

Import Data in Free Format Option

The ***Import Data in Free Format*** option on the ***File*** menu is associated with the ***Open Data File*** dialog box. A data set, stored in free format, can be converted to a PRELIS system file (*.**psf**) by selecting this option. See Section 3.11 for an example and Section 7.1 for a description of PRELIS system files.

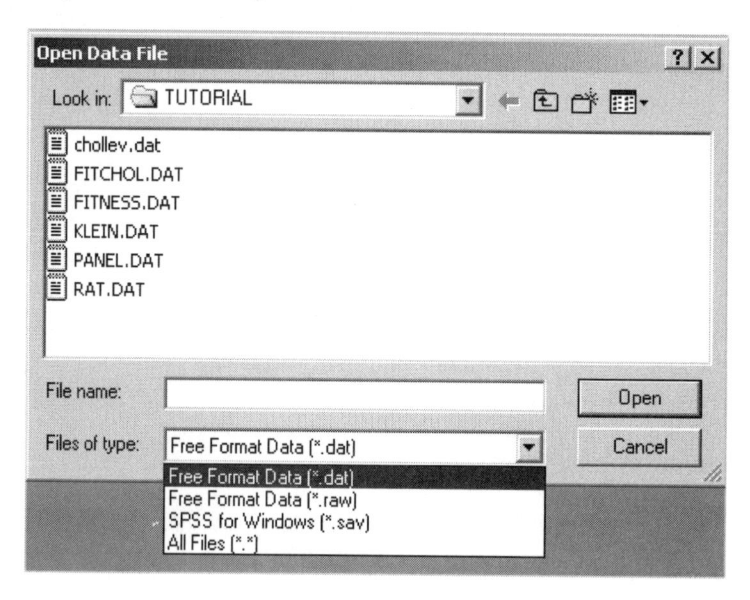

Import External Data in Other Formats Option

The ***Import External Data in Other Formats*** option on the ***File*** menu is associated with the ***Input Database*** dialog box. LISREL can import files from many software packages and convert these files to PRELIS system files.

See Sections 3.2 and 3.3 for examples and Chapter 7 for a description of the PRELIS system file.

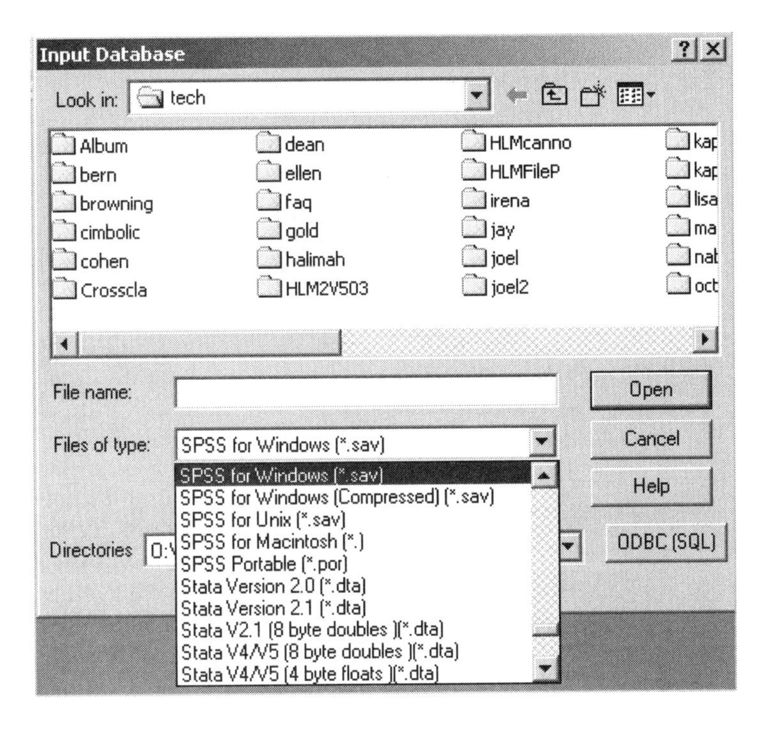

Print Setup Option

If this option is selected, the standard Windows print setup dialog box is displayed. This enables the user to select paper size, orientation and other properties of the installed printer.

Previously Opened Files Option

This portion of the ***File*** menu shows a list of the most recently opened files. To open one of the files listed, move the mouse cursor to the file name and click.

Exit Option

If this option is selected, all files opened in LISREL are closed and the user exits the application. If a file has been modified, the user will be prompted to save it before the application is closed.

2.1.5 View Menu

The ***View*** menu provides access to the ***Toolbar*** and ***Status Bar*** options discussed below.

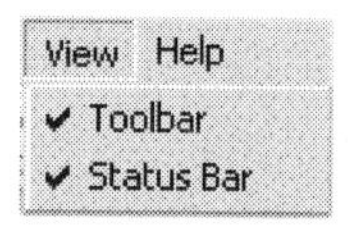

Toolbar Option

If the ***Toolbar*** option is checked, the LISREL main menu bar is displayed as shown below.

By opening a syntax file, all the ***Edit*** toolbar functions are activated as shown below:

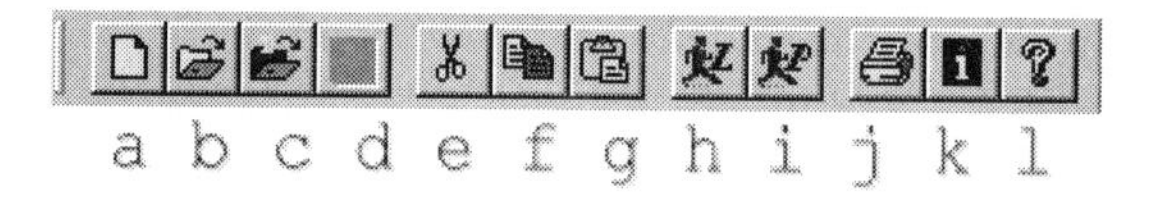

An explanation of the icon buttons designated by letters of the alphabet are as follows:

- **New File (a)**
- **Open File (b)**
- **Import Data (c)**: Import data from other software applications and convert such files to ***.psf** format.
- **Save Text (d)**: This is the standard Windows function to save changes made to a file.
- **Cut Text (e)**: A marked block of text is moved to the clipboard. Move the cursor to the position in the existing file where you want to put this text, or open a file if text is to be moved to it. On the ***Edit*** menu, click ***Paste*** or use the ***Paste*** icon button on the ***Edit*** toolbar.
- **Copy Text (f)**: A marked block of text is copied to the clipboard. Move the cursor to the position where you want to copy the text or open a file if text is to be copied to it. On the ***Edit*** menu, click ***Paste*** or use the ***Paste*** icon button on the ***Edit*** toolbar.
- **Paste Text (g)**: The clipboard contents are pasted into a file at the position of the cursor.

- **Run LISREL Syntax (h)**: Click this icon button on the LISREL main toolbar to start a LISREL analysis based on the contents of the current file. This file must be a text file that contains either SIMPLIS or LISREL syntax.
- **Run PRELIS Syntax (i)**: Click this icon button on the LISREL main menu bar to start a PRELIS analysis based on the contents of the current file. This file must be a text file that contains PRELIS syntax.
- **Print Document (j)**: If this icon button is clicked, a standard Windows dialog box is displayed which enables the user to select an alternative printer, cancel the print job, etc.
- **Help menu (k)**: Click this icon button to display the contents of the help file.

Status Bar Option

When the ***Status Bar*** option is checked, information regarding the program status appears in a bar at the bottom of the LISREL window as demonstrated below:

Change the printer and printing options

2.2 The PSF Window

Most of the statistical procedures described in the *PRELIS 2: User's Reference Guide* can be performed by clicking the appropriate buttons. When a PRELIS system file (*.**psf**) is opened, the PRELIS system file toolbar appears. The ***Edit*** menu has the usual cut, paste, select, delete, find, etc. options.

2.2.1 File Menu

An additional option on the ***File*** menu is to save the contents of a *.**psf** file as ASCII data.

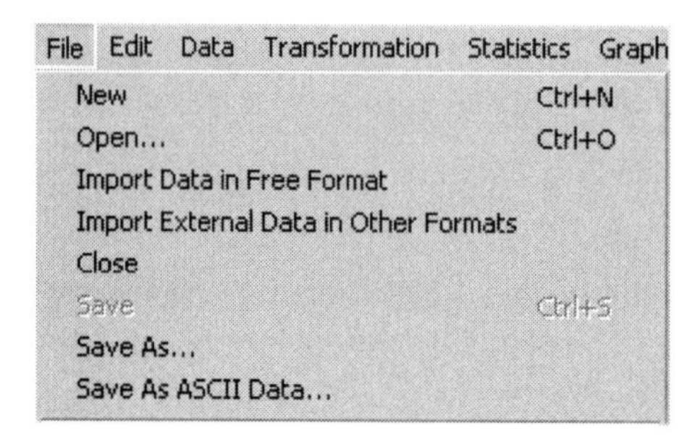

The input for this dialog box is associated with the RA command in PRELIS syntax.

2.2.2 Edit Menu

The ***Edit*** menu provides standard Windows functions and access to the ***Data Format*** dialog box through the ***Format*** option.

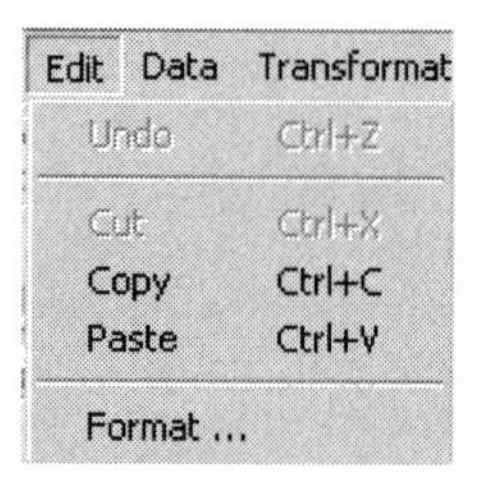

Format Option

The ***Format*** option on the ***Edit*** menu is associated with the ***Data Format*** dialog box. Use this dialog box to change the column widths and the number of decimals displayed. One can also align the data values by selecting ***Left***, ***Center*** or ***Right***.

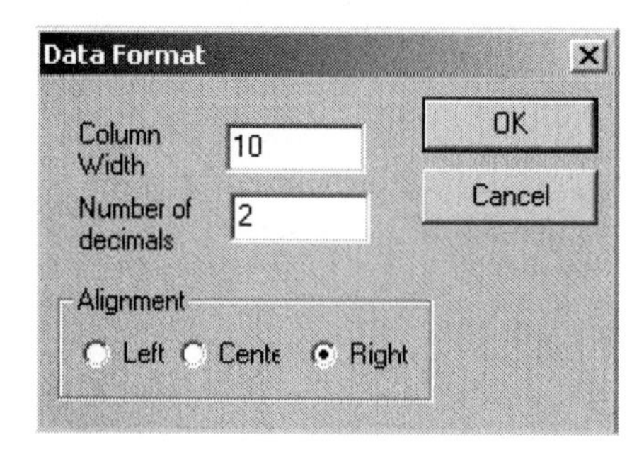

2.2.3 Data Menu

The ***Data*** menu shown below enables one to modify the contents of the PRELIS system file by, for example, changing variable types, inserting new variables, assigning category labels and defining missing value codes.

In this menu, the ***Define Variables*** option is associated with the `CO` and `OR` commands in PRELIS and the ***Select Variables/Cases*** option is associated with the `SE` and `SC` commands in PRELIS respectively.

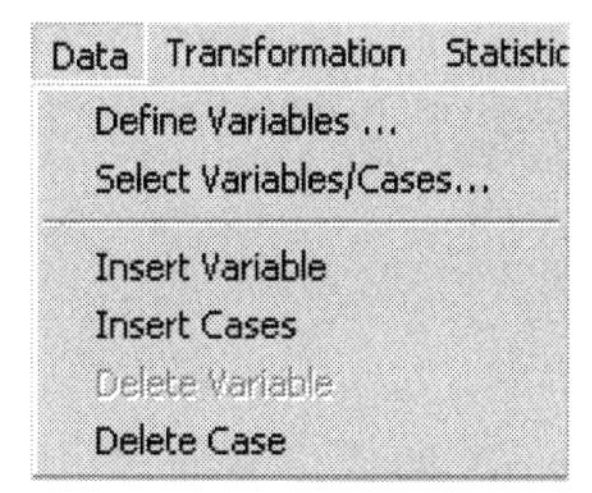

Examples of the use of most options are given in Chapter 3.

Define Variables Option

The ***Define Variables*** option on the ***Data*** menu is associated with the ***Define Variables*** dialog box.

When a variable is highlighted, *i.e.* selected, it can be renamed, its type can be changed, category labels can be assigned to ordinal variables, and missing values can be defined, as illustrated in Chapter 3.

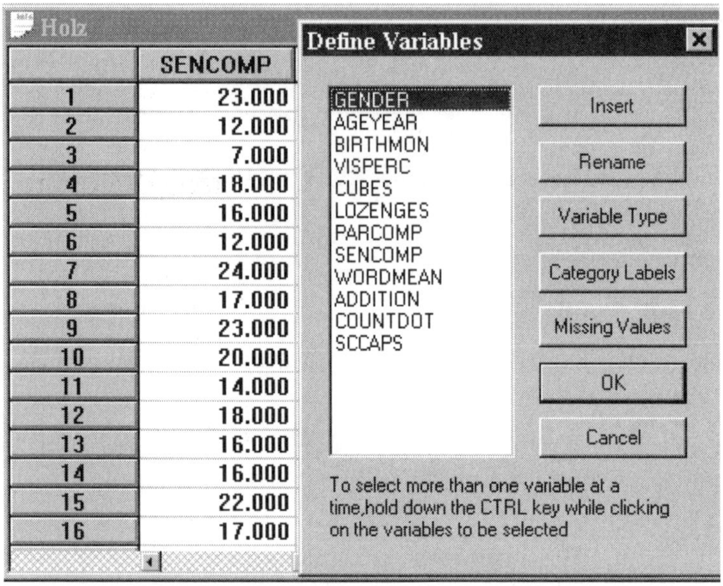

Buttons on this dialog box are associated with the following PRELIS syntax:

- ***Insert*** : `NE` command
- ***Variable Type***: `CO`, `OR`, `CA`, `CB`, `CE` commands
- ***Missing Values***: `MI` command.

Select Variables/Cases Option

The ***Select Variables/Cases*** option on the ***Data*** menu provides access to two dialog boxes: the ***Select Variables*** dialog box and the ***Select Cases*** dialog box.

Select Variables Dialog Box

Use the ***Select Variable*** dialog box to select any subset of variables. Select ***Output Options*** to save the new data set as a *.**psf** or *.**dat** file.

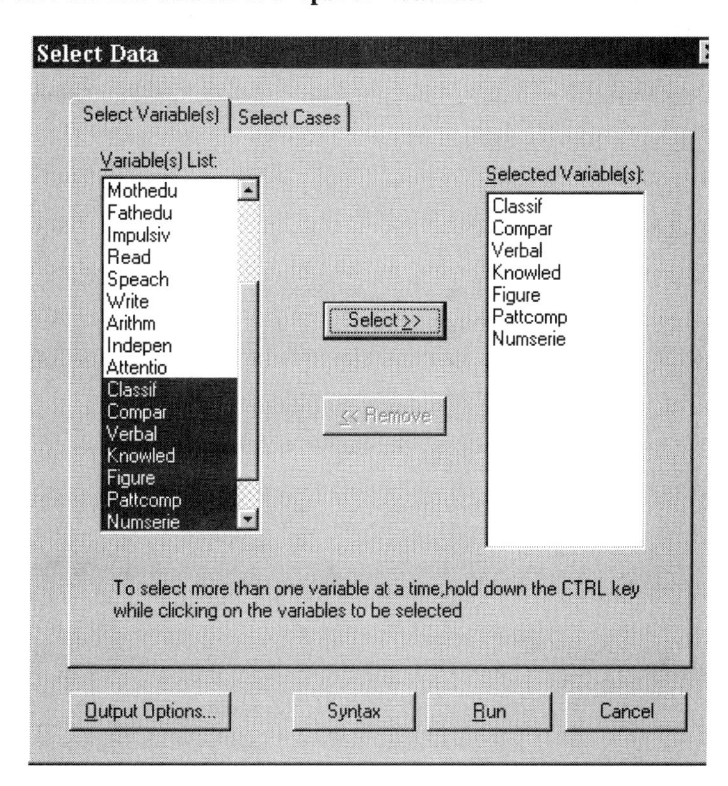

Select Cases Dialog Box

Create a new data set by selecting only odd or even cases, etc. Select ***Output Options*** to save the data to a new *.**psf** file.

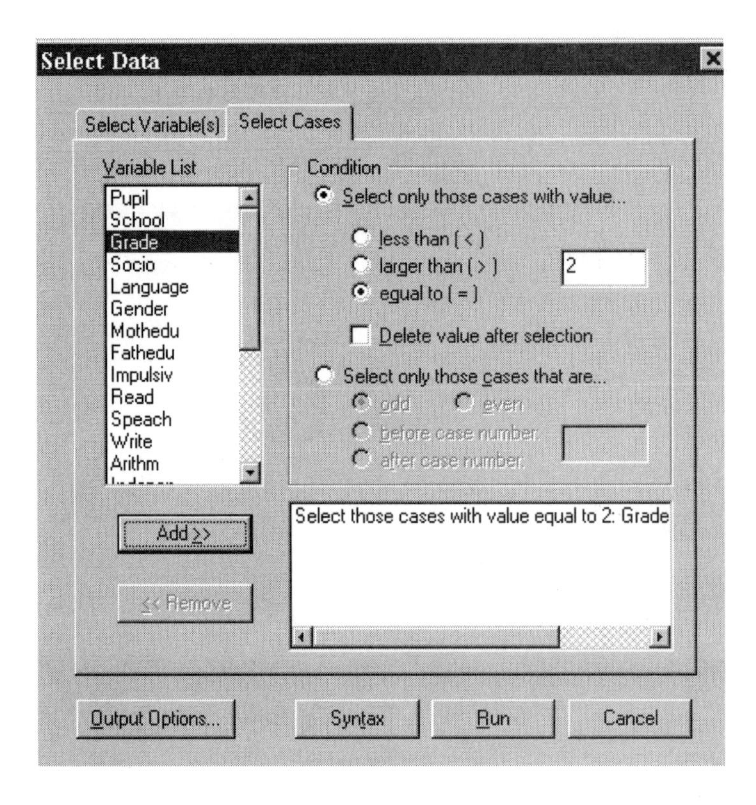

Insert Variable Option

When this option is selected, a new variable is inserted into the PRELIS spreadsheet. All cases will be assigned a value of 0 by default. These numbers can be changed manually, or by using the ***Compute*** option on the ***Transformation*** menu.

To insert a variable between the variables *CONTIN1* and *ORDINAL1* in the PRELIS system file **data100.psf**, click on the label *ORDINAL1* to highlight the column.

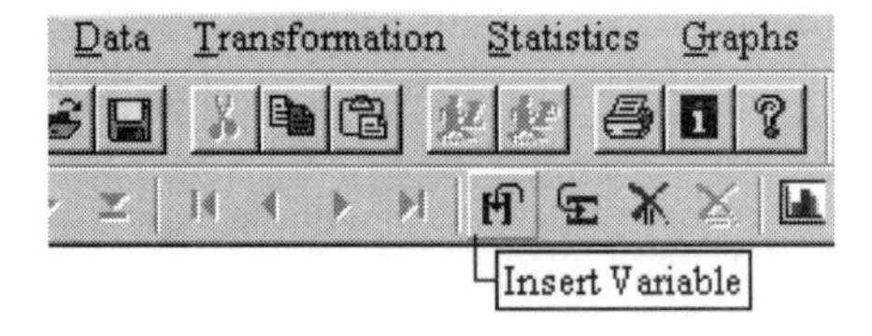

Select the ***Insert Variable*** option from the ***Data*** menu or click on the ***Insert Variable*** icon button to insert the new variable, shown as *var7* in the illustration below. Note that all the values in the *var7* column are equal to zero. Replace the zeros by entering the appropriate values or by using the ***Compute*** option on the ***Transformation*** menu.

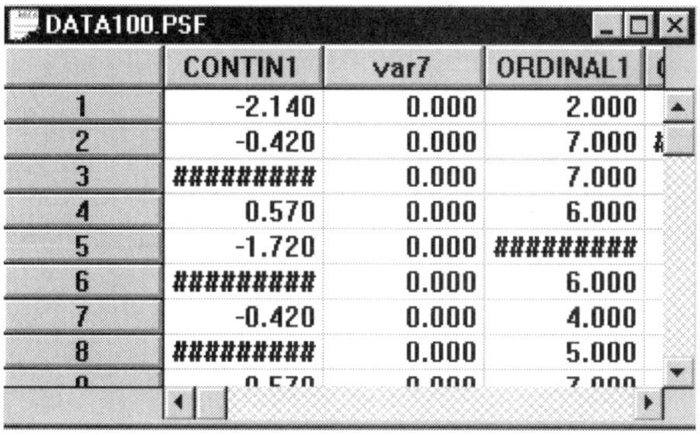

DATA100.PSF

	CONTIN1	var7	ORDINAL1
1	-2.140	0.000	2.000
2	-0.420	0.000	7.000
3	#########	0.000	7.000
4	0.570	0.000	6.000
5	-1.720	0.000	#########
6	#########	0.000	6.000
7	-0.420	0.000	4.000
8	#########	0.000	5.000
9	0.570	0.000	7.000

Note:

This option is associated with the NE command in PRELIS.

Insert Cases Option

When a case in the PRELIS system file is highlighted before this option is selected, a new case is inserted into the PRELIS spreadsheet. All variables will be assigned a value of 0 for this case. These numbers can be changed manually, or by using the ***Compute*** option on the ***Transformation*** menu. If, however, no case is highlighted before selection of this option, the ***Insert Cases*** dialog box will open to allow the user to specify the number of cases to be inserted. This dialog box is used to insert a number of new cases into the PRELIS spreadsheet. All variables will be assigned a value of 0 for this case. These numbers can be changed manually, or by using the ***Compute*** option on the ***Transformation*** menu. To insert a new record (case) between cases 13 and 14, click on the case number 14 button to highlight the case, as shown in the illustration below:

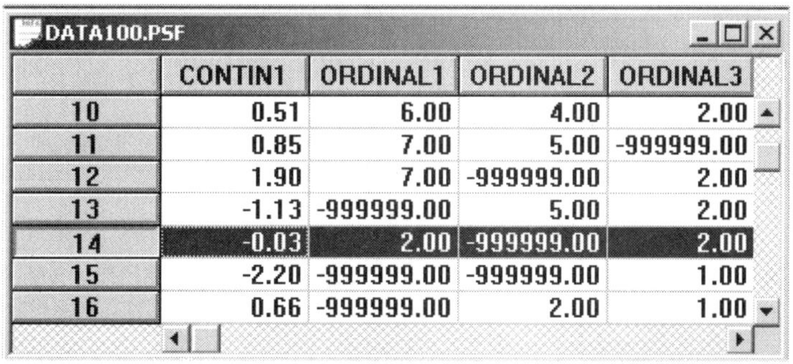

DATA100.PSF

	CONTIN1	ORDINAL1	ORDINAL2	ORDINAL3
10	0.51	6.00	4.00	2.00
11	0.85	7.00	5.00	-999999.00
12	1.90	7.00	-999999.00	2.00
13	-1.13	-999999.00	5.00	2.00
14	-0.03	2.00	-999999.00	2.00
15	-2.20	-999999.00	-999999.00	1.00
16	0.66	-999999.00	2.00	1.00

Select the ***Data***, ***Insert Cases*** option or, alternatively, click the ***Insert Case*** icon button and enter 1 in the ***Insert Cases*** dialog box.

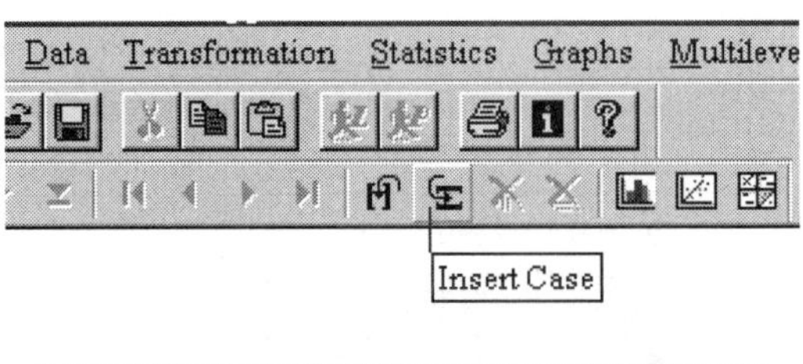

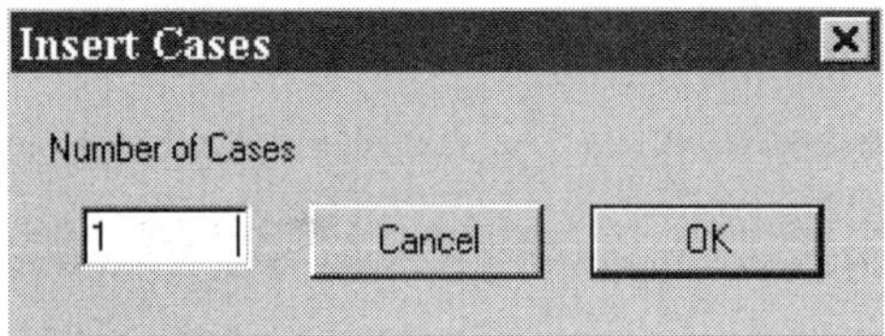

The image below shows that case 14 is the new record. The previous case 14 becomes case 15. Replace the zeros by entering the appropriate values.

DATA100.PSF

	CONTIN1	ORDINAL1	ORDINAL2	ORDINAL3
10	0.51	6.00	4.00	2.00
11	0.85	7.00	5.00	-999999.00
12	1.90	7.00	-999999.00	2.00
13	-1.13	-999999.00	5.00	2.00
14	0.00	0.00	0.00	0.00
15	-0.03	2.00	-999999.00	2.00
16	-2.20	-999999.00	-999999.00	1.00

Delete Variable Option

To delete a variable from the spreadsheet, click on the variable label and then select this option with the ***Data***, ***Delete Variable*** option or the ***Delete Variable*** icon button. In the PRELIS system file shown below, we wish to delete the variable *CONTIN2*.

DATA100.PSF

	ORDINAL3	CONTIN2	ORDINAL4
11	2.000	0.880	2.000
12	4.000	4.000	2.000
13	2.000	1.540	1.000
14	0.000	0.000	0.000
15	2.000	2.450	3.000
16	2.000	-0.750	3.567
17	1.000	-2.260	1.000
18	1.000	1.060	2.000

Select the ***Data***, ***Delete Variable*** option or click the ***Delete Variable*** icon button

to remove the selected variable. The effect of this action is illustrated below:

DATA100.PSF

	ORDINAL3	ORDINAL4
11	2.000	2.000
12	4.000	2.000
13	2.000	1.000
14	0.000	0.000
15	2.000	3.000
16	2.000	3.567
17	1.000	1.000
18	1.000	2.000

Note:

This dialog box is associated with the `SE` and `SD` command in PRELIS.

Delete Case Option

To delete a case from the spreadsheet, click on the case number and then select this option or, alternatively, click the ***Delete Case*** icon button. In the PRELIS system file shown below, we wish to remove cases 13, 14 and 15. To mark these cases, click on case 13 and, with the mouse button held down, drag the cursor to case number 15.

Data100

	ORDINAL1	ORDINAL2	ORDINAL3
10	6.000	4.000	2.00
11	7.000	5.000	-999999.00
12	7.000	-999999.000	2.00
13	-999999.000	5.000	2.00
14	2.000	-999999.000	2.00
15	-999999.000	-999999.000	1.00
16	-999999.000	2.000	1.00

Select the ***Data***, ***Delete Case*** option or click the ***Delete Cases*** icon button to remove the selected case(s).

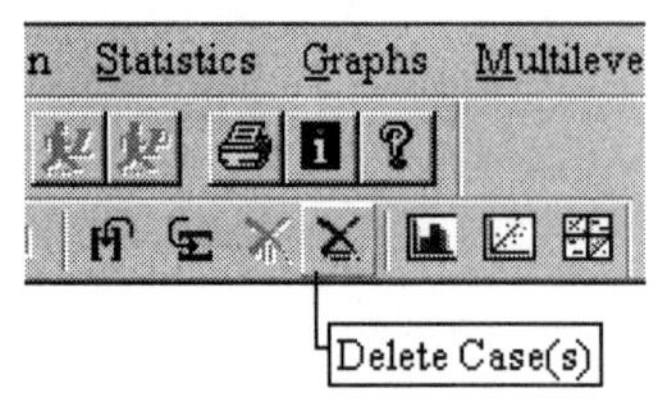

Note:

This dialog box is associated with the `SC` command in PRELIS.

2.2.4 Transformation Menu

The ***Transformation*** menu can be used to recode the values of a variable, to define a weight variable, or to compute functions of variables.

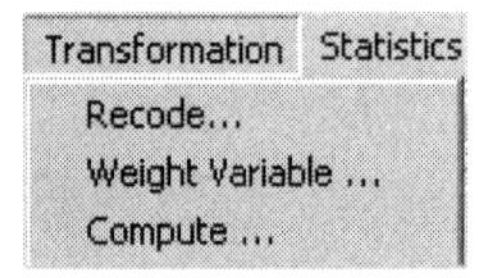

Examples of the use of most of these options can be found in Chapter 3.

Note:

Options on this menu are associated with the following PRELIS syntax:

- ***Recode*** : `RE` command
- ***Weight Variable*** : `WE` command
- ***Compute*** : `NE` command

Recode Option

The ***Recode*** option on the ***Transformation*** menu is associated with the ***Recode Variables*** dialog box. The ***Recode Variables*** dialog box shown below shows the recoding of the variable *GENDER*. All values of *GENDER* equal to zero are to be replaced with a value of -1. Once the old and new values are entered, click the ***Add*** button to add the recoding to the syntax window.

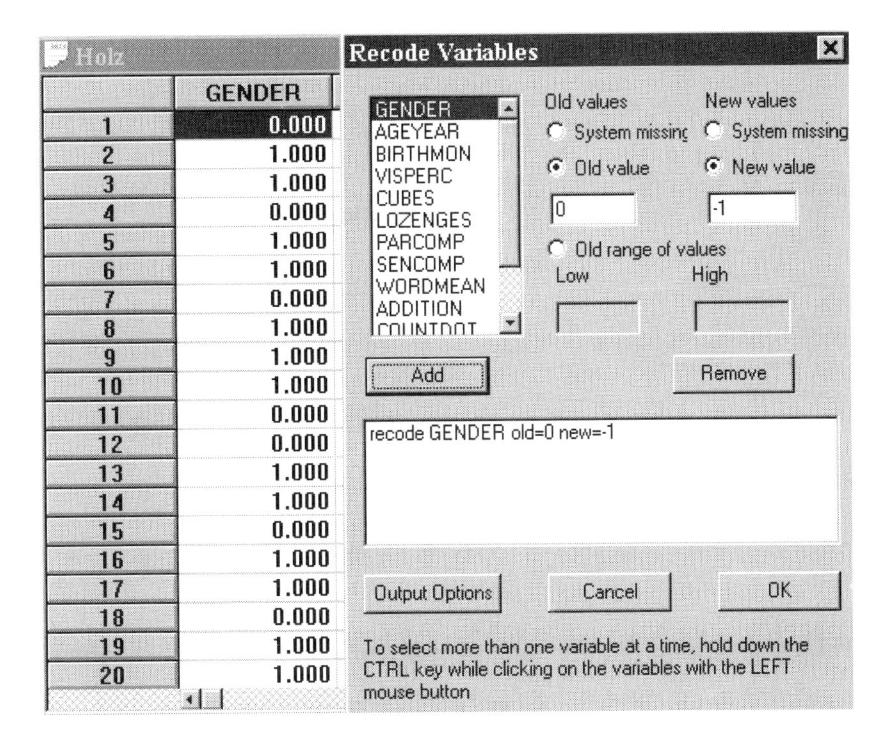

Weight Option

The ***Weight*** option on the ***Transformation*** menu is associated with the ***Weight Variable*** dialog box.

PRELIS uses a weight variable to weight each case. A variable is defined as a weight variable in order to specify that the *i*-th case is to be counted n_i times. The ***Weight Variable*** dialog box below shows the selection of the variable *FREQ* as a weight variable.

See the *PRELIS 2: User's Reference Guide*, pp. 82-84 for more information.

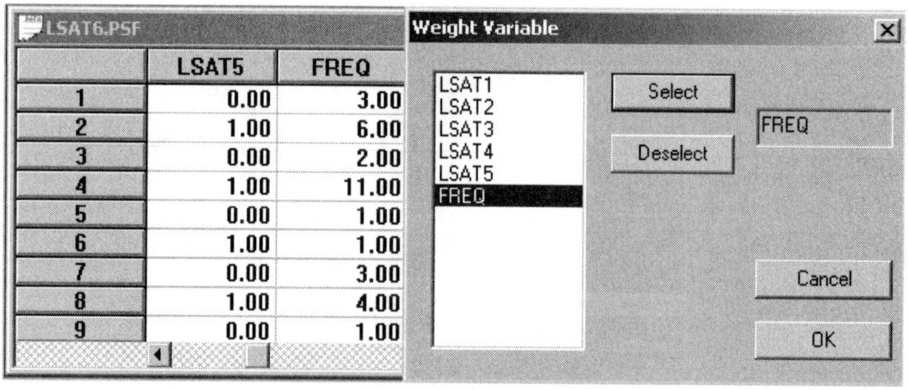

LSAT6.PSF

	LSAT5	FREQ
1	0.00	3.00
2	1.00	6.00
3	0.00	2.00
4	1.00	11.00
5	0.00	1.00
6	1.00	1.00
7	0.00	3.00
8	1.00	4.00
9	0.00	1.00

Weight Variable

LSAT1
LSAT2
LSAT3
LSAT4
LSAT5
FREQ

Select
Deselect
FREQ
Cancel
OK

Compute Option

The ***Compute*** option on the ***Transformation*** menu is associated with the ***Compute*** dialog box.

This dialog box is used to add or create new variables and calculate functions of variables. A new variable may be added by clicking the ***Add*** button, and the rest of the equation is completed by dragging variable names to the syntax text box or clicking the *, +, etc. buttons. The ***Output Options*** may be used to change the random number seed when simulating data.

As an illustration, the ***Compute*** dialog box is used to add a new variable, *AGEMON*, and to define the new variable as a combination of other variables in the *.**psf** file.

See Sections 3.5.4 and 3.10.4 for additional examples.

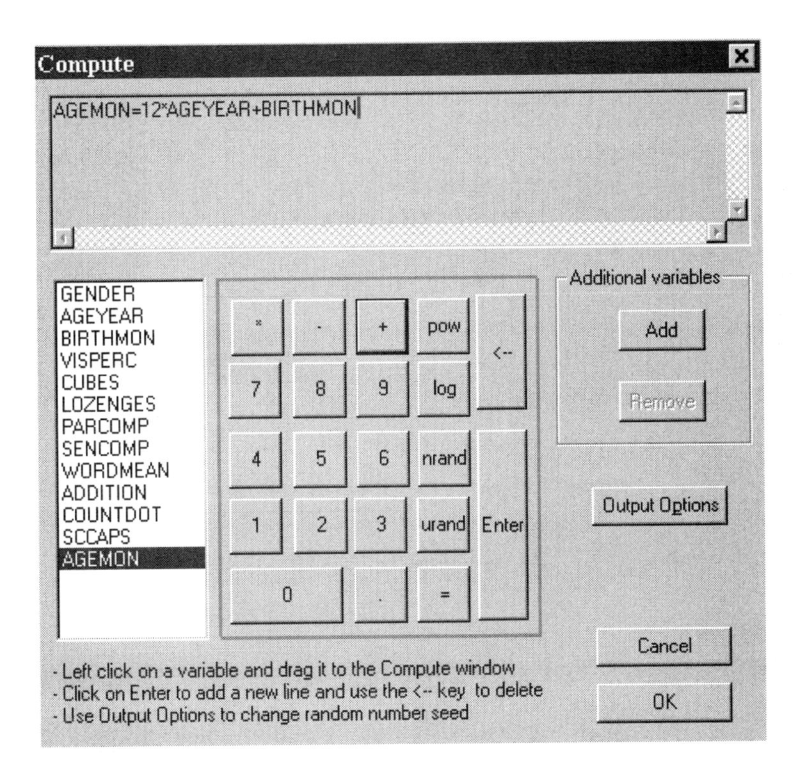

2.2.5 Statistics Menu

The ***Statistics*** menu may be used to perform a variety of statistical procedures.

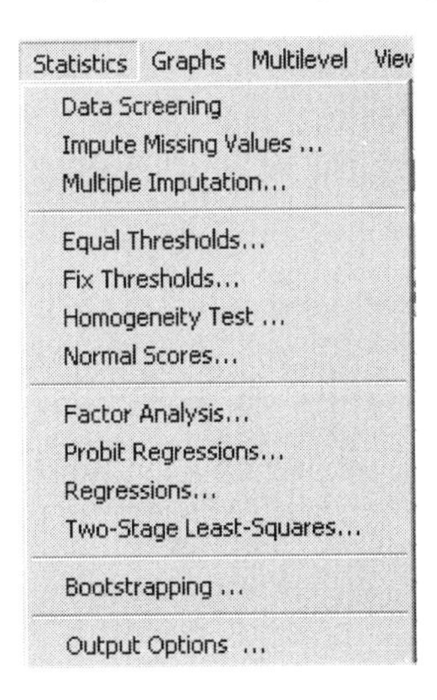

Note:

The options on this menu are associated with the following PRELIS syntax:

- ***Impute Missing Values*** : `IM` command
- ***Multiple Imputation*** : `EM` and `MC` commands
- ***Equal Thresholds*** : `ET` command
- ***Fix Thresholds*** : `FT` command
- ***Homogeneity Test*** : `HT` command
- ***Normal Scores*** : `NS` command
- ***Factor Analysis*** : `FA` command
- ***Regressions*** : `RG` command
- ***Two-Stage Least-Squares*** : `RG` command
- ***Bootstrapping*** : `BS` keyword
- ***Output Options*** : `OU` command.

Data Screening Option

For each variable in the spreadsheet, PRELIS lists all data values found in the raw data and the number and relative frequency of each data value, and it gives a bar chart showing the distribution of the data values.

For more information on data screening, see the *PRELIS 2: User's Reference Guide* pages 145 and 146. An example of data screening can be found in the Section 3.2 of this manual.

Impute Missing Values Option

The ***Impute Missing Values*** option on the ***Statistics*** menu is associated with the ***Impute Missing Values*** dialog box.

Missing values can be imputed from a set of matching variables. Select the list of variables to be imputed and click the ***Add*** (imputed variables) button. Similarly, a list of matching variables is selected.

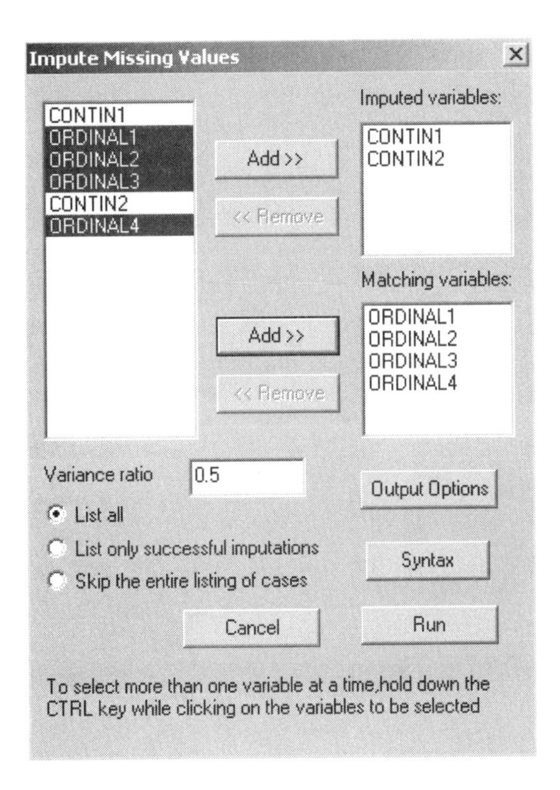

For more information on imputing, see the *PRELIS 2: User's Reference Guide* pages 155-160. An example of imputation can be found in Section 3.2.5 of this manual.

Note:

This dialog box is associated with the `IM` command in PRELIS.

Multiple Imputation Option

The ***Multiple Imputation*** option on the ***Statistics*** menu is associated with the ***Multiple Imputation*** dialog box.

Missing values can be imputed using either the Expected Maximization (EM) or the Monte Carlo Markov Chain (MCMC) method of imputation described in Chapter 7.

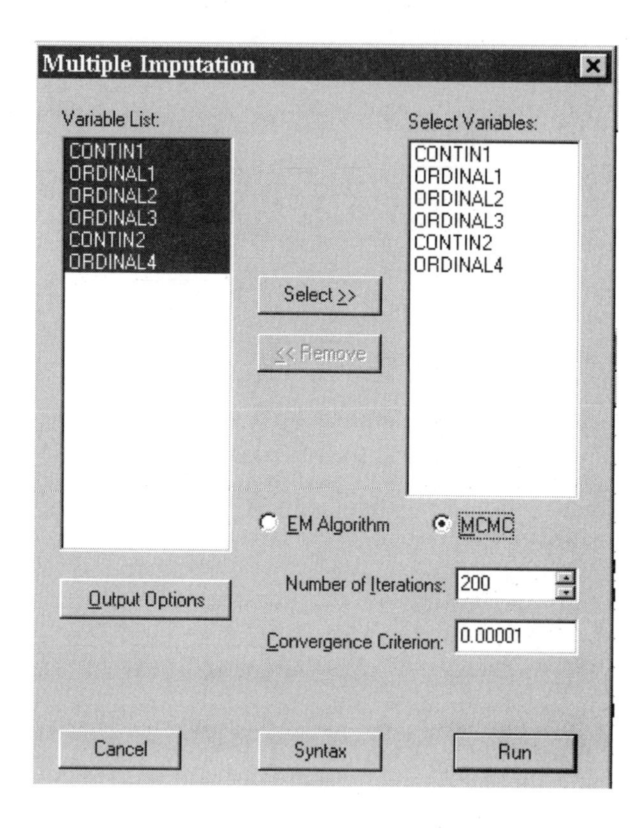

Note:

- By default, if no variables are selected from the ***Variable List***, all variables are used.
- The ***Output Options*** button may be used to save the imputed data set, change the random number seed when simulating data, and simulate a specified number of data sets.
- The ***Syntax*** button may be used to generate PRELIS syntax for modification or saving of the relevant commands generated here.
- The ***Number of Iterations*** and ***Convergence Criterion*** fields can be used to control the iterative procedure.

Section 3.11 of this manual contains a detailed example.

Equal Thresholds Option

The ***Equal Thresholds*** option on the ***Statistics*** menu is associated with the ***Equal Threshold Test*** dialog box.

To perform an equal thresholds test, select ordinal variables with the same number of categories. Click ***Add*** for each selection. When done, click ***Syntax*** to generate syntax or ***Run*** to run the example. The ***Output Options*** button may be used to control the PRELIS output.

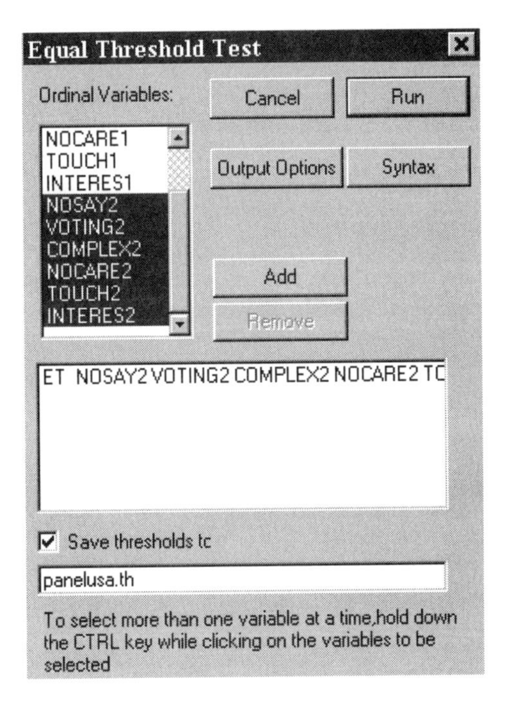

The command

```
ET varlist
```

will instruct PRELIS to estimate the thresholds under the condition that they are *equal* for all ordinal variables in the `varlist`. Underlying normality is assumed.

For more information on this command, see the *PRELIS 2: User's Reference Guide* pp. 176-8.

Note:

This dialog box is associated with the `ET` command.

Fix Thresholds Option

The ***Fix Thresholds*** option on the ***Statistics*** menu is associated with the ***Fix Thresholds*** dialog box.

Select the variables to be fixed by clicking on a variable and then the ***Add*** button. Specify the name of the file that contains the thresholds and, when done, click ***Syntax*** to generate syntax or ***Run*** to run the example. The ***Output Options*** button may be used to control the PRELIS output. The thresholds of variables can be fixed by reading in threshold values for the variables from an external file created by first estimating the thresholds to be equal and saving the estimated thresholds to a file.

Polychoric correlations and means and variances of the variables underlying the ordinal variables can then be estimated for each group separately by including the commands

```
FT = filename varlist1
FT varlist2
....
```

and reading the fixed thresholds from the file `filename`. Thresholds given on the first line of the file will be assigned to variables in `varlist1`, thresholds in the second line to `varlist2`, etc. The thresholds for all variables in a particular `varlist` will be set equal to the values given in `filename`.

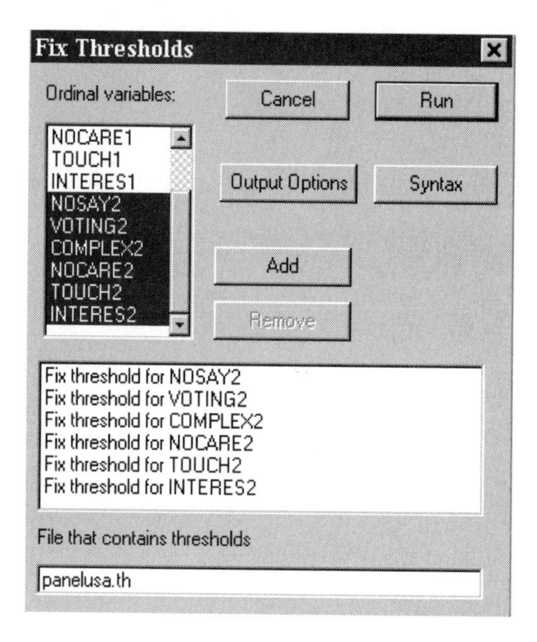

For more information on this command, see the *PRELIS 2: User's Reference Guide* pp. 176-8.

Note:

This dialog box is associated with the `FT` command in PRELIS.

Homogeneity Test Option

The ***Homogeneity Test*** option on the ***Statistics*** menu is associated with the ***Homogeneity Test*** dialog box.

The homogeneity test is a test of the hypothesis that the marginal distributions of two categorical variables with the same number of categories *k* are the same. This test can be applied to nominal as well as ordinal variables and is represented by the

```
HT var1 var2
```

command in the PRELIS syntax file.

The homogeneity test (`HTest`) differs from the equal threshold test (`ETest`) in two ways:

- It does not assume underlying normal variables.
- If underlying normality is assumed, the homogeneity hypothesis implies the equal threshold hypothesis.

To perform a homogeneity test, select a pair of ordinal variables with the same number of categories. Click ***Add*** for each pair. When done, click ***Syntax*** to generate syntax or ***Run*** to run the example. The ***Output Options*** button may be used to control the PRELIS output.

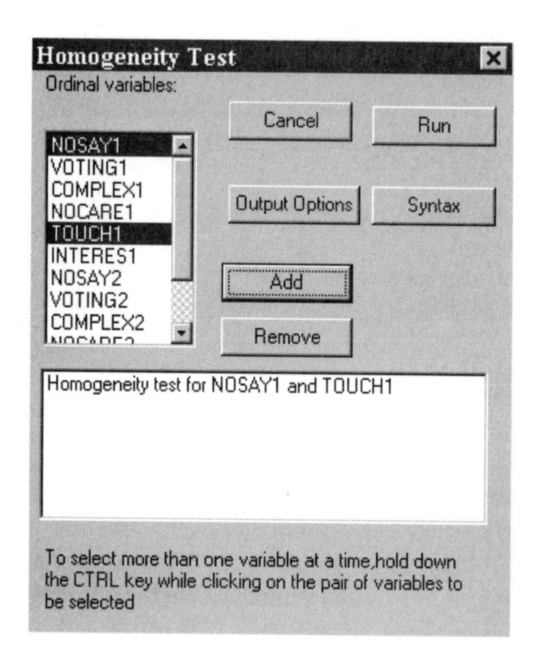

For an example of a homogeneity test, see Section 3.2.6. For more information on this command, see the *PRELIS 2: User's Reference Guide* pp. 173-176.

Note:

This dialog box is associated with the `HT` command in PRELIS.

Normal Scores Option

The ***Normal Scores*** option on the ***Statistics*** menu is associated with the ***Normal Scores*** dialog box.

A possible solution to non-normality is to normalize the variables before analysis. Normal scores offer an effective way of normalizing a continuous variable for which the origin and unit of measurement have no intrinsic meaning, such as test scores. Normal scores may be computed for ordinal and continuous variables.

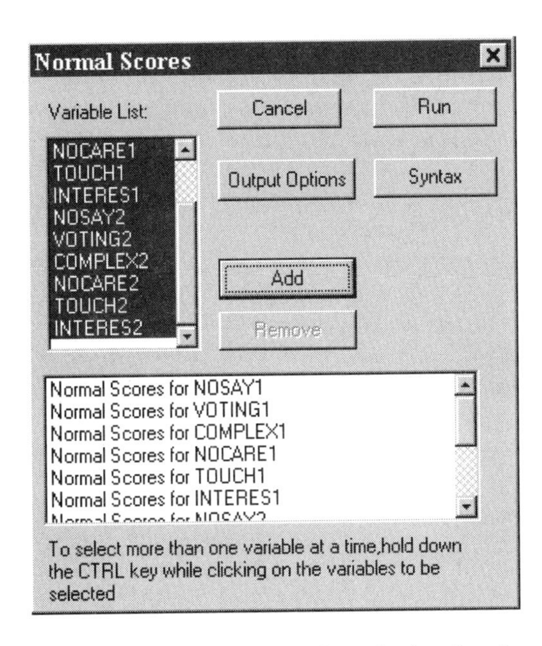

Select variables for which normal scores are to be calculated and use the ***Add*** button to add the selected variables to the text box. When done, click ***Syntax*** to generate syntax or ***Run*** to run the example. The ***Output Options*** button may be used to control the PRELIS output. An example of the calculation of normal scores is given in Section 3.6 of this manual.

Note:

This dialog box is associated with the `NS` command in PRELIS. For more information, see Section 3.4 in *LISREL 8: New Statistical Features.*

Factor Analysis Option

The ***Factor Analysis*** option on the ***Statistics*** menu is associated with the ***Factor Analysis*** dialog box. It can also be used to perform principal component analysis.

Exploratory factor analysis is useful in the early stages of investigation to study the measurement properties of the observed variables. ML estimates of factor loadings (unrotated, VARIMAX rotated or PROMAX rotated) are provided. The number of factors may be specified by the user or will be determined automatically by the program.

Select a subset of variables and use the ***Select*** button to add the selection to the ***Select a subset of Variables*** list box. When done, click ***Syntax*** to generate syntax or ***Run*** to run the example. The ***Output Options*** button may be used to control the PRELIS output.

If no subset of variables is selected, all variables will be used.

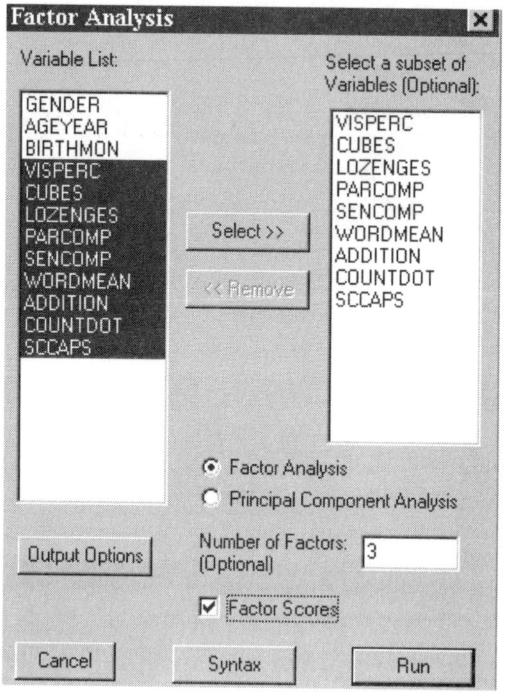

Note:

This dialog box is associated with the `FA` and `PC` commands in PRELIS. For more information, see Sections 3.2 and 3.3 in *LISREL 8: New Statistical Features*.

Probit Regressions Option

The ***Probit Regressions*** option on the ***Statistics*** menu is associated with the ***Probit Regressions*** dialog box.

For each ordinal *y*-variable, the univariate probit regression of y^* on the *x*-variables may be obtained, where y^* is the variable underlying *y*. For each pair of ordinal variables, the conditional polychoric correlation for a given *x* will also be estimated.

There are three ways of handling the thresholds in the probit regressions.

- The default alternative is to estimate the thresholds jointly with the regression coefficients in the probit regressions.

- To estimate thresholds from the marginal distributions of the ordinal variables, the marginal threshold command (MT) may be used.
- To estimate regression coefficients for fixed thresholds specified by the user, the fixed threshold command (FT) may be used.

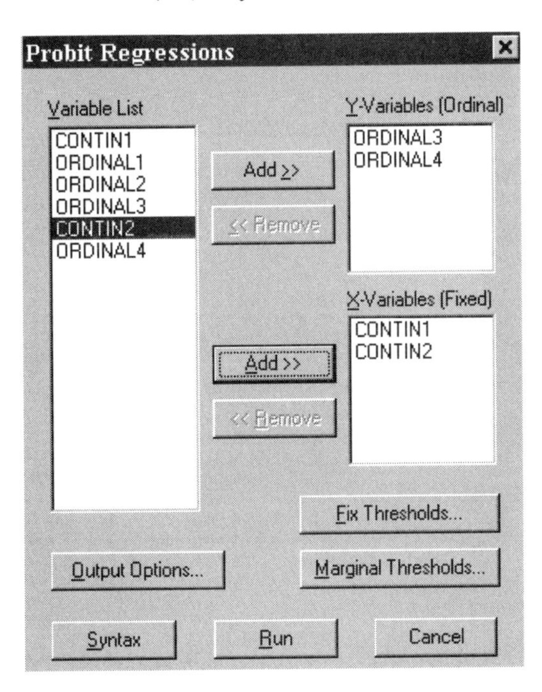

Select variables as *y*-variables or *x*-variables and use the ***Add*** button to add the selected variables to the respective fields. When done, click ***Syntax*** to generate syntax or ***Run*** to run the example. The ***Output Options*** button may be used to control the PRELIS output. Note that the ***Fix Thresholds*** and ***Marginal Thresholds*** buttons provide access to the ***Fix Thresholds*** and ***Marginal Thresholds*** dialog boxes discussed earlier in this section.

From the variable list, select a list of *y*-variables. Each of the *y*-variables must be of ordinal type. Select a set of fixed *x*-variables, for example *CONTIN1* and *CONTIN2*. Univariate probit regressions will be performed for each remaining variable on *CONTIN1* and *CONTIN2*.

PRELIS makes a distinction between *y*- and *x*-variables. The *x*-variables can be fixed or random variables. If they are random, their joint distribution is unspecified and assumed not to contain any parameters of interest. To specify *x*-variables, the command

```
FI varlist
```

must be included in the syntax file, where `varlist` is a list of the *x*-variables. All other variables will be assumed to be ordinal, unless they are declared continuous or censored.

For more information on the thresholds, see pp. 173-178 of the *PRELIS 2: User's Reference Guide*. Additional information on probit regression may be found on pp. 178 - 182 of the same guide.

Note:

This dialog box is associated with the `FI`, `FT` and `MT` commands in PRELIS.

Marginal Thresholds Dialog Box

The ***Marginal Thresholds*** dialog box is used to estimate thresholds for ordinal variables from their marginal distributions and is accessed from the ***Probit Regressions*** dialog box discussed previously. See the *PRELIS 2: User's Reference Guide* pages 80 and 180 for more information.

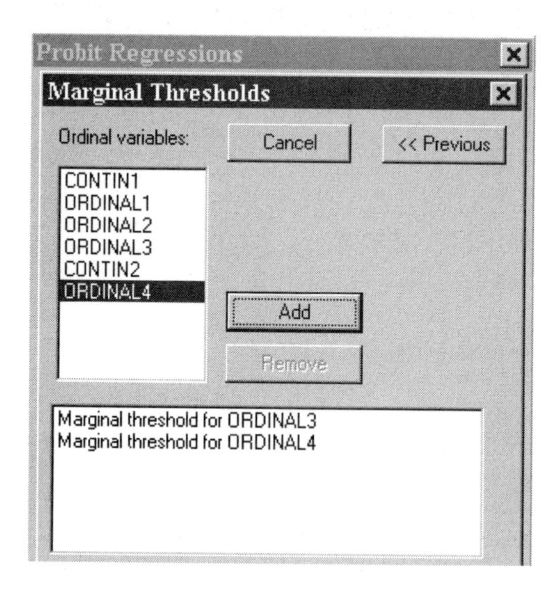

Note:

This dialog box is associated with the `MT` command in PRELIS.

Regressions Option

The ***Regressions*** option on the ***Statistics*** menu is associated with the ***Regressions*** dialog box. This dialog box is used to estimate the regression of any variable on any set of variables. For each regression, the following output is produced:

- estimated regression coefficients based on the moment matrix requested with the `MA` keyword.
- standard errors of the estimated regression coefficients
- *t*-values of the estimated regression coefficients
- residual variance
- squared multiple correlation.

The basic form of the `RG` command is:

```
RG y-varlist ON x-varlist
```

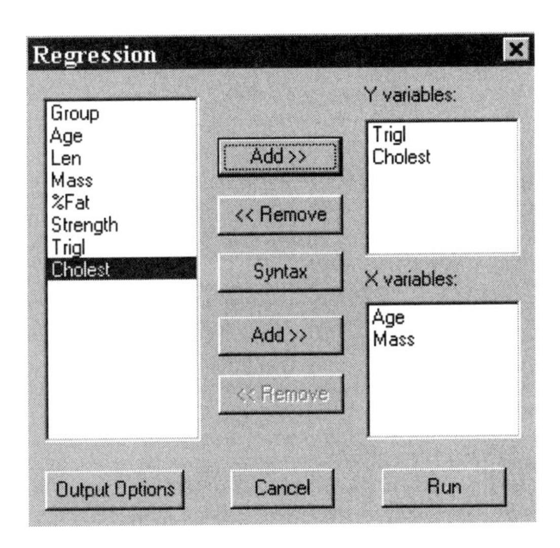

Y-variables, *Trigl* and *Cholest*, and two predictors (*X*-variables), *Age* and *Mass*, are shown in the dialog box. PRELIS will estimate two regression equations, these being (1) *Trigl* on *Age* and *Mass* and (2) *Cholest* on *Age* and *Mass*.

For more information on the `RG` command, see the *PRELIS 2: User's Reference Guide* pp. 90-91. See Section 3.4.4 for an example.

Note:

This dialog box is associated with the `RG` command in PRELIS.

Two Stage Least-Squares Option

The ***Two-Stage Least-Squares*** option on the ***Statistics*** menu is associated with the ***Two-Stage Least-Squares*** dialog box.

Two-Stage Least-Squares (TSLS) often provide sufficient information to judge whether a structural equation model is reasonable or not. TSLS estimates and their standard errors may be obtained quickly, without iterations, for several typical structural equation models and is useful especially in the early stages of investigation.

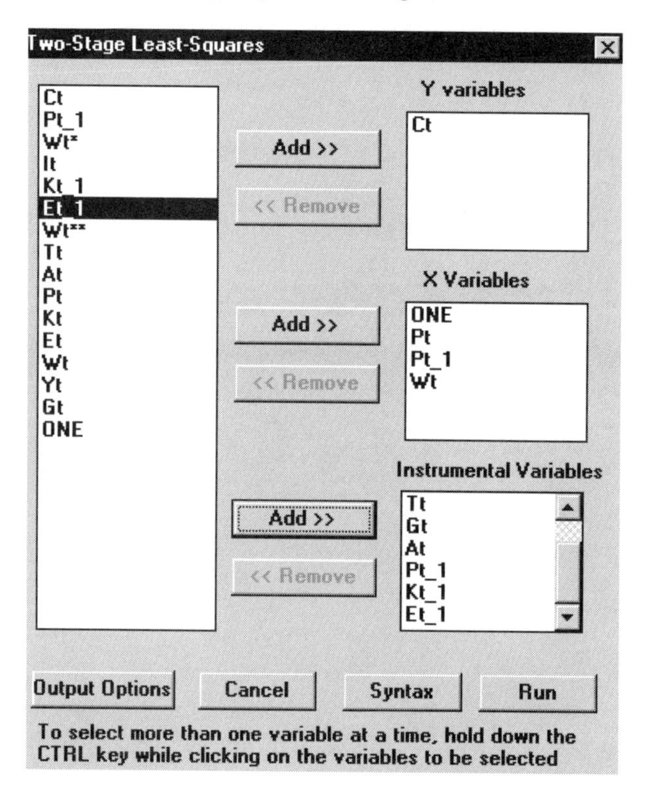

See *LISREL 8: New Statistical Features*, Chapter 3, for a detailed discussion. An example is given in Section 3.7 of this manual.

Note:

This dialog box is associated with the RG command in PRELIS.

Bootstrapping Option

The ***Bootstrapping*** option on the ***Statistics*** menu is associated with the ***Bootstrapping*** dialog box.

Bootstrap samples of an existing raw data set can be obtained with PRELIS. The user may specify the number of bootstrap samples, the sample fraction and the file names in which moment matrices are to be stored. Use the ***Output Options*** button for more control. Click ***Syntax*** to generate syntax or ***Run*** to run the example.

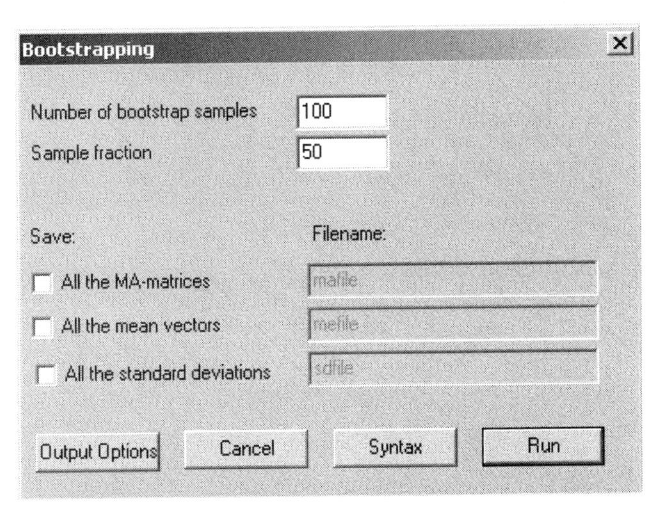

For more information on bootstrapping, see the *PRELIS 2: User's Reference Guide* pages 184-188. An example of bootstrapping can be found in the Section 3.9.

Note:

This dialog box is associated with the `BS` keyword in PRELIS.

Output Options Option

The ***Output Options*** option on the ***Statistics*** menu is associated with the ***Output*** dialog box.

The ***Output*** dialog box controls list output and is used to save results in external files. The following options are available:

- Specifying the type of moment matrix to be estimated and the filename to which the matrix is to be saved (`CM`, `KM`, `MM` keywords).

- Files to which transformed raw data, asymptotic covariance matrix of estimated variances, covariances or correlations, mean vectors and standard deviations will be saved.
- Specifying the integer starting value for the random number generator (`IX` keyword).
- Specifying the width and number of decimals in each field for the raw data file to be saved (`WI`, `ND` keywords).
- Printout of summary statistics (`XM`, `XB`, `XT` options).
- Number of repetitions (used in data simulations; `RP` keyword).

Examples of the use of this dialog box can also be found in Chapter 3 of this guide. For complete information on the `OU` command and other available options, see the *PRELIS 2: User's Reference Guide* pp. 92-100.

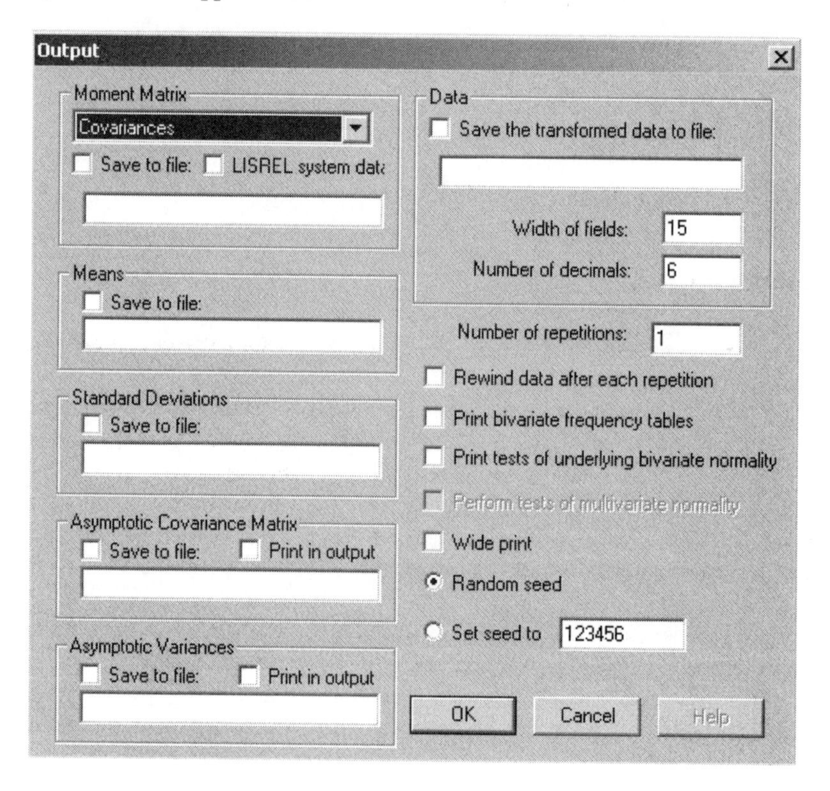

Note:

This dialog box is associated with the `OU` command in PRELIS.

2.2.6 Graphs Menu

The ***Graphs*** menu is used to make univariate, bivariate and multivariate plots of the data in a PRELIS system file (*.**psf**).

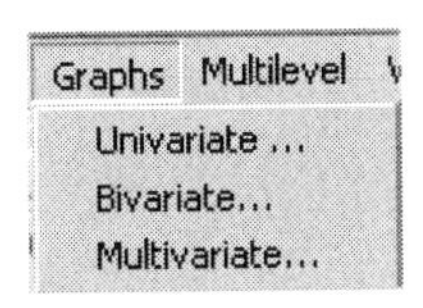

The following graph types are available:

- ***Univariate***: Bar chart, pie chart and histogram.
- ***Bivariate***: Box-and-whisker plot, 3-D bar chart, scatter plot, and line plot.
- ***Multivariate***: Scatter plot matrix.

Attributes of the graph can be changed by using the ***Image*** dialog box.

Univariate Option

The ***Univariate*** option on the ***Graphs*** menu is associated with the ***Univariate Plots*** dialog box.

A histogram is displayed if the selected variable has more than 15 distinct values, otherwise a bar chart will be displayed. By checking ***Normal curve overlay***, the normal distribution approximation to the histogram is displayed in the form of a smooth curve. Pie charts or bar charts can also be created using the ***Univariate Plots*** dialog box, provided that the selected variable is ordinal.

A pie chart of the variable *BIRTHMON* (a variable in the PRELIS system file **holz.psf** in the **tutorial** folder) is obtained by checking the ***Pie chart*** radio button:

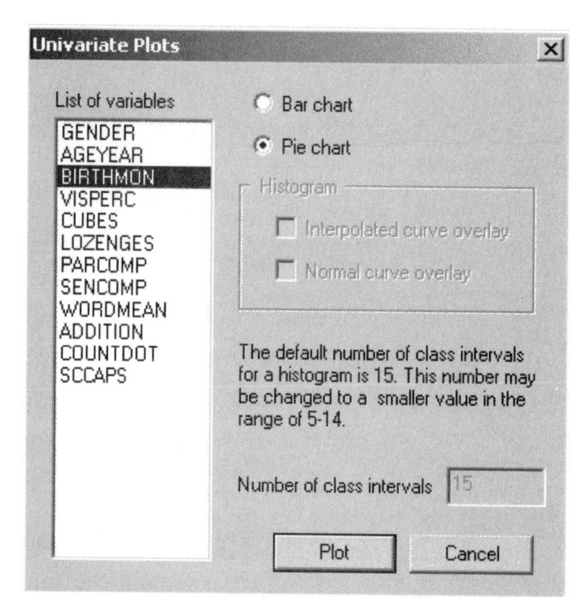
Univariate Plots
List of variables
GENDER
AGEYEAR
BIRTHMON
VISPERC
CUBES
LOZENGES
PARCOMP
SENCOMP
WORDMEAN
ADDITION
COUNTDOT
SCCAPS
Bar chart
Pie chart
Histogram
Interpolated curve overlay
Normal curve overlay
The default number of class intervals for a histogram is 15. This number may be changed to a smaller value in the range of 5-14.
Number of class intervals
15
Plot
Cancel

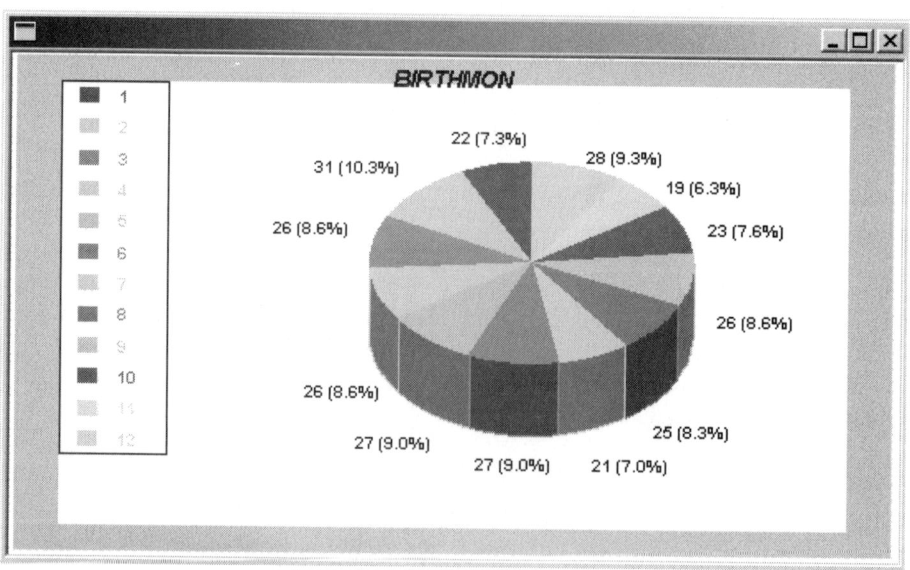
BIRTHMON
1
2
3
4
5
6
7
8
9
10
11
12
22 (7.3%)
28 (9.3%)
19 (6.3%)
23 (7.6%)
26 (8.6%)
25 (8.3%)
21 (7.0%)
27 (9.0%)
27 (9.0%)
26 (8.6%)
26 (8.6%)
31 (10.3%)

Bivariate Option

The ***Bivariate*** option on the ***Graphs*** menu is associated with the ***Bivariate Plots*** dialog box.

The selected *Y*-variable is plotted against the selected *X*-variable. A ***box-and-whisker plot***, ***3-D bar chart***, ***scatter plot*** or ***line plot*** can be selected. The line plot is obtained by taking the arithmetic average of all the *Y*-values that correspond to the same *X*-value. A 3D-bar chart of two ordinal variables may also be produced.

Box-and-Whisker Plot

A box-and-whisker plot is useful for depicting the locality, spread and skewness of a data set. It also offers a useful way of comparing two or more data sets with each other with regard to locality, spread and skewness.

A box-and-whisker plot of the variables *LOZENGES* and *BIRTHMON* (available in the PRELIS system file **holz.psf** in the **tutorial** folder) is obtained by checking the ***Box-and-Whisker Plot*** check box.

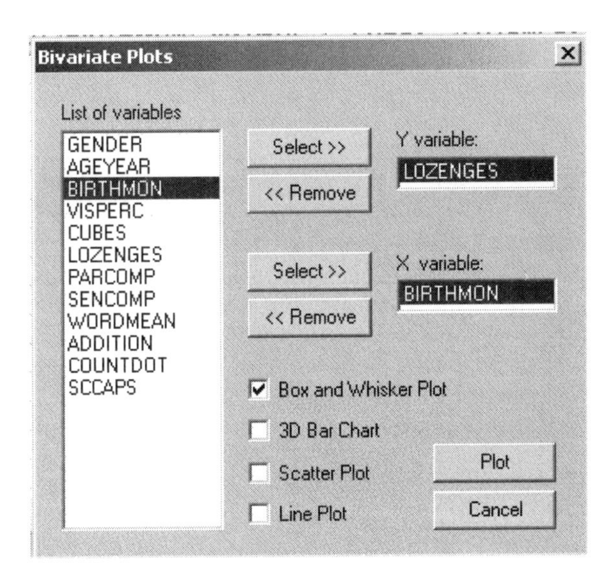

A description of box-and-whisker plots is given in Chapter 7. From the graphical display below, it is evident that there is little, if any, relationship between the *LOZENGES* scores and the birth month of a person.

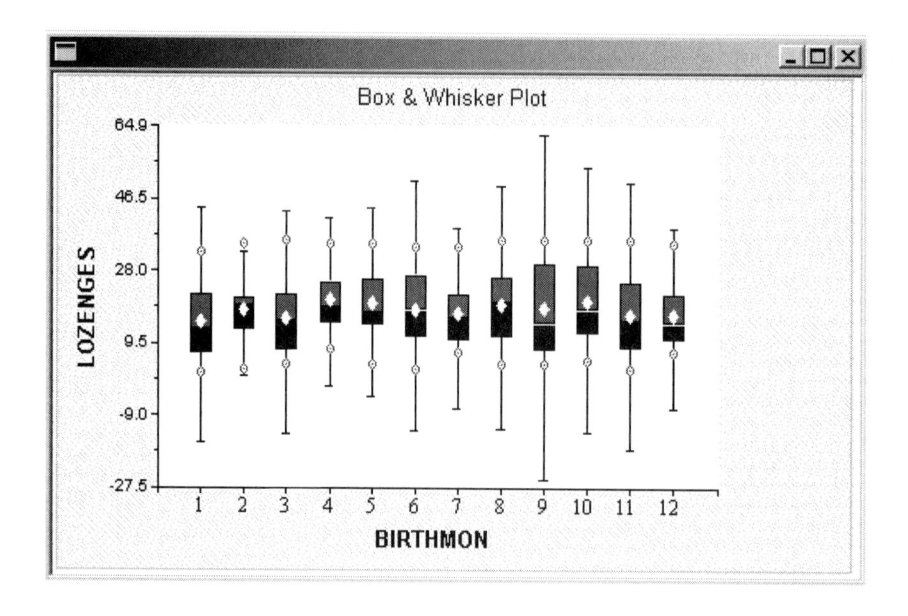

Scatter and Line Plots

A scatter plot and line plot of the variables *PARCOMP* and *SENCOMP* (available in the PRELIS system file **holz.psf** in the **tutorial** folder) are obtained by checking the ***Scatter Plot*** and ***Line Plot*** check boxes after selecting the variables using the ***Select*** buttons:

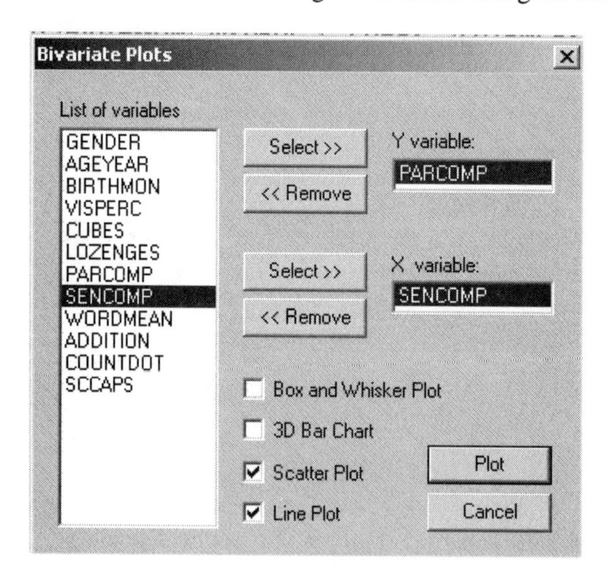

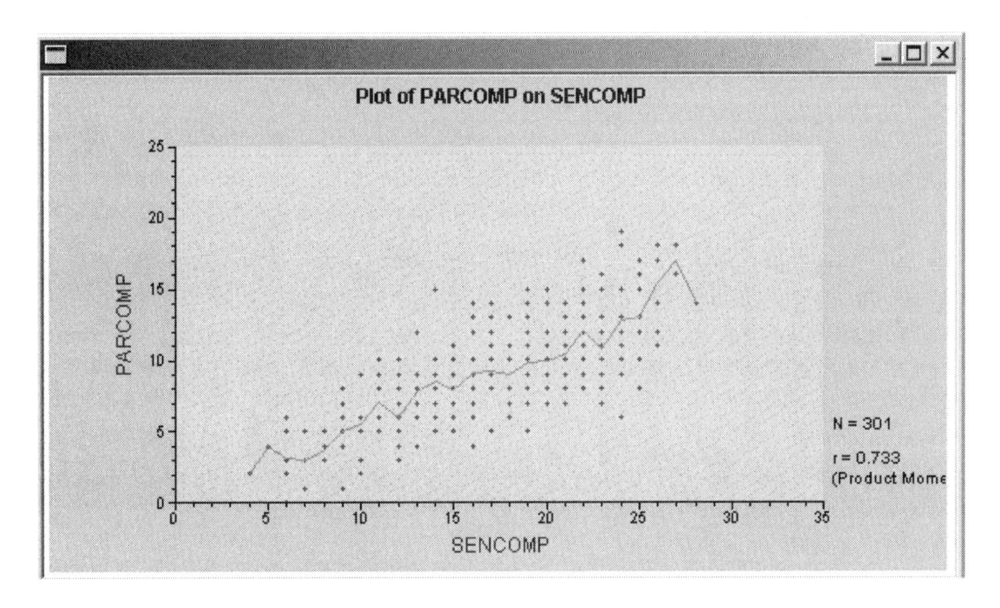

The plot reveals a relatively strong linear correlation ($r = 0.733$) between these variables.

3-D Bar Chart

A three-dimensional bar chart of the variables *BIRTHMON* and *GENDER* (available in the PRELIS system file **holz.psf** in the **tutorial** folder) is obtained by checking the ***3-D Bar Chart*** check box after selecting the variables using the ***Select*** buttons:

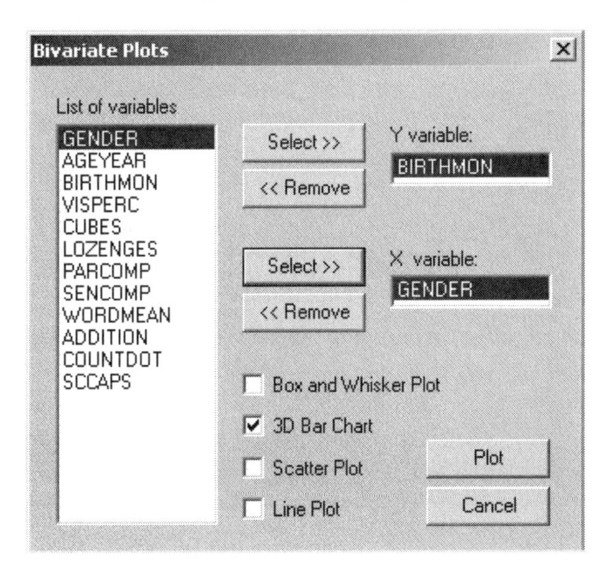

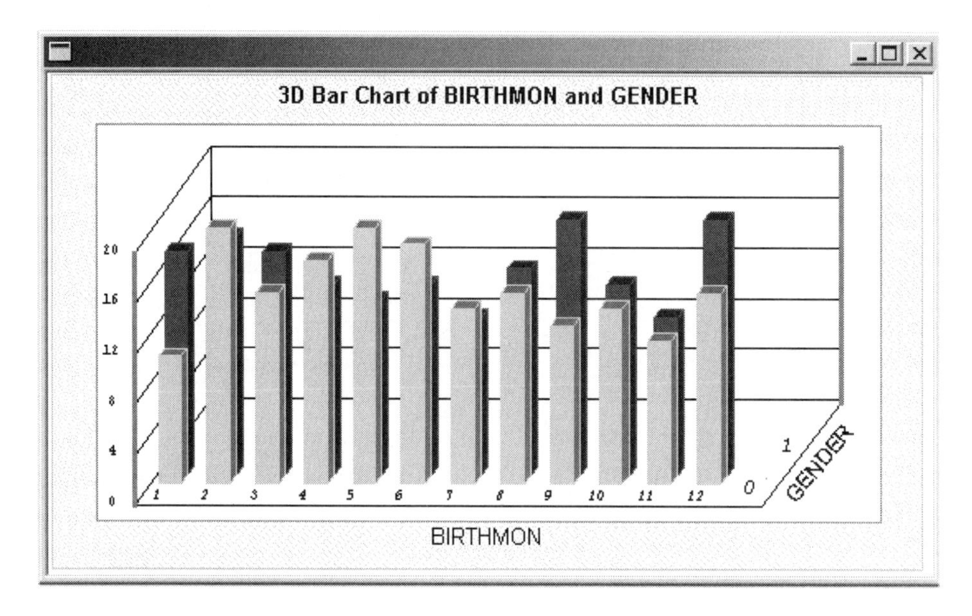

From this plot, it may be deduced that there is an approximately even split of males and females across month of birth.

Multivariate Option

The ***Multivariate*** option on the ***Graphs*** menu is associated with the ***Scatter Plot Matrix*** dialog box.

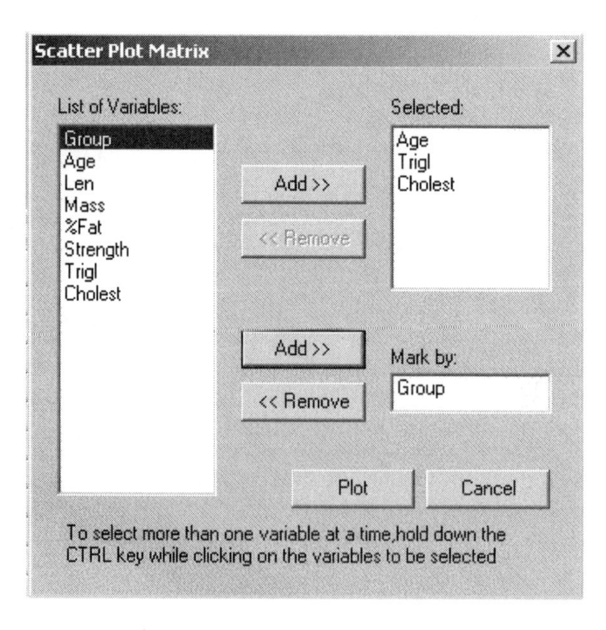

Suppose $(X_i, Y_i), i = 1, 2, \ldots, n$ indicate the paired measurements of two variables *X* and *Y*. A two-dimensional representation of these pairs of observations is known as a scatter plot. Such plots are particularly useful tools in an exploratory analysis conveying information about the association between *X* and *Y*, the dependence of *Y* on *X* where *Y* is a response variable, the clustering of the points, the presence of outliers, etc.

If only one predictor *X* is considered, a plot of *Y* against *X* provides a useful summary of a regression problem. A second predictor can be included by adding a third dimension to the two-dimensional plot of *Y* on *X*. However, if more than 2 predictors are to be included, the data cannot be viewed in total, because they are in too many dimensions. One helpful graphical method to study the relationships between variables in such a case is the scatter plot matrix, which enables the researcher to obtain multiple two-dimensional plots of higher dimensional data.

In the image below, data from the PRELIS system file **fitchol.psf** are used. The variables *Age*, *Trigl* and *Cholest* are plotted against each other for the four groups of adult males. The groups were weightlifters, students, marathon athletes and coronary patients respectively.

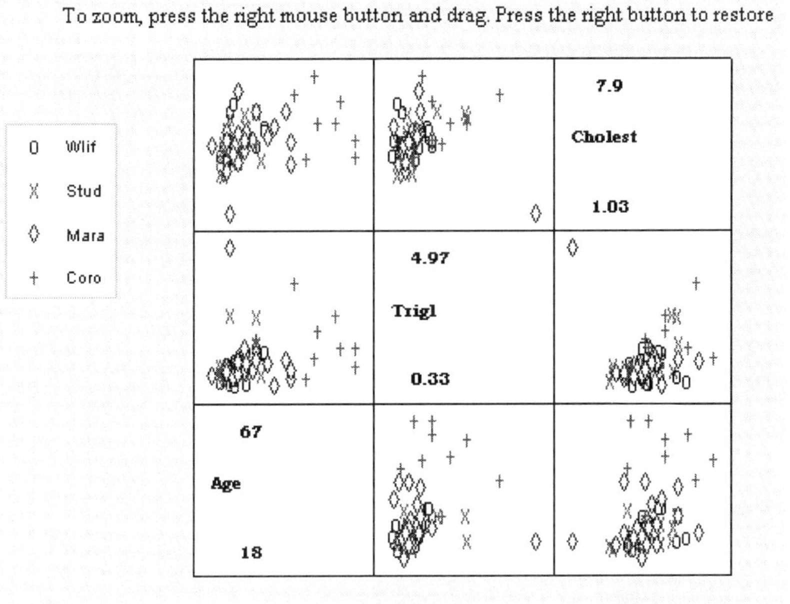

From the scatter plot matrix we see that, by looking at the (*Age*, *Trigl*) and (*Age*, *Cholest*) plots, that the coronary patients, denoted with a "+" symbol, are clustered together, away from the main cluster of points formed by the other three groups. In the (*Age*, *Age*) segment, for example, the minimum and maximum values of the *Age* variables are given.

Options are provided to zoom in and out of any segment of the scatter plot matrix and to change the plot symbols, legends, format of text and colors used in the graphical display. For more information, see the scatter plot example in Chapter 7.

2.2.7 Image Menu

When a graph is displayed, the ***Image*** menu is activated. This menu enables one to change parameters or the scale of either the *Y* or *X* axis or both from linear to logarithmic.

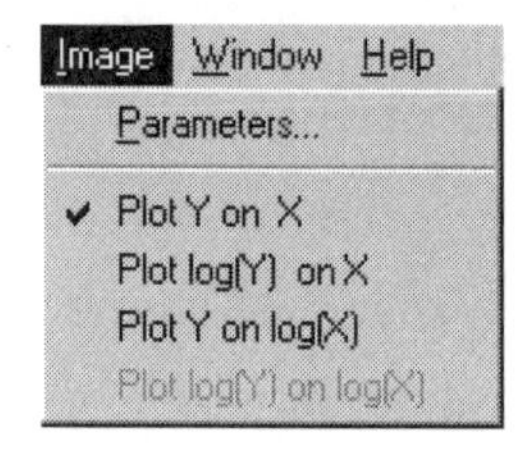

Parameters Option

The ***Parameters*** option is used to change attributes of the graph displayed and is associated with the ***Graph Parameters*** dialog box. This dialog box is used to change the position, size and color of the currently selected graph and its plotting area.

The following functions are defined:

- The ***Left***, ***Top***, ***Width*** and ***Height*** edit controls allow the user to specify a new position and size of the graph (relative to the page window) and of the plotting area (relative to the graph window).
- The ***Color*** drop-down list box is used to specify the graph window color and the color of the graph's plotting area.
- If the ***Border*** check box is checked, the graph will have a border around it.
- If the ***Border*** check box is checked, the ***Border Attributes*** button leads to another standard dialog box (the ***Line Parameters*** dialog box discussed elsewhere in this section) that allows specification of the thickness, color, and style of the border line.

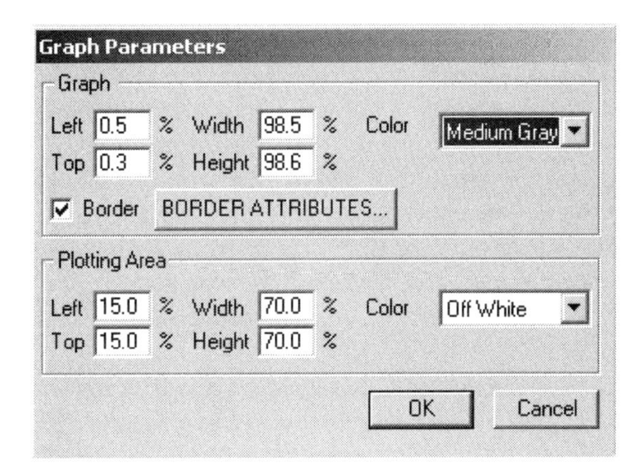

In addition to the ***Graphs Parameters*** dialog box, a number of other dialog boxes may be used to change attributes of graphs. The dialog boxes accessible depend on the type of graph displayed. The dialog boxes are:

- ***Axis Labels*** dialog box
- ***Text Parameters*** dialog box
- ***Bar Graph Parameters*** dialog box
- ***Line Parameters*** dialog box
- ***Legend Parameters*** dialog box
- ***Pie Chart*** dialog box
- ***Pie Slice Parameters*** dialog box

The user may access any of these dialog boxes by double-clicking in the corresponding section of the graph. For example, double-clicking in the legend area of the graph will activate the ***Legend Parameters*** dialog box. Double-clicking on the title of the graph, on the other hand, will provide access to the ***Text Parameters*** dialog box.

These dialog boxes will now be discussed in turn.

Axis Labels Dialog Box

This dialog box is used for editing axis labels.

The following functions are defined:

- The ***Labels Position*** group box controls the position of the labels relative to the axis or plotting area.
- The ***Last Label*** group box allows manipulation of the last label drawing options. If ***On*** is selected, the last label is displayed like the others. If ***Off*** is selected, it is

not displayed. If ***Text*** is selected, the text string entered in the edit box below will be displayed instead of the last numerical label.

- The format of the numerical labels can be specified using the radio buttons in the ***Format*** group box.
- The ***Date Parameters*** group box becomes active once the ***Date*** radio button is checked. The ***Date Format*** box selects the date format to use for labels, while the ***Date Time Base*** box selects the time base (minute, hour, day, week, month, year) for the date calculations. The ***Starting Date*** drop-down list boxes specify the starting date that corresponds to the axis value of 0. All dates are calculated relative to this value.
- If the ***Set Precision*** check box is not checked, the labels' precision is determined automatically. If it is checked, the number entered into the ***#Places*** field specifies the number of digits after the decimal point.
- The ***Text Parameters*** button leads to another dialog box (see the next section) that controls the font, size, and color of labels.

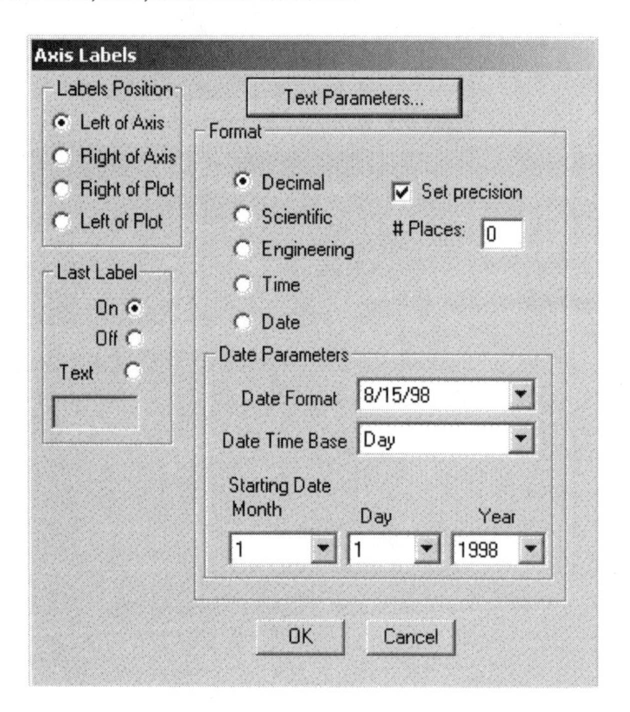

Text Parameters Dialog Box

This dialog box is used for editing text strings, labels, titles, etc. It can be called from some of the other dialog boxes controlling the graphic features in LISREL.

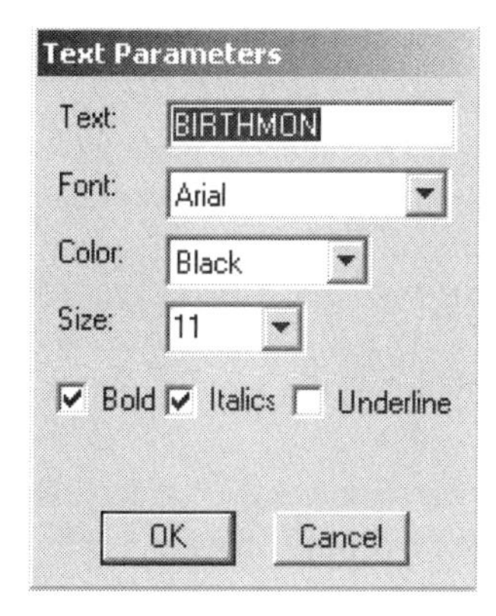

The following functions are defined:

- The ***Text*** edit control allows the user to edit the text string.
- The ***Font*** drop-down list box allows control of the font typeface.
- The text color can be selected from the ***Color*** drop-down list box.
- The size of the fonts, in points, is controlled by the ***Size*** drop-down list box.
- The ***Bold***, ***Italic*** and ***Underline*** check boxes control the text style.

Bar Graph Parameters Dialog Box

This dialog box is used for editing the parameters of all bars in a regular bar graph, or a selected group member of grouped bar graphs.

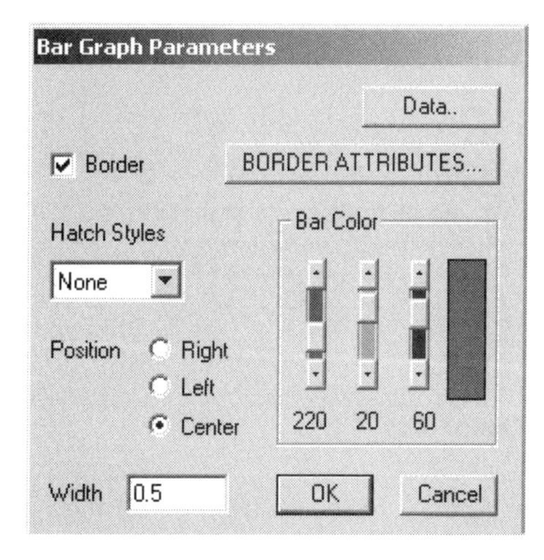

It operates as follows:

- If the ***Border*** check box is checked, the bars have a border around them. In this case, the ***Border Attributes*** button leads to another standard dialog box that controls border thickness, color and style.
- The ***Data*** button leads to the spreadsheet-style window for editing plotted data points.
- The ***Hatch Style*** drop-down list box allows the user to choose the hatch style for bars.
- The ***Bar Color*** scrolling bars control the bar RGB color.
- The ***Position*** radio buttons control the bar position relative to the independent variable values.
- The ***Width*** string field allows the user to enter the bar width in units of the independent variable.

Line Parameters Dialog Box

This dialog box is used for editing lines in the graph.

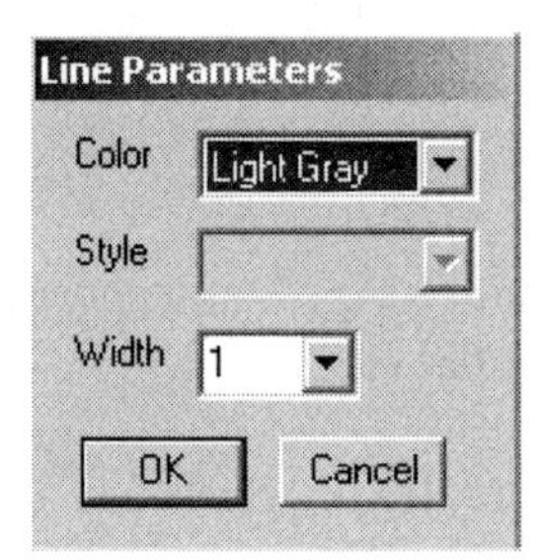

It has the following functions:

- The ***Color*** drop-down list box controls the line color.
- The ***Style*** drop-down list box, visible when activated, allows selection of a line style.
- The ***Width*** control specifies the line width, in screen pixels.

Legend Parameters Dialog Box

This dialog box allows the editing of legends. It opens when the user double-clicks while the cursor is anywhere inside the legend box, except over a symbol representing a plotting object.

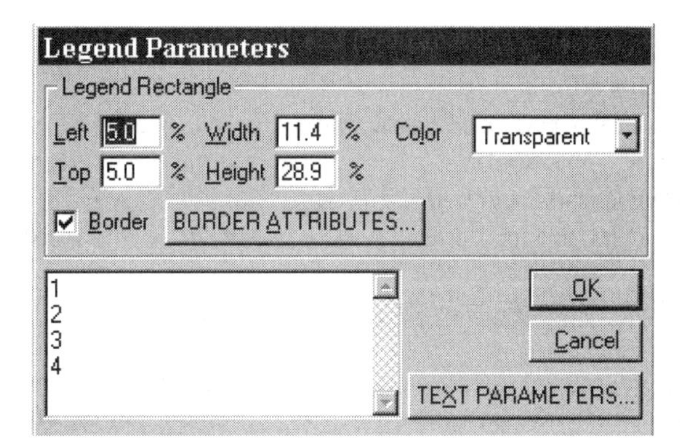

This dialog box operates as follows:

- The ***Left***, ***Top***, ***Width*** and ***Height*** string fields allow the user to specify a new position and size of the legend bounding rectangle relative to the graph window.
- The ***Color*** drop-down list box specifies the legend rectangle background color.
- If the ***Border*** check box is checked, the rectangle will have a border. In this case, the ***Border Attributes*** button leads to another standard dialog box that controls border thickness, color and style of the border line.
- The multi-line text box in the lower left corner lists and allows editing of each of the legend text strings.
- The ***Text Parameters*** button leads to the ***Text Parameters*** dialog box discussed earlier.

Pie Chart Parameters Dialog Box

This dialog box is used for editing pie charts.

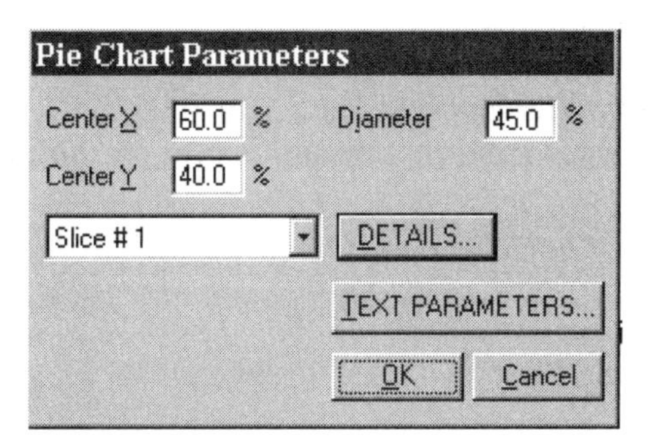

It operates as follows:

- The string fields ***Center X***, ***Center Y***, and ***Diameter*** allow the user to enter the pie position and size in graph normalized coordinates.

- The drop-down list box contains labels for individual slices. It allows the user to select a slice for editing.
- The ***Details*** button leads to the dialog box for editing the selected pie slice parameters.
- The ***Text Parameters*** button leads to the ***Text Parameters*** dialog box and allows the user to specify the font, size, and color of the pie chart labels.

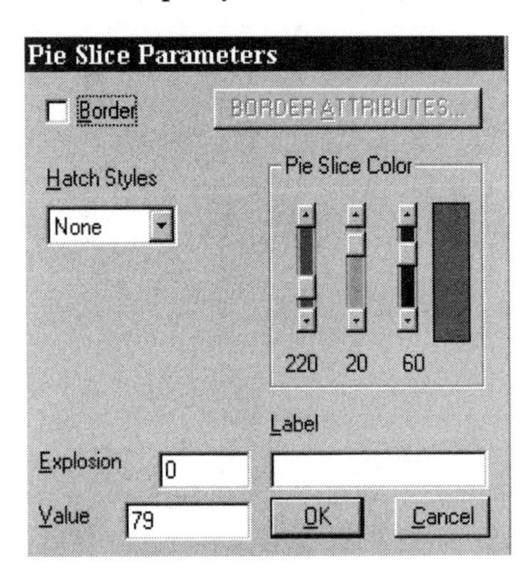

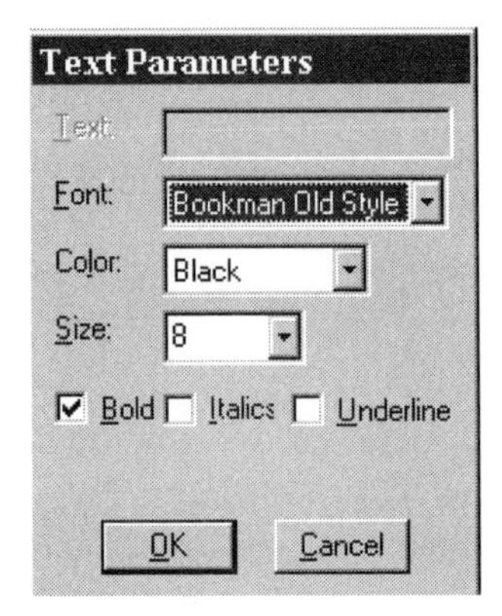

Pie Slice Parameters Dialog Box

This dialog box (see above) is used for editing a selected pie slice.

It operates as follows:

- If the ***Border*** check box is checked, the slice will have a border. In this case, the ***Border Attributes*** button leads to another standard dialog box that controls border thickness, color and style of the border line.
- The ***Hatch Style*** drop-down list box allows the user to choose the hatch style for the slice.
- The ***Pie Slice Color*** scrolling bars control the RGB color of the slice.
- The ***Explosion*** string field allows the user to enter a value between 0 and 1 for the explosion coefficient. The default is 0.
- The ***Value*** string field allows the user to enter a data value.
- The ***Label*** string field allows the user to edit the pie slice text label.

Plot Y on X / Plot log(Y) on X Options

The next set of options on the ***Image*** menu (see below) are only activated once a graph is displayed. By selecting any of these options, the graph can be changed from one form to the other as shown below, where the ***Plot Y on X*** and ***Plot log(Y) on X*** options are used.

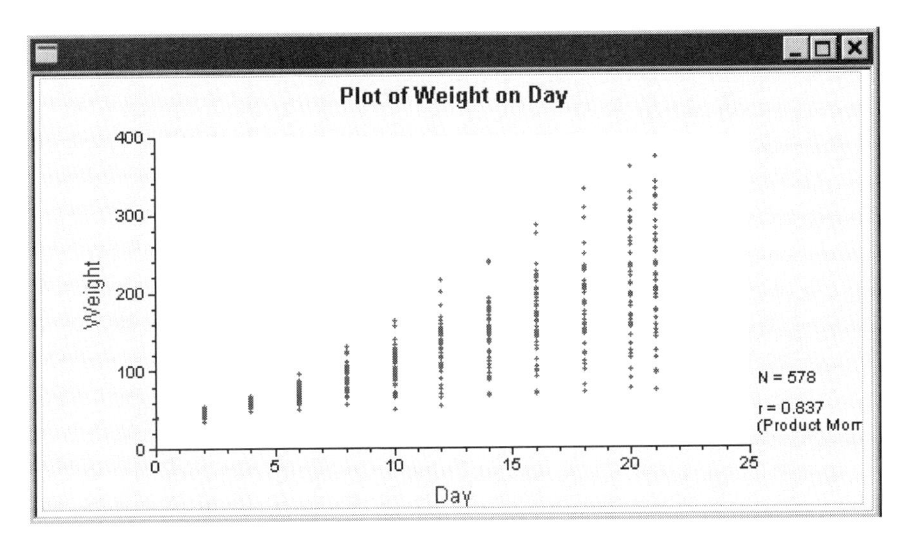

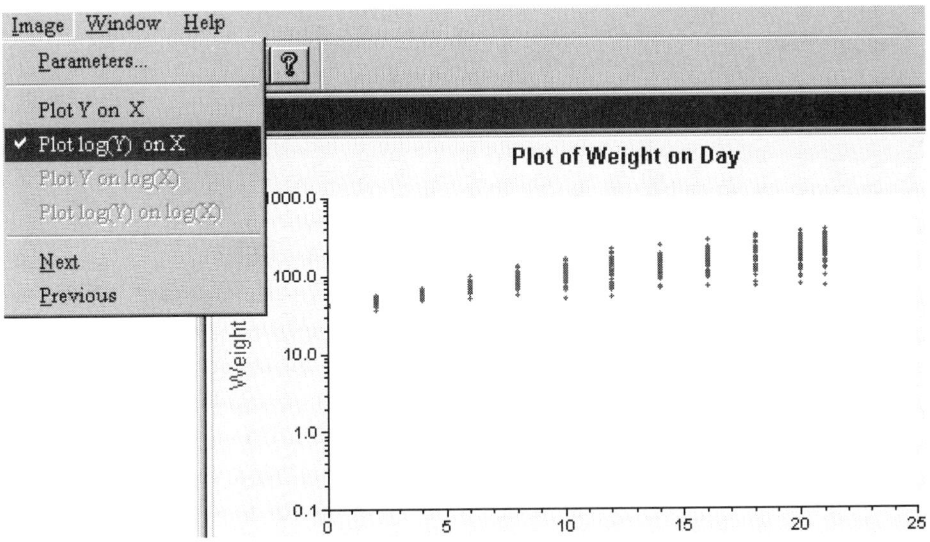

2.2.8 Multilevel Menu

Multilevel models may be fitted to data by using the options on the ***Multilevel*** menu. This menu is only available when a PRELIS system file (*.**psf**) is opened.

The ***Multilevel*** menu has two submenus, for linear and non-linear models respectively, as shown below.

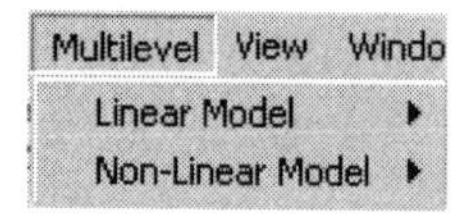

For more information on multilevel models, please see Chapters 7 and 8. Examples can be found in Chapter 5.

Linear Model Option

The ***Linear Model*** option allows access to four dialog boxes as shown below. These dialog boxes enable one to build syntax for the most commonly used hierarchical linear models. Chapter 7 gives an overview of the more specialized commands that can be included by adding the syntax file.

The four dialog boxes accessed from this menu will now be discussed in turn.

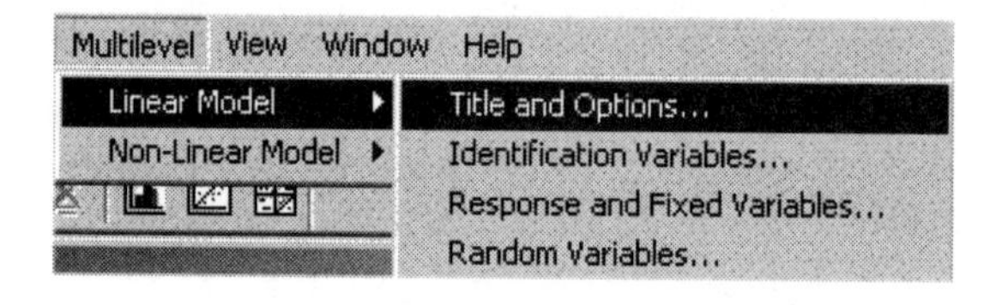

Linear Model: Title and Options Option

The ***Title and Options*** option on the ***Multilevel, Linear Model*** menu is associated with the ***Title and Options*** dialog box shown below.

The ***Title and Options*** dialog box is used to provide a title for the analysis and options concerning the iterative procedure. For an illustration of the use of this dialog box, please see Chapter 5. Click ***Next*** to go to the ***Identification Variables*** dialog box when constructing a syntax file.

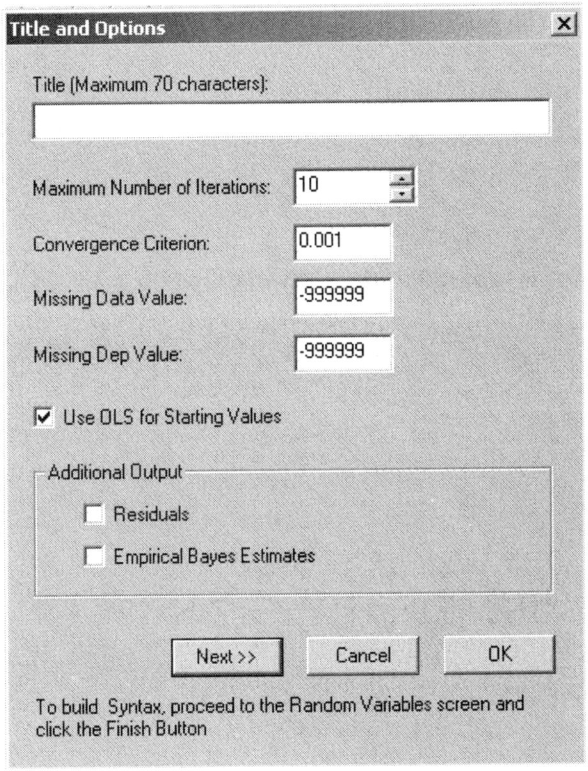

Title and Options

Title (Maximum 70 characters):

Maximum Number of Iterations: 10

Convergence Criterion: 0.001

Missing Data Value: -999999

Missing Dep Value: -999999

Use OLS for Starting Values

Additional Output

Residuals

Empirical Bayes Estimates

Next >> Cancel OK

To build Syntax, proceed to the Random Variables screen and click the Finish Button

The default values for the maximum number of iterations, convergence criterion, etc. are shown above.

Linear Model: Identification Variables Option

The ***Identification Variables*** option on the ***Multilevel, Linear Model*** menu is associated with the ***Identification Variables*** dialog box shown below.

The ***Identification Variables*** dialog box is used to select the variables in the PRELIS system file (***.psf**) that identify the various levels of the hierarchy. For an illustration of the use of this dialog box, please see Chapter 5. Click ***Next*** to go to the ***Select Response and Fixed Variables*** dialog box when constructing a syntax file.

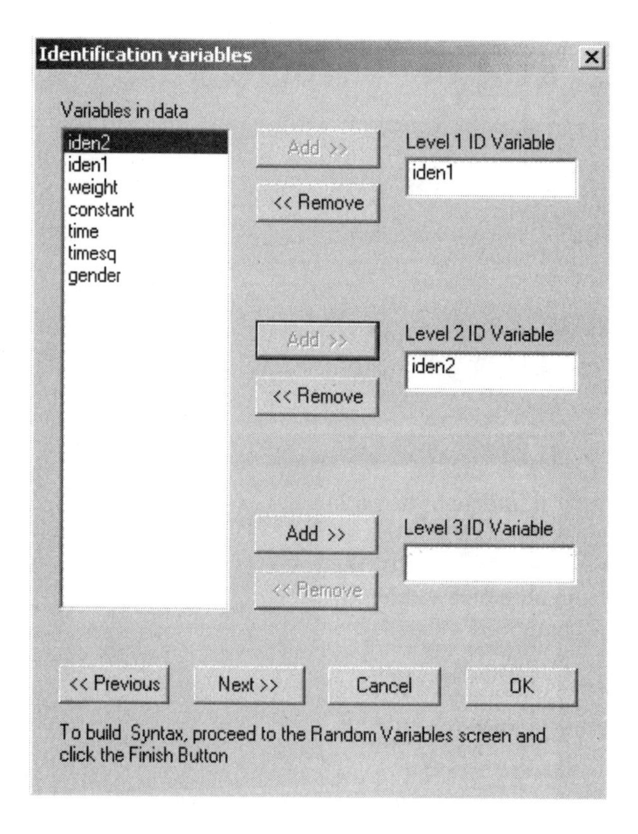

The dialog box above corresponds to a typical selection for a level-2 model.

Linear Model: Select Response and Fixed Variables Option

The ***Response and Fixed Variables*** option on the ***Multilevel, Linear Model*** menu is associated with the ***Select Response and Fixed Variables*** dialog box shown below.

The ***Select Response and Fixed Variables*** dialog box is used to select the outcome and fixed variables to be included in the model from the PRELIS system file (***.psf**). For an illustration of the use of this dialog box, please see Chapter 5. Click ***Next*** to go to the ***Random Variables*** dialog box when constructing a syntax file. If the creation of dummy variables is requested, the `DUMMY` command will be included in the generated syntax file.

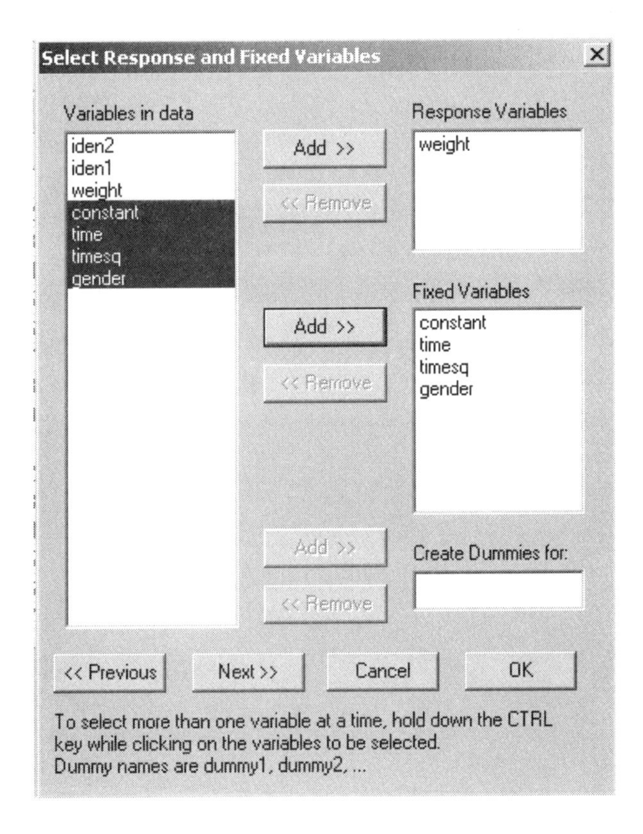

See Section 8.4 for more information on the `DUMMY` command.

Linear Model: Random Variables Option

The ***Random Variables*** option on the ***Multilevel, Linear Model*** menu is associated with the ***Random Variables*** dialog box shown below.

The ***Random Variables*** dialog box is used to select the random variables to be included in the model from the PRELIS system (*.**psf**) file. Clicking the ***Finish*** button will generate the syntax.

For an illustration of the use of this dialog box, please see Chapter 5.

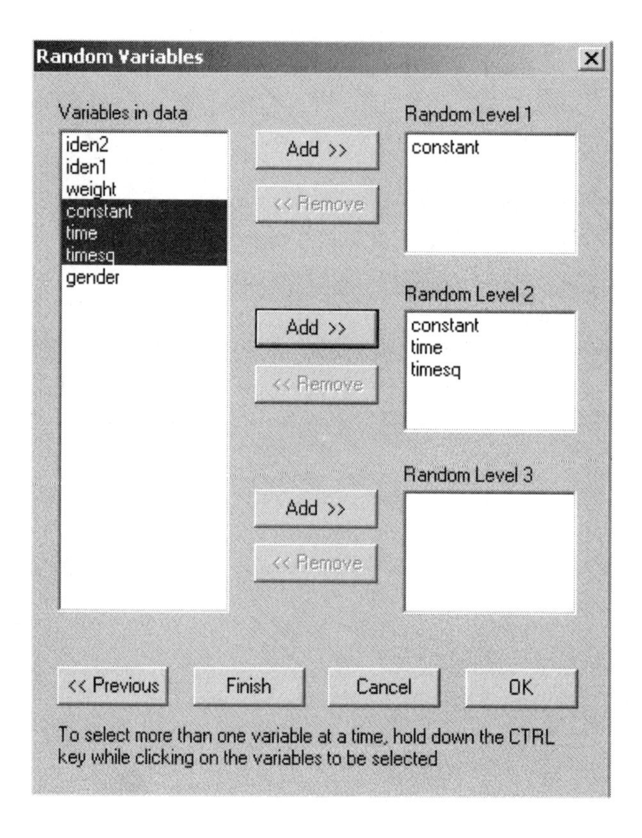

Non-Linear Model Option

The ***Non-Linear Model*** option allows access to five dialog boxes as shown below. The five dialog boxes listed will now be discussed in turn.

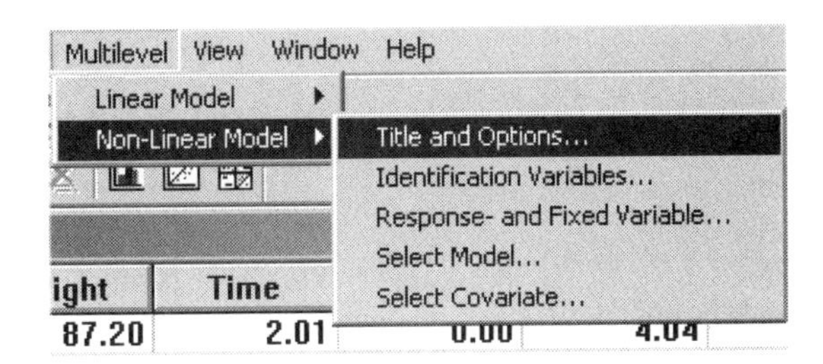

This option is used to fit appropriate models to repeated measurement data that exhibit a non-linear trend over measurement occasions. Suitable candidates for this modeling approach are, for example:

- height measurement of humans
- weights of chickens on two types of diets
- chest measurement of boys and girls from ages 5 to 14.

We consider models that are linear in the regression parameters (for example, any polynomial model) as linear. On the other hand, the model

$$y = b_1 x^{b_2} + e$$

is non-linear in the parameters b_1 and b_2 is therefore considered a non-linear model.

More examples are given in the **nonlinex** folder. Presently, there are five curve types available that can be selected or used in combination when a two-component model is fitted.

Non-Linear Model: Title and Options Option

The ***Title and Options*** option on the ***Multilevel, Non-Linear Model*** menu is associated with the ***Title and Options*** dialog box shown below.

The ***Title and Options*** dialog box is used to provide a title for the analysis and options concerning the iterative procedure.

The ***Number of Quadrature Points*** determines the accuracy of the numerical integration procedure when the Maximum Likelihood method of estimation is selected. This number can be high (20-80) for regression models with 3 or fewer coefficients, but should be lower (6-12) for models with more than 3 coefficients, since high dimensional numerical integration are computationally intensive.

For more details, see Chapter 7. An example of the use of this dialog box is given in Chapter 5.

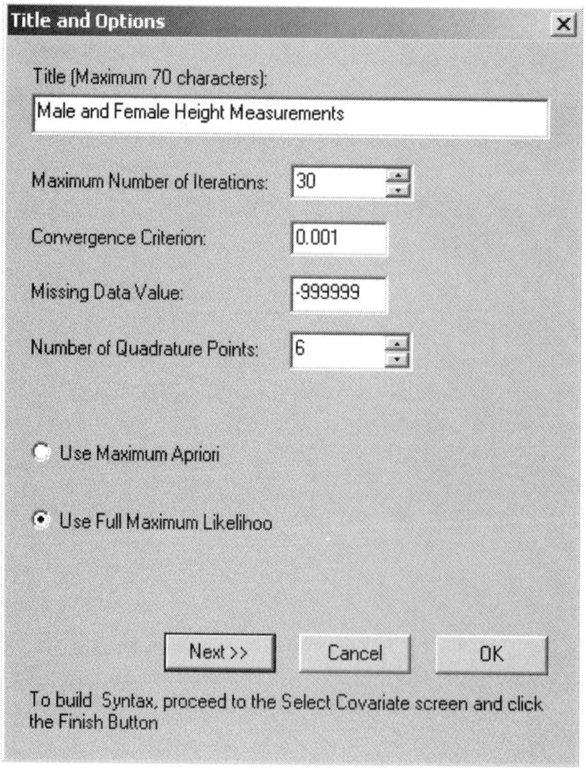

Title and Options

Title (Maximum 70 characters):

Male and Female Height Measurements

Maximum Number of Iterations: 30

Convergence Criterion: 0.001

Missing Data Value: -999999

Number of Quadrature Points: 6

Use Maximum Apriori

Use Full Maximum Likelihoo

Next >> | Cancel | OK

To build Syntax, proceed to the Select Covariate screen and click the Finish Button

Non-Linear Model: Identification Variables Option

The ***Identification Variables*** option on the ***Multilevel, Non-Linear Model*** menu is associated with the ***Identification Variables*** dialog box shown below.

The ***Identification Variables*** dialog box is used to select the variables in the PRELIS system file (***.psf**) that identify the various levels of the hierarchy. Click ***Next*** to go to the ***Response and Fixed Variable*** dialog box when constructing a syntax file.

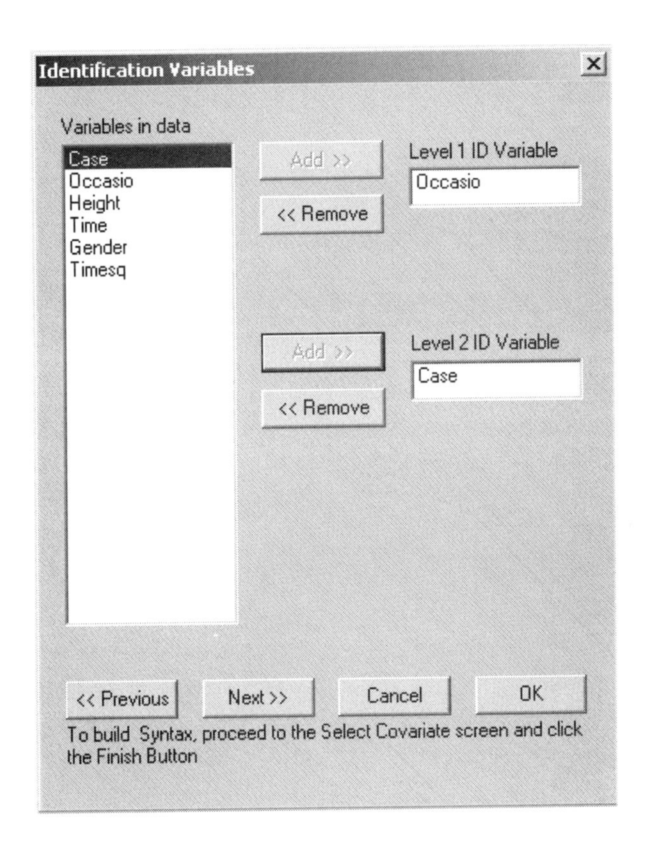

For an illustration of the use of this dialog box, please see Chapter 5.

Non-Linear Model: Response and Fixed Variable Option

The ***Response and Fixed Variable*** option on the ***Multilevel, Non-Linear Model*** menu is associated with the ***Response and Fixed Variable*** dialog box shown below.

This dialog box is used to select the outcome (dependent) variable and fixed variable. Note that only one fixed variable is allowed (see ***Select Model*** dialog box).

See the multilevel examples in Chapter 5 for more information.

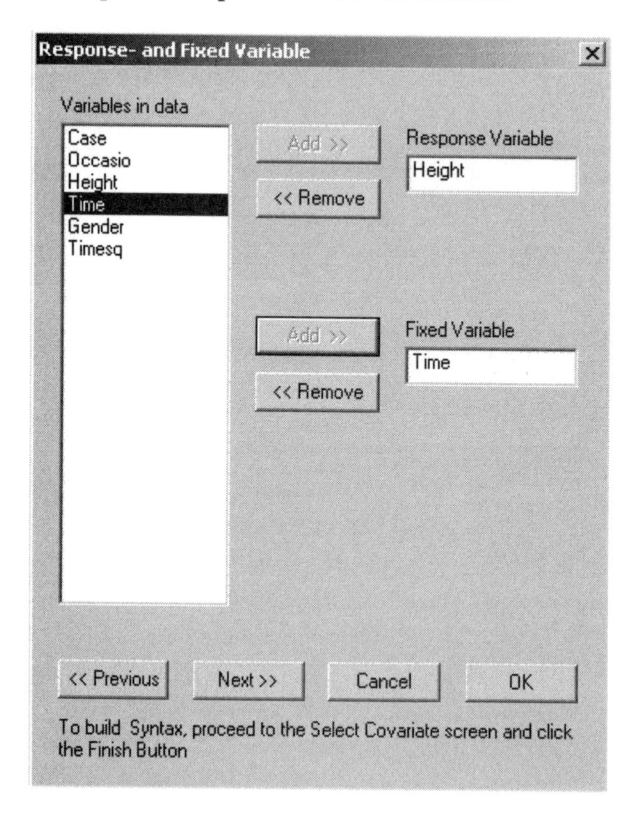

Non-Linear Model: Select Model Option

The ***Select Model*** option on the ***Multilevel, Non-Linear Model*** menu is associated with the ***Select Model*** dialog box shown below.

Models of the type

$$y = f_1(x) + f_2(x) + e$$

may be selected in this dialog box. Function types available are:

- logistic,
- Gompertz,
- Monomolecular,
- power, and
- exponential.

For the definitions of these models, see Chapter 7. An example of a non-linear model is given in Chapter 5.

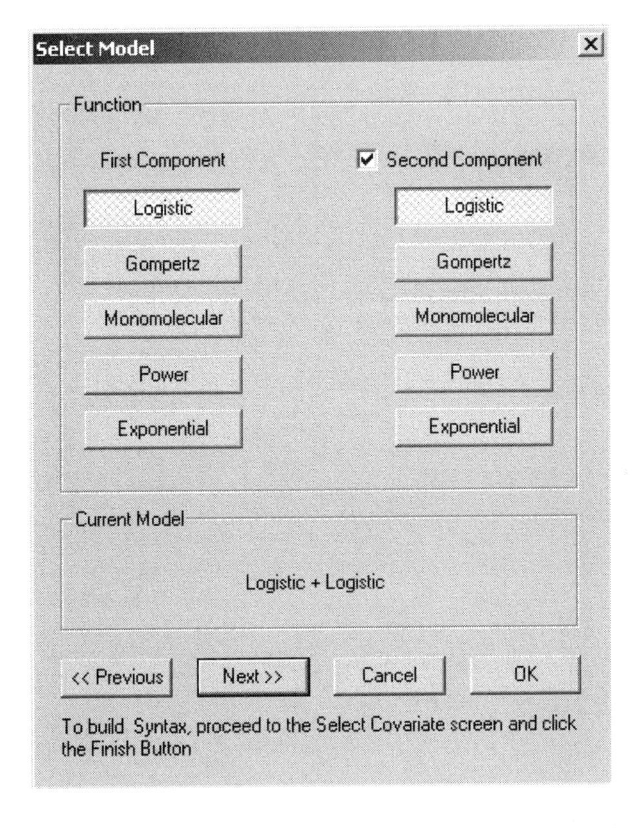

Non-Linear Model Select Covariate Option

The ***Select Covariate*** option on the ***Multilevel, Non-Linear Model*** menu is associated with the ***Select Covariate*** dialog boxes shown below.

These two dialog boxes are used to select a covariate for the non-linear multilevel model. The model coefficients, $b_1, b_2, \ldots, c_6$ can be regarded as outcome variables at level-2 of the hierarchy. As an example, the use of these boxes for the double-logistic model

$$y = \frac{b_1}{1+\exp(-b_2 b_3 x)} + \frac{c_1}{1+\exp(-c_2 c_3 x)} + e$$

where

$$\begin{aligned} b_1 &= \beta_1 + \gamma_1 gender + u_1 \\ b_2 &= \beta_2 + \gamma_2 gender + u_2 \\ &\vdots \\ c_3 &= \beta_6 + \gamma_6 gender + u_6 \end{aligned}$$

is shown below. See Chapter 5 for an example and Chapter 7 for more information on the non-linear models.

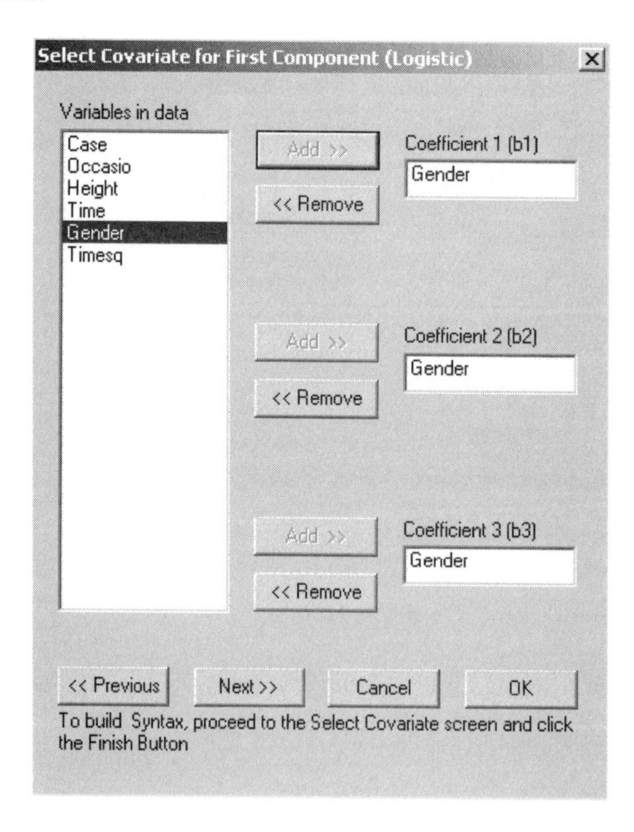

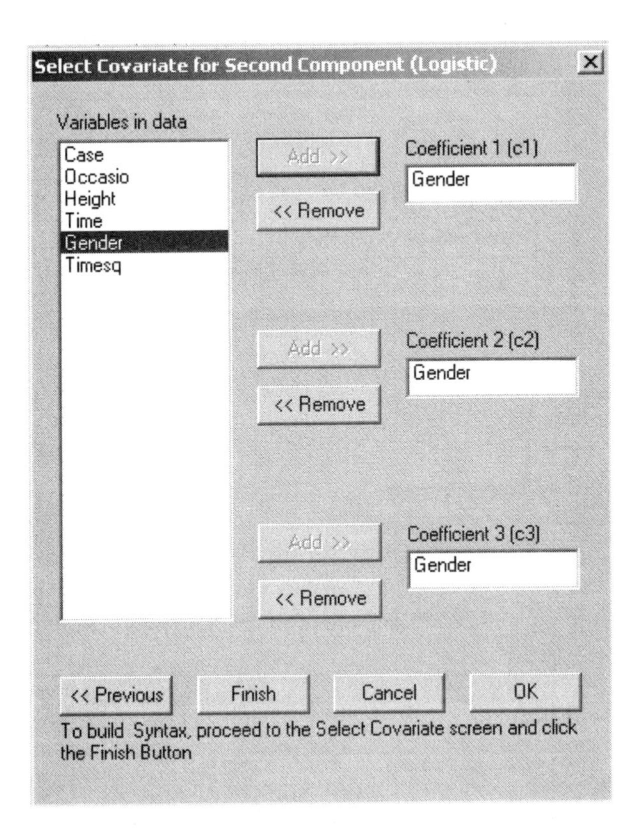

The PRELIS syntax generated by these dialog boxes is shown below.

HEIGHT.PR2

```
OPTIONS METHOD = ML CONVERGE = 0.001000 MAXIT
TITLE = Male and Female Height Measurements;
 SY='D:\lisrel850\Nonlinex\HEIGHT.PSF';
ID1 = Occasio;
ID2 = Case;
RESPONSE = Height;
FIXED = Time;
MODEL = Logistic + Logistic;
COVARIATES b1 = Gender
           b2 = Gender
           b3 = Gender
           c1 = Gender
           c2 = Gender
           c3 = Gender;
```

2.2.9 View Menu

When a PRELIS system file (*.**psf**) is open, the ***Data Toolbar*** option is available from the ***View*** menu, along with the options to ***Scroll Down One Page*** and to move to the ***Bottom*** of the file.

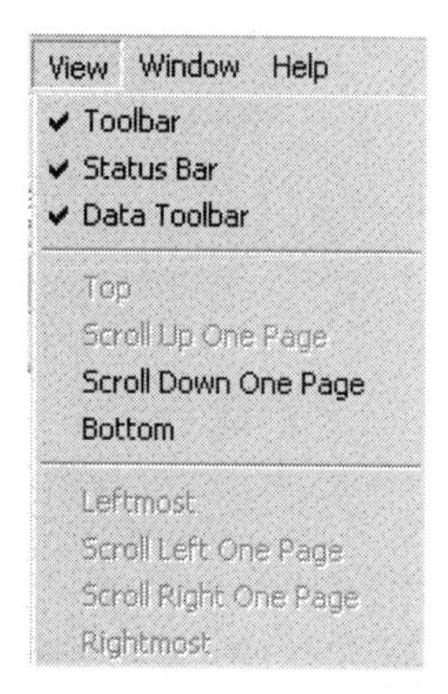

Data Toolbar Option

Checking the ***Data Toolbar*** option from the ***View*** menu adds the ***Scroll Data*** toolbar to the main menu bar. Some of the functions available are only applicable when large *.**psf** files are opened, in which case the spreadsheet displays segements of 50 variables and 5000 cases at a time.

The functions of the buttons are (from left to right):

- Scroll to the top
- Scroll up 5000 lines
- Scroll down 5000 lines
- Scroll to the bottom
- Scroll to the extreme left
- Scroll 50 variables to the left
- Scroll 50 variables to the right
- Scroll to the extreme right

2.2.10 Window Menu

The Window menu allows the user to arrange windows and switch between windows.

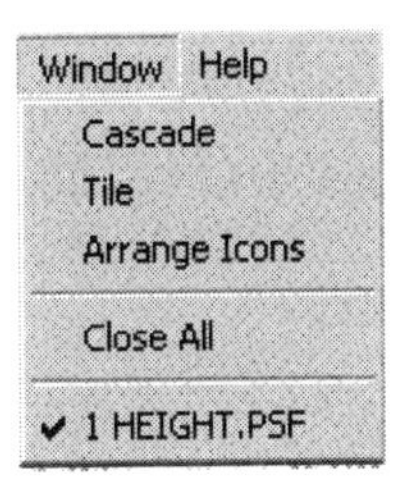

2.3 Text Editor Window

When a text file is opened, the main menu bar changes to:

2.3.1 File Menu

The ***File*** menu now includes options to run LISREL and PRELIS syntax files, except when the file opened is of type *.**out**.

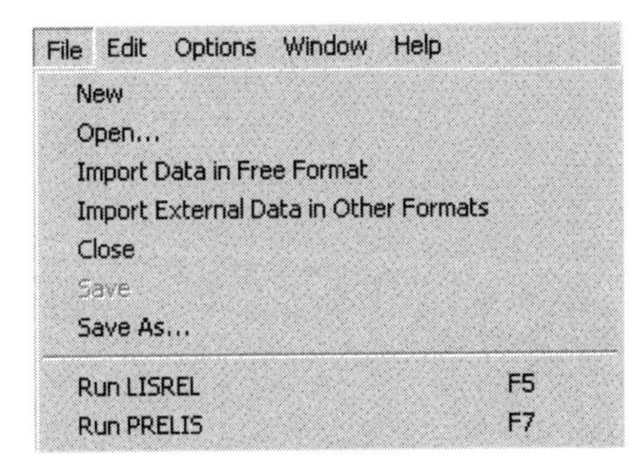

2.3.2 Edit Menu

The ***Edit*** menu available in the text editor window adds standard Windows functionality to the manipulation of text files opened in the text editor window, as shown below.

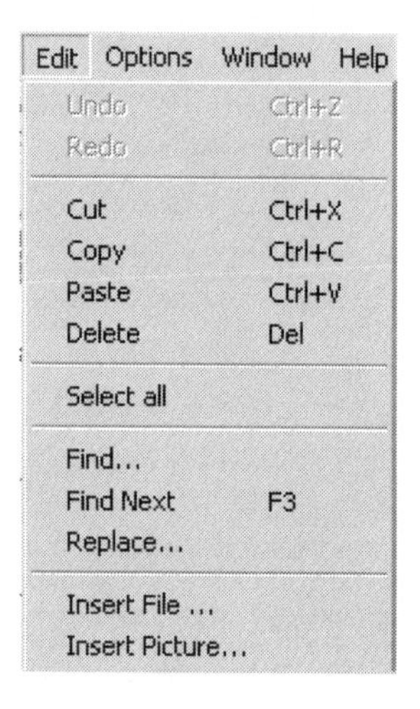

2.3.3 Options Menu

When a text file is opened, the ***Options*** menu is activated. This menu is shown below. It controls the ***Toolbar***, ***Status Bar*** and ***Edit Bar*** as well as the general appearance of the text displayed.

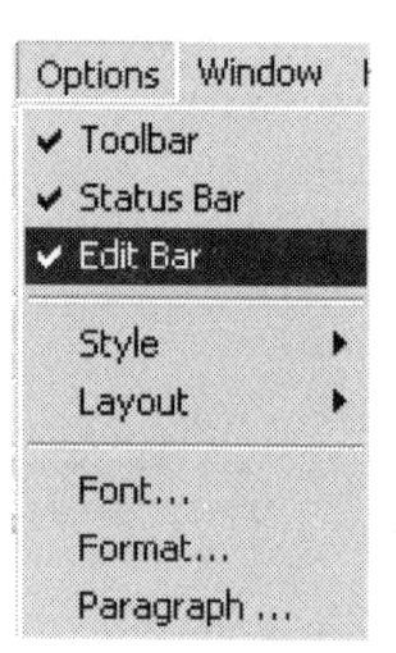

Edit Bar Option

The ***Edit Bar*** is activated whenever a text file is displayed in the text editor window. Examples, apart from standard text files, are:

- PRELIS syntax or output files,
- LISREL project files (*.**lpj**), or
- SIMPLIS project files (*.**spj**).

The functions of the button icons are described below. To select the text to be changed, click and drag the cursor over the text that you want to change and then click on the appropriate icon button of choice.

- **Bold Text**: Change marked text to bold face.
- **Italic Text**: Change marked text to italic type.
- **Underline Text**: Underline marked text.
- **Strike Through**: Draw a horizontal line through marked text.
- **New Font and Color**: Change font and color of marked text or of document. By clicking this icon, a pull-down menu is activated which provides a list of fonts and colors that are available on your system. If you change fonts, keep in mind that LISREL and PRELIS output require fonts with fixed pitch (so-called monospace fonts since each character has the same width) to keep tables and matrices vertically aligned.
- **Align Left**: Left align selected text. Note that text with blanks inserted in the beginning of each line will not left align since blanks are also interpreted as valid text characters.
- **Center Text**: Center selected text. Note that text with blanks inserted in the beginning of each line will not center since blanks are also interpreted as valid text characters.
- **Align Right**: Right align selected text.
- **Justify Text**: Change the paragraph justification from the default left justification (or "ragged" right).
- **Single Space Text**: Marked text is displayed with single vertical space, that is, no blank lines between successive lines of text.
- **1.5 Spacing**: Marked text is displayed with about half the width of one line of text additional space between lines.
- **Double Space Text**: Marked text is displayed with double vertical space, that is, a blank line is inserted between any two successive lines of text.

The options described above can also be accessed via the ***Options*** menu as shown below. The ***Style*** option controls the fonts, while the ***Layout*** option controls the adjustment and spacing of text.

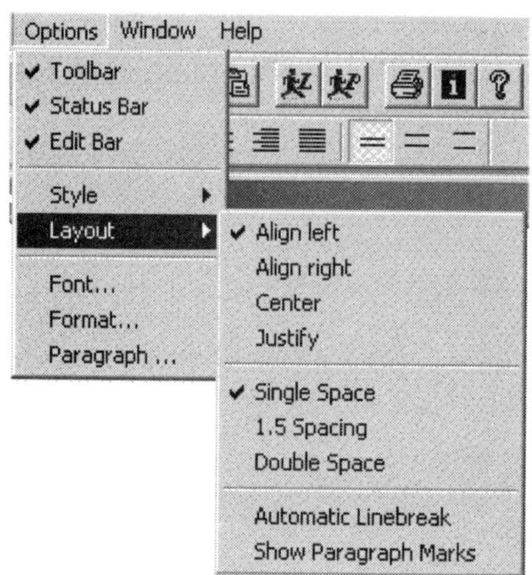

2.4 The Path Diagram Window

When a path diagram is displayed or a *.**pth** file is opened, the main menu for the path diagram window appears. The Path Diagram menu bar has the ***File***, ***Edit***, ***Setup***, ***Draw***, ***View***, ***Image***, ***Output***, ***Window*** and ***Help*** menus as shown in the image below.

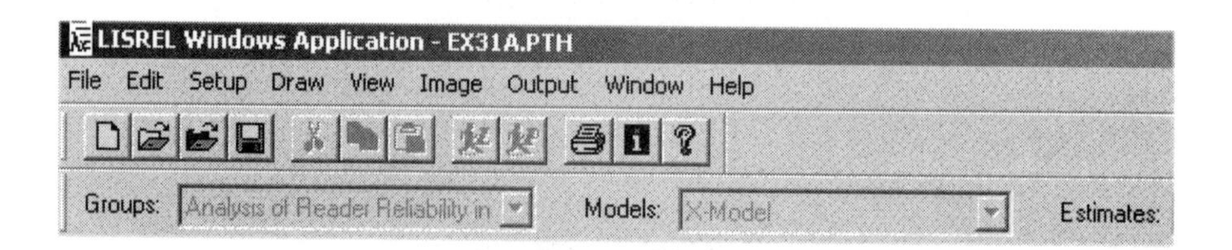

2.4.1 File Menu

An addition to the ***File*** menu when a path diagram is displayed is the ability to export a *.**pth** file to a *.**wmf** or *.**gif** format.

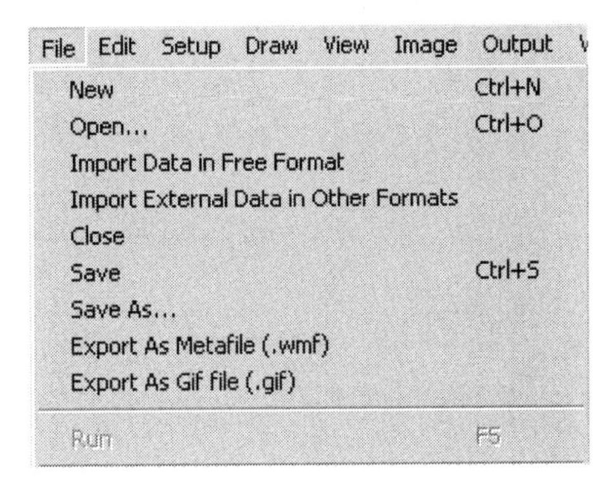

2.4.2 Edit Menu

In the path diagram window, the ***Edit*** menu has only three options, all standard Windows functions.

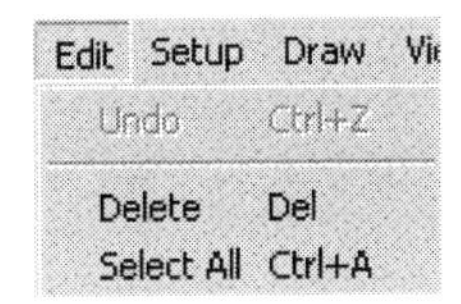

2.4.3 Setup Menu

The ***Setup*** menu is displayed below.

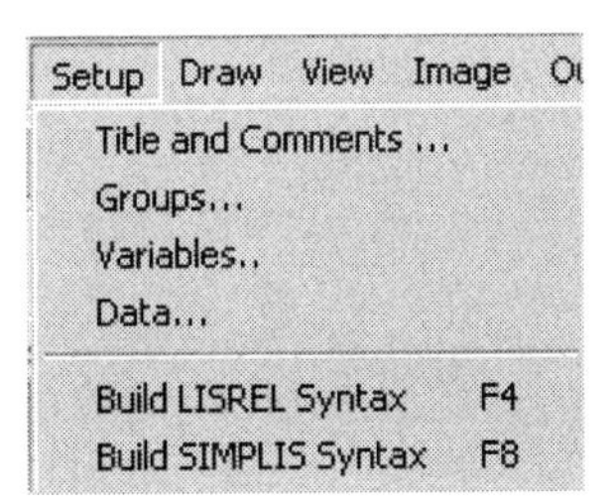

When building LISREL or SIMPLIS syntax from a path diagram, the options displayed are used to:

- Provide a title and comments
- Provide group names (multi-sample analysis)
- Create a list of observed and latent variables
- Specify type and location of the input data, for example: covariance matrix, **c:\temp\ex10.cov**
- Build LISREL or SIMPLIS syntax

Note:

The options on the ***Setup*** menu are associated with the following LISREL syntax:

- ***Title and Comments*** : `TI` command
- ***Groups*** : `NG` keyword
- ***Variables*** : `LA`, `LE`, `LK` commands
- ***Data*** : `DA` command

The options on the ***Setup*** menu are illustrated in detail in Chapter 4.

Title and Comments Option

The ***Title and Comments*** option on the ***Setup*** menu is associated with the ***Title and Comments*** dialog box shown below.

The ***Title and Comments*** dialog box is used to provide a title and additional comments. Use of it is optional. Click ***Next*** to go to the ***Group Names*** dialog box.

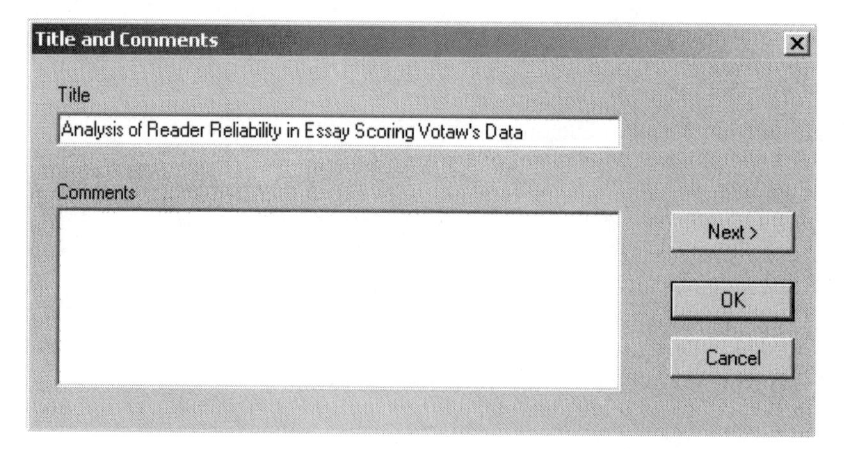

Groups Option

The ***Groups*** option on the ***Setup*** menu is associated with the ***Group Names*** dialog box shown below.

The ***Group Names*** dialog box is used to provide labels in the case of multiple-group analysis. If only one group is analyzed, this dialog box is not used. Click ***Next*** to go to the ***Labels*** dialog box. ***Click Previous*** to go to the ***Title and Comments*** dialog box.

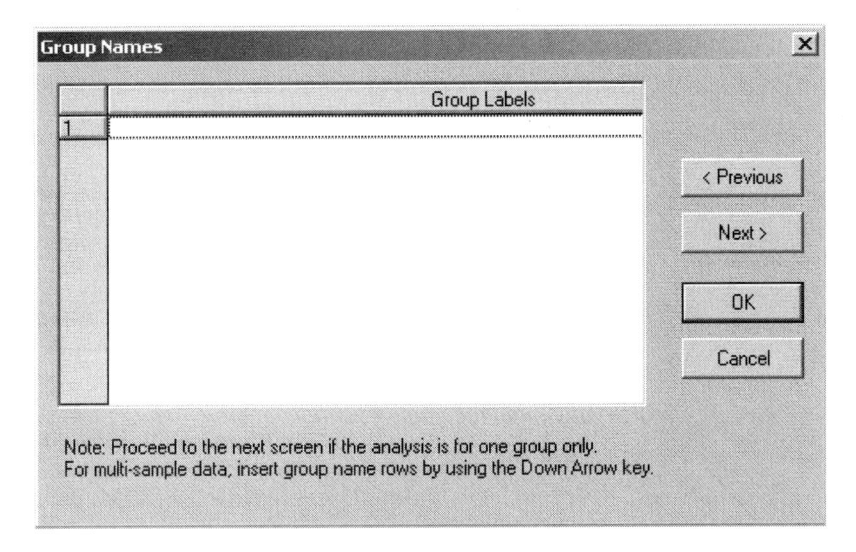

See Sections 4.4, 4.8.2, and 4.10 in Chapter 4 for examples.

Variables Option

The ***Variables*** option on the ***Setup*** menu is associated with the ***Labels*** dialog box shown below.

The ***Labels*** dialog box is used to provide labels for variables. Click ***Next*** to go to the ***Data*** dialog box. Click ***Previous*** to go to the ***Group Names*** dialog box. The ***Add/Read Variables*** button provides access to the ***Add/Read Variables*** dialog box. The ***Add Latent Variables*** button is used to go to the ***Add Variables*** dialog box. The ***Add/Read Variables*** and ***Add Variables*** dialog box are discussed next.

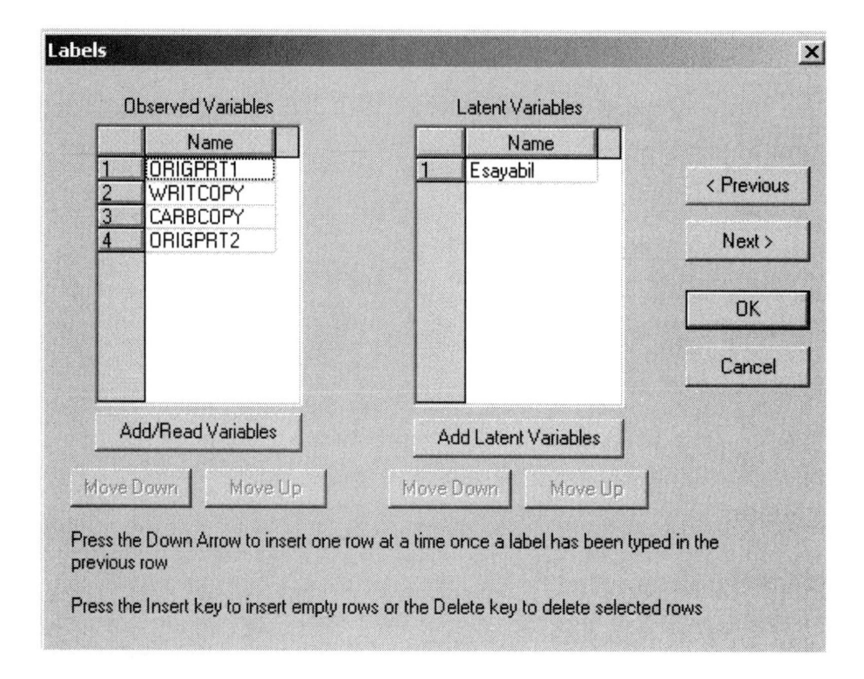

Add/Read Variables Dialog Box

This dialog box is used to read variables (labels) from an external LISREL data system file (*.**dsf**) or a PRELIS system file (*.**psf**), or to type in names.

- The ***Read from file*** radio button and drop-down list box may be used to select the type of file to be read.
- The ***Browse*** button allows the user to browse for the file to be used.
- The ***Add list of variables*** radio button is used to access the ***Add Variables*** dialog box.

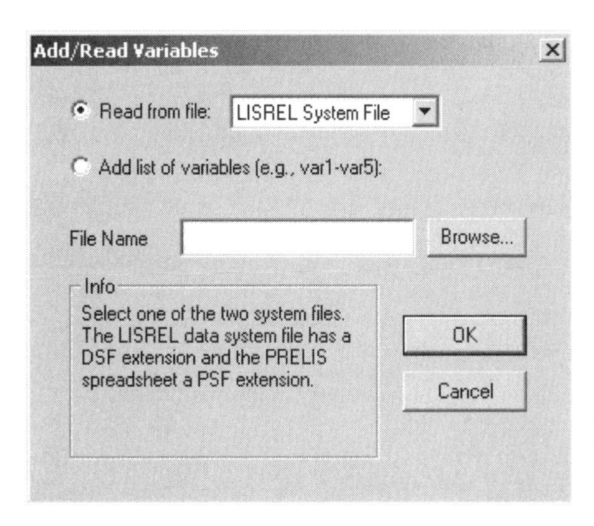

Add Variables Dialog Box

One or more variables can be added to the PRELIS system file (spreadsheet). If *V1-V3* is entered, for example, the variable names *V1*, *V2*, and *V3* are generated.

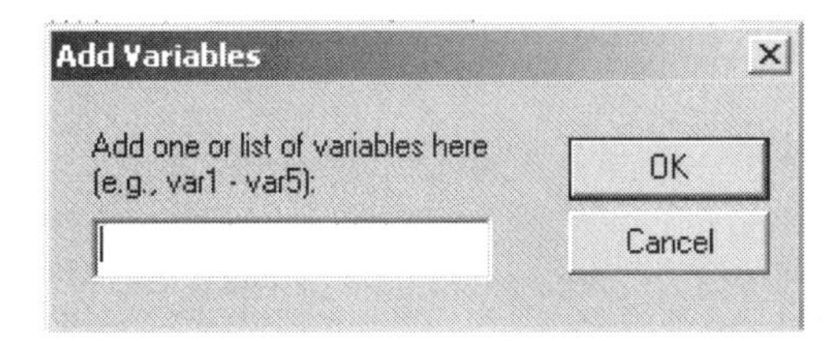

This dialog box is also activated if one clicks the ***Add Latent Variables*** button on the ***Labels*** dialog box. An example of adding variables can be found in Sections 4.3 and 4.4.

Data Option

The ***Data*** option on the ***Setup*** menu is associated with the ***Data*** dialog box shown below.

The ***Data*** dialog box is used to specify, for each group, the file type. For example, a *.**dsf** file is associated with covariances/correlations (see ***Statistics from*** drop-down list box) and a *.**psf** file with raw data. The ***Number of Observations*** field is used to indicate the number of observations to be used in the analysis. The ***Matrix to be analyzed*** drop-down list box is used to select the type of matrix to be analyzed, while the ***Weight*** check box allows the specification of a weight matrix, such as, for example, an asymptotic covariance matrix. See the examples in Chapter 4 for a more detailed description.

Click ***Previous*** to go to the ***Labels*** dialog box. Click ***OK*** to return to the path diagram window.

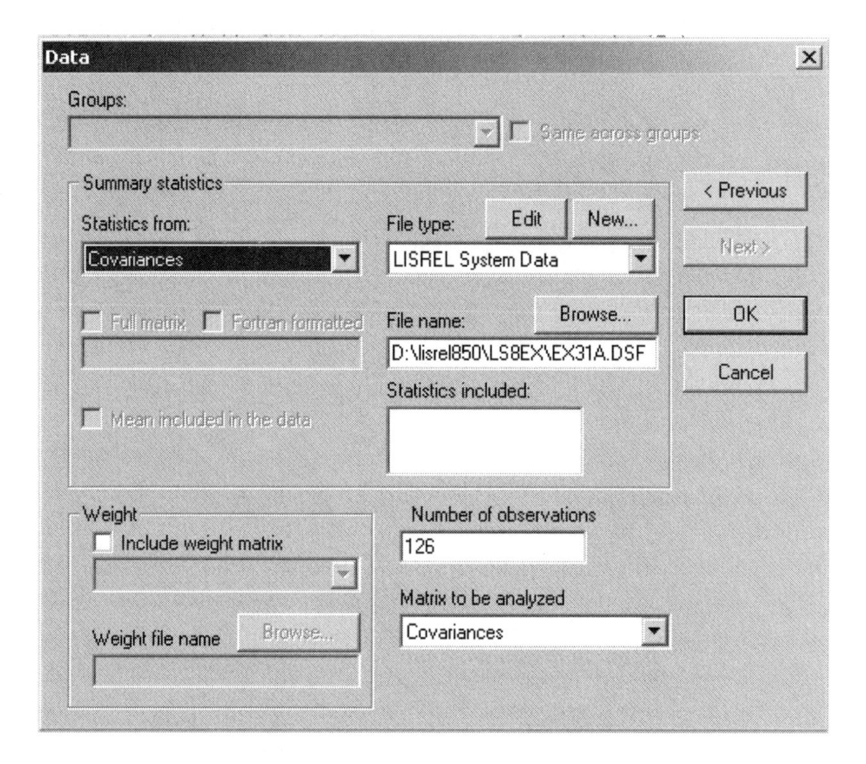

Build LISREL/SIMPLIS Syntax Options

The final two options on the ***Setup*** menu allows the users to build LISREL or SIMPLIS syntax from the path diagram.

2.4.4 Draw Menu

When creating LISREL/SIMPLIS syntax from a path diagram or whenever a path diagram is displayed on the screen, the ***Draw*** menu is activated. One can either use the ***Draw*** toolbox shown at the bottom of the image below or select an option from the ***Draw*** menu. If an item is selected from this menu, the corresponding toolbar icon button is highlighted.

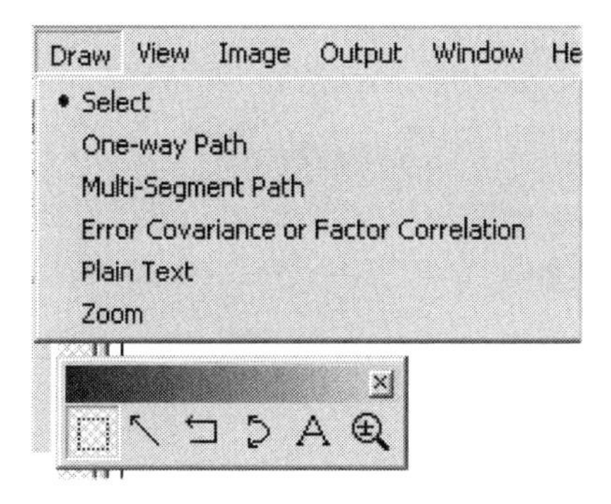

The options and corresponding icon buttons are discussed in turn below.

Select Object Option

The ***Select Object*** option is used to select one or more objects in the path diagram. A selected object may be moved or resized. The color, font style, line thickness, etc. can also be adjusted. This is done by clicking the right mouse button once an object is selected.

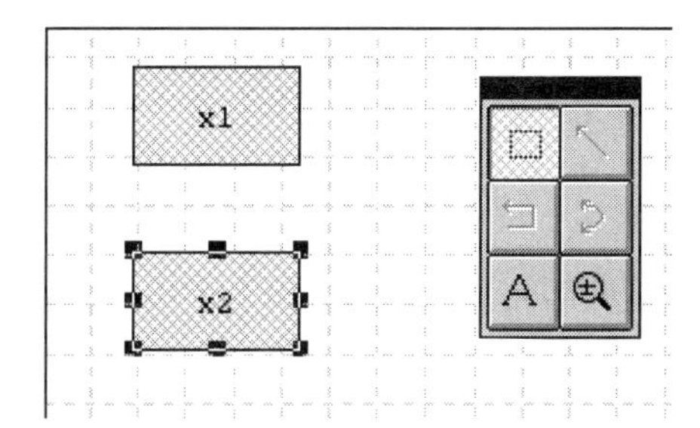

One-way Path Option

The ***One-way Path*** option is used to draw a path from one variable to another. The diagram below demonstrates the drawing of a path from the latent variable *ksi1* to the observed variable *x2*.

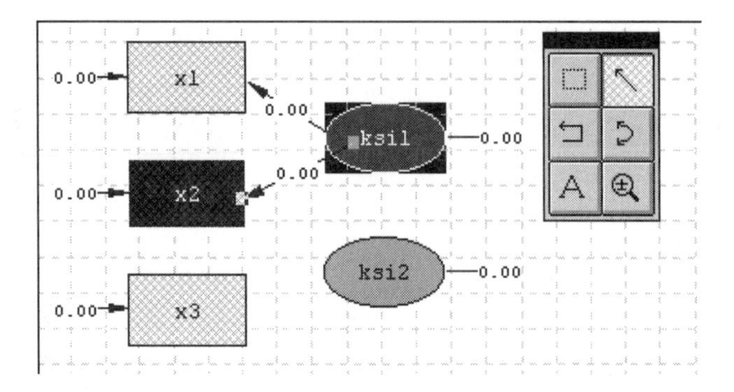

To draw the path, click on the ***One-way Path*** icon button (single-headed arrow icon button on the toolbox to the right in the image above). Click with the left mouse button inside the *ksi1* ellipse and drag the mouse to within the *x2* rectangle. Release the mouse button. To deactivate the arrow function, click on the ***Select Object*** icon button.

Two-way Path Option

The ***Two-way Path*** option (double-headed arrow icon button on the toolbox to the right in the image below) is used to denote a correlation between two variables. In the diagram shown below, a covariance path is drawn between the variables *ksi1* and *ksi2*.

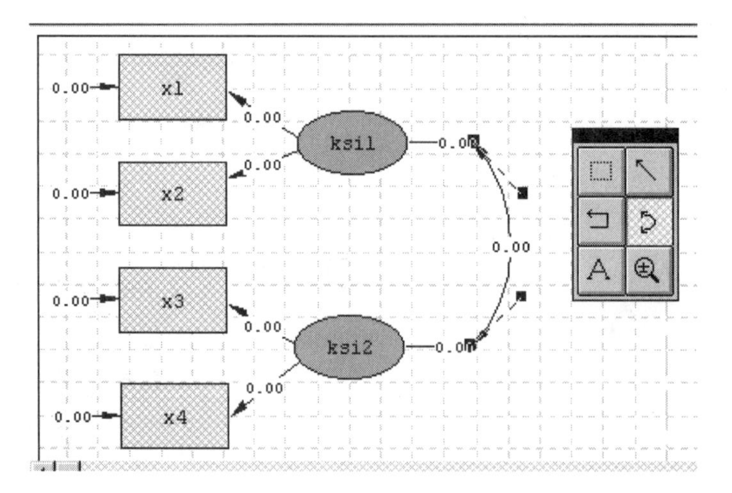

To draw this path, click on the ***Two-way Path*** icon button to activate this choice. Click on the horizontal line indicating the *ksi1* variance and drag the mouse cursor to the horizontal line indicating the *ksi2* variance.

Multi-Segment Line Option

The ***Multi-segment Line*** option, represented by the icon button highlighted on the toolbox to the right in the image below, is used to draw a path as a series of segments.

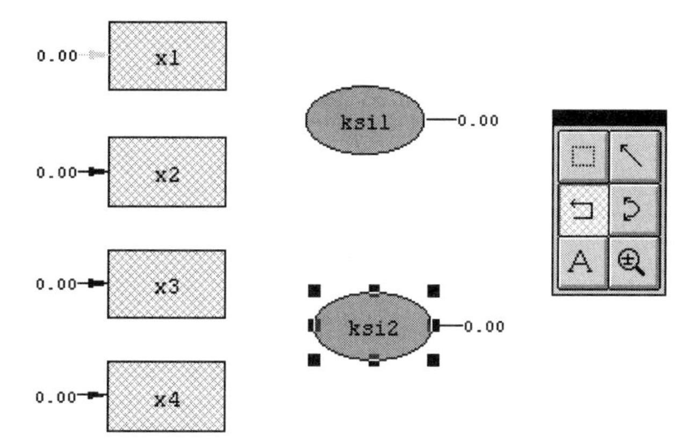

Note:

Do not use this tool for drawing a single-segment line.

Click the ***Multi-segment Line*** icon button and move the mouse button to within the ellipse (*ksi1*). Without releasing the left button, drag a vertical line to the desired position and release the mouse button.

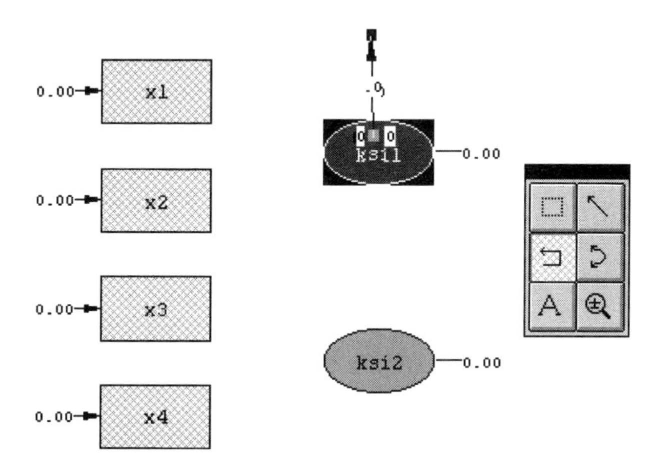

While at this position, click the left mouse button and drag a horizontal line to the desired position, then briefly release the mouse button while remaining at this position.

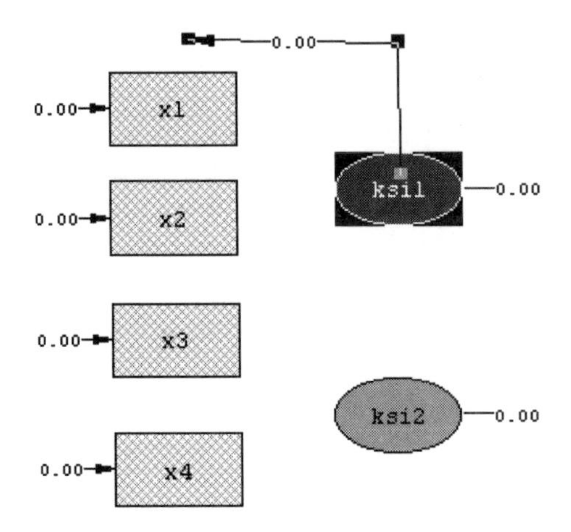

Finally, drag a vertical line down to within the rectangle (*x1*) and release the mouse button. Undo the ***Multi-Segment Line*** option by clicking on the ***Select Object*** icon button (highlighted in the image below) of the draw toolbar.

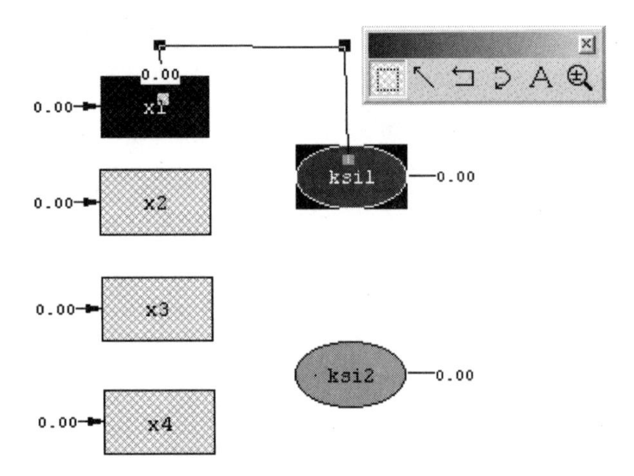

Plain-Text Option

The ***Plain-Text*** option may be used to add additional text to the path diagram prior to printing it.

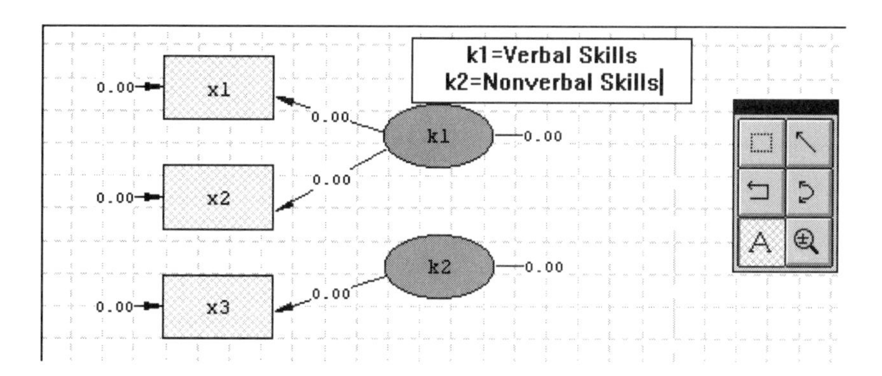

Click on the ***Plain-Text*** icon button (see highlighted icon button in the image above) and then draw a rectangle on the path diagram by dragging the mouse button diagonally from top left to bottom right. Release the mouse button and enter the text. Deactivate the ***Plain-Text*** icon button by clicking the ***Select Object*** icon button. The font and style of the text may be changed by selecting the text area as an object. By clicking the right mouse button, a pop-up dialog box appears. From this menu select ***Options***.

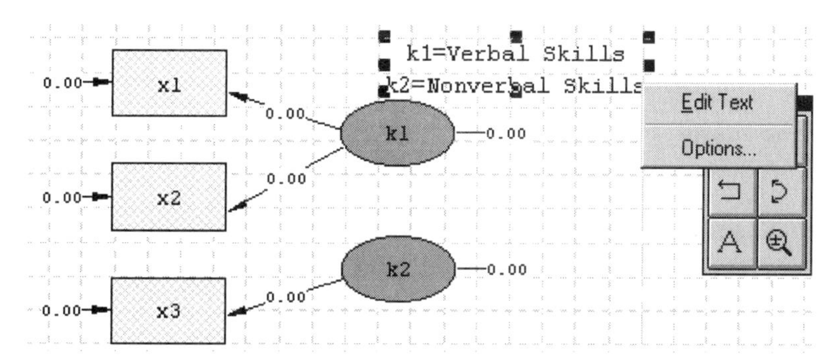

Use the ***Options*** dialog box discussed elsewhere in this section to change the font style and color.

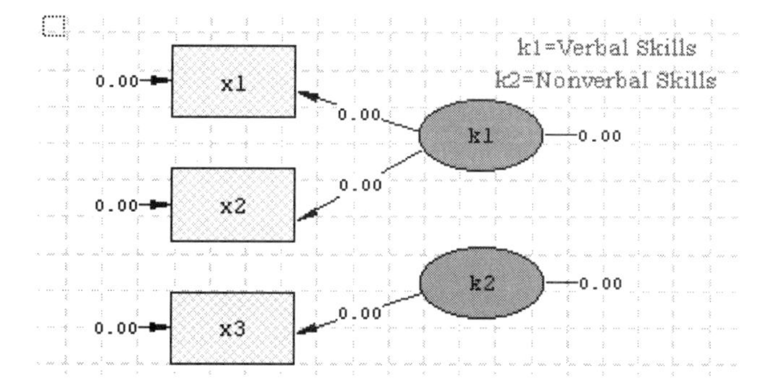

Zoom Option

Use the ***Zoom*** icon button (highlighted in the toolbox shown below) to reduce or increase the size of the path diagram.

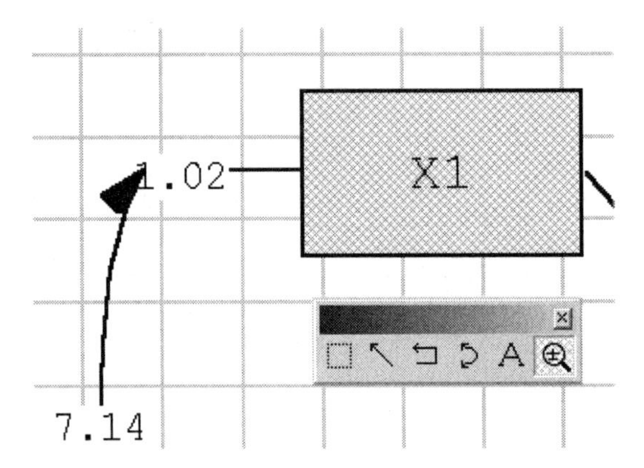

Use the right mouse button to increase and the left mouse button to decrease the size of the path diagram.

2.4.5 View Menu

When a path diagram is open in the path diagram window, the ***View*** menu is expanded with options specific to path diagrams, as shown below.

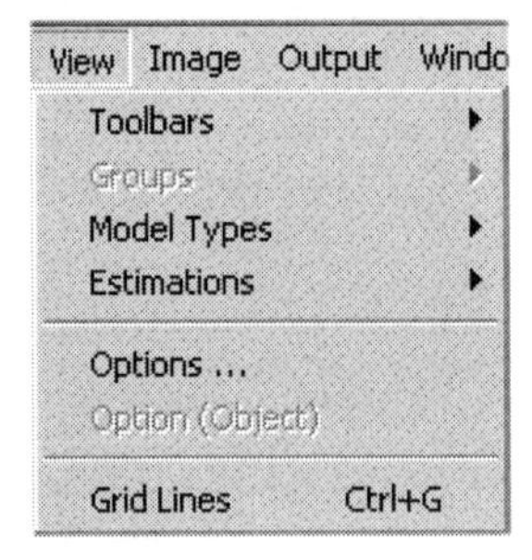

Toolbars Option

The ***Toolbars*** option provides access to the ***Toolbar***, ***Status Bar***, ***Typebar***, ***Variables*** and ***Drawing Bar*** options.

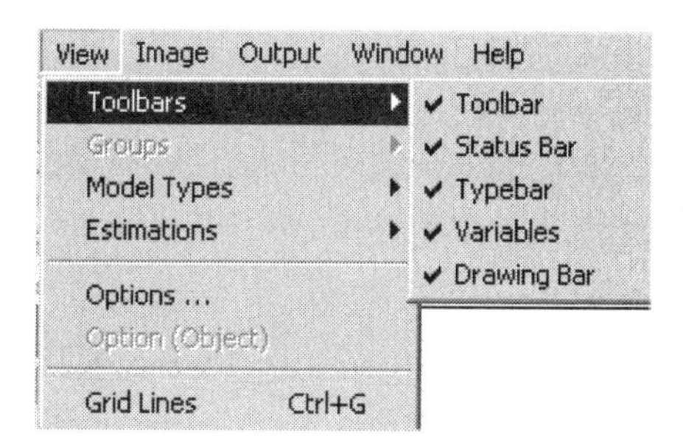

Four of the five toolbars that can be selected through use of the ***Toolbars*** options are shown below. In the first image, the ***Toolbar*** is displayed at the top of the image, with the ***Type Bar*** right below it. In the second image, the ***Variables*** bar, which is added to the left of the path diagram window, is shown. The ***Drawing Bar***, also known as the ***Draw*** toolbox, is shown to the right of the ***Variables*** bar.

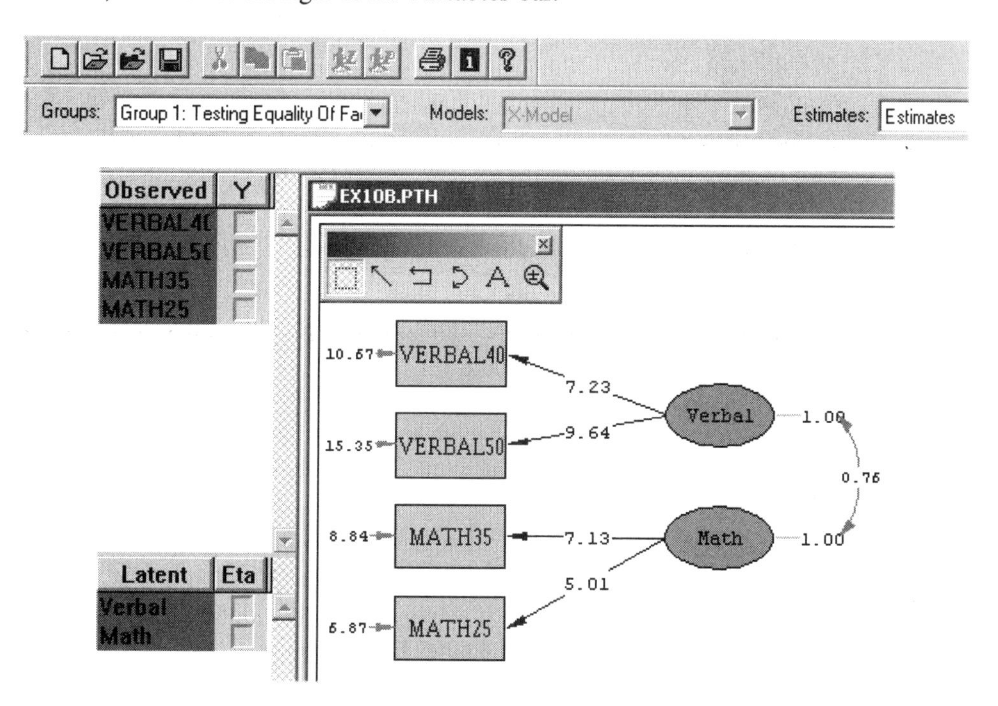

Model Types Option

The ***Model Types*** option provides access to options allowing the user to view, in turn,

- The basic model
- The X-model

- The Y-model
- The structural model
- Correlated errors
- The mean model.

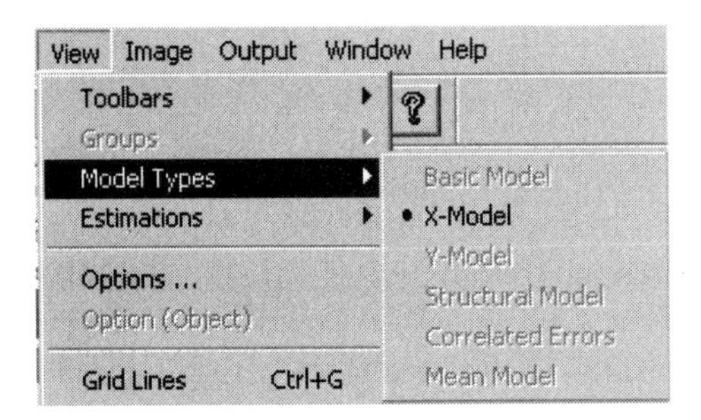

Selecting any of these options will lead to the display of the values associated with the selected option on the path diagram.

For examples of the various types, see Section 4.5.5.

Estimations Option

The ***Estimations*** option provides access to options allowing the user to view, in turn,

- The estimates
- The standardized solution
- A conceptual diagram
- *t*-values
- The modification indices
- The expected changes.

Selecting any of these options will lead to the display of the values associated with the selected option on the path diagram.

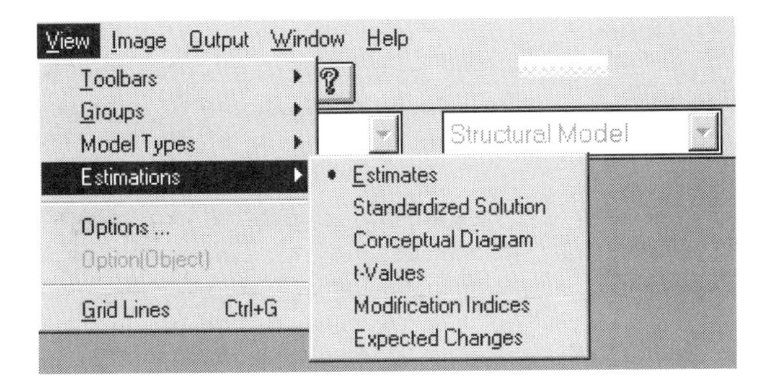

For examples of these various options, see Chapter 4.

Options Option

The ***Options*** option on the ***View*** menu is associated with the ***Options*** dialog box.

This dialog box is used to control the appearance of objects on the path diagram.

- The ***Outline***, ***Outline Color***, ***Fill Color*** and ***Font*** (of the label) for any type of object on the path diagram can be changed here.
- The type of object to which these changes should be applied is selected using the ***Type of Object*** drop-down list box.
- The color of non-significant *t*-values may be adjusted using the ***Non-significant t Color*** button which provides access to a color palette from which to select a new color.
- Background color may be adjusted using the ***Background Color*** button which provides access to a color palette from which to select a new color.

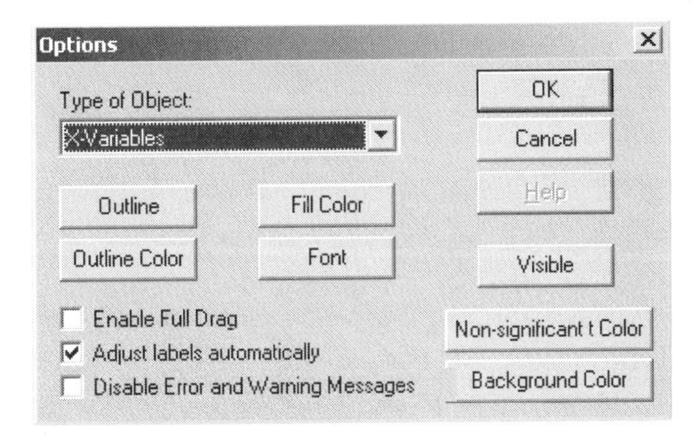

Option (object) Option

The ***Option (object)*** option on the ***View*** menu is associated with the ***Options for Object*** dialog box shown below.

This dialog box may be used to change the attributes on any selected object. An object in the path diagram window may be selected using the ***Select Object*** icon on the ***Draw*** toolbox.

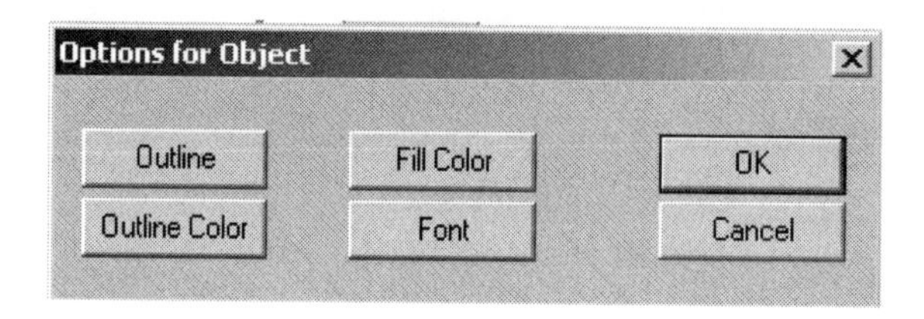

Grid Lines Option

The ***Grid Lines*** option on the ***View*** menu can also be selected to display grid lines when drawing a new path diagram. For further information, see Chapter 4.

2.4.6 Image Menu

If the active window is a path diagram window, the ***Image*** menu is activated. From the ***Edit*** menu, choose ***Select all***. The path diagram can then be rotated, moved, etc. A black and white path diagram, useful for using in publications, can be created using the ***Black and White*** option on this menu.

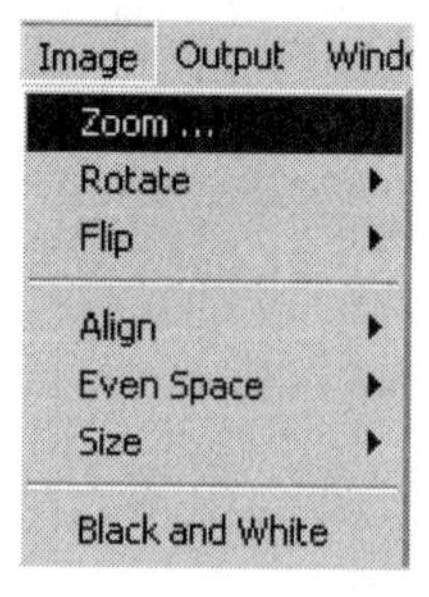

Use of these options is illustrated in the examples given in Chapter 4.

2.4.7 Output Menu

The ***Output*** menu enables one to change the default SIMPLIS or LISREL output syntax.

For example, the number of decimals displayed and the method of estimation may be changed by using the menus that are activated when an ***Output*** selection is made. If a model has been fitted to the data, one may view the ***Fit Indices*** file by selecting this option on the ***Output*** menu.

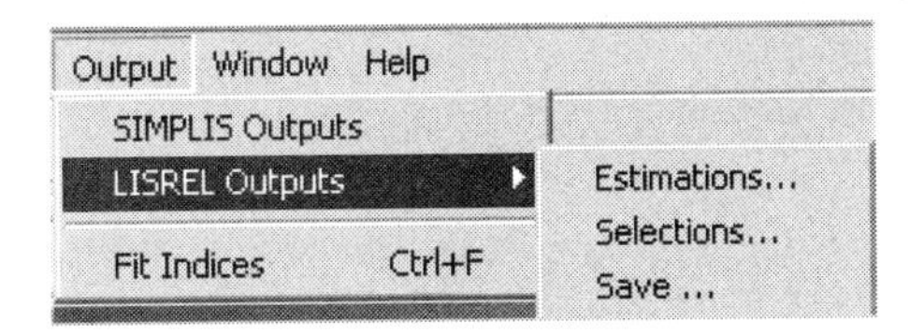

SIMPLIS Outputs Option

The ***SIMPLIS Outputs*** option on the ***Output*** menu is associated with the ***SIMPLIS Outputs*** dialog box shown below. This dialog box may be used to control the method of estimation and other options associated with the iterative procedure. It may also be used to change the number of decimals in the output, the printing of residuals, etc.

SIMPLIS Outputs

Method of Estimation

Maximum Likelihood
Generalized Least Squares
Two-stage Least Squares
Generally Weighted Least Squares
Instrument Variables
Diagonally Weighted Least Squares
Unweighted Least Squares

Set Check Admissibility to 20 Iterations
OK
Cancel
Maximum Number of Iterations -1
Number of Decimals (0-8) in the Output 2
Default
Print Residuals
Wide Print
Save Sigma (fitted matrix) .sis
Select LISREL Outputs
Crossvalidate File .cvf
Invoke Path Diagram

Note:

The ***SIMPLIS Outputs*** option is associated with the `Output` command in SIMPLIS.

LISREL Outputs Option

The ***LISREL Outputs*** option on the ***Output*** menu is associated with three dialog boxes:

- ***Estimations***
- ***Selections***, and
- ***Save***.

These three dialog boxes are discussed next.

Note:

The ***LISREL Outputs*** option is associated with the `OU` command in LISREL.

Estimations Dialog Box

The ***Estimations*** dialog box may be used to control the method of estimation and other options associated with the iterative procedure.

When building LISREL/SIMPLIS syntax from a path diagram, it is important to ensure that the appropriate method of estimation is selected. For example, suppose the observed variables are ordinal and that the asymptotic covariance matrix of the polychoric coefficients was computed. In this case, the appropriate estimation procedure is Weighted Least-Squares (WLS) which corresponds to the ***Generally Weighted Least Squares*** radio button.

See *LISREL 8: User's Reference Guide* pp. 91-99 for a description of this command.

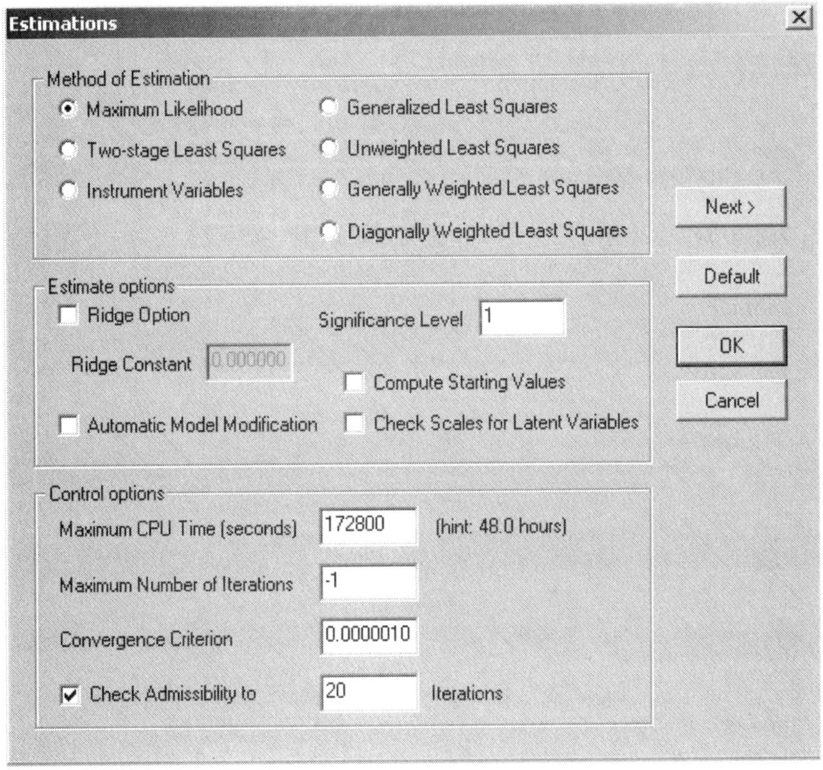

Note:

This dialog box is associated with the OU command in LISREL.

Selections Dialog Box

The ***Selections*** dialog box allows the user to invoke a path diagram, control the number of decimals in the output, and specify the type of output required.

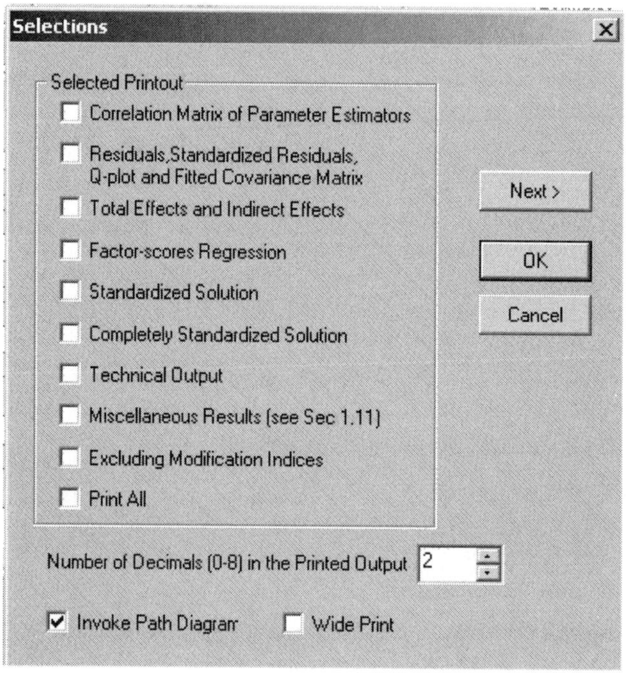

Note:

This dialog box is associated with the OU command in LISREL.

Save Matrices Dialog Box

The ***Save Matrices*** dialog box allows the user to specify external files to which, for example, the asymptotic covariance matrix of parameter estimates and the *t*-values can be saved. The user is referred to page 96 of *LISREL 8:User's Reference Guide* for a detailed description.

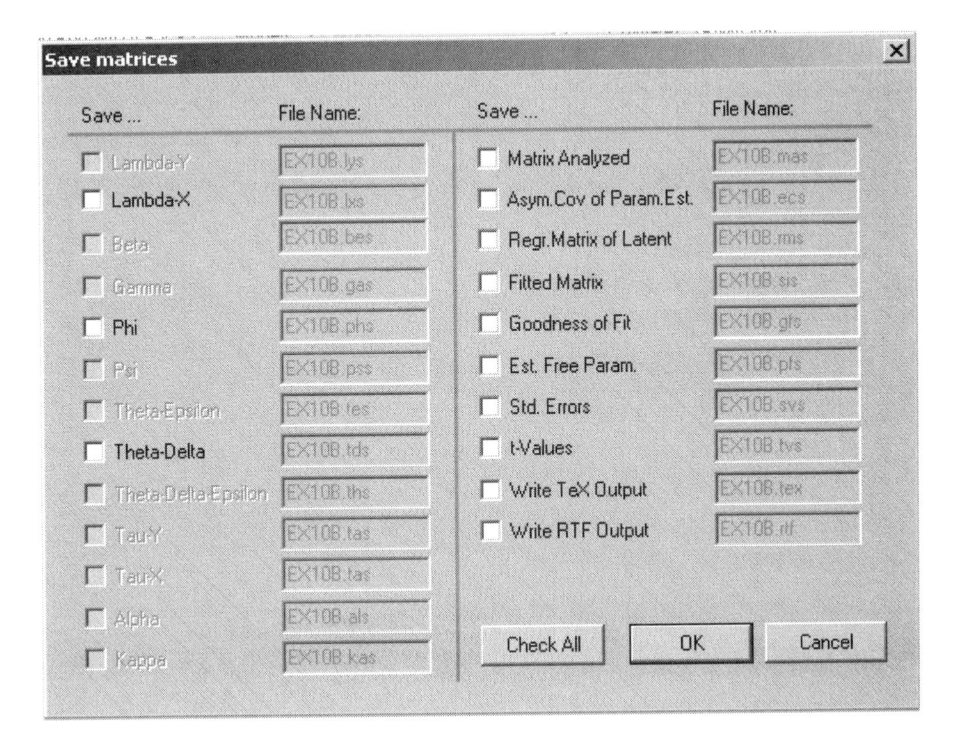

Note:

This dialog box is associated with the OU command in LISREL.

2.5 SPJ Window

When a SIMPLIS project (*.**spj**) is created or opened, the main menu bar changes to:

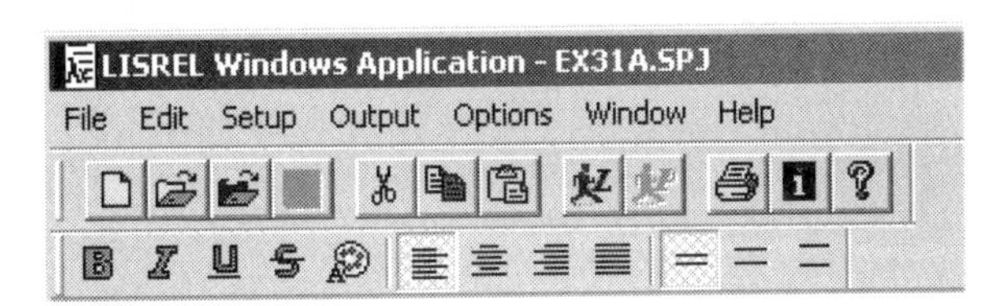

In addition, the ***Relationships*** keypad for use in changing syntax is displayed below the opened SIMPLIS project file as shown below. The only menu with differences in functionality is the ***Setup*** menu. For all other menus, the user is referred to the detailed discussion of these in previous sections of this chapter.

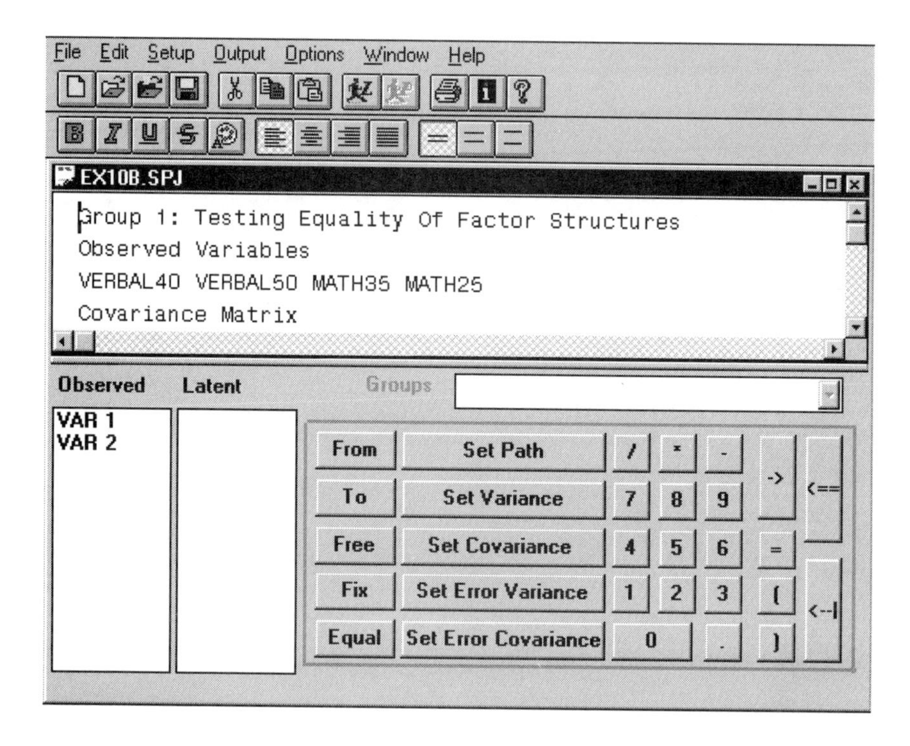

2.5.1 Setup Menu

In the case of the SPJ window, the Setup menu includes the following options that have already been discussed in the section dealing with options available in the case of the PSF window: ***Title and Comments***, ***Groups***, ***Variables***, and ***Data***.

New options available in the SPJ window are:

- ***Build SIMPLIS Syntax***: This option allows the user to generate SIMPLIS syntax based on the SIMPLIS project currently open.
- ***Build LISREL Syntax***: This option allows the user to generate LISREL syntax based on the SIMPLIS project currently open.

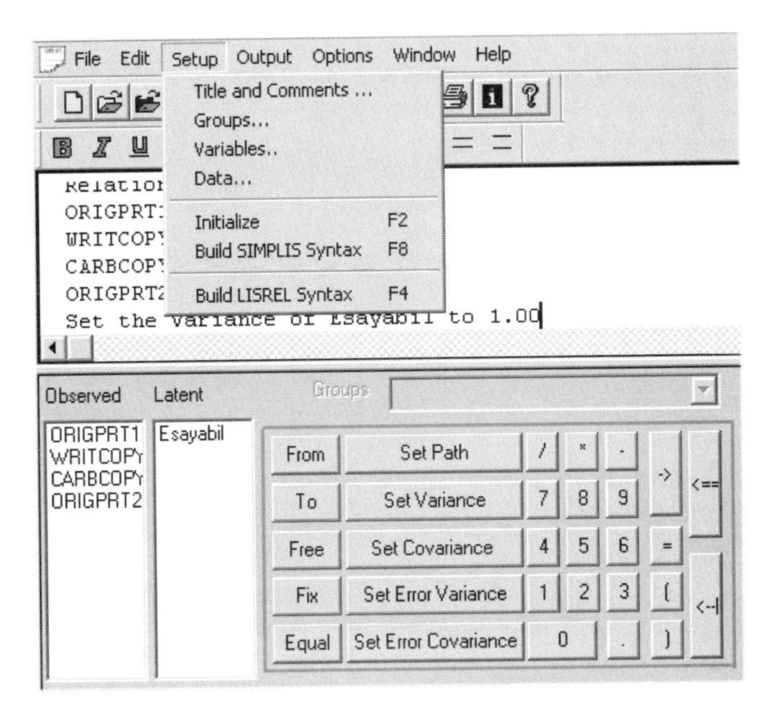

2.6 The LPJ Window

The main menu bar displayed when a LISREL project is opened is shown below.

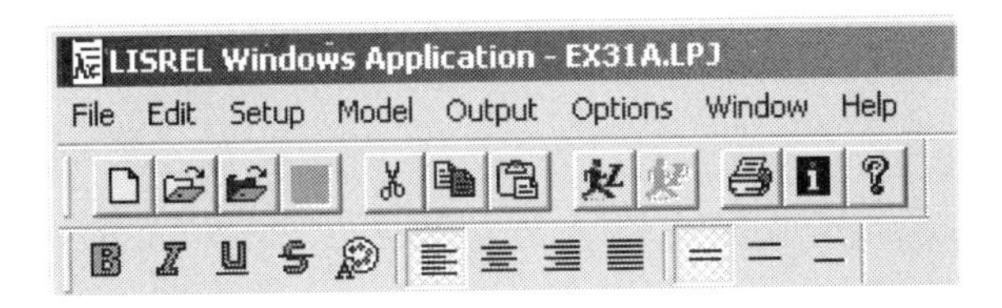

One may use the ***Setup***, ***Model*** or ***Output*** options on the main menu bar to add, modify or delete existing code. The ***File*** and ***Edit*** menus for the LPJ window are the same as those for the SPJ window discussed earlier.

2.6.1 Setup Menu

The ***Setup*** menu for the LPJ window has an additional option in the ***Data*** dialog box. This dialog box is accessed when the ***Data*** option is selected from the ***Setup*** menu. In the ***Data*** dialog box shown below, the ***Next*** button is activated.

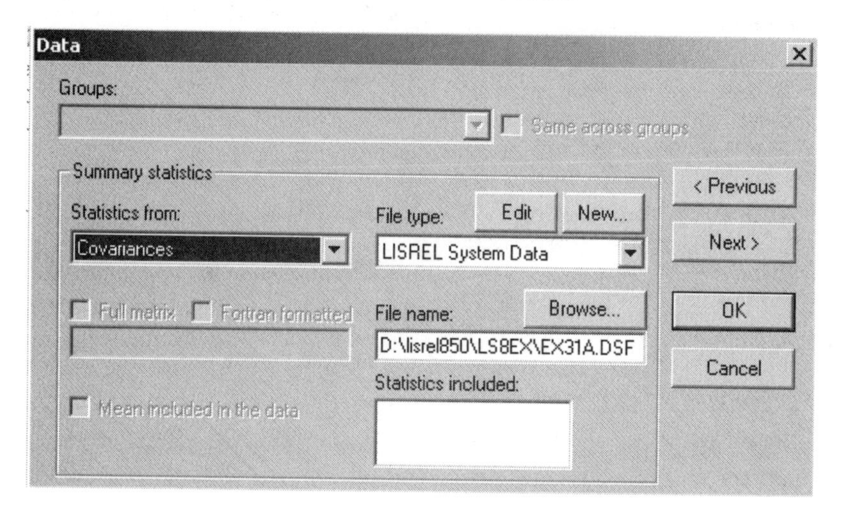

This button provides sequential access to the following dialog boxes:

- ***Define Observed Variables***
- ***Define Latent Variables***
- ***Model Parameters***
- ***Constraints***
- ***Selections***
- ***Save Matrices***

The ***Selections*** and ***Save Matrices*** dialog boxes are discussed in the previous section on the path diagram window. The use of these dialog boxes is illustrated and discussed in Section 4.7.

2.6.2 Model Menu

The ***Model*** menu is activated when a LISREL project (***.lpj**) file is opened. For more information on how to use the ***Observed Variables***, ***Latent Variables***, ***Parameters*** and ***Constraints*** dialog boxes, see the examples in Chapter 4. Some of these dialog boxes may also be accessed using the ***Data*** dialog box discussed earlier.

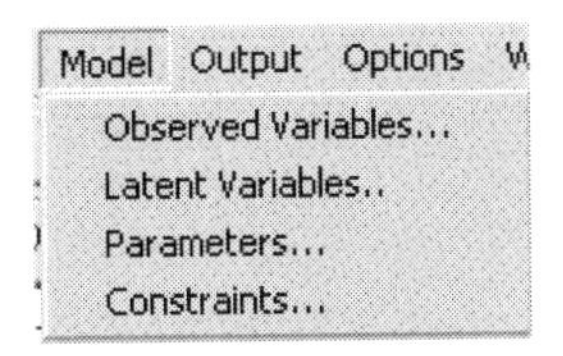

The four options correspond to the following dialog boxes, which are discussed in detail in Section 4.7:

- ***Define Observed Variables***
- ***Define Latent Variables***
- ***Model Parameters***
- ***Constraints***

Note:

The options on the ***Model*** menu are associated with the following LISREL syntax:

- ***Observed Variables*** : `LA` command
- ***Latent Variables*** : `LK` and `LE` commands
- ***Constraints*** : `CO` command.

2.7 Output Window

When an output file is displayed, it is shown in the text editor window. There is only one additional option that is available to the user in the case of an output file, and this is to be found on the ***File*** menu. For all other options available, the user is referred to the section describing the functionality of the text editor window.

2.7.1 File Menu

The option to convert the output file to a Rich Text Format file (*.**rtf**), a LaTeX file (*.**tex**) or to a Webpage (*.**htm**) is available to the user. Three options, one or each of the available output formats, are accessible through the ***File*** menu.

3 PRELIS Examples

3.1 Running PRELIS Syntax Files

The examples given in the *PRELIS 2: User's Reference Guide* are contained in the subfolder **pr2ex** of LISREL.

From the main menu bar, select the ***Open*** option from the ***File*** menu to activate the ***Open*** dialog box. Select the appropriate drive, folder, and file name. Select the file **ex2.pr2** as shown below. When done, click ***Open***.

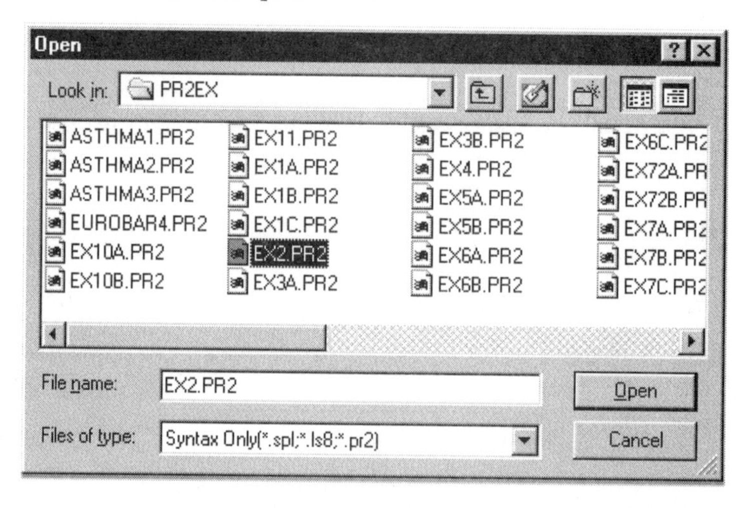

The contents of the file **ex2.pr2** are displayed in a text editor window. This file may be edited, portions of it may be copied to the clipboard, etc. with the same procedures used in a text editor program.

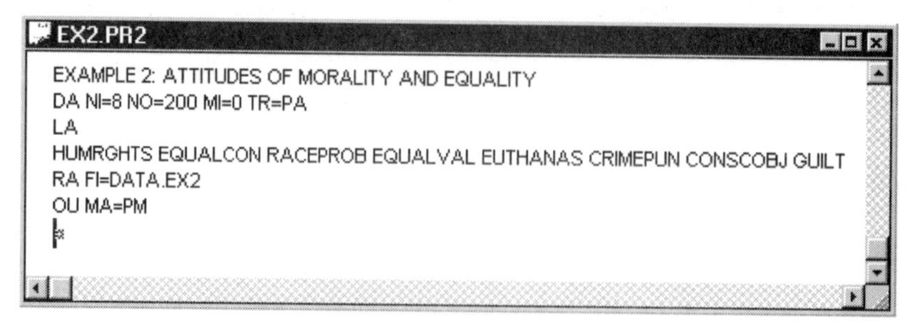
EX2.PR2

```
EXAMPLE 2: ATTITUDES OF MORALITY AND EQUALITY
DA NI=8 NO=200 MI=0 TR=PA
LA
HUMRGHTS EQUALCON RACEPROB EQUALVAL EUTHANAS CRIMEPUN CONSCOBJ GUILT
RA FI=DATA.EX2
OU MA=PM
```

To run PRELIS, click on the ***Run PRELIS*** icon button shown below.

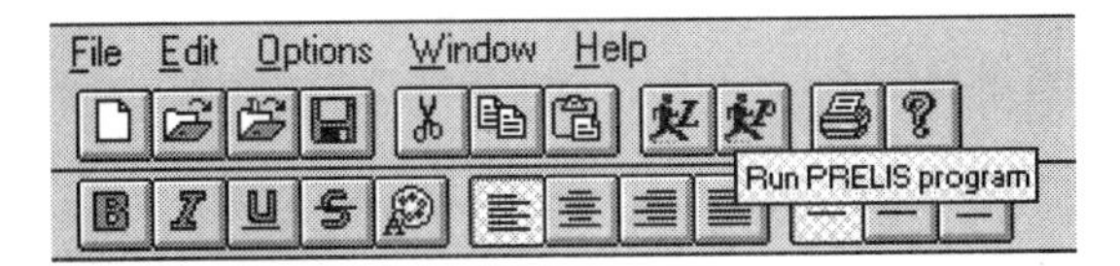

A selection of the output is displayed below:

```
EX2.OUT
    GUILT Frequency Percentage Bar Chart
      1       15         7.5     0000000
      2       40        20.1     0000000000000000000
      3      104        52.3     000000000000000000000000
      4       40        20.1     0000000000000000000

 Bivariate Distributions for Ordinal Variables (Frequ

                   EQUALCON              RACEPROB
              ------------------    ------------------
 HUMRGHTS      1    2    3    4      1    2    3    4
              ------------------    ------------------

    1          0    0    2    1      2    0    0    1
    2          3    4    4    1      0    2    7    3
    3          0   13   39   21      6    7   36   24
    4          2    7   41   60      7    8   41   52
```

The file **ex2.out** may be converted to *.**rtf**, *.**htm** or *.**tex** format. This is accomplished by selecting the ***Convert Output to RTF***, or ***Convert Output to HTM*** or ***Convert Output to TEX*** options from the ***File*** menu.

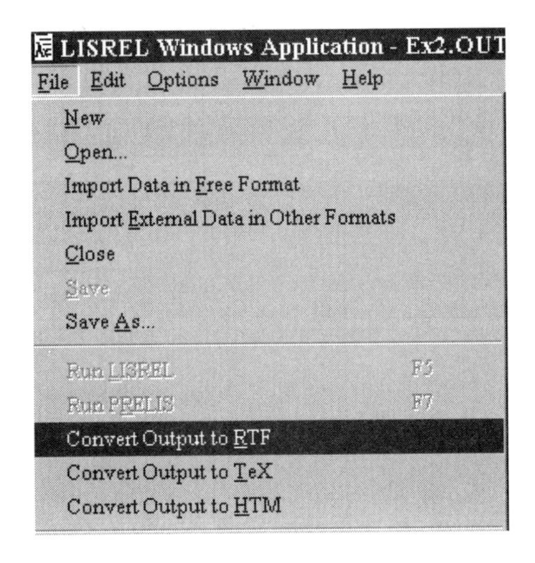

A selection of the output contained in **ex2.rtf** is displayed below. Compare this with the corresponding selection in **ex2.out**.

GUILT	Freq.	Perc.	
1	15	7.5	□□□□□□□
2	40	20.1	□□□□□□□□□□□□□□□□□□□□
3	104	52.3	□□□□□□□□□□□□□□□□□□□□□□□□□□
4	40	20.1	□□□□□□□□□□□□□□□□□□□□

Bivariate Distributions for Ordinal Variables (Fre

	EQUALCON				RACEPROB			
HUMRGHTS	1	2	3	4	1	2	3	4
1	0	0	2	1	2	0	0	1
2	3	4	4	1	0	2	7	3
3	0	13	39	21	6	7	36	24
4	2	7	41	60	7	8	41	52

3.2 Exploratory Analysis of Political Survey Data

The SPSS file **dataex7.sav** contains six variables on a subset of cases from the first cross-section of a political survey (see *PRELIS 2: User's Reference Guide* pages 145-162). Use is made of a number of dialog boxes to

- import **dataex7.sav** and save it as a PRELIS (*.**psf**) spreadsheet file
- do data screening
- assign labels (names) for categories of the ordinal variables
- define missing values
- impute missing values and calculate polychoric coefficients and the asymptotic covariance matrix
- perform a homogeneity test for pairs of ordinal variables
- graphically display the univariate distributions of the variables.

3.2.1 Importing Data into PRELIS

From the ***File*** menu, select ***Import External Data in Other Formats***:

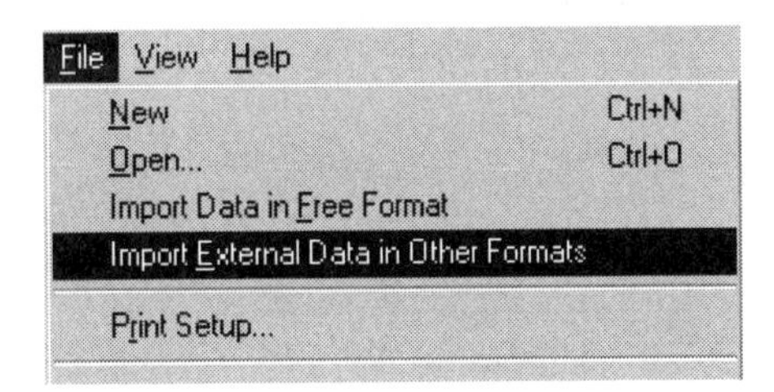

The user is prompted to select the drive and path where the data file to be imported is stored.

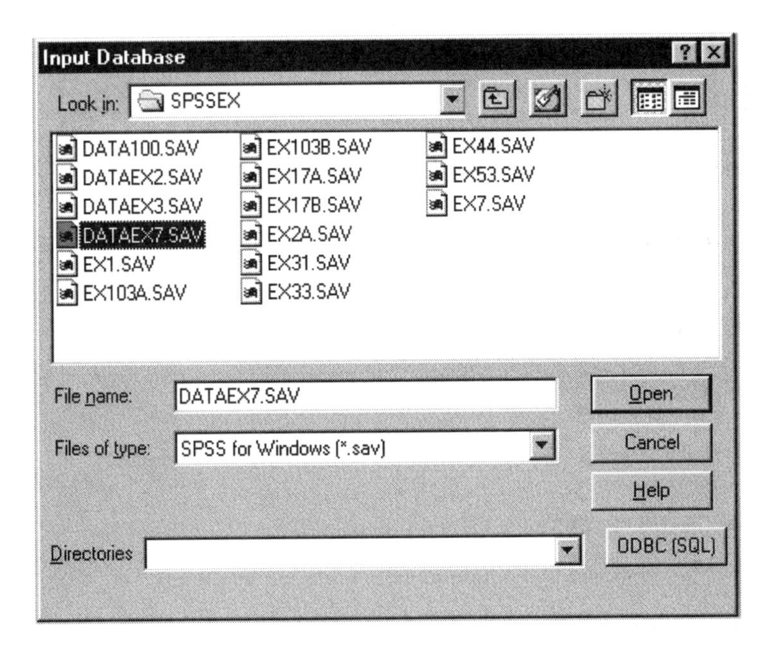

We would like to import an SPSS for Windows (*.**sav**) file named **dataex7.sav**. From the ***Files of Type*** drop-down menu, select SPSS for Windows (*.**sav**). Once **dataex7.sav** is selected from the **spssex** folder, click ***Open*** to proceed.

Hint: Click (left mouse button) on the ***Files of Type*** drop-down list box and type in the letter "S" to go directly to all files types starting with the letter "S".

Select a file name and folder in which the PRELIS system file should be saved. Note that the extension has to be *.**psf**. When done, click ***Save*** to display the system file as a spreadsheet.

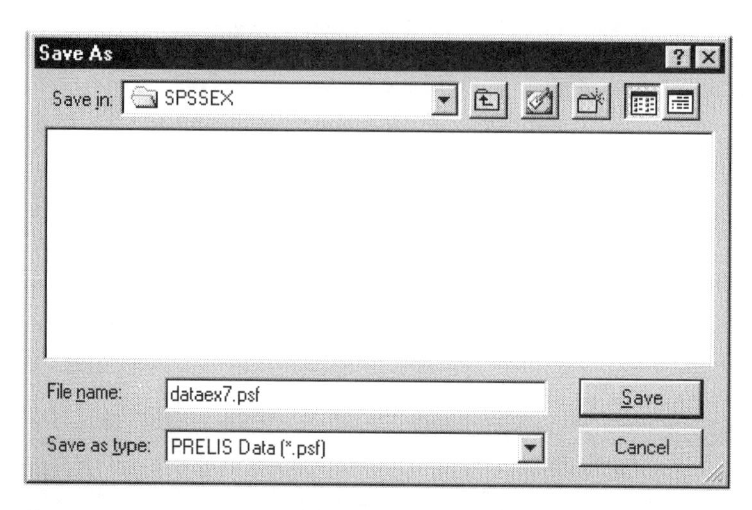

DATAEX7.PSF

	NOSAY	VOTING	COMPLEX	NOCARE	TOUCH
1	2.000	2.000	1.000	1.000	1.000
2	2.000	3.000	3.000	3.000	2.000
3	3.000	2.000	2.000	3.000	3.000
4	3.000	3.000	2.000	3.000	2.000
5	2.000	2.000	1.000	2.000	2.000
6	2.000	2.000	1.000	1.000	2.000
7	3.000	2.000	2.000	3.000	3.000
8	2.000	2.000	2.000	2.000	1.000
9	3.000	1.000	2.000	2.000	2.000
10	2.000	1.000	2.000	2.000	1.000
11	2.000	2.000	1.000	2.000	2.000

The file **dataex7.psf** contains data for six variables, these being *NOSAY*, *VOTING*, *COMPLEX*, *NOCARE*, *TOUCH* and *INTEREST*. Use the ***Edit*** toolbar to view the contents of this file.

Note that the ***Scroll data*** toolbar shown below can be used to view the contents of a very large data set. The first eight buttons are activated if the file contains more than 50 variables and/or more than 5000 cases.

The functions of the icons from left to right are:

- Scroll to the top
- Scroll up 5000 cases
- Scroll down 5000 cases
- Scroll to the bottom
- Scroll to the extreme left
- Scroll 50 variables to the left
- Scroll 50 variables to the right
- Scroll to the extreme right
- Insert a variable
- Insert a case
- Delete selected variable
- Delete selected case

3.2.2 Data Screening

The simplest form of data screening is performed by selecting ***Data Screening*** from the ***Statistics*** menu:

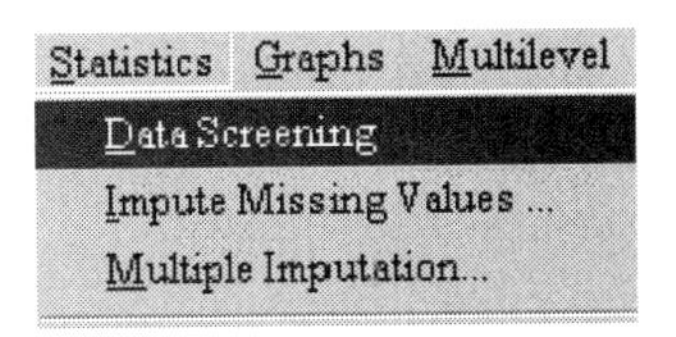

For each variable, **dataex7.out** lists the number and relative frequency of each data value, and it gives a bar chart showing the distribution of the data values. For the variable *INTEREST*, it is as follows:

```
INTEREST Frequency Percentage Bar Chart
     1        44         15.6    □□□□□□□□□□□□□□□□
     2       137         48.6    □□□□□□□□□□□□□□□□□□□□□□□□□□□□□□□□□□□□□□□□□□□□□□□□□
     3        95         33.7    □□□□□□□□□□□□□□□□□□□□□□□□□□□□□□□□□□
     4         6          2.1    □□
```

3.2.3 Assigning Labels to Categories

Each of the 6 political efficacy items has four distinct numerical values. To assign category labels `AS`, `A`, `D` and `DS` to the numerical values `1`, `2`, `3`, and `4` (`AS` = Agree strongly, `A` = Agree, `D` = Disagree, `DS` = Disagree Strongly) we proceed as follows.

Select the ***Define Variables*** option from the ***Data*** menu to activate the ***Define Variables*** dialog box.

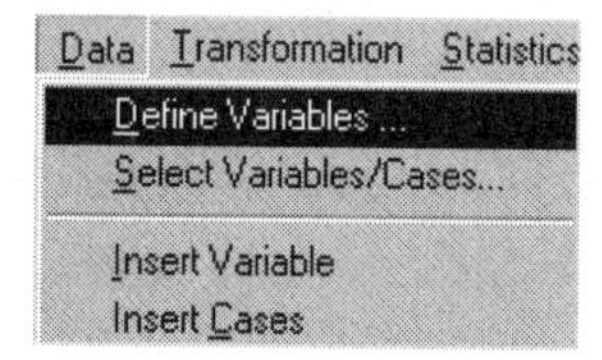

Select all six variables by holding down the ***CTRL*** key on the computer keyboard while clicking on each variable (left mouse button). Alternatively one can drag the cursor over the six variables with the left mouse button held down. The selected variables will be highlighted as shown below:

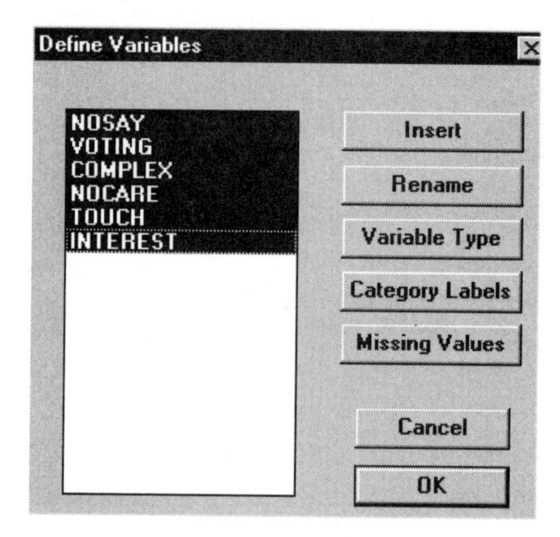

Click ***Category Labels*** to obtain the ***Category Labels*** dialog box:

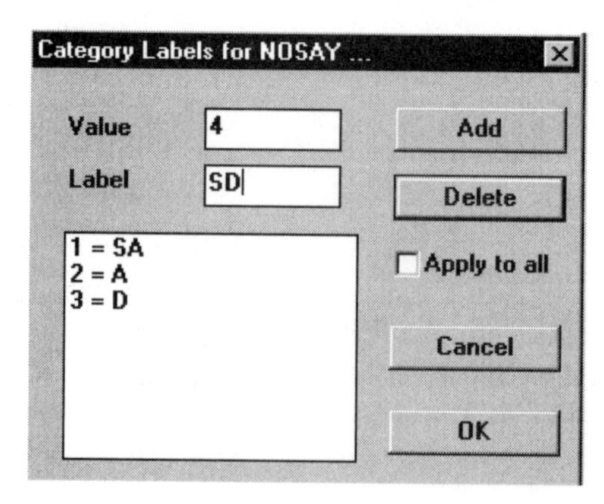

Use this dialog box to enter a numerical value and its corresponding label, then click ***Add*** to display the result in the string field at the bottom. Finally, when the last value and label (`4` and `SD`) have been entered in the ***Value*** and ***Label*** fields (see above), click ***Add*** to obtain the following dialog box.

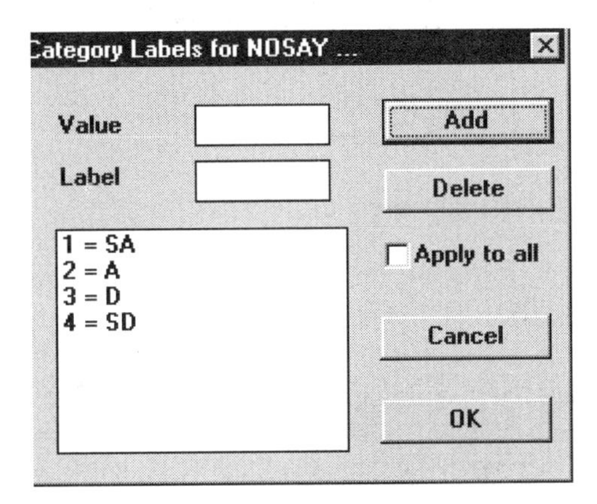

Any of these choices can be deleted by highlighting it (left mouse button) and then clicking on the ***Delete*** button. When done, click ***OK*** to return to the ***Define Variables*** dialog box.

3.2.4 Define Missing Values

Suppose that the data set contains values 8 and 9, which are actually codes to indicate missing responses. From the ***Define Variables*** dialog box, we select the ***Missing Values*** option to activate the ***Missing Values*** dialog box shown below:

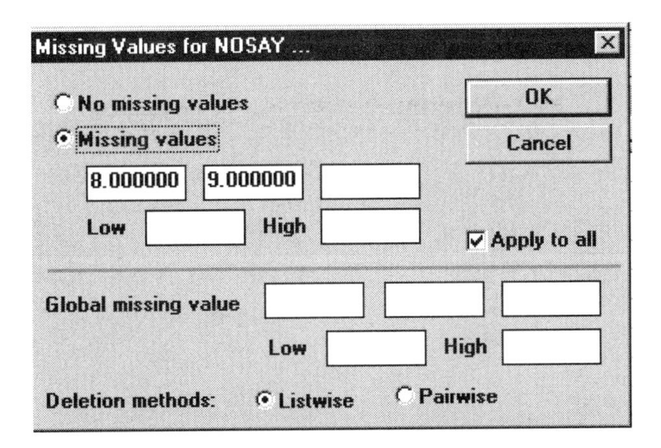

Click the ***Missing values*** radio button. Enter the values 8 and 9 in the first two string fields as shown above. Check the ***Apply to all*** check box and the ***Listwise*** deletion radio button. When this is done, click ***OK*** to return to the ***Define Variables*** dialog box. Click ***OK*** to exit this dialog box and return to the PSF window.

Note:

It is important to select ***Save*** from the ***File*** menu for the changes to take effect.

3.2.5 Imputation of Missing Values

Missing values can be handled by choosing pairwise or listwise deletion. PRELIS offers another possible way to handle missing values, namely by imputation, i.e. by substitution of real values for missing values. The value to be substituted for the missing value for a case is obtained from another case that has a similar response pattern over a set of matching variables (see the *PRELIS 2: User's Reference Guide* pages 153-8).

To do this, select the ***Impute Missing Values*** option from the ***Statistics*** menu. By clicking on ***Impute Missing Values***, the dialog box shown below appears. Suppose that we would like to impute the missing values on each variable using all the other variables as matching variables. Cases with missing values will be eliminated after imputation.

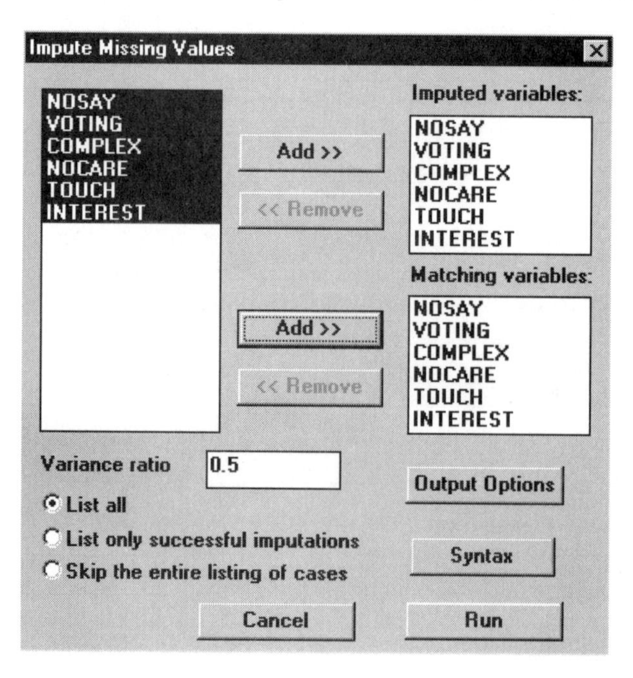

Hold the ***CTRL*** key down while selecting each of the 6 variables (left mouse button). Click the ***Add*** button to add the selected list to Imputed variables and then click the ***Add*** to matching variables button.

To obtain moment matrices, click ***Output Options*** in the dialog box shown above. The ***Output*** dialog box appears.

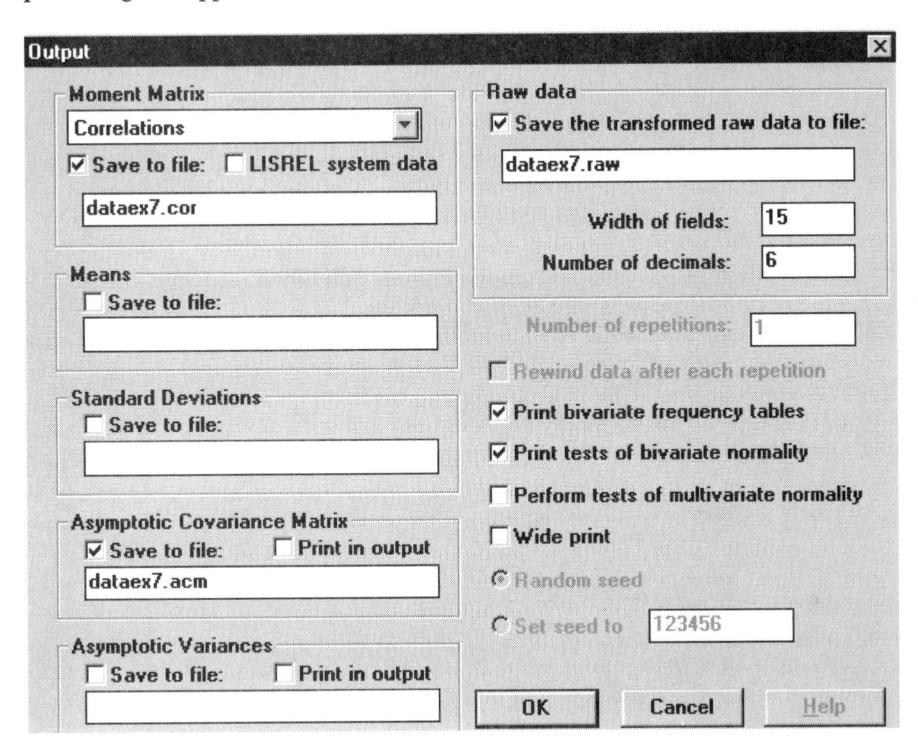

From the ***Moment Matrix*** drop-down list box, select ***Correlations***. Select ***Save to File*** and type in the name of the file. To save the newly created data set check the ***Save the transformed raw data to file*** check box and enter a file name. Note that if the file extension is ***.psf**, the file will be saved as a PRELIS system file. Also check the ***Save to file*** check box under ***Asymptotic Covariance Matrix*** and enter **dataex7.acm** in the string field. Click ***OK*** to return to the ***Impute Missing Values*** dialog box. If satisfied with the selections made, click ***Run*** to invoke PRELIS. On execution, an output file named **dataex7.out** is produced. Alternatively, click ***Syntax*** to create a PRELIS syntax file.

3.2.6 Homogeneity Tests

The homogeneity test (see the *PRELIS 2: User's Reference Guide* pages 173-176) is a test of the hypothesis that the marginal distributions of two categorical variables with the same number of categories (k) are the same. The χ^2 statistic for testing this hypothesis has k -1 degrees of freedom.

The homogeneity test is particularly useful in longitudinal studies to test the hypothesis that the distribution of a variable has not changed from one occasion to the next.

To test for homogeneity, select ***Homogeneity Test*** from the ***Statistics*** menu to open the ***Homogeneity Test*** dialog box.

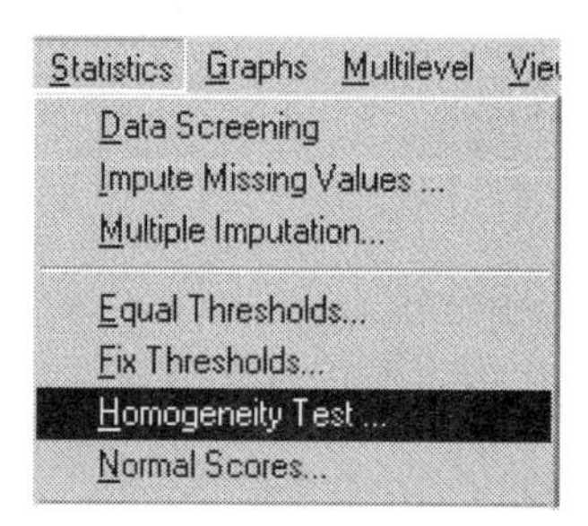

Hold the ***CTRL*** keyboard button down while selecting two variables from the list of ***Ordinal variables*** as shown in the dialog box below.

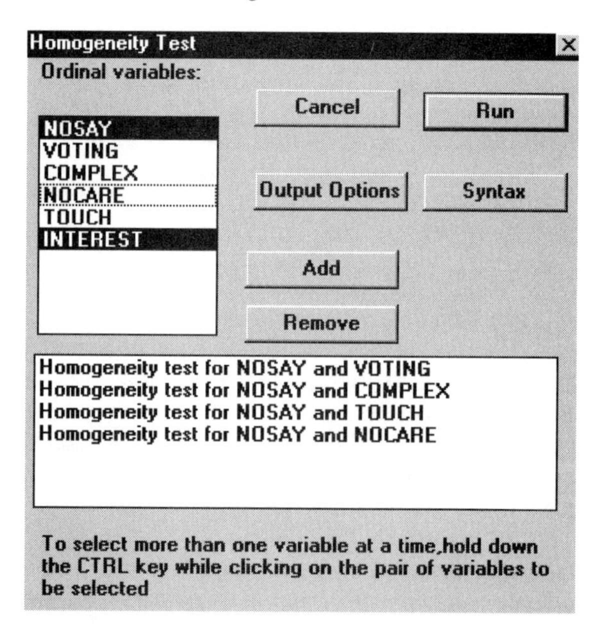

Once two variables are highlighted, the ***Add*** button is activated, and the selection can be entered by clicking on this button. The dialog box above shows that the variables *NOSAY* and *INTEREST* are highlighted. By clicking ***Add***, this test will be added to the 4 tests shown in the bottom window of the dialog box.

When all the tests are entered, click ***Output Options*** for additional output or, if not required, the ***Run PRELIS*** icon button to start the PRELIS analysis. A summary of the test results is given below:

Homogeneity Tests

Variable	vs.	Variable	Chi-Squ.	D.F.	P-Value
NOSAY	vs.	VOTING	53.220	3	0.000
NOSAY	vs.	COMPLEX	134.792	3	0.000
NOSAY	vs.	NOCARE	29.270	3	0.000
NOSAY	vs.	TOUCH	94.407	3	0.000
NOSAY	vs.	INTEREST	51.844	3	0.000

3.2.7 Graphical Display of Univariate Distributions

A bar chart of the univariate distribution of each variable may be obtained by clicking on a variable name to highlight the column of data values as shown below.

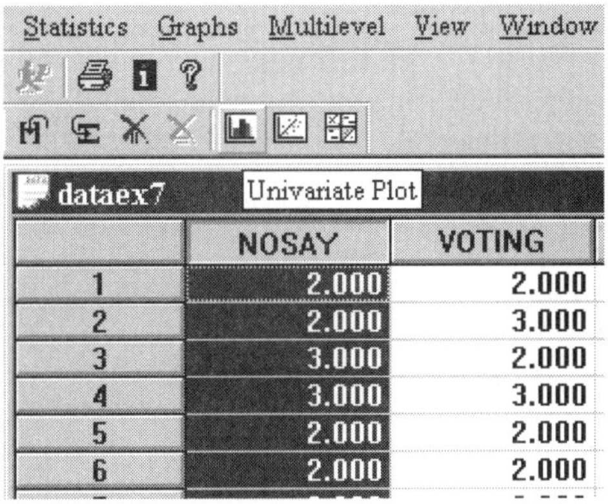

Once this is done, click the ***Univariate Plot*** icon button (third icon from the right on the toolbar) to create the bar chart. Alternatively, one may select the ***Univariate*** option from the ***Graphs*** menu.

The bar chart given below shows that the majority of respondents selected category 2 (Agree) whilst only a small number selected the "Strongly Disagree" category.

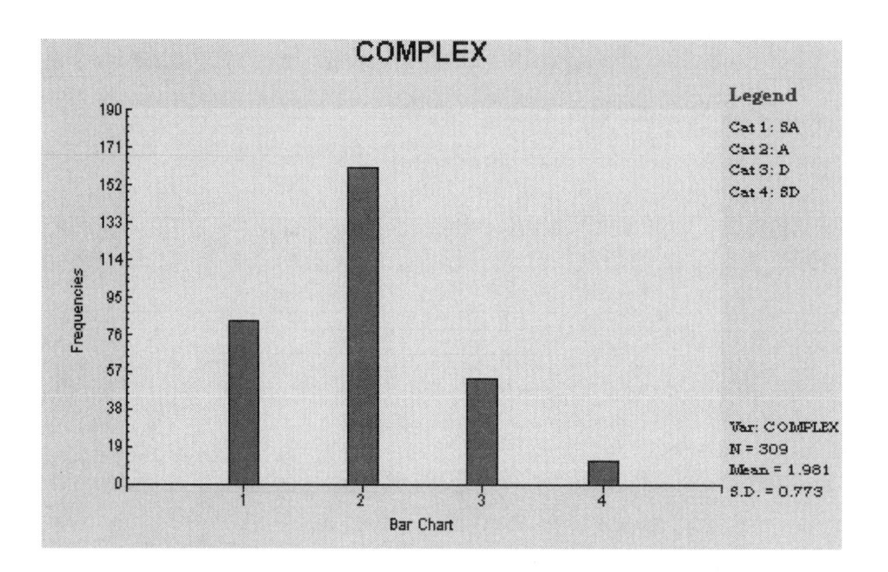

3.3 Import Data and Compute Polychoric Coefficients

LISREL provides the user with numerous features when importing data into PRELIS. The user can, for example, delete variables, select cases and compute various sample statistics interactively by making use of a set of menus.

As an illustration, an SPSS data set, **data100.sav**, which contains two continuous and four ordinal variables will be imported. The interactive PRELIS environment will be then used to

- delete the continuous variables
- select odd numbered cases (that is cases 1, 3, ...)
- calculate a matrix of polychoric correlation coefficients
- save this and the new data set to external files.

From the ***File*** menu, select ***Import External Data in Other Formats***:

The ***Input Database*** dialog box that subsequently appears will prompt the user to select the drive and path where the data file to be imported is stored. From the ***List Files of Type*** drop-down list box select SPSS for Windows (*.**sav**) and then select the SPSS for Windows file named **data100.sav** from the **spssex** folder. Click ***Open*** to go to the ***Save As*** dialog box.

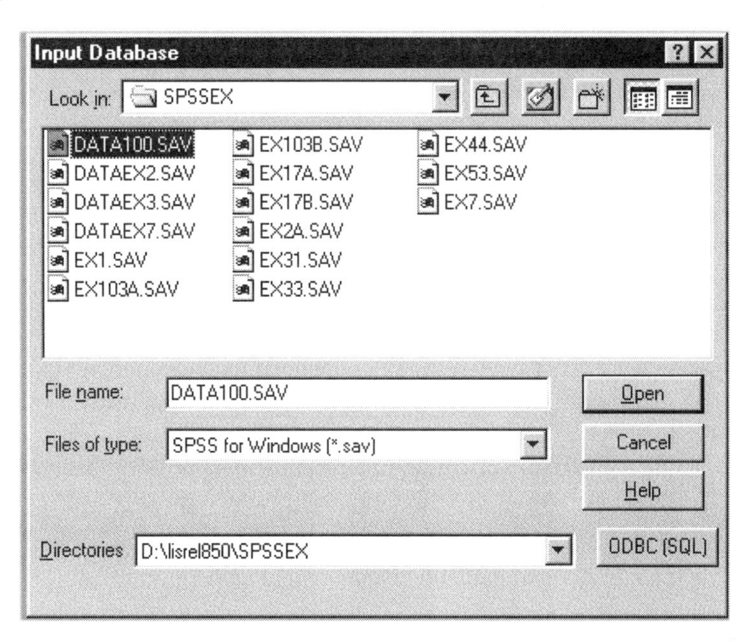

Select a file name and folder in which the PRELIS data should be saved. For the present example, the folder chosen is **spssex** and the file name selected is **data100.psf**. Note that the extension has to be *.**psf**. When done, click ***Save***.

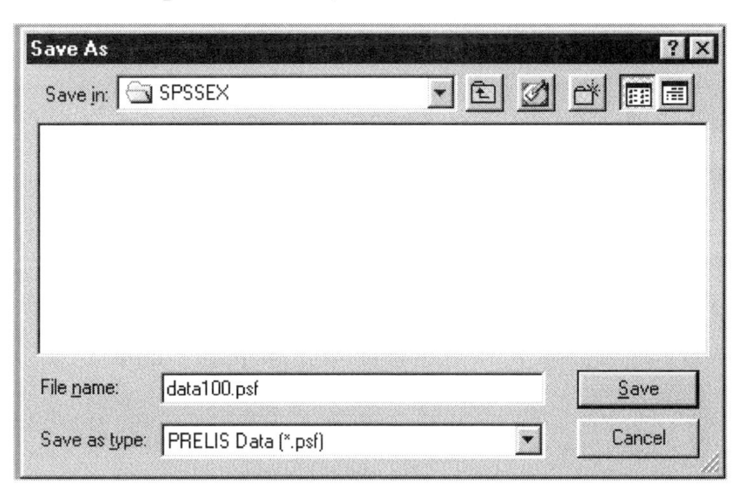

The PRELIS data spreadsheet will show up in the PSF window. Note that the file **data100.psf** contains data values for six variables, these being *CONTIN1*, *ORDINAL1*, *ORDINAL2*, *ORDINAL3*, *CONTIN2* and *ORDINAL4*.

Since we are only interested in the ordinal variables, *CONTIN1* and *CONTIN2* may be deleted from the data set. To delete the variable *CONTIN1*, click on the rectangle containing the name *CONTIN1*. The column containing the data values will change in color as shown below. This variable may now be deleted by clicking on the ***Delete Variable(s)*** button (the X button shown below) or, alternatively, by selecting the ***Delete Variable*** option from the ***Data*** menu.

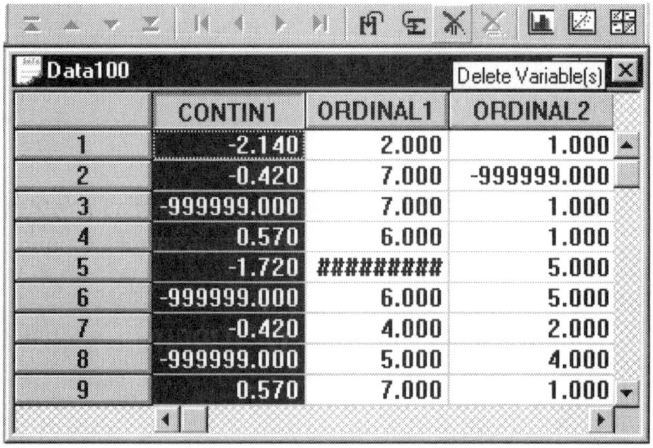

Data100 — Delete Variable(s)

	CONTIN1	ORDINAL1	ORDINAL2
1	-2.140	2.000	1.000
2	-0.420	7.000	-999999.000
3	-999999.000	7.000	1.000
4	0.570	6.000	1.000
5	-1.720	#########	5.000
6	-999999.000	6.000	5.000
7	-0.420	4.000	2.000
8	-999999.000	5.000	4.000
9	0.570	7.000	1.000

The variable *CONTIN2* is subsequently deleted by clicking on the rectangle containing the name *CONTIN2* and then on ***Delete Variable(s)*** icon button. Note that one should use the ***File, Save*** option to ensure that these changes take effect.

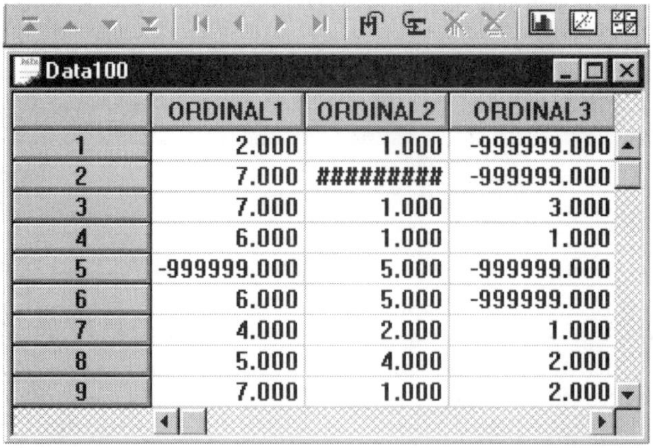

Data100

	ORDINAL1	ORDINAL2	ORDINAL3
1	2.000	1.000	-999999.000
2	7.000	#########	-999999.000
3	7.000	1.000	3.000
4	6.000	1.000	1.000
5	-999999.000	5.000	-999999.000
6	6.000	5.000	-999999.000
7	4.000	2.000	1.000
8	5.000	4.000	2.000
9	7.000	1.000	2.000

The resulting data set consists of 4 variables and 100 cases. Note that the ####### symbol in a data field indicates a missing value, or a value that contains too many digits to be displayed correctly. One can view such values by changing the ***Width*** and ***Number of decimals*** defaults by using the ***Edit*** menu to open the ***Data Format*** dialog box.

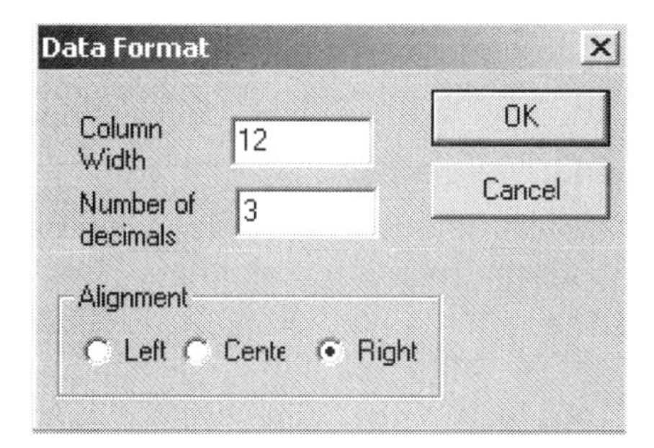

The next step in the data analysis procedure is to select the odd-numbered cases. Select the ***Select Variables/Cases*** option from the ***Data*** menu to obtain the dialog box shown below.

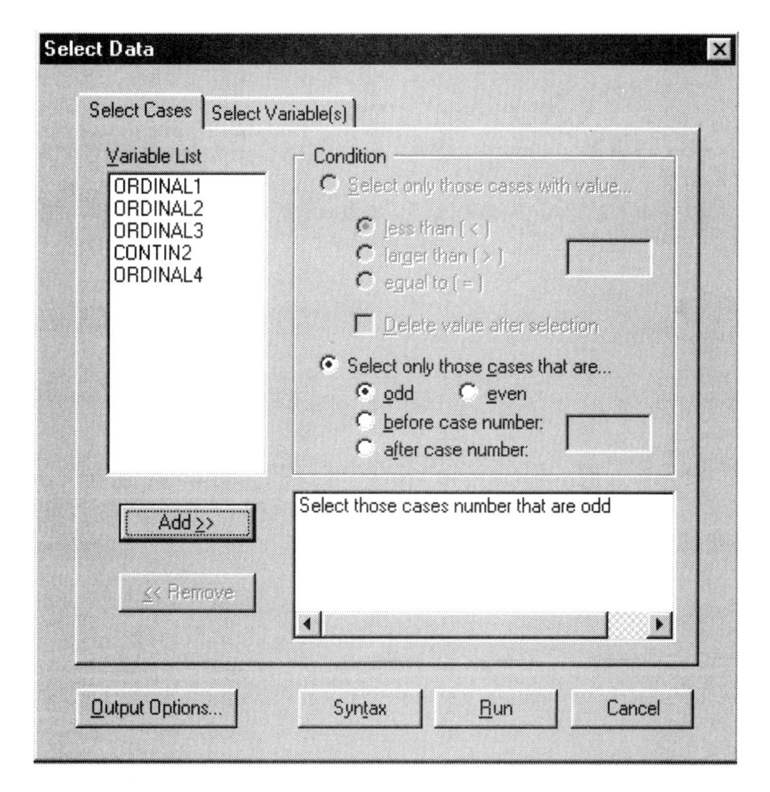

Click the ***Select only those cases*** radio button on the ***Select Cases*** tab and then click the ***odd*** radio button. This instruction is transferred to the text box by clicking the ***Add*** button. In doing so, the words `select those case numbers that are odd`

appear in the text box. Any command in this box can be removed by highlighting the specific command before clicking ***Remove***. Once done, click ***Output Options*** to obtain the ***Output*** dialog box.

Enter `10` in the ***Width of fields*** string field and `1` in the ***Number of decimals*** string field. Using the ***Moment Matrix*** drop-down list box, select ***Correlations***. Check the ***Save to file*** check box and type in the name of the file (**data100.cor**). Save the data set to be created by checking the ***Save the transformed data to file*** check box and entering a file name. For the present example, the name **data100.raw** is used.

Notes:

- Tests of multivariate normality are only performed on continuous variables.
- PRELIS computes a polychoric correlation matrix if the variables are ordinal.
- When saving data under a filename with the extension *.**psf**, a PRELIS system file with that name is created.

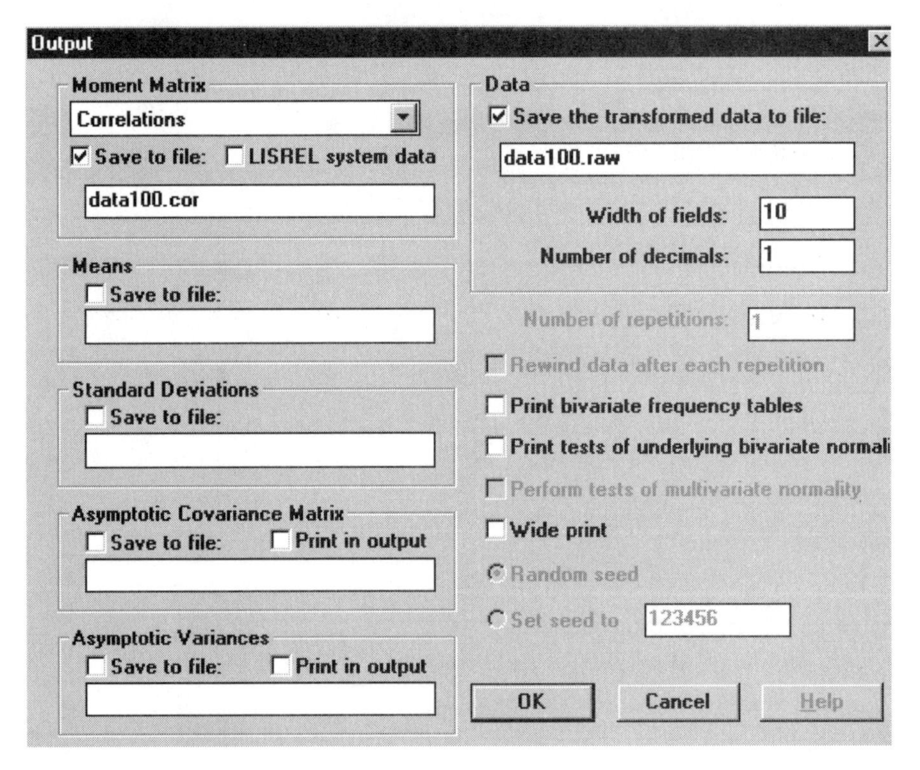

Click ***OK*** when done. This action will return control to the ***Select Data*** dialog box. If satisfied with the selections made, click ***OK*** to invoke the PRELIS program. On execution, an output file named **data100.out** is produced. A selection of the output is given.

```
SY=C:\lisrel850\TUTORIAL\Data100.PSF
SC CASE ODD
OU XM SM=data100.cor RA=data100.raw WI=10 ND=1 XM

Number of Missing Values per Variable

 ORDINAL1  ORDINAL2  ORDINAL3  ORDINAL4
 --------  --------  --------  --------
       10        11        14        11

Distribution of Missing Values

Total Sample Size =     50

Number of Missing Values     0     1     2     3
         Number of Cases    20    17    10     3
Listwise Deletion

Total Effective Sample Size =     20
```

By closing the output file to return to the PSF window, bar charts can be produced for each of the ordinal variables. Do so by clicking on the rectangle containing the name of the variable and then on the ***Univariate Plot*** icon button.

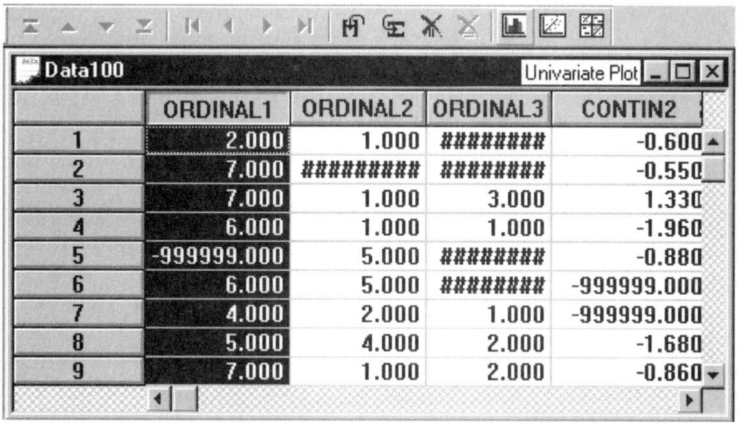

Data100 — Univariate Plot

	ORDINAL1	ORDINAL2	ORDINAL3	CONTIN2
1	2.000	1.000	########	-0.600
2	7.000	#########	########	-0.550
3	7.000	1.000	3.000	1.330
4	6.000	1.000	1.000	-1.960
5	-999999.000	5.000	########	-0.880
6	6.000	5.000	########	-999999.000
7	4.000	2.000	1.000	-999999.000
8	5.000	4.000	2.000	-1.680
9	7.000	1.000	2.000	-0.860

A bar chart similar to the one shown next can be obtained by clicking on a vertical bar in the graph to invoke the ***Bar Graph Parameters*** dialog box. Change the bar type from ***Vertical Bars*** to ***Vertical 3D Bars*** (see Section 3.4.4 for more details).

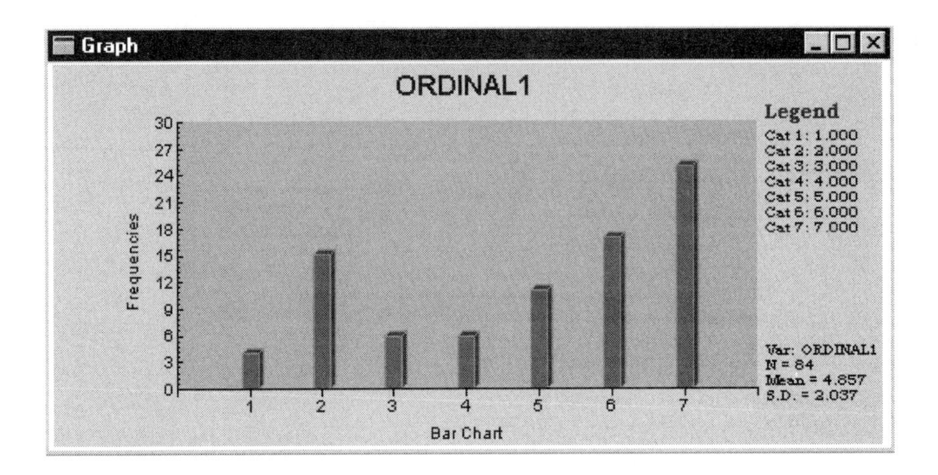

3.4 Exploratory Analysis of Fitness Data

Once a *.**psf** file is opened, the PRELIS toolbar appears and a number of additional dialog boxes becomes accessible to the user. PRELIS syntax can be generated by making use of these dialog boxes. As an illustration, we use a data set called **fitchol.dat** (Du Toit, Steyn and Stumpf, 1986.)

3.4.1 Description of the Fitness/Cholesterol Data

Two of the many factors that are known to have some influence or relevance on the condition of the human heart are physical fitness and blood cholesterol level. In a related research project, four different homogeneous groups of adult males were considered. A number of plasma lipid parameters were measured on each of the 66 individuals and fitness parameters were also measured on three of the four groups.

The groups are:

Group	Description
1	Weightlifter ($n_1 = 17$)
2	Student (control; $n_2 = 20$)
3	Marathon athlete ($n_3 = 20$)
4	Coronary patient ($n_4 = 9$)

The characteristics that we will consider here are:

Variable	Label
X1 = Age (years)	Age
X2 = Length (cm)	Len
X3 = Mass (kg)	Mass
X4 = Percentage fat	%Fat
X5 = Strength-breast (lb)	Strength
X6 = Triglycerides	Trigl
X7 = Cholesterol (total)	Cholest

The first 5 lines of the data set are given below:

OBS	GROUP	X1	X2	X3	X4	X5	X6	X7
1	1	22	179.2	107.1	15.2	3.0	0.58	4.44
2	1	30	183.0	112.2	20.3	4.6	1.51	4.88
3	1	26	175.7	78.0	17.5	3.7	1.20	4.33
4	1	23	182.5	79.7	16.1	3.3	0.75	3.66
5	1	29	178.0	81.8	14.1	2.7	0.75	4.57

3.4.2 Exploratory Data Analysis

From the ***File*** menu, select ***Import External Data in Other Formats***.

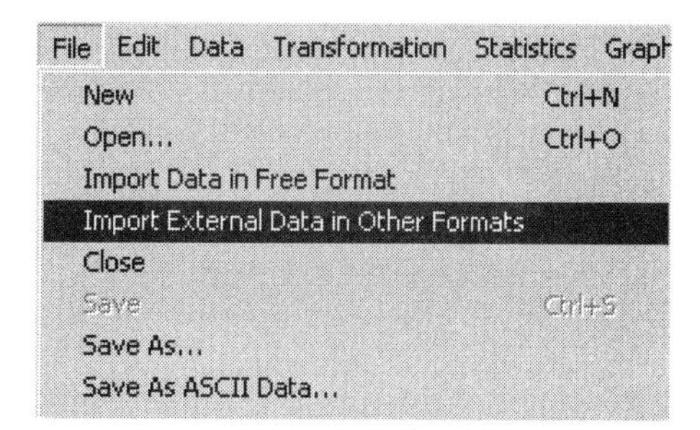

The ***Input Database*** dialog box that appears will prompt the user to select the drive and path where the raw data is stored. For the present example, the folder name is **pr2ex** and the file name is **fitchol.dat**.

For this data set, it would be less time consuming to use the ***Import Data in Free Format*** option as illustrated in Section 3.11. In this section, however, we illustrate how ASCII files in fixed or free format with specific field delimiters can be imported into a *.**psf**.

Select ***ASCII*** **(*.dat, *.txt, *.csv,*.lst,*.prn)** from the ***Files of Type*** drop-down list box and select the file **filtchol.dat** from the **tutorial** folder. Click ***Open*** to go to the ***ASCII Input Format Options*** dialog box shown on the next page.

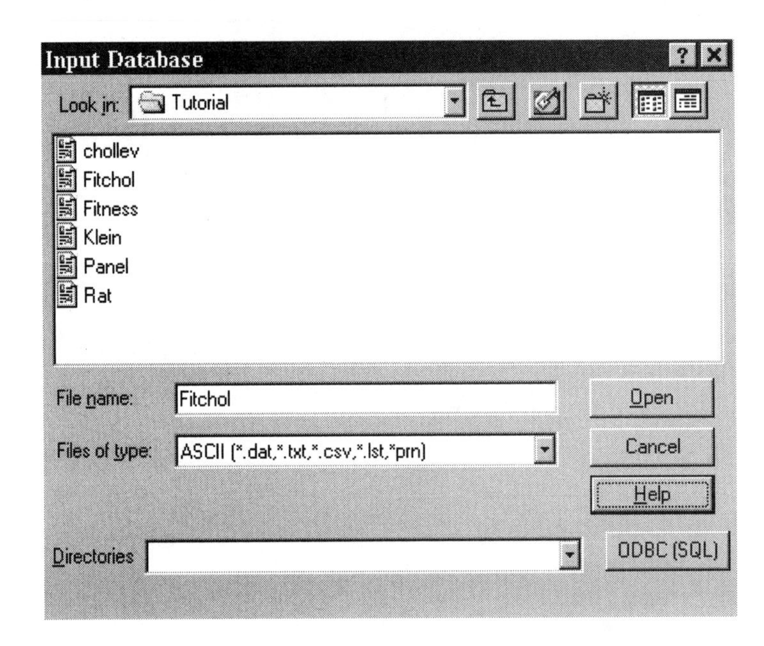

Since the data are in free format, make sure that the ***File is Fixed Format*** and ***Variable Names on Row#1*** box are not checked. Note that the ***Variable Names on Row#1*** is only checked when importing data from statistical packages like SPSS or SAS and spreadsheet or database programs like LOTUS or MS ACCESS.

Also note that data values in the file **fitchol.dat** are separated by blanks, replace the default ***Field Separator*** (a comma) by a blank (keyboard space bar). The surround character is used with variable labels that contain blanks, for example `Class A` or `Type III`.

Click ***OK*** to activate the ***ASCII Dictionary Builder*** dialog box shown on the next page.

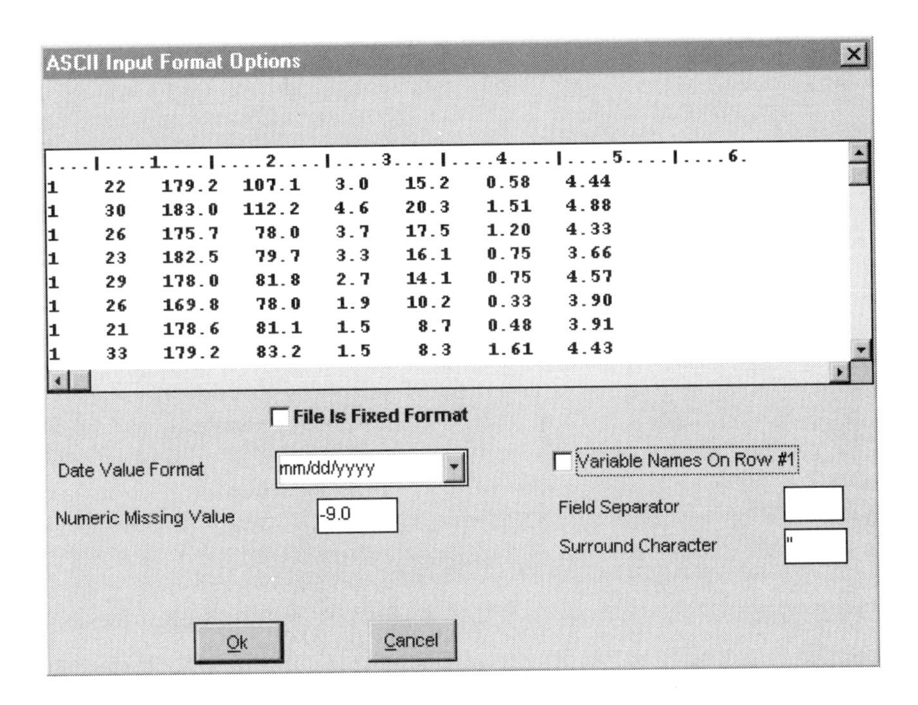

In the ***# Len Type Field Name*** field, click anywhere on the first line. Then click on the ***Type*** drop-down list box to activate the possible data types. Select ***Float (4 bytes)***. The default variable name *f1* can be changed to the name *Group* by selecting the first row in the ***Len Type Field Name*** field and providing information in the ***Field Information*** section of this dialog box (see dialog box below). Repeat the steps outlined above for the variables *Age*, *Len*, . . . , *Cholest*.

Notes:

- If a character variable is to be read from a free-format file, click on the ***Length*** selector button (up arrow) until the maximum length of the character variable in question is indicated.
- When creating a PRELIS system file, all the character (non-numeric) variables in the data set will be skipped.
- Contextual help on the ***Input Database*** dialog box may be accessed by clicking the ***Help*** button.

When done, click ***Finished*** as shown on the bottom of the dialog box given below.

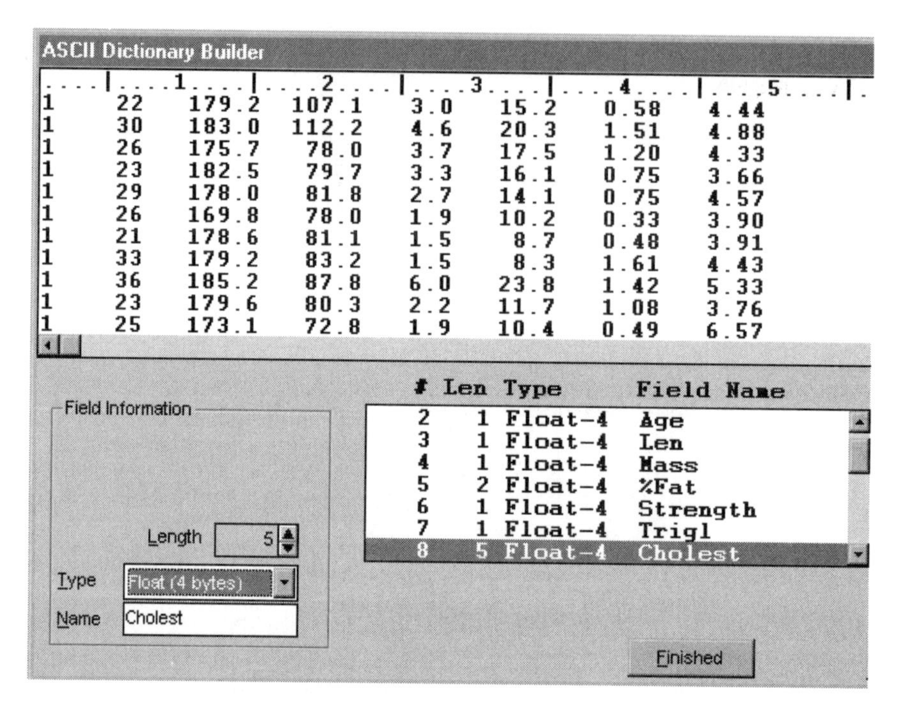

Select a filename and folder in which the PRELIS data should be saved. For the present example the folder chosen is **tutorial** and the filename selected is **fitchol.psf.** Note that the extension has to be *.**psf**. When done, click ***OK*** to open the PSF window in which the newly create *.**psf** file is displayed.

FITCHOL.PSF

	Group	Age	Len	Mass	%Fat
1	1.000	22.000	179.200	107.100	3.000
2	1.000	30.000	183.000	112.200	4.600
3	1.000	26.000	175.700	78.000	3.700
4	1.000	23.000	182.500	79.700	3.300
5	1.000	29.000	178.000	81.800	2.700
6	1.000	26.000	169.800	78.000	1.900
7	1.000	21.000	178.600	81.100	1.500
8	1.000	33.000	179.200	83.200	1.500
9	1.000	36.000	185.200	87.800	6.000
10	1.000	23.000	179.600	80.300	2.200

Since the fitness data set (**fitchol.dat**) contains missing values that are denoted by `-9.0`, we would like to remove these values prior to doing any exploratory analyses.

To do so, select the ***Define Variables*** option from the ***Data*** menu to obtain the ***Define Variables*** dialog box.

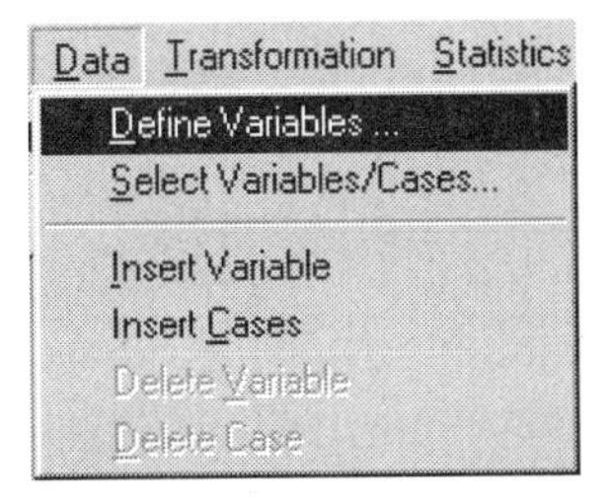

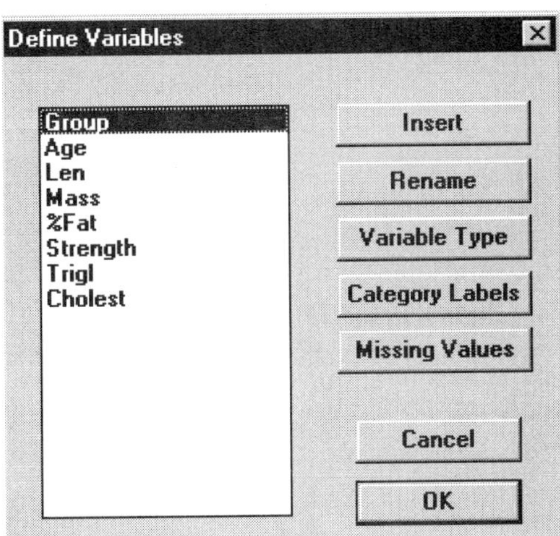

Select the variable in the first row. This action will activate the ***Variable Type***, ***Category Labels***, ***Missing Values*** and ***Labels*** buttons. Click ***Missing Values*** to obtain the ***Missing Values for*** dialog box. On this dialog box, type the value of `-9.0` in the ***Global missing value*** field. Click ***OK*** to return to the ***Define Variables*** dialog box.

Missing Values for Group ...
No missing values
Missing values
Low
High
OK
Cancel
Apply to all
Global missing value -9.0
Low
High
Deletion methods: Listwise Pairwise

Suppose that we would like to specify all variables as continuous. On the ***Define Variables*** dialog box, select the variable *Age* in the second row and then click ***Variable Type*** to activate the ***Variable Type for*** dialog box. Check the ***Continuous*** radio button and check the ***Apply to all*** check box. Finally click ***OK*** to return to the ***Define Variables*** dialog box.

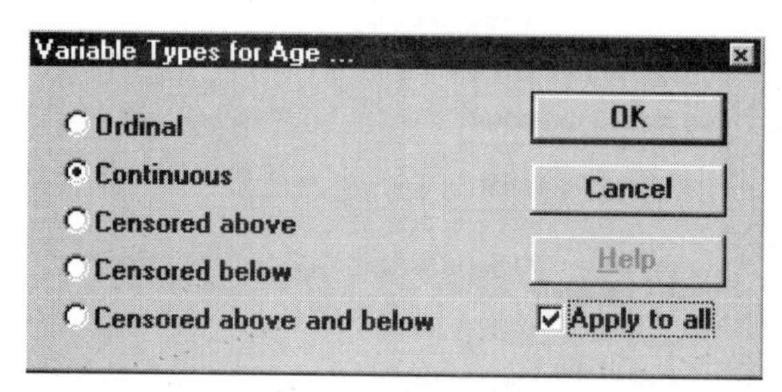

Since the ***Apply to all*** check box was selected, all variables are defined as continuous. The variable *Group*, however, has only 4 categories. We therefore repeat the procedure by selecting ***Define Variables*** option from the ***Data*** menu. On the ***Define Variables*** dialog box, select *Group* (first row), then click on ***Variable Type*** and select ***Ordinal*** from the list of variable types.

The dialog box may also be used to change the variable labels by clicking on a variable and then ***Rename***. In this example, we changed the name *Len* to *Length*.

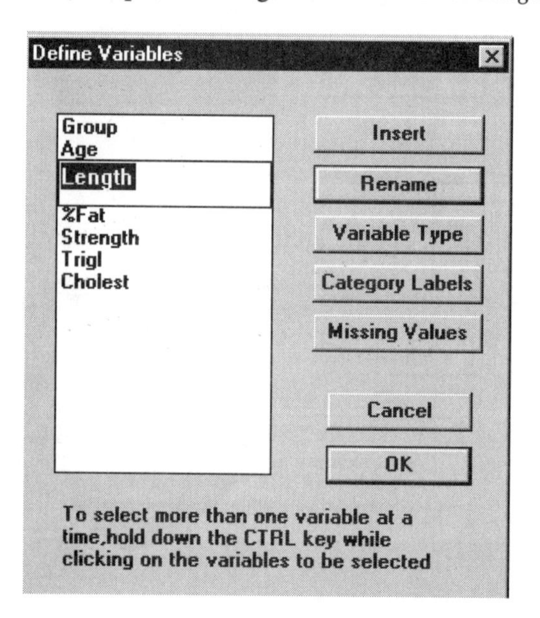

Click ***OK*** to obtain the following PSF window. The data spreadsheet illustrates the effect of the change in the variable label.

FITCHOL.PSF

	Group	Age	Length	Mass	%Fat
1	1.000	22.000	179.200	107.100	3.000
2	1.000	30.000	183.000	112.200	4.600
3	1.000	26.000	175.700	78.000	3.700
4	1.000	23.000	182.500	79.700	3.300
5	1.000	29.000	178.000	81.800	2.700
6	1.000	26.000	169.800	78.000	1.900
7	1.000	21.000	178.600	81.100	1.500
8	1.000	33.000	179.200	83.200	1.500
9	1.000	36.000	185.200	87.800	6.000
10	1.000	23.000	179.600	80.300	2.200

Before fitting an appropriate model to data, it may be wise to run a data screening procedure. The PRELIS data screening procedure provides information on the distribution of missing values, univariate summary statistics and test of univariate normality for continuous variables. The procedure also provides information on the distribution of variables over a number of class intervals.

To do data screening, select ***Statistics*** from the main menu bar and then the ***Data Screening*** option from the ***Statistics*** menu.

A selection of the output file obtained is shown below:

Number of Missing Values	0	1	2	3	4
Number of Cases	57	0	0	0	9

Listwise Deletion
Total Effective Sample Size = 57

Univariate Summary Statistics for Continuous Variables

Variable	Mean	St. Dev.	T-Value	Skewness	Kurtosis	Min.	Freq.	Max.	Freq.
AGE	27.807	6.075	34.556	1.145	0.952	18.000	1	45.000	2
LEN	178.760	5.641	239.235	0.198	-0.510	168.200	1	192.600	1
MASS	75.830	10.524	54.398	1.256	2.392	58.800	1	112.200	1
%FAT	3.146	1.068	22.232	0.748	0.173	1.400	1	6.000	1
STRENGTH	15.081	3.753	30.339	0.299	-0.267	7.600	1	24.000	1
TRIGL	1.024	0.708	10.917	3.614	16.362	0.330	1	4.970	1
CHOLEST	4.573	1.048	32.939	-0.268	1.392	1.030	1	7.190	1

Test of Univariate Normality for Continuous Variables

Variable	Skewness		Kurtosis		Skewness and Kurtosis	
	Z-Score	P-Value	Z-Score	P-Value	Chi-Square	P-Value
AGE	3.590	0.000	1.600	0.110	15.449	0.000
LEN	0.622	0.534	-0.663	0.507	0.827	0.661
MASS	3.937	0.000	2.648	0.008	22.516	0.000
%FAT	2.346	0.019	0.668	0.504	5.948	0.051
STRENGTH	0.936	0.349	-0.104	0.917	0.887	0.642
TRIGL	11.329	0.000	5.369	0.000	157.177	0.000
CHOLEST	-0.841	0.400	1.985	0.047	4.648	0.098

3.4.3 Graphical Displays

To produce a histogram of the data in the ***Length*** column, click on the rectangle containing the name ***Length***. The column containing the data values will change in color as shown below.

FITCHOL.PSF

	Group	Age	Length	Mass	%Fat
1	1.000	22.000	179.200	107.100	3.000
2	1.000	30.000	183.000	112.200	4.600
3	1.000	26.000	175.700	78.000	3.700
4	1.000	23.000	182.500	79.700	3.300
5	1.000	29.000	178.000	81.800	2.700
6	1.000	26.000	169.800	78.000	1.900
7	1.000	21.000	178.600	81.100	1.500
8	1.000	33.000	179.200	83.200	1.500
9	1.000	36.000	185.200	87.800	6.000
10	1.000	23.000	179.600	80.300	2.200

On the main LISREL menu bar, select the ***Univariate Plot*** icon button. Note that, if you move the mouse to this icon, the words `Univariate Plot` will be displayed in the proximity of the icon button.

A histogram of the variable *Length* is displayed below. The colors of the background and the bars can be changed by double-clicking on the appropriate area. This action will produce a dialog box in which the necessary change may be made. Likewise, one can change the color and appearance of the axes and labels by double clicking on a specific label, for example on ***Histogram***. The dialog box allows one to select a font and font size and to change the color of the text. The use of various dialog boxes to edit graphs are discussed in Chapter 2.

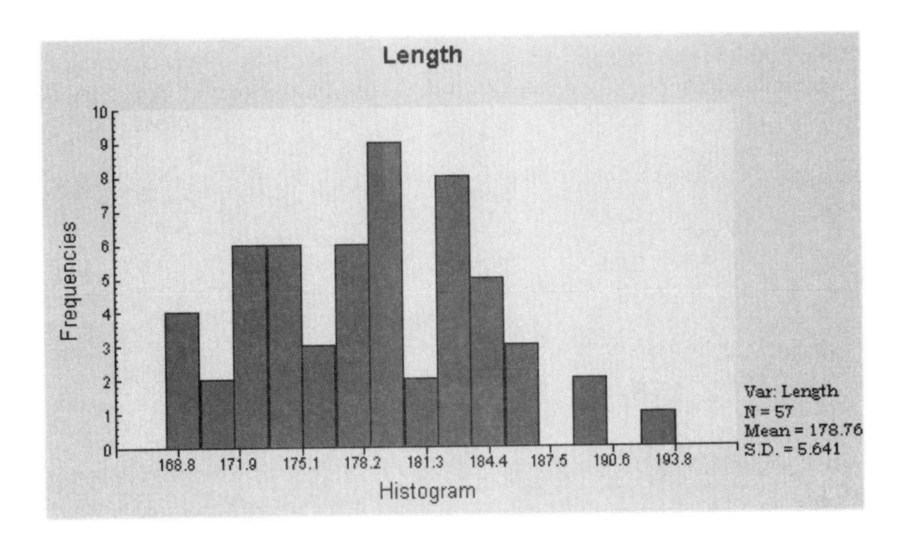

As an alternative to producing a histogram by clicking on the ***Univariate plot*** icon button, one may select the ***Graphs, Univariate*** option on the main menu bar as shown below. Before doing so, make sure that a *variable* (column) in the spreadsheet is not selected. If a column is selected, click on the rectangle containing the corresponding variable label to deselect it.

The selection of the ***Graphs, Univariate*** option will activate the ***Univariate Plots*** dialog box. Click on the variable to be displayed and check, for example, the ***Normal curve overlay*** check box. Click ***Plot*** to proceed.

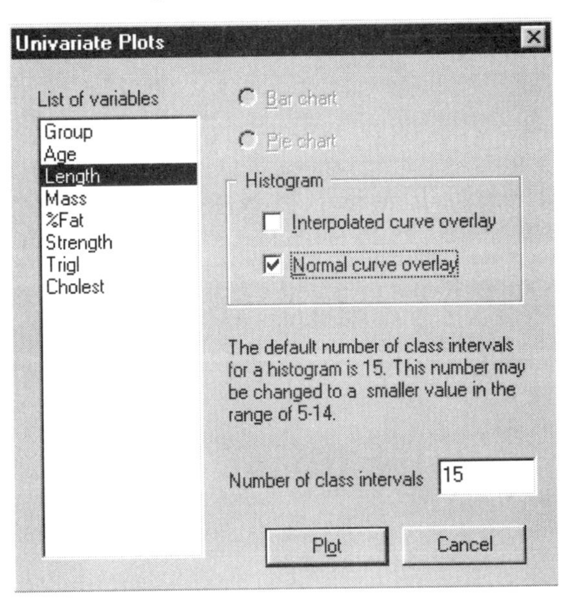

The histogram of the variable *Length* is displayed. The normal curve is superimposed on the histogram. This simultaneous display of a histogram and normal curve is useful in visually assessing whether the normality assumption for a given variable is realistic.

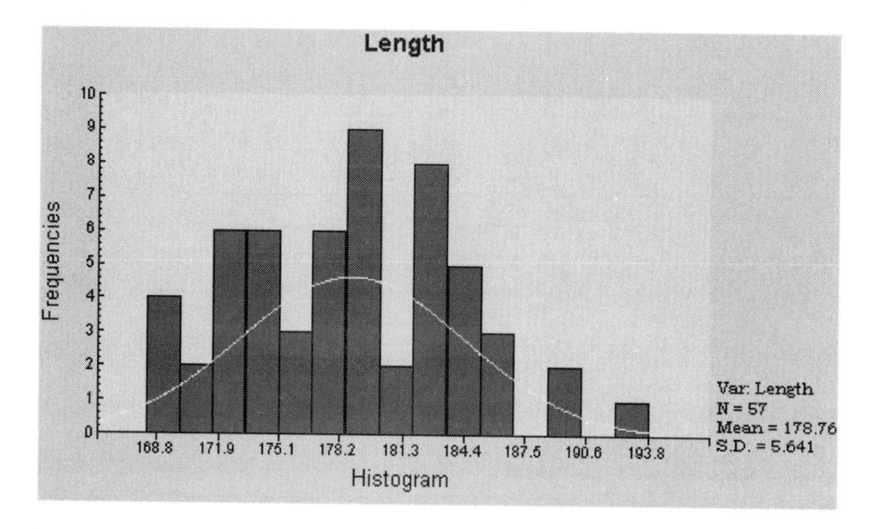

If a variable has less than 16 distinct categories, the ***Draw histogram*** option produces a bar chart representation of the data. This is the case for the variable *Group,* which has 4 distinct values, corresponding to

1 = Weightlifter
2 = Student
3 = Marathon athlete
4 = Coronary patients

The bar chart below shows that the number of individuals in the first three groups are almost equal (n = 17, 20 and 20), but that the sample contains a relatively small number of coronary patients (n = 9).

The default legend shows the categories of the variable *Group* and the values assigned to each category. By double clicking on the `Cat1:1` description, the ***Text Parameters*** dialog box is activated.

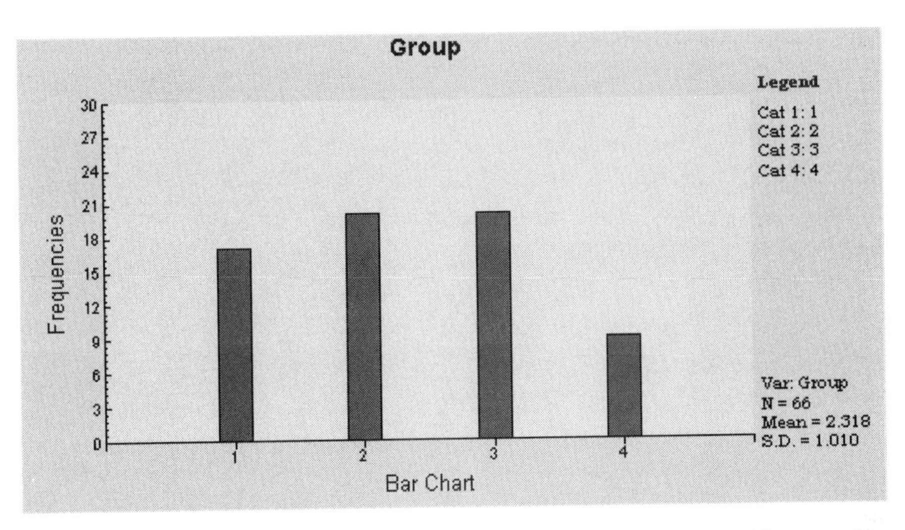

Change `Cat1:1` to `1 = Weightlifter` and click ***OK***. Repeat this procedure to change the descriptions of categories 2, 3, and 4 to `Student`, `Marathon` and `Coronary`, respectively.

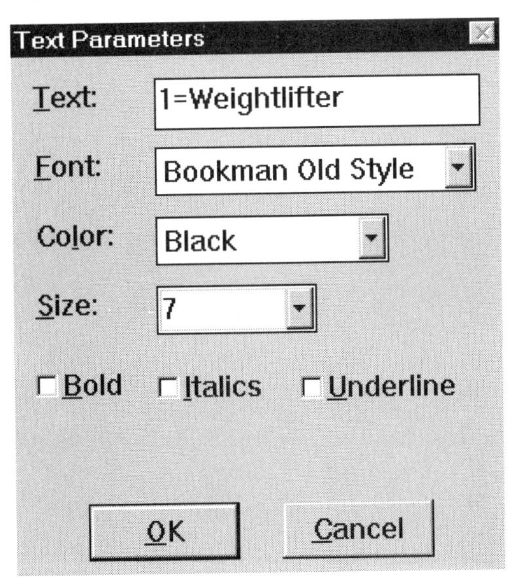

The appearance of the vertical bars can be changed by double-clicking on the top portion of one of highest bars until the ***Bar Graph Parameters*** dialog box is activated. Change ***Type*** to ***Vert. 3-D bars*** and use the ***bar color*** sliders to change to the desired color. Click ***OK*** when done.

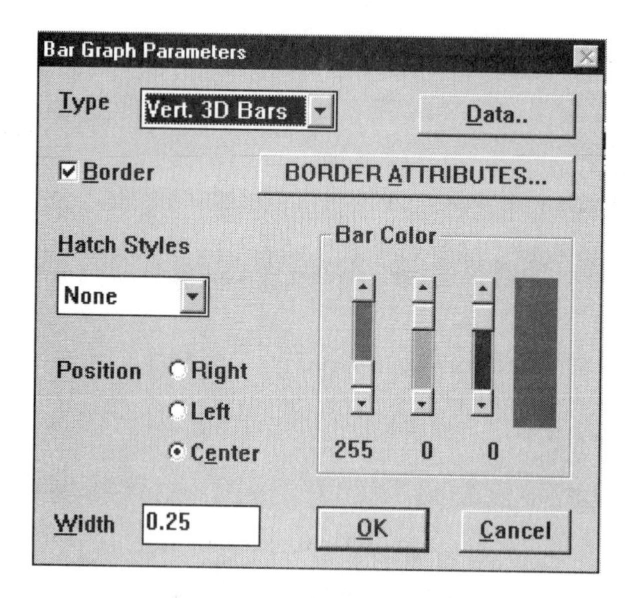

The bar chart shown below reflects these changes in attributes. The bar chart for the variable *Group* may now be printed. This is done by selecting the ***Print*** option from the ***File*** menu. Alternatively, the bar chart can be exported as a Windows metafile (*.**wmf**) by selecting the ***Export as Metafile*** option from the ***File*** menu.

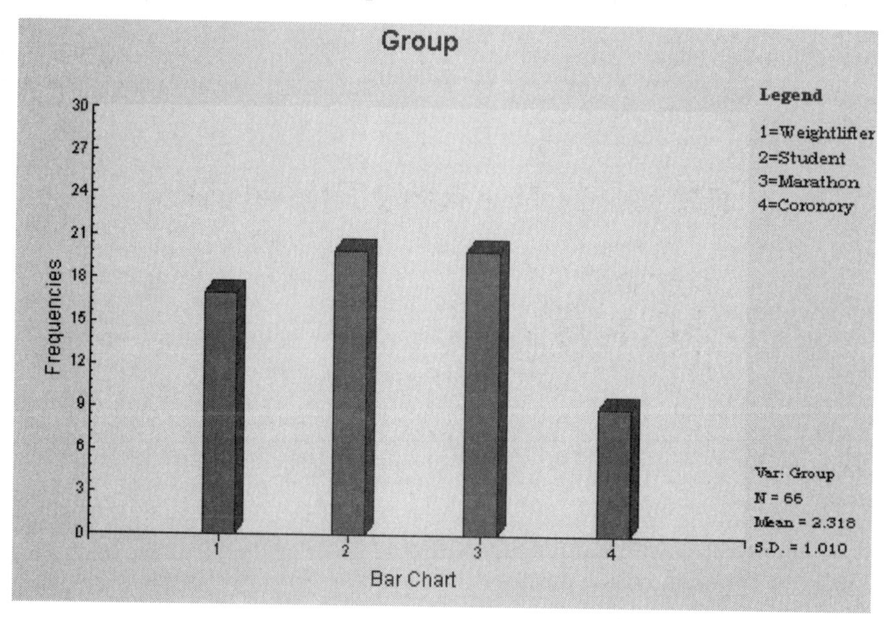

3.4.4 Regression Analysis

We may be interested in the prediction of *Cholesterol* (total) by the variables age, length, percentage fat and strength. To obtain the regression equation, select the ***Regressions*** option from the ***Statistics*** menu.

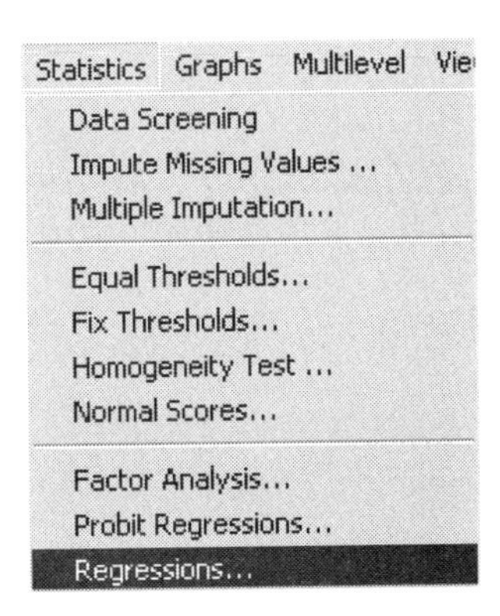

The ***Regression*** dialog box is shown below. This dialog box allows the user to select variables as either *Y* or *X*-variables. Click on a variable, for example *Cholest*, to activate the ***Add*** button. Click the top ***Add*** button to add this variable to the list of *Y*-variables. Select the variables *Age*, *Length*, *Mass*, *%Fat*, and *Strength* and click the bottom ***Add*** button to add the variable to the list of *X*-variables. When the selection procedure is complete, click on ***Output Options*** to activate the ***Output*** dialog box.

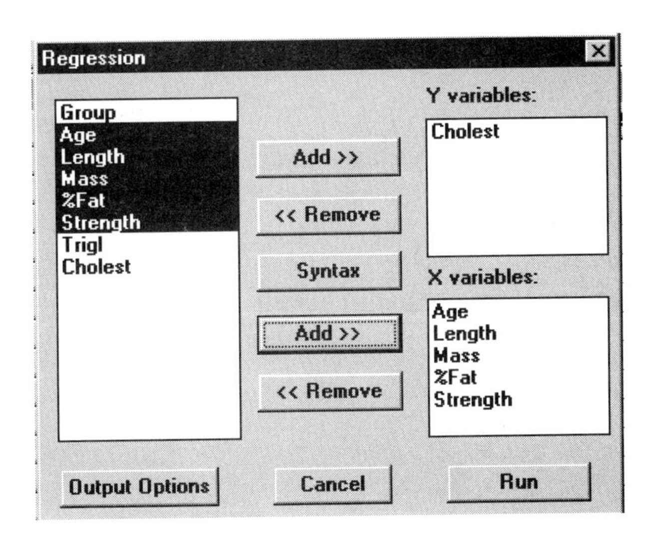

A selection of the output is given below:

Estimated Regressions

CHOLEST on	AGE	LEN	MASS	%FAT	STRENGTH
Regr. Coeff.	0.060	-0.055	-0.006	0.559	-0.119
Std. Error.	0.023	0.025	0.014	0.914	0.258
T - Values	2.626	-2.176	-0.412	0.611	-0.463

Residual Variance = 0.875D+00
Squared Multiple Correlation =0.203

3.5 Data Manipulation and Bivariate Plots

In this example, use is made of the Fitness/Cholesterol data described in detail in the previous section. The data set contains 8 variables, these being *Group*, *Age*, *Len*, *Mass*, *%Fat*, *Strength*, *Trigl* and *Cholest*.

The categorical variable *Group* has 4 values, these being 1, 2, 3, and 4, denoting weight lifters (n_1 = 17), students (n_2 = 20), marathon athletes (n_3 = 20) and coronary patients (n_4 = 9) respectively.

The PRELIS environment will be used to

- Assign labels to the categories of *Group*
- Change the variable types of the variables *Age*, *Len*, . . . , *Cholest* from the default type of ordinal to continuous
- Insert a new variable called *Totchol* into the data set
- Assign the value *Trigl* + *Cholest* to the new variable *Totchol*
- Create a data subset consisting of the 20 marathon athletes
- Display bivariate plots for the marathon athlete subgroup
- Calculate covariances and means for groups 1 to 3

3.5.1 Assigning Labels to Categories of Ordinal Variables

To display the fitness/cholesterol data, the ***File***, ***Open*** option is used. Select ***Files of type*** ***.psf** from the drop-down list box. The file **fitness.psf** can be found in the **tutorial** folder.

From the ***Data*** menu, select ***Define Variables*** and highlight the variable *Group* as illustrated in the next image.

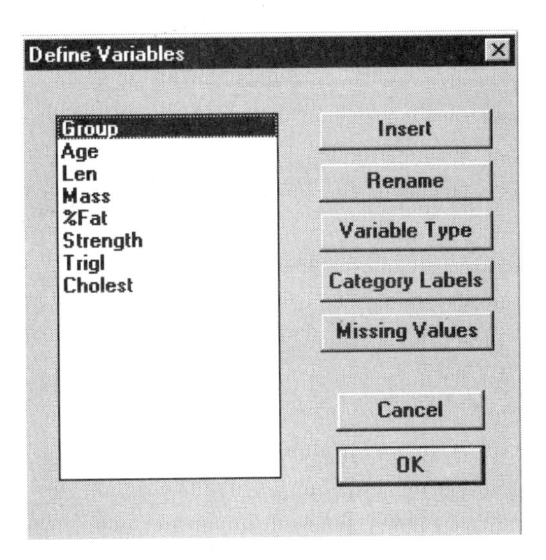

Once this is done, click on the ***Category Labels*** button.

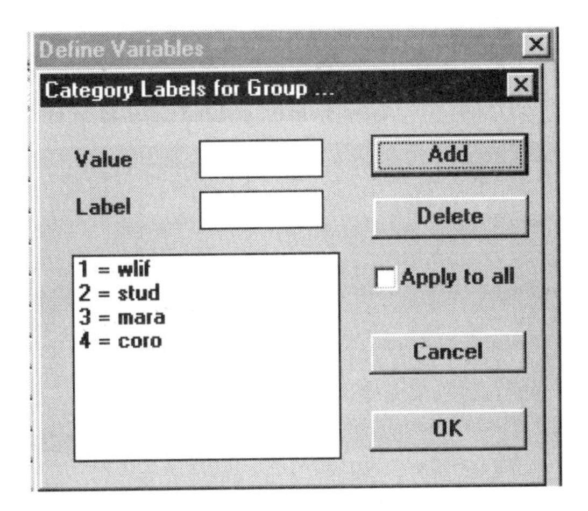

In the value string field, type the value `1` and in the label string field type the value `wlif`. Click the ***Add*** button to obtain the code `1 = wlif` in the syntax window. Proceed in a similar fashion until `4 = coro` has been entered in the code string field. When done, click ***OK*** to return to the ***Define Variables*** dialog box.

3.5.2 Change Variable Type

As default, PRELIS assumes that all variables are ordinal (see page 146-7 of the *PRELIS 2: User's Reference Guide*). Variables with more than 15 categories will be classified as continuous. To change the default variable type of the variables *Age*, *Len*, . . . , *Cholest*, we use the ***Define Variable*** option on the ***Data*** menu. From the ***Define Variables*** dialog box, select the variables *Age* to *Cholest* (use the ***CTRL*** keyboard button to make a multiple selection).

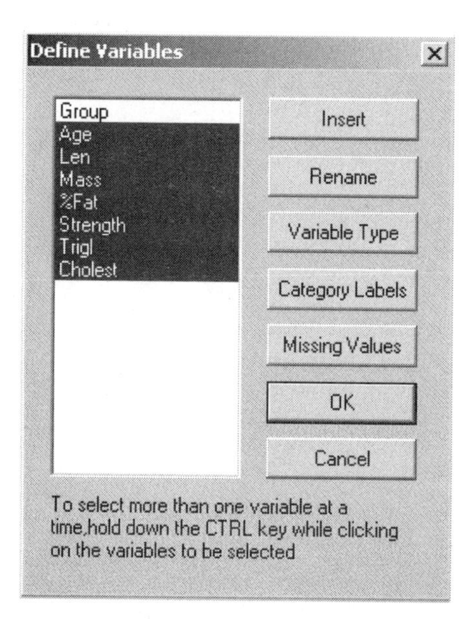

Click on the ***Variable Type*** button to activate the dialog box shown below.

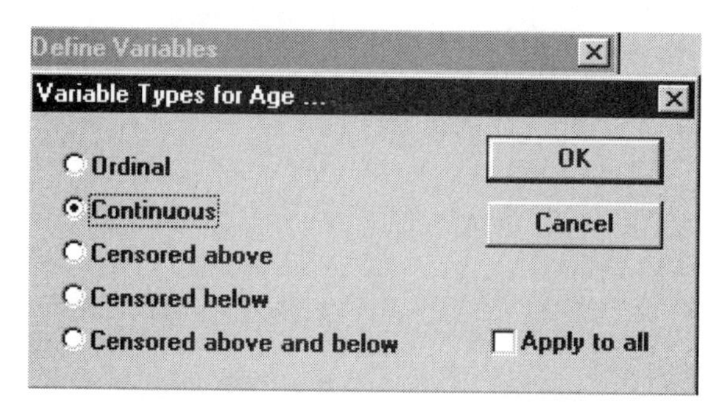

Select ***Continuous*** and click ***OK***. One can save the PRELIS system file under a new name, e.g. **fitness1.psf**, by using the ***Save As*** option on the ***File*** menu.

3.5.3 Insert New Variables

To insert new variables in a *.**psf** file, use the ***Insert*** button on the ***Define Variables*** dialog box. The ***Define Variables*** dialog box is obtained by clicking on the ***Data*** menu on the main menu bar. Note that the ***Data***, ***Transformations***, ***Statistics***, and ***Graphs*** items are only activated when a PRELIS system file (*.**psf**) is opened.

To insert the new variable *TotChol*, click on the ***Insert*** button to activate the dialog box labeled ***Add one or a list of variables***. Type in *TotChol*, as shown in the dialog box below and, when done, click the ***OK*** button.

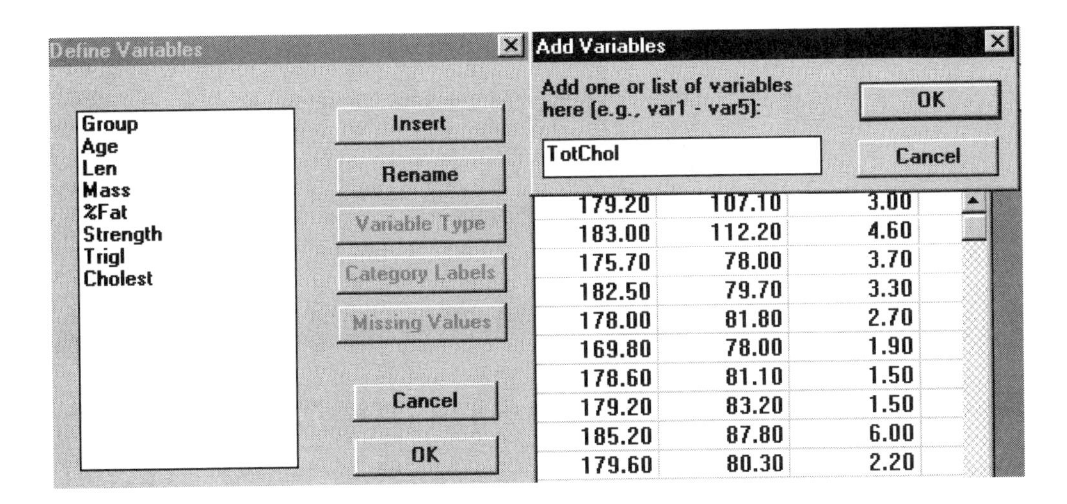

The dialog box below shows that the variable *Totchol* has been added to the list. *Totchol* will appear in the spreadsheet as an additional column with the value 0 assigned to each individual. Click ***OK*** when done.

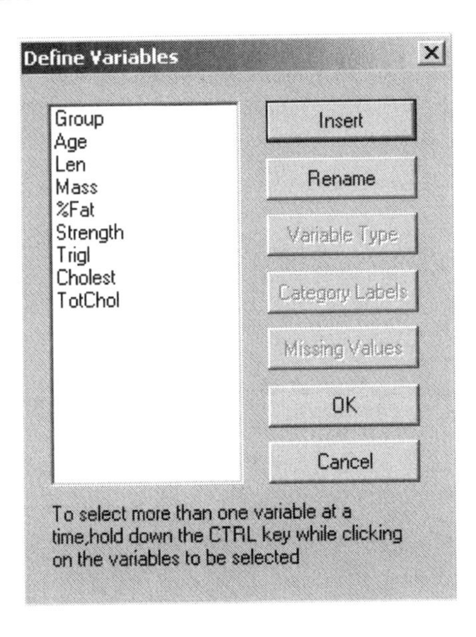

A portion of the spreadsheet, containing the variable *Totchol*, is shown below:

FITNESS.PSF

	%Fat	Strength	Trigl	Cholest	TotChol
1	3.000	15.200	0.580	4.440	0.000
2	4.600	20.300	1.510	4.880	0.000
3	3.700	17.500	1.200	4.330	0.000
4	3.300	16.100	0.750	3.660	0.000
5	2.700	14.100	0.750	4.570	0.000

3.5.4 Compute Values for a New Variable

One can use the items on the ***Transformation*** menu to, for example, recode or compute new values for the variables contained in the PRELIS spreadsheet.

In the previous section a new variable *TotChol* was added to **fitness.psf** and we now wish to enter the values `TotChol = Trigl + Cholest` for this variable. Open the ***Compute*** dialog box shown below by selecting the ***Compute*** option from the ***Transformation*** menu.

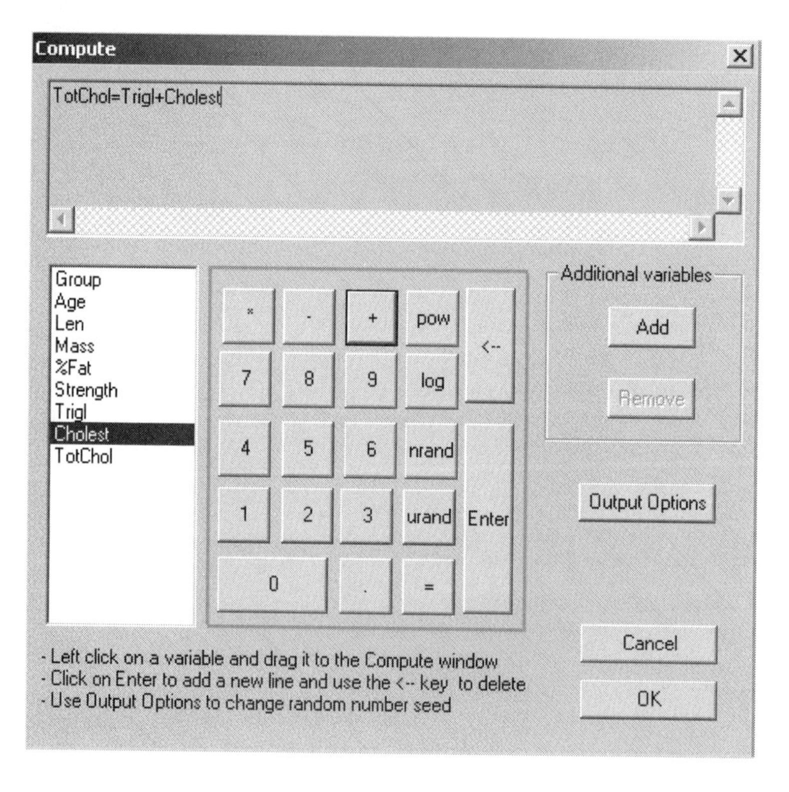

The ***Compute*** dialog box displays the new variable as a function of *Trigl* and *Cholest*. To enter this equation, click on *TotChol* (left mouse button). While holding the mouse button down, drag this item to the position shown in the ***Compute*** dialog box and release the mouse button. Click on the "=" sign, then drag *Trigl* to the position next to the "=" sign. When this is done, click on the "+" sign and finally drag *Cholest* to the position shown. Click ***OK*** to update the *.**psf** file. The result of this computation is shown below.

Fitchol

	Trigl	Cholest	TotChol
1	0.58	4.44	5.02
2	1.51	4.88	6.39
3	1.20	4.33	5.53
4	0.75	3.66	4.41
5	0.75	4.57	5.32
6	0.33	3.90	4.23

It is also possible to save the data under a new name, e.g. **fitness2.psf** by clicking the ***Output Options*** button on the ***Compute*** dialog box and then checking the ***Save the transformed data to file*** check box.

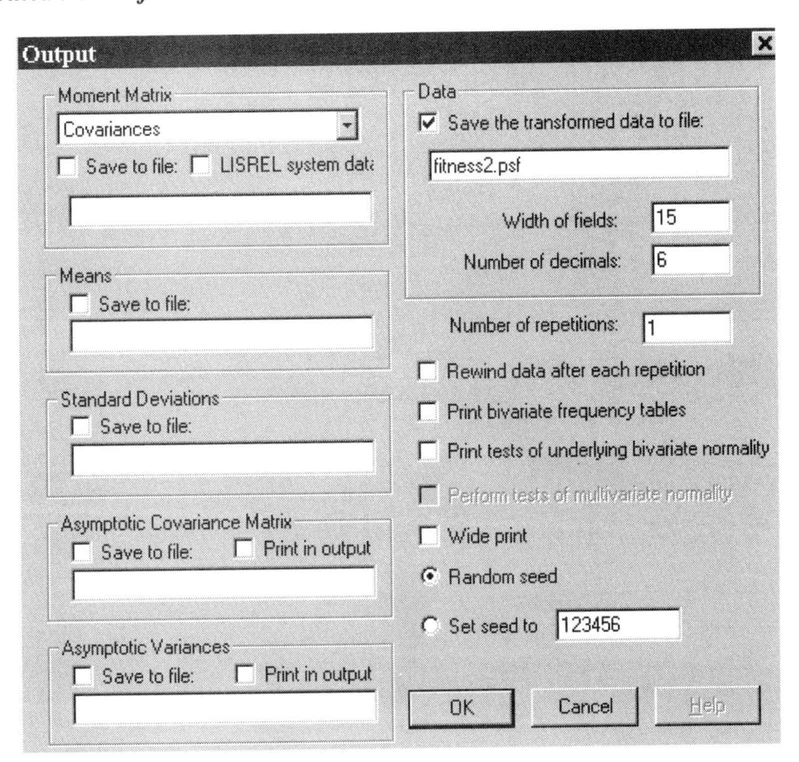

3.5.5 Creating a Subset of the Data

Suppose that we want to focus on the subgroup of marathon athletes (*Group* = 3). To create a data set consisting of this subgroup, select the ***Data, Select Variables/Cases*** option as shown below.

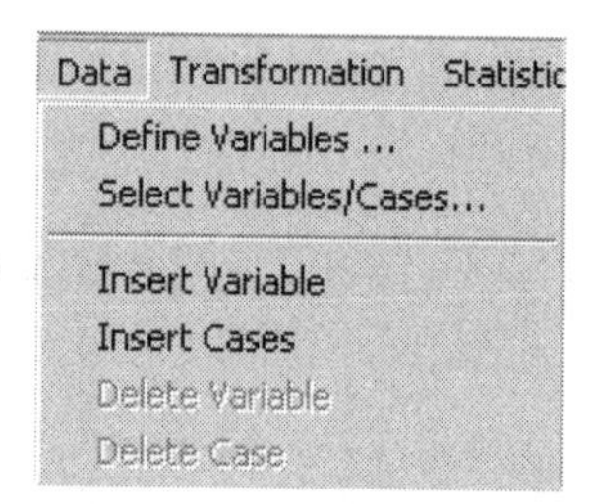

The ***Select Data*** dialog box appears.

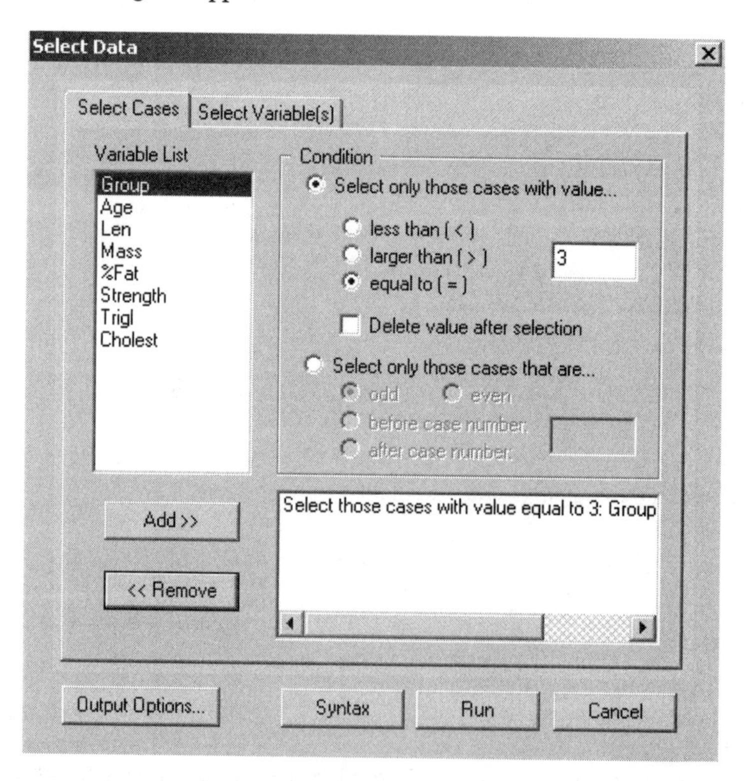

Highlight the variable *Group* and select the ***Select only those cases with value*** and ***equal to*** options. Type in the value of `3` and then click the ***Add*** button to obtain the words `select those cases with value equal to 3:`Group shown in the window below the ***Add*** and ***Remove*** buttons. This syntax can be removed if it is highlighted and the ***Remove*** button is clicked. Choose the ***Output Options*** button to save the file under a new name, for example **fitnessm.psf**.

Note:

If the file is not saved under a new name, the existing *.**psf** will be overwritten.

When done, click ***OK*** to start PRELIS. The data subset consisting of `Group=3` values only will be created. This spreadsheet, **fitnessm.psf**, is shown below.

FITNESSM.PSF

	Group	Age	Len	Mass	%Fat
1	3.00	45.00	182.40	89.50	5.30
2	3.00	23.00	176.40	58.80	1.90
3	3.00	27.00	169.10	70.70	5.80
4	3.00	37.00	173.70	76.40	3.60
5	3.00	18.00	190.70	70.70	2.70

3.5.6 Bivariate Plots

A bivariate plot can be obtained by clicking on the ***Bivariate Plot*** icon button shown next to the ***Univariate Plot*** icon button on the second tool bar box below.

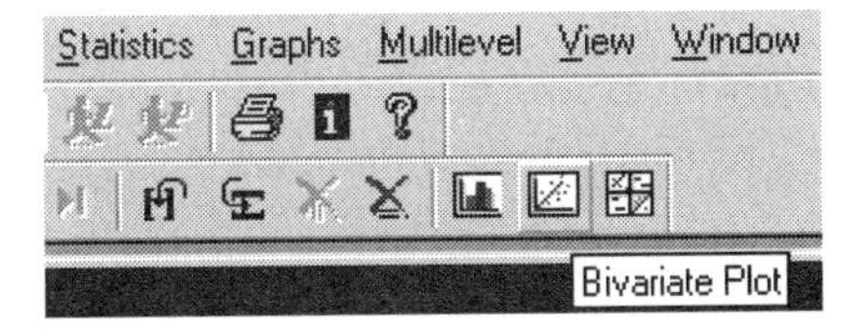

One can also obtain a bivariate plot of two variables by selecting the ***Graphs***, ***Bivariate*** option to obtain the ***Bivariate Plots*** dialog box.

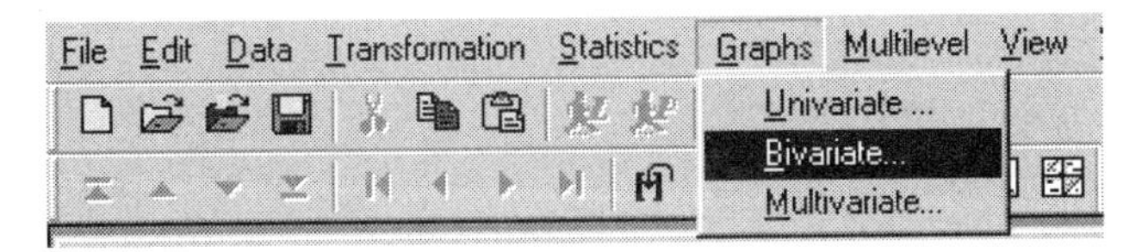

Click on *Cholest* and then click on the ***Select (Y variable)*** button. Once this is done, highlight *Age*, as shown below, and click on the ***Select (X variable)*** button. Also select the ***Scatter Plot*** and ***Line Plot*** options. To produce a bivariate graph, click ***Plot***.

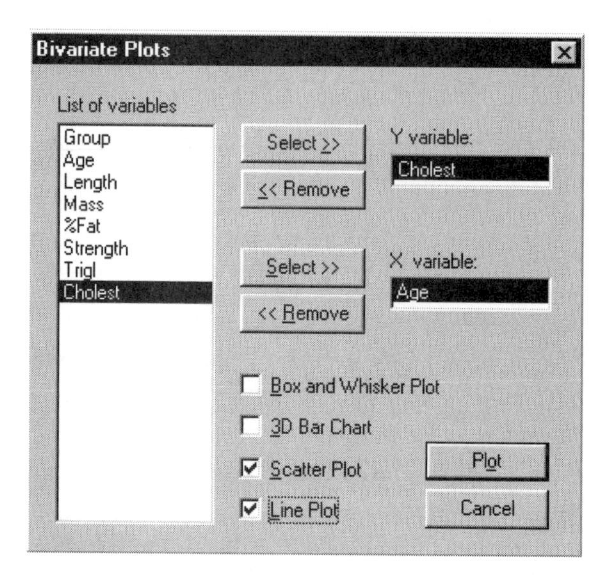

A plot of *Cholest* on *Age* is obtained, as shown below:

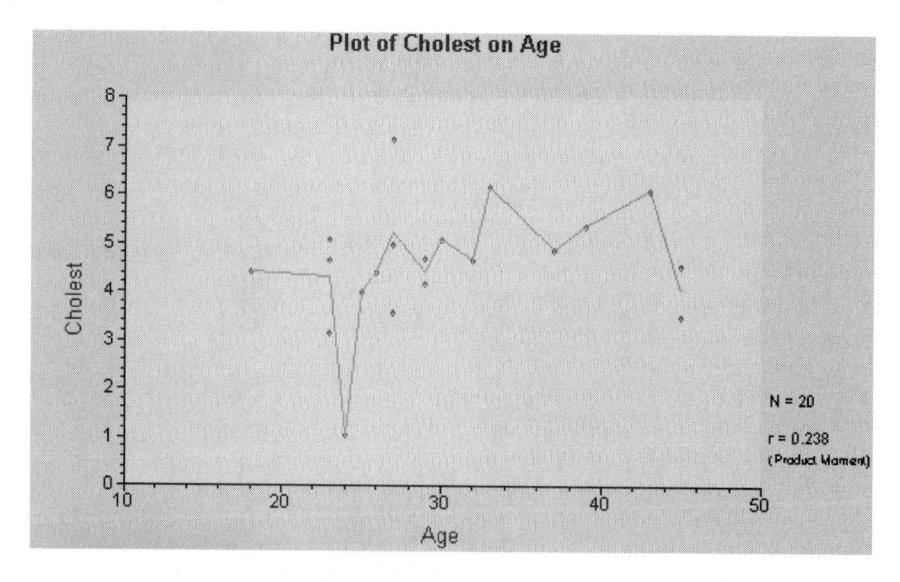

Note that the graph shows the product moment correlation coefficient and the sample size. The line plot is obtained by taking the arithmetic average of all *Y*-values that correspond to the same *X*-value. As such, the line plot shows the general trend in the *Y*-variable with increase in *X*.

In practice, one often finds that two variables are not linearly related, but a transformation on one or both the variables may exhibit a stronger linear relationship.

The Y and X axes can therefore be changed from linear to logarithmic to investigate the Log(Y), X or Y, log(X) or log(Y), log(X) relationships.

To plot *Cholest* on log(*Age*), for example, select the ***Plot Y on log(X)*** option from the ***Image*** menu.

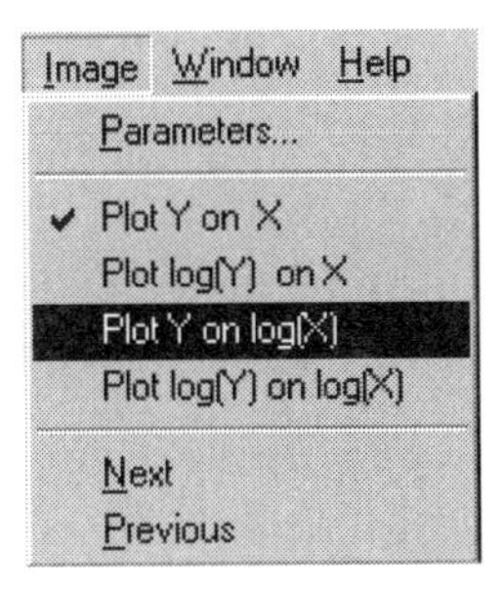

The resulting graph is shown below. Note that this particular transformation does not exhibit a stronger association between the two variables.

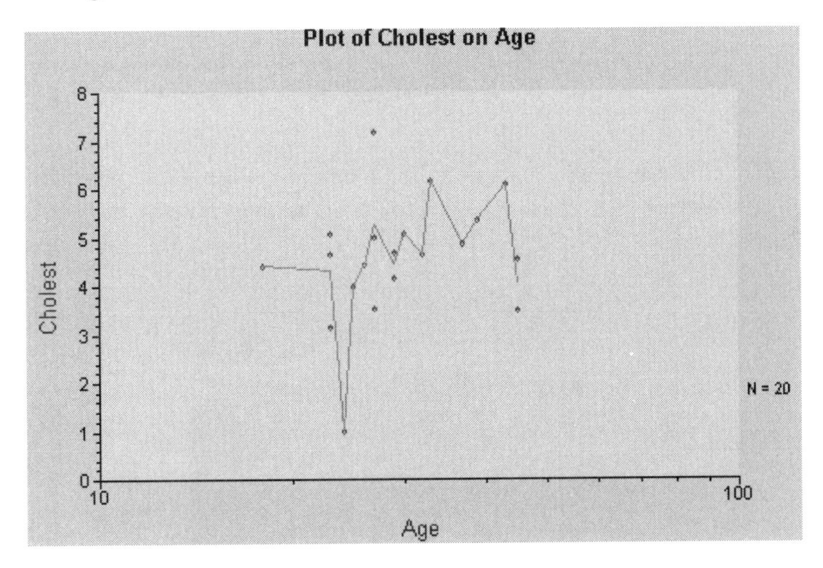

As with path diagrams, histograms, bar charts and bivariate plots can be exported as Windows Metafiles (*.**wmf** extension). These files can be imported into any of the Microsoft Office products, and can also be opened with various graphics software packages. This, together with the fact that PRELIS or LISREL output can be converted to a *.**rtf** , *.**tex** or *.**htm** file, should greatly assist researchers in producing reports or publications.

To save a bivariate plot as a *.**wmf** graphics file, select the ***Export as Metafile*** option from the ***File*** menu.

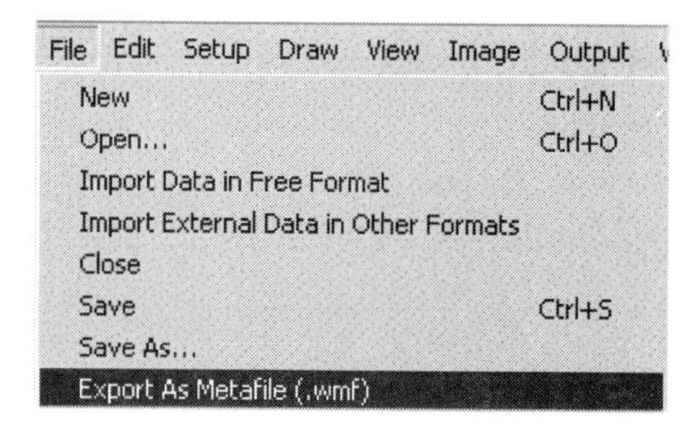

By clicking on ***Export as Metafile***, the user is prompted to save the *.**wmf** file under any name and in any folder.

3.5.7 Calculate and Save Covariances and Means for Subgroups

Suppose that we want to calculate covariances, means and standard deviations for each of the first 3 groups of the fitness/cholesterol data.

The procedure is illustrated by selecting *Group* = 2 (2 = students). Select the ***Select Variables/Cases*** option from the ***Data*** menu to obtain the ***Select Data*** dialog box.

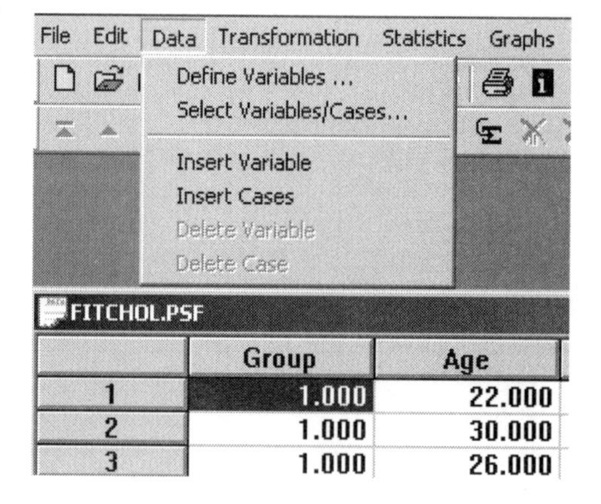

Click the ***Select Cases*** tab on the ***Select Data*** dialog box:

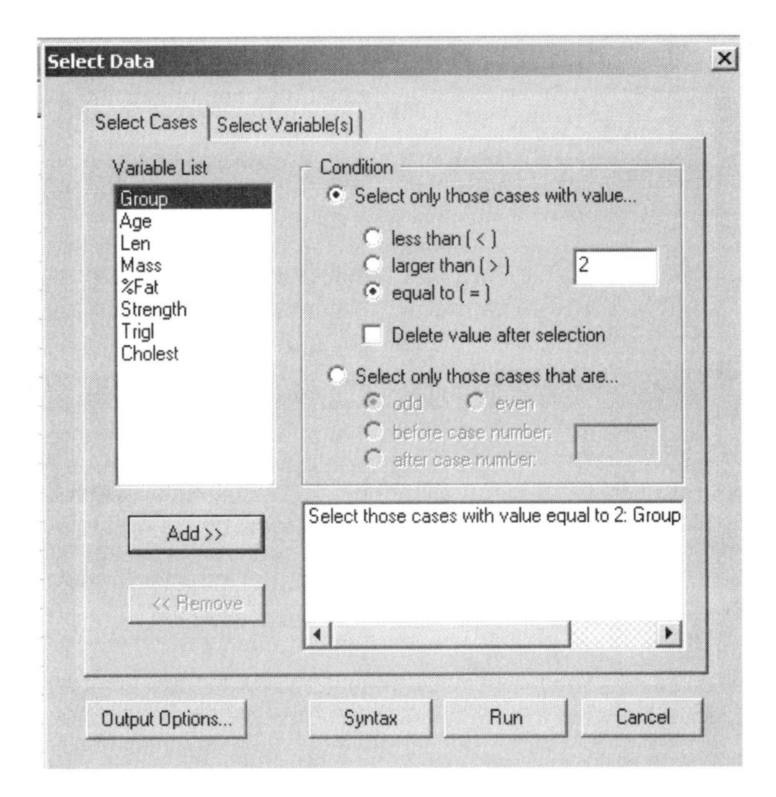

Select *Group* (click on *Group* with the left mouse button) and then complete the ***Select only those cases with value*** portion of the dialog box. Since the variable *Group* will now assume values of 2 only, it should not be included in the calculation of the means and covariances.

If this variable is to be deleted, check the ***Delete variables after selection*** option. Click the ***Add*** button to generate the necessary syntax.

Once this is done, the ***Output Options*** button should be clicked to produce the ***Output*** dialog box given below.

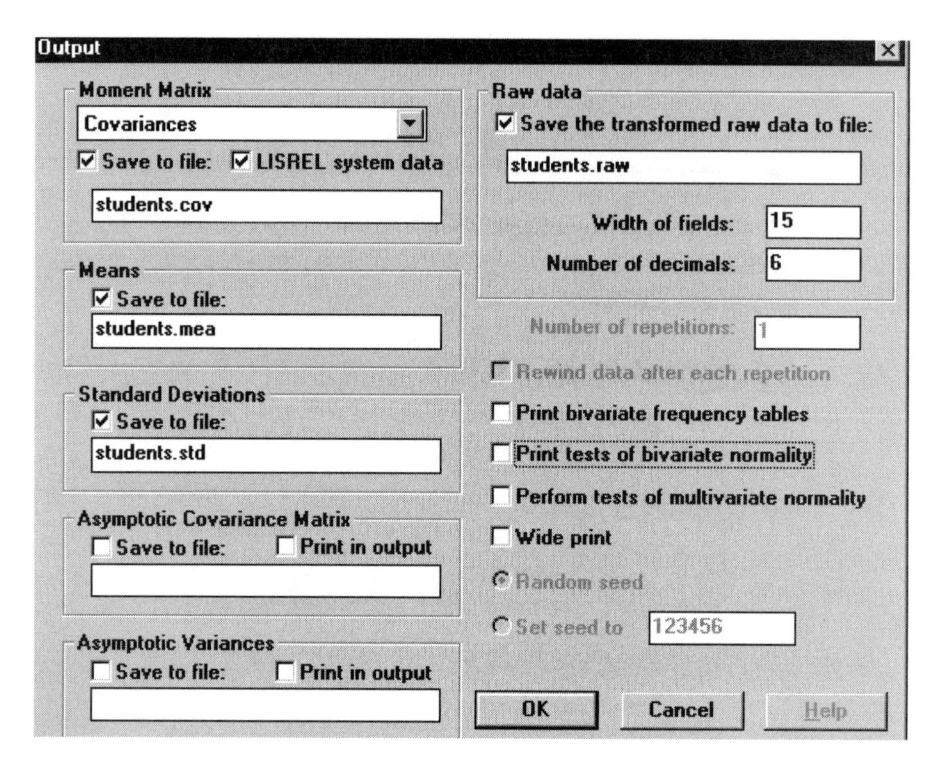

Complete the dialog box as shown. When done, click ***OK*** to return to the ***Select Cases*** dialog box. Click ***OK*** to run PRELIS.

A portion of the PRELIS output is given below:

```
SY = F:\LISREL850\PR2EX\FITNESS.PSF
OU MA=CM SM=students.cov RA=students.raw WI=15 ND=6
ME=students.mea
SD=students.std XM

Total Sample Size =     20
```

Covariance Matrix

	AGE	LEN	MASS	%FAT	STRENGTH	TRIGL
AGE	19.734					
LEN	8.576	38.234				
MASS	12.507	26.173	69.178			
%FAT	-0.924	-1.056	2.492	0.912		
STRENGTH	-2.922	-4.754	9.334	3.326	12.350	
TRIGL	0.880	-0.376	0.358	0.063	0.282	0.386
CHOLEST	2.412	-2.394	0.222	0.254	1.045	0.397
TOTCHOL	3.292	-2.770	0.580	0.317	1.327	0.783

```
Covariance Matrix

            CHOLEST    TOTCHOL
           --------   --------
CHOLEST       1.064
TOTCHOL       1.461      2.244


Means

      AGE        LEN       MASS       %FAT   STRENGTH      TRIGL
 --------   --------   --------   --------   --------   --------
   26.050    179.760     73.070      3.260     15.615      1.052


Means

  CHOLEST    TOTCHOL
 --------   --------
    4.507      5.559
```

3.6 Normal Scores

The analysis of continuous non-normal variables in structural equation models is problematic in several ways. If the maximum likelihood (ML) method is used, standard errors and χ^2 may be incorrect. In theory, weighted least squares (WLS or ADF) with a correct weight matrix should produce correct estimates of standard errors and chi-squares, but this requires a very large sample. Sometimes a reasonable compromise is to use ML despite the non-normality and correct for the bias in standard errors, but this too requires a very large sample.

Another solution to non-normality is to normalize the variables before analysis. Normal scores offer an effective way of normalizing a continuous variable for which the origin and unit of measurement have no intrinsic meaning, such as test scores.

Example: Normalizing nine Psychological Variables

A detailed description of this example is given in *LISREL 8: New Statistical Features*. To normalize the raw scores of the nine psychological variables in file **npv.psf**, we proceed as follows.

Select the ***File, Open*** option and select ***PRELIS system file*** (*.**psf**) from the ***Files of type*** drop-down list box. Select the file **npv.psf** from the **lis850ex** folder. Click ***Open*** to proceed.

Note that normal scores may be calculated for both ordinal and continuous variables. From the ***Statistics*** menu, select the ***Normal Scores*** option to obtain the ***Normal Scores*** dialog box.

Statistics Graphs Multilevel Models View Window Help

Data Screening
Impute Missing Values ...
Equal Thresholds...
Fix Thresholds...
Homogeneity Test ...
Normal Scores
Factor Analysis
Probit Regressions...
Regressions...
Two-Stage Least-Squares
Bootstrapping ...
Output Options ...

VIS PERC	CUBES	LOZENGES
23.000	19.000	4.000
33.000	22.000	17.000
34.000	24.000	22.000
29.000	23.000	9.000
16.000	25.000	10.000
30.000	25.000	20.000
36.000	33.000	36.000
28.000	25.000	9.000
30.000	25.000	11.000

Select all 9 variables and click ***Add***. A variable may be removed by clicking on the specific variable in the bottom dialog box and then using the ***Remove*** button. When all the variables for which normal scores are to be computed are selected, click ***Output Options*** to save the normal scores to an external file.

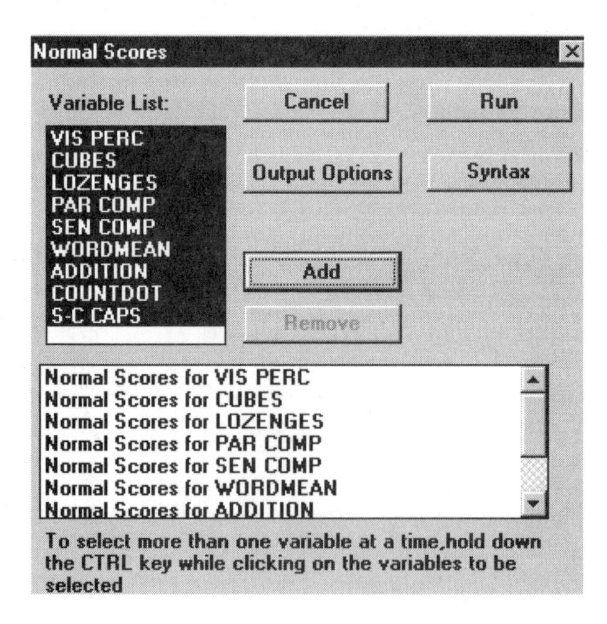

On the ***Output*** dialog box, check the ***Save the transformed data to file*** check box and enter the file name **npv_ns.raw**. The data can also be saved as **npv_ns.psf**, in which case a PRELIS system file is created.

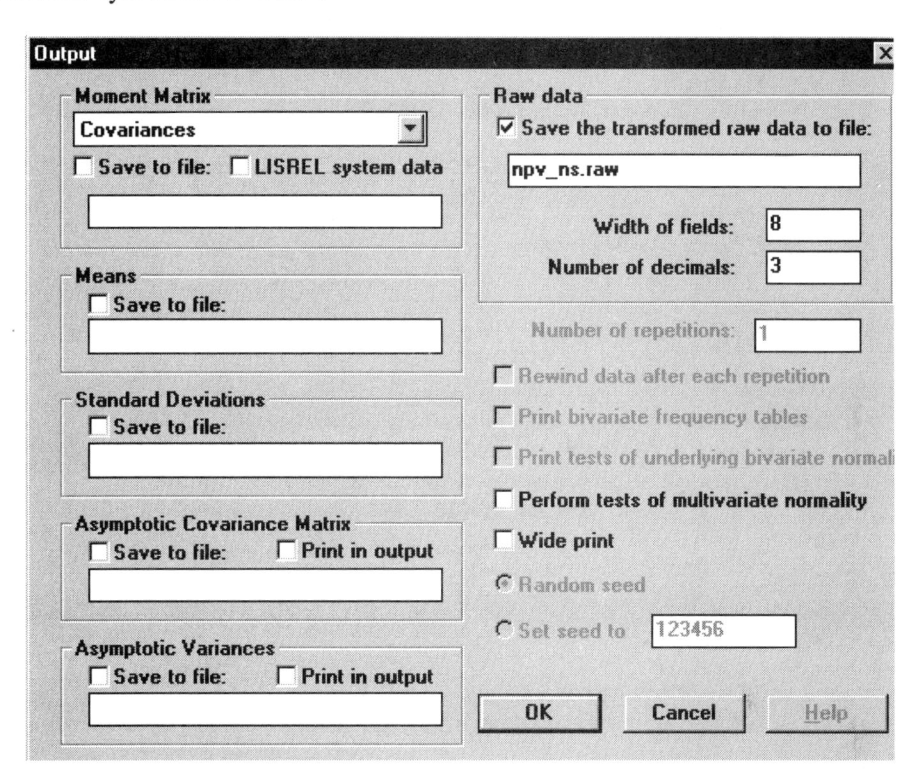

When done, click ***OK*** and then ***Run*** on the ***Normal Scores*** dialog box. The first ten rows of normal scores for the nine variables are shown below.

NPV_NS.RAW

```
  -0.931  -1.742  -1.887   0.201  -0.448  -1.047  -0.793  -1.302  -0.918
   0.447  -0.556   0.317  -0.621  -0.448  -1.047  -1.044  -0.595   0.043
   0.596  -0.122   0.747   0.506   0.000   0.409  -1.821  -1.074   1.000
  -0.130  -0.326  -0.736  -0.202   0.000  -0.784   0.958  -0.362  -1.243
  -2.070   0.130  -0.585  -0.621   1.421   0.892   0.905   0.747  -0.805
   0.017   0.130   0.627   0.201   0.773   0.262   0.165   0.174   0.236
   0.893   1.538   2.789   1.737   1.421   2.789   1.410   1.484   2.789
  -0.308   0.130  -0.736   0.201  -0.254  -0.784   0.307  -0.713  -0.515
   0.017   0.130  -0.437   0.506   0.400  -1.303   0.585   0.244   0.069
  -1.207   0.130  -1.347  -0.202   0.400  -0.009   0.000  -0.456  -0.418
```

3.7 Two-Stage Least-Squares

Two-Stage Least-Squares (TSLS) is particularly useful for estimating econometric models of the form

$$\mathbf{y} = \mathbf{B}\mathbf{y} + \mathbf{\Gamma}\mathbf{x} + \mathbf{u},$$

where $\mathbf{y} = (y_1, y_2, ..., y_p)$ is a set of endogenous or jointly dependent variables, $\mathbf{x} = (x_1, x_2, ..., x_q)$ is a set of exogenous or predetermined variables uncorrelated with the error terms $\mathbf{u} = (u_1, u_2, ..., u_p)$, and $\mathbf{B}$ and $\mathbf{\Gamma}$ are parameter matrices. A typical feature of the above model is that not all y-variables and not all x-variables are included in each equation.

A necessary condition for identification of each equation is that, for every y-variable on the right side of the equation, there must be at least one x-variable excluded from that equation. There is also a sufficient condition for identification, the so-called *rank condition*, but this is often difficult to apply in practice. For further information on identification of interdependent systems, see, *e.g.*, Goldberger (1964, pp. 313-318).

Example: Klein's Model I of US Economy

Klein's (1950) Model I is a classical econometric model which has been used extensively as a benchmark problem for studying econometric methods. It is an eight-equation system based on annual data for the United States in the period between the two World Wars. It is dynamic in the sense that elements of time play important roles in the model.

The data set, **klein.dat**, consists of the following 15 variables:

Ct	Aggregate Consumption
Pt_1	Total Profits, previous year
*Wt**	Private Wage Bill
It	Net Investment
Kt_1	Capital Stock, previous year
Et_1	Total Production of Private Industry, previous year
*Wt***	Government Wage Bill
Tt	Taxes
At	Time in Years from 1931
Pt	Total Profits
Kt	End-of-year Capital Stock
Et	Total Production of Private Industry
Wt	Total Wage Bill
Yt	Total Income
Gt	Government Non-Wage Expenditure

To estimate the consumption function, we use *Ct* as the *y*-variable, *Pt*, *Pt_1* and *Wt* as *x*-variables and *Wt***, *Tt*, *Gt*, *At*, *Pt_1*, *Kt_1* and *Et_1* as the z-variables. An intercept term in the equation can be estimated by introducing a variable denoted in this example by *ONE*, which is a constant equal to 1 for each year. When an intercept term is introduced, the moment matrix (`MM`) is used instead of the covariance matrix (`CM`).

To start the analysis, select the ***Import Data in Free Format*** option from the ***File*** menu.

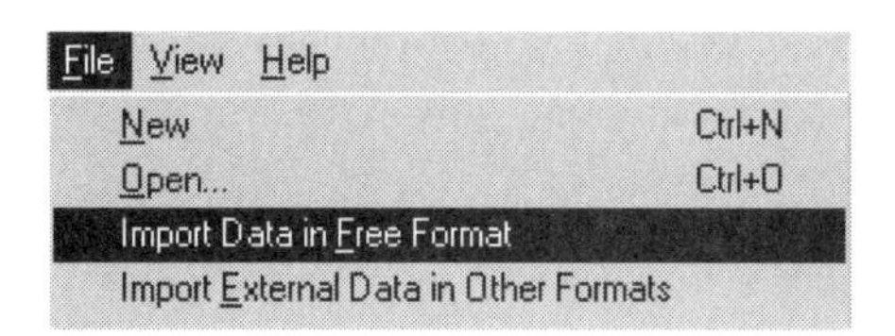

Select the data file **klein.dat** from the **lis850ex** folder. When the ***Open*** button is clicked in the ***Open Data File*** dialog box, the user is prompted for the number of variables. Enter `15` in the ***Number*** string field.

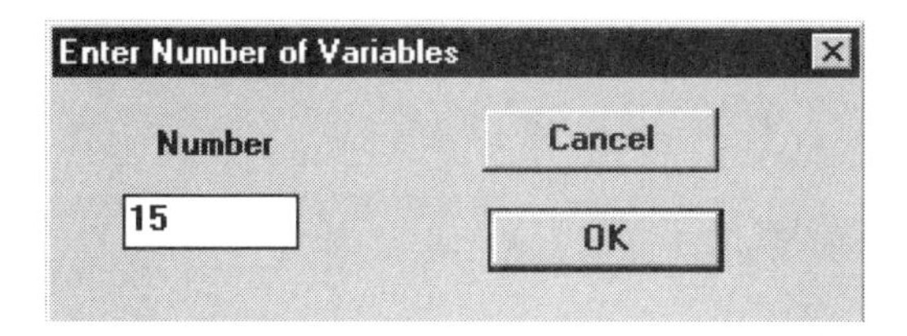

Click ***OK*** to obtain the PRELIS system file **klein.psf**. Note that default variable names are assigned: *VAR 1*, *VAR 2*, … , *VAR 15*.

KLEIN.PSF

	VAR 1	VAR 2	VAR 3	VAR 4
1	41.900	12.700	25.500	-0.200
2	45.000	12.400	29.300	1.900
3	49.200	16.900	34.100	5.200
4	50.600	18.400	33.900	3.000
5	52.600	19.400	35.400	5.100
6	55.100	20.100	37.400	5.600
7	56.200	19.600	37.900	4.200

To change the variable names, select ***Define Variables*** from the ***Data*** menu.

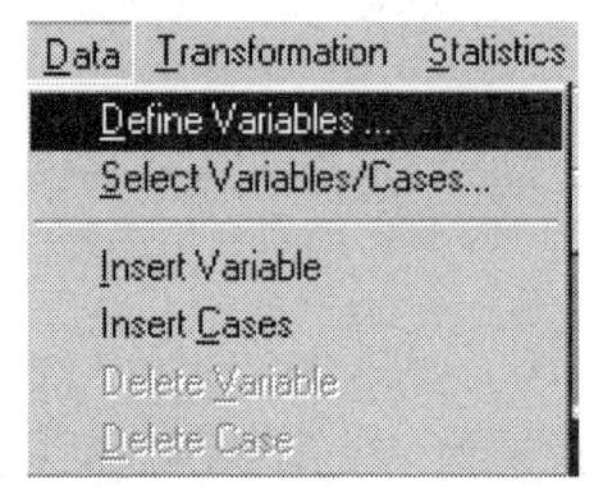

Click on the ***Rename*** button or double click on a variable name to invoke an edit box. When the new variable name is entered, press the ***Enter*** key on the keyboard or click once outside the edit box.

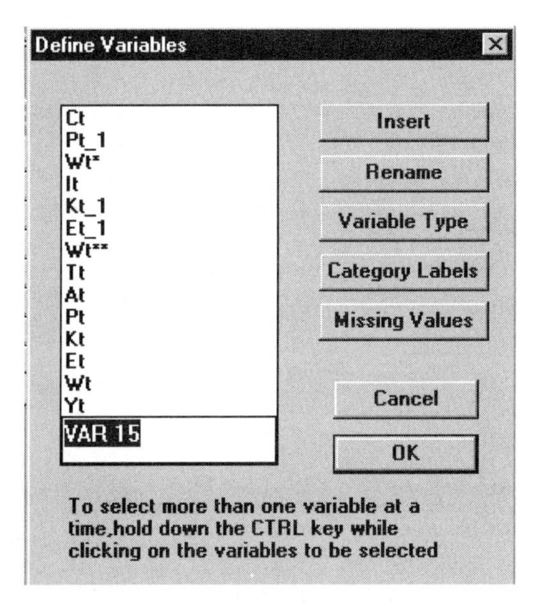

To change the variable type from the default of ordinal to continuous, select the variable *Ct* and click on ***Variable Type*** on the ***Define Variables*** dialog box to go to the ***Variable Type for*** dialog box.

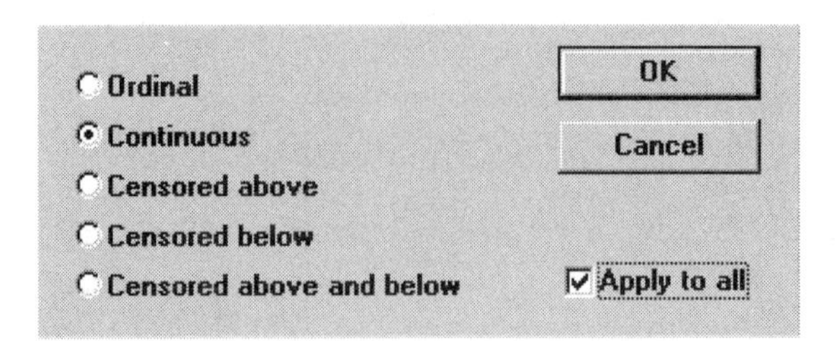

Check the ***Apply to all*** check box, then click ***OK*** to return to the ***Define Variables*** dialog box. On this dialog box, also click ***OK***. The spreadsheet now contains the renamed variables.

To add the intercept term, select the ***Compute*** option from the ***Transformation*** menu. This will invoke the ***Compute*** dialog box shown below. Click ***Add*** to add one or more variables to the existing list.

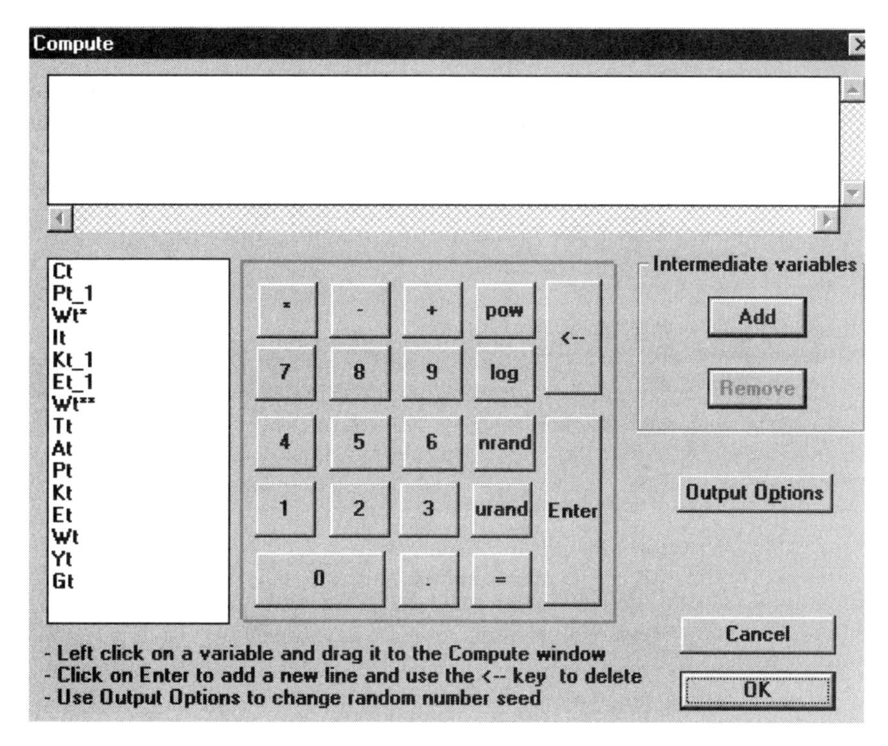

Type `ONE` in the text box of the ***Add Variables*** dialog box and click ***OK*** when done.

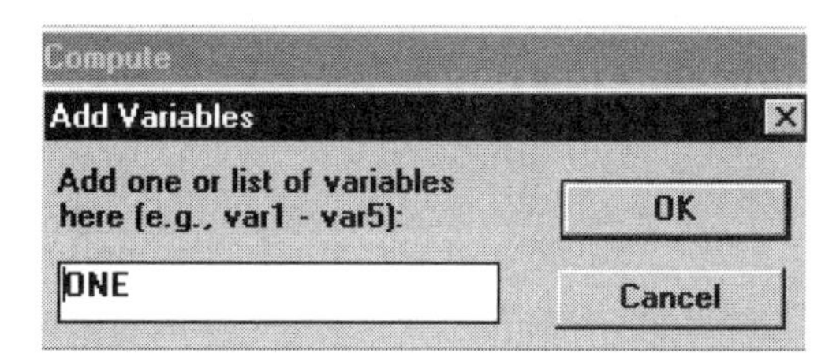

Click on the variable *ONE* and drag it to the ***Compute*** window. Next, click on the equal sign and then on the number `1`.

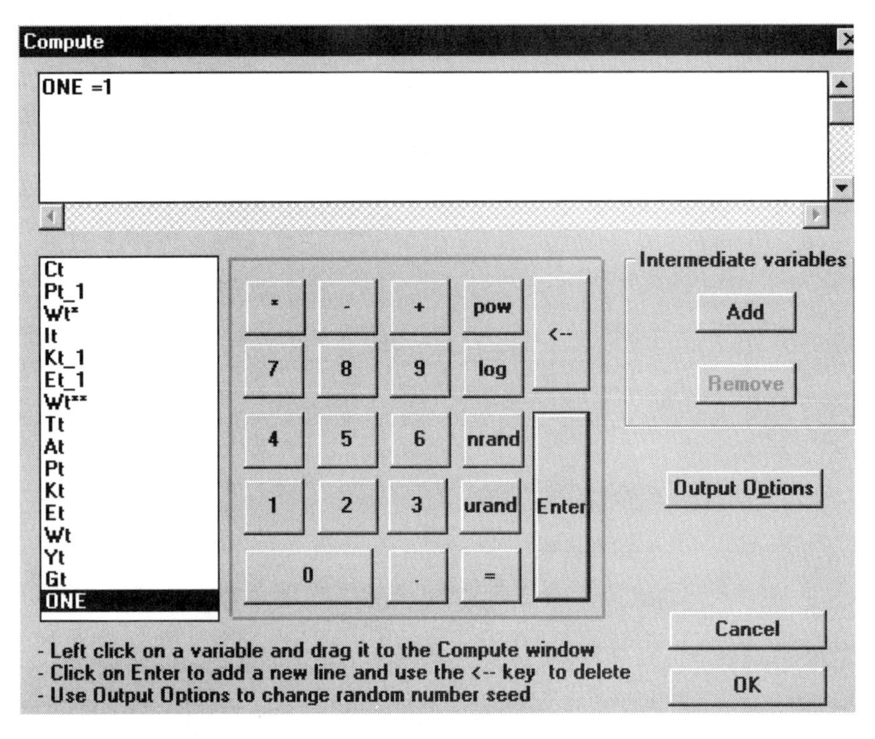

When done, click ***OK*** to generate the new spreadsheet with the variable *ONE* added as the last column. Finally, select ***Two-Stage Least-Squares*** from the ***Statistics*** menu to obtain the ***Two-Stage Least-Squares*** dialog box.

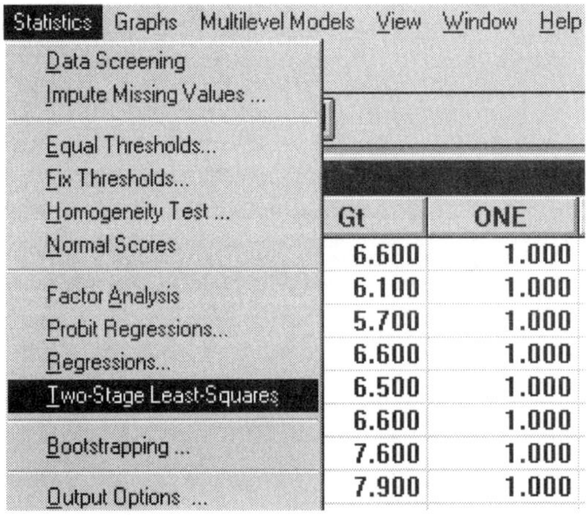

Statistics Graphs Multilevel Models View Window Help

Data Screening
Impute Missing Values ...
Equal Thresholds...
Fix Thresholds...
Homogeneity Test ...
Normal Scores
Factor Analysis
Probit Regressions...
Regressions...
Two-Stage Least-Squares
Bootstrapping ...
Output Options ...

Gt	ONE
6.600	1.000
6.100	1.000
5.700	1.000
6.600	1.000
6.500	1.000
6.600	1.000
7.600	1.000
7.900	1.000

Enter *Ct* as the *Y*-variable, *ONE*, *Pt*, *Pt_1* and *Wt* as the *X*–variables and *ONE*, *Wt***, *Tt*, *Gt*, *At*, *Pt_1*, *Kt_1* and *Et_1* as the instrumental variables.

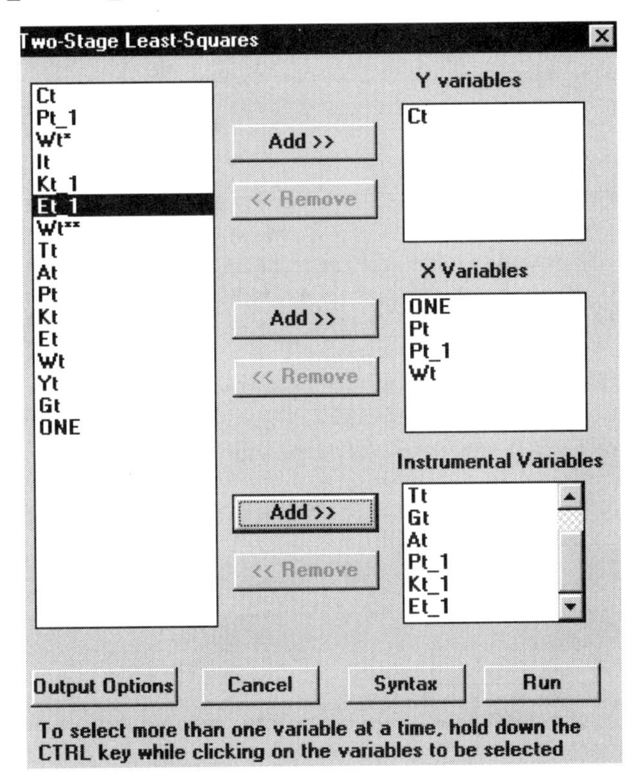

Click on ***Output Options*** and select ***Moments about zero*** from the ***Moment Matrix*** drop-down list box. At this stage one could also save the new data set under a different name. If the file extension is *.**psf**, a PRELIS system file is created. When done, click ***OK*** to return to the ***Two-Stage Least-Squares*** dialog box and then select ***Run*** to do the analysis or ***Syntax*** to view the newly created PRELIS syntax file (**klein.pr2**).

Consult the *LISREL 8: New Statistical Features* for additional information regarding this example.

3.8 Exploratory Factor Analysis

In an exploratory factor analysis, one wants to explore the empirical data to discover and detect characteristic features and interesting relationships without imposing any definite model on the data. An exploratory factor analysis may be structure generating, model generating, or hypothesis generating. In confirmatory factor analysis, on the other hand, one builds a model assumed to describe, explain, or account for the empirical data in terms of relatively few parameters.

Exploratory factor analysis is a technique often used to detect and assess latent sources of variation and covariation in observed measurements. It is widely recognized that exploratory factor analysis can be quite useful in the early stages of experimentation or test development.

Example: Exploratory Factor Analysis of Nine Psychological Variables

To illustrate exploratory factor analysis we use a classical data set. Holzinger and Swineford (1939) collected data on twenty-six psychological tests administered to 145 seventh- and eighth-grade children in the Grant-White school in Chicago. Nine of these tests are selected for this example. The nine selected variables and their intercorrelations are given in the table below.

Table 3.1: Correlation Matrix for Nine Psychological Variables

VIS PERC	1.000								
CUBES	0.326	1.000							
LOZENGES	0.449	0.417	1.000						
PAR COMP	0.342	0.228	0.328	1.000					
SEN COMP	0.309	0.159	0.287	0.719	1.000				
WORDMEAN	0.317	0.195	0.347	0.714	0.685	1.000			
ADDITION	0.104	0.066	0.075	0.209	0.254	0.178	1.000		
COUNTDOT	0.308	0.168	0.239	0.104	0.198	0.121	0.587	1.000	
S-C CAPS	0.487	0.248	0.373	0.314	0.356	0.222	0.418	0.528	1.000

From the ***File*** menu, select ***Import Data in Free Format***. From the ***Open Data File*** dialog box, select file of the type *.**raw** and choose the file **npv.raw** from the **lis850ex** folder. This file contains 145 observations on the nine psychological measurements.

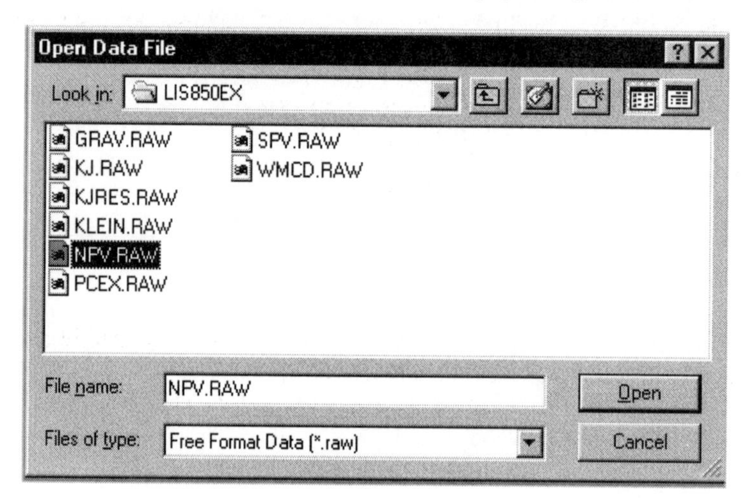

Click ***Open*** to obtain the ***Enter Number of Variables*** dialog box and enter 9 in the ***Number*** string field.

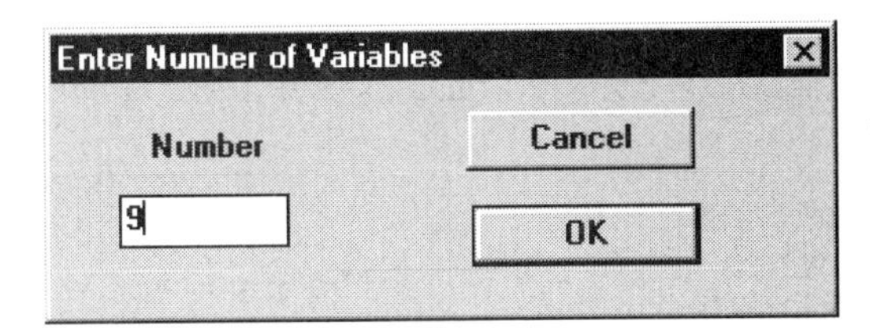

Click ***OK*** to display the PRELIS system file with default variable names *VAR 1*, *VAR 2*, …, *VAR 9*.

File Edit Data Transformation Statistics Graphs Multilevel Mode

NPV.PSF

	VAR 1	VAR 2	VAR 3
1	23.000	19.000	4.000
2	33.000	22.000	17.000
3	34.000	24.000	22.000
4	29.000	23.000	9.000
5	16.000	25.000	10.000
6	30.000	25.000	20.000

To change the default names to *VIS PERC*, *CUBES*, *LOZENGES*, *PAR COMP*, *SEN COMP*, *WORDMEAN*, *ADDITION*, *COUNTDOT* and *S-C CAPS*, select the ***Define Variables*** option from the ***Data*** menu to obtain the ***Define Variables*** dialog box.

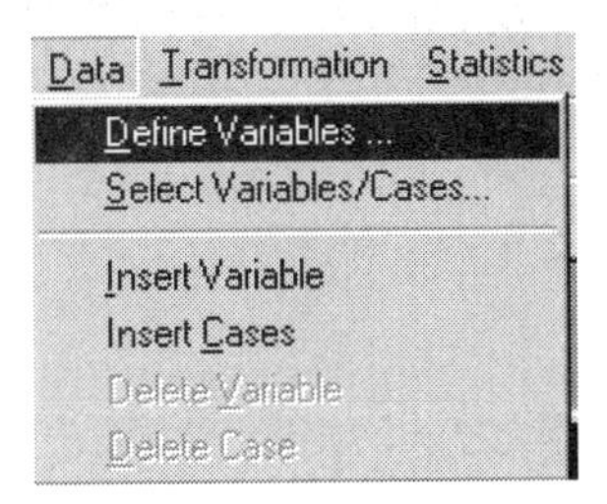

Names may be changed by clicking ***Rename*** or, alternatively, by double clicking on an existing variable name. Once a variable is entered, click once just below the box in which the variable name is entered (for example *S-C CAPS*) in the dialog box shown below or press the ***Enter*** key on the keyboard.

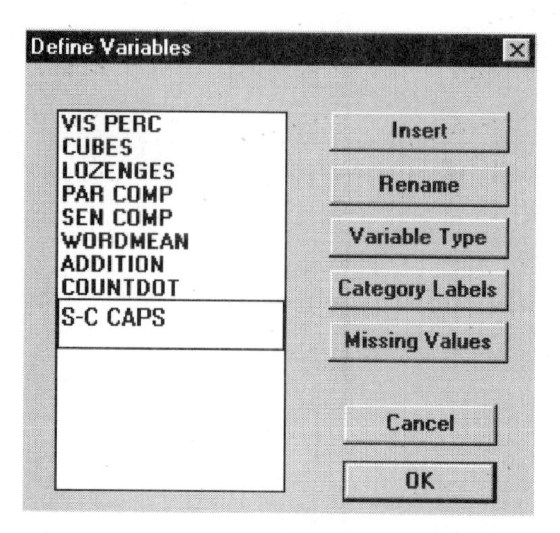

The nine psychological variables should be treated as continuous variables. The PRELIS default variable type, however, is ordinal. To change the default variable type, select all the variables in the variable list and click the ***Variable Type*** button. All the variables can be selected simultaneously by clicking on the first variable and then, with the left mouse button held down, dragging the cursor over the remaining variables.

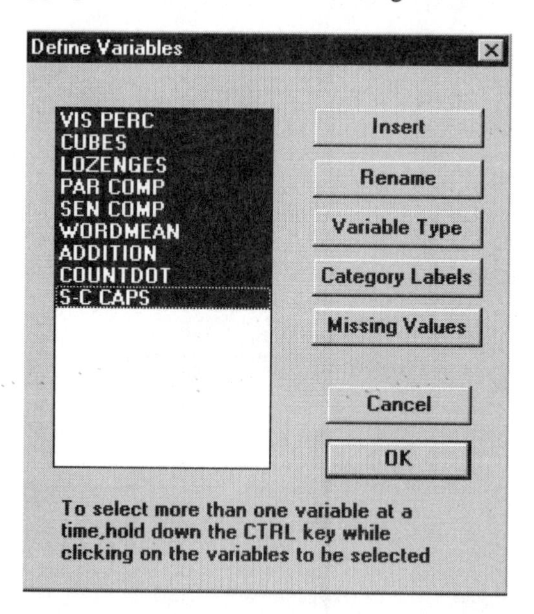

Change the variable type to continuous and click ***OK*** when done. If only one variable had been selected, use of the ***Apply to all*** check box would have changed all the variables in the list from ordinal to continous variables.

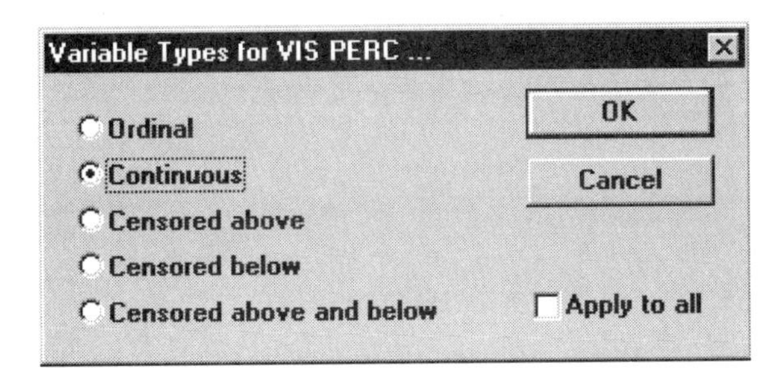

Although LISREL can determine the number of factors to be extracted if this is not specified by the user, we shall assume that the number of factors equals three.

An exploratory factor analysis is done by selecting the ***Factor Analysis*** option from the ***Statistics*** menu to obtain the ***Factor Analysis*** dialog box. Not all the variables in a data set may be suitable for a factor analysis, since the data may contain demographic variables such as marital status, gender, etc. A subset of the list of variables may therefore be selected. For the present example, all nine variables are used and therefore we do not have to use the ***Select*** button. Enter 3 for the number of Factors and click ***Run*** to produce output or ***Syntax*** to view (and possibly modify) the syntax file before running PRELIS. Factor scores will be saved in the file **npv.fsc**. Note that one can also use the ***Factor Analysis*** dialog box to perform principal component analysis.

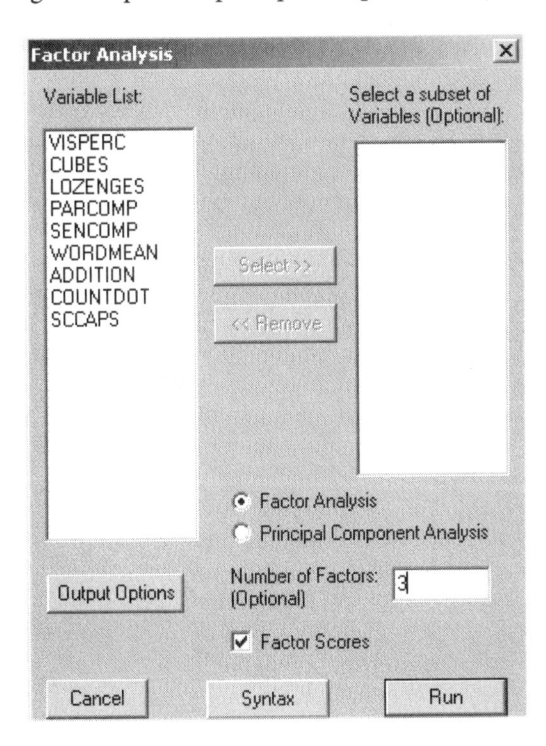

A portion of the factor analysis output is shown below. This output and example are described in detail in *LISREL 8: New Statistical Features*.

NPV.OUT

Promax-Rotated Factor Loadings

	Factor 1	Factor 2	Factor 3	Unique Var
VIS PERC	0.678	0.039	0.031	0.499
CUBES	0.531	-0.013	-0.051	0.740
LOZENGES	0.670	0.064	-0.059	0.535
PAR COMP	0.074	0.848	-0.045	0.241
SEN COMP	0.007	0.805	0.086	0.302
WORDMEAN	0.083	0.796	-0.049	0.322
ADDITION	-0.184	0.131	0.793	0.388
COUNTDOT	0.200	-0.145	0.774	0.317
S-C CAPS	0.431	0.041	0.446	0.456

3.9 Bootstrap Samples from Political Survey Data

The *PRELIS 2: User's Reference Guide* provides a description of the bootstrapping technique (see pp. 184 - 188). To illustrate how this can be done using PRELIS menus and dialog boxes, suppose 100 bootstrap samples of size 148 each are to be drawn from the PRELIS system file **dataex7.psf**. This data set was obtained by importing the SPSS ***.sav** file, **dataex7.sav** from the **spssex** folder. See Section 3.2 for more information on how to import data.

To open the file **dataex7.psf**, select the ***Open*** option on the ***File*** menu to obtain the ***File Open*** dialog box. Select from the ***Files of type*** drop-down list box ***PRELIS Data*** **(*psf)** and the **spssex** folder. When **dataex7.psf** is selected, clicked ***OK*** to display the data set.

Select the ***Bootstrapping*** option from the ***Statistics*** menu to obtain the ***Bootstrapping*** dialog box.

Bootstrapping

Number of bootstrap samples 100

Sample fraction 50

Save: Filename:

All the MA-matrices ex7boot.cov

All the mean vectors mefile

All the standard deviations sdfile

Output Options Cancel Syntax Run

Since the sample fraction is 50, each bootstrap sample will be 50% of the original sample size of 297, that is, of size 148. Select the ***Save: all the MA-matrices*** option and choose a file name for a file that will contain 100 correlation matrices.

Click ***Output Options*** on the ***Bootstrapping*** dialog box and choose ***Correlations*** for the matrices to be computed.

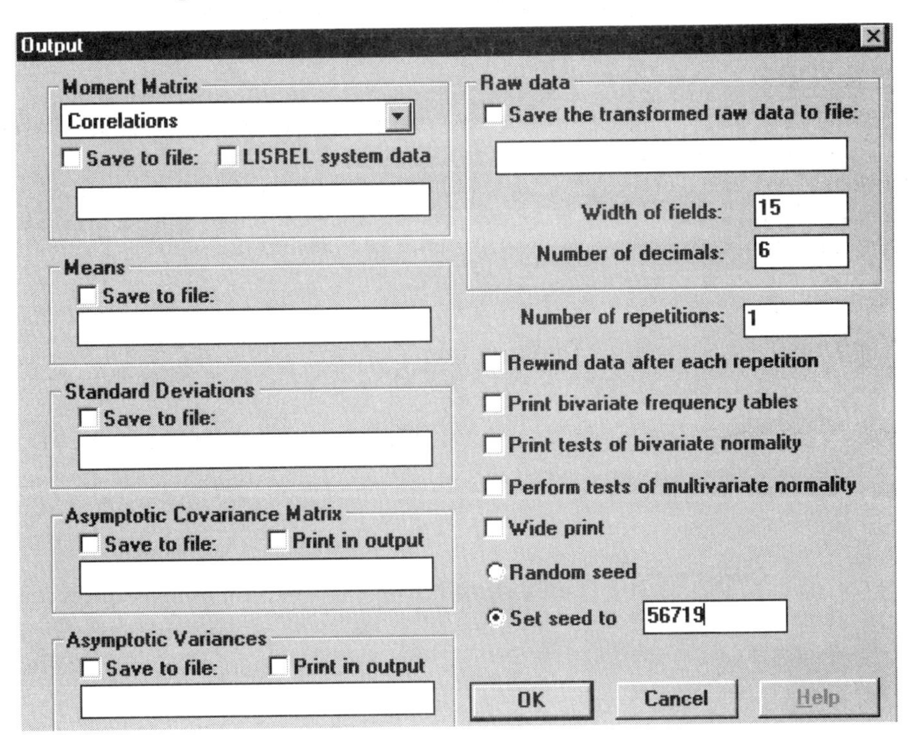

The ***Output*** dialog box also allows you to set the random number generator seed, so that this run can be replicated exactly. For this example, a value of `56719` was chosen. Click ***OK*** to return to the ***Bootstrapping*** dialog box and click ***Run*** to start the PRELIS analysis.

After this run, the file **bootex7.cor** contains 100 correlation matrices, each with 21 items (the lower triangular part of a 6 x 6 corrrelation matrix). The correlation matrices for the first two bootstrap samples are:

```
0.10000D+01  0.30610D+00  0.10000D+01  0.35929D+00  0.40094D+00  0.10000D+01
0.45201D+00  0.25276D+00  0.55734D+00  0.10000D+01  0.41889D+00  0.26356D+00
0.38436D+00  0.58064D+00  0.10000D+01  0.55686D+00  0.18972D+00  0.45974D+00
0.71211D+00  0.62033D+00  0.10000D+01
0.10000D+01  0.26861D+00  0.10000D+01  0.43686D+00  0.29063D+00  0.10000D+01
0.55195D+00  0.27254D+00  0.33373D+00  0.10000D+01  0.43043D+00  0.38556D+00
0.22970D+00  0.59980D+00  0.10000D+01  0.54469D+00  0.25376D+00  0.42777D+00
0.82663D+00  0.50061D+00  0.10000D+01
```

Note that the numbers are given in so-called scientific notation, where D+01 indicates that the number is to be multiplied by 10, D-01 indicates multiplication by 0.1, etc.

A 3-D bar chart of *VOTING* and *INTEREST* can be obtained by selecting the ***Bivariate*** option from the ***Graphs*** menu.

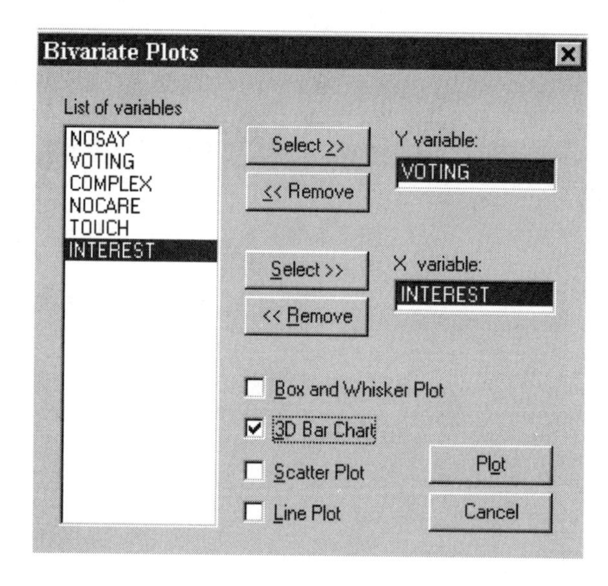

After completing the ***Bivariate Plots*** dialog box as shown above, click ***Plot***. The resulting plot is shown below:

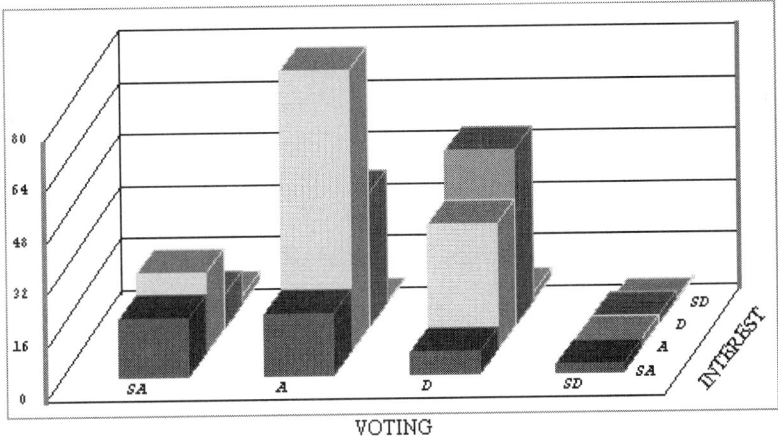

3.10 Monte Carlo Studies

The interactive PRELIS environment may be used to simulate data from uniform or normal distributions, or from mixtures of these distributions. To illustrate, suppose we want to simulate 50 observations on three variables, *V1*, *V2*, and *V3*. Suppose further that *V1* has a normal distribution with a mean of 100 and standard deviation of 15; *V2* = *V1* + *e2*, where *e2* has a normal distribution with mean zero and standard deviation of 15; and finally *V3* = *V1* + *V2* + *e3* where *e3* has a uniform distribution with a mean of 10.

This may be accomplished by specifying

```
V1 = 100 + 15 * NRAND
V2 = V1 + 15 * NRAND
V3 = V1 + V2 + 20 * URAND,
```

where (see pages 189-204 of the *PRELIS 2: User's Reference Guide*) NRAND has a normal (0;1) distribution and URAND a uniform (0;1) distribution.

The next sections illustrate

- The creation of a new PRELIS system file
- Insertion of variables into the PRELIS system file
- Insertion of cases
- Simulation of data values
- Graphical displays

3.10.1 New PRELIS (*.psf) Data

From the ***File*** menu, select ***New*** to obtain the ***New*** dialog box. Select ***PRELIS Data*** and click ***OK*** when done.

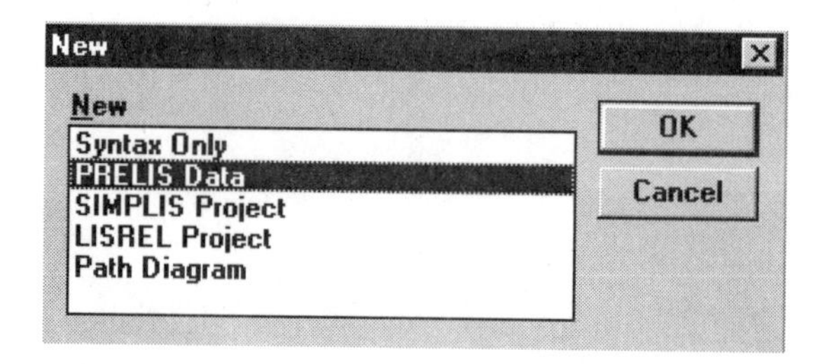

A new (but empty) PRELIS system file is displayed in the PSF window.

In order to simulate 50 observations of the form

```
V1 = 100 + 15 * NRAND,
V2 = V1 + 15 * NRAND, and
V3 = V1 + V2 + 20 * URAND,
```

we need to create an empty data matrix with 3 columns and 50 rows, where each column represents a variable and each row an observation (case).

3.10.2 Inserting Variables into the Spreadsheet

We start by inserting the three variables *V1* to *V3*. To do so, select the ***Define Variables*** option on the ***Data*** menu to display the ***Define Variables*** dialog box.

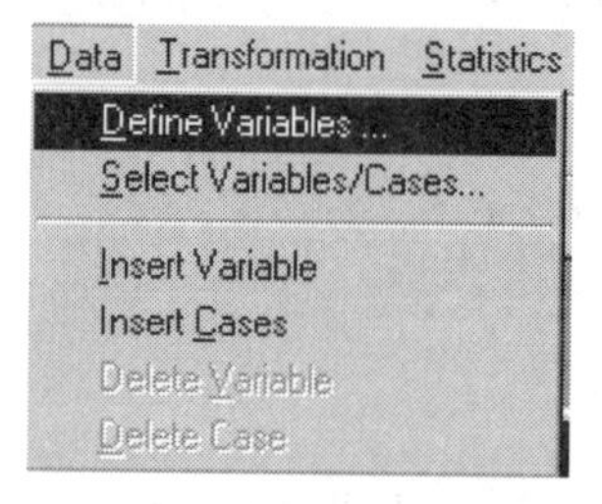

On the ***Define Variables*** dialog box, click ***Insert*** to invoke the ***Add Variables*** dialog box. In this box, enter `V1-V3` and click ***OK***. Then also click ***OK*** on the ***Define Variables*** dialog box.

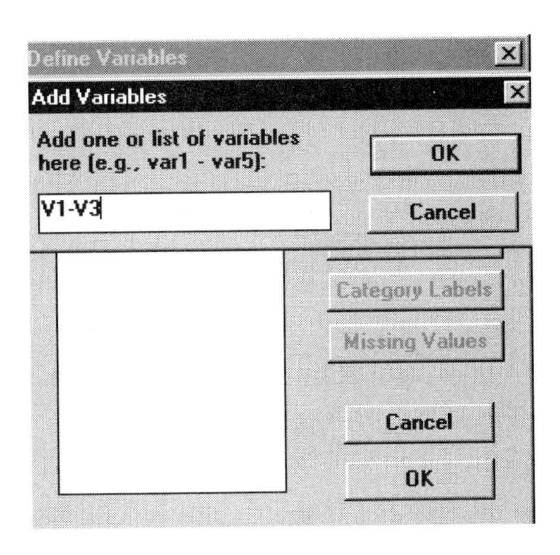

The PRELIS system file will now contain 3 columns of zero values corresponding to the variables *V1-V3*.

DATA1

	V1	V2	V3
	0.00	0.00	0.00
	0.00	0.00	0.00
	0.00	0.00	0.00
	0.00	0.00	0.00
	0.00	0.00	0.00
	0.00	0.00	0.00

At this stage, this file should be saved as a ***.psf**. This is accomplished by clicking the ***File, Save as*** option to open the ***Save As*** dialog box. Select a filename and folder, for example **data1.psf**. Click ***Save***. In the next section, we will insert 50 cases into **data1.psf**.

3.10.3 Inserting Cases into the Spreadsheet

From the ***Data*** menu, select the ***Insert Cases*** option. A dialog box will appear in which the ***Number of cases*** to be inserted may be entered. Note that the default number of cases is 1. This is the number of cases that will be inserted if the ***Number of Cases*** field is left blank. Click ***OK*** when done.

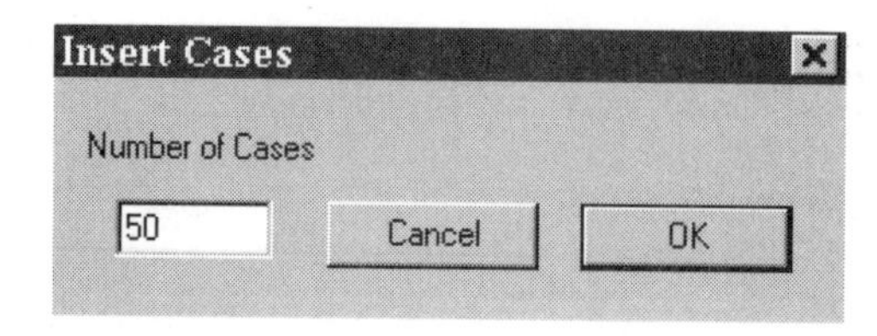

The spreadsheet shown below now contains 50 cases. Note that, prior to inserting the simulated data values into 3 columns, the file should be saved. To do so, select the ***Save*** option from the ***File*** menu.

DATA1

	V1	V2	V3
39	0.00	0.00	0.00
40	0.00	0.00	0.00
41	0.00	0.00	0.00
42	0.00	0.00	0.00
43	0.00	0.00	0.00
44	0.00	0.00	0.00
45	0.00	0.00	0.00
46	0.00	0.00	0.00
47	0.00	0.00	0.00
48	0.00	0.00	0.00
49	0.00	0.00	0.00
50	0.00	0.00	0.00

3.10.4 Simulating Data Values

From the ***Transformation*** menu, select the ***Compute*** option. This activates the ***Compute*** dialog box.

The three lines of syntax

```
V1 = 100 + 15 * NRAND,
V2 = V1 + 15 * NRAND, and
V3 = V1 + V2 + 20 * URAND,
```

are entered as follows. Click on *V1* and with the left mouse button held down, drag it to the position shown in the transformation window. When done, click the "=" sign, then enter `100 + 15 * NRAND` by clicking the corresponding buttons on the compute pad. To start the next syntax line, click on the ***Enter*** button. Proceed as described above until the 3 lines have been added.

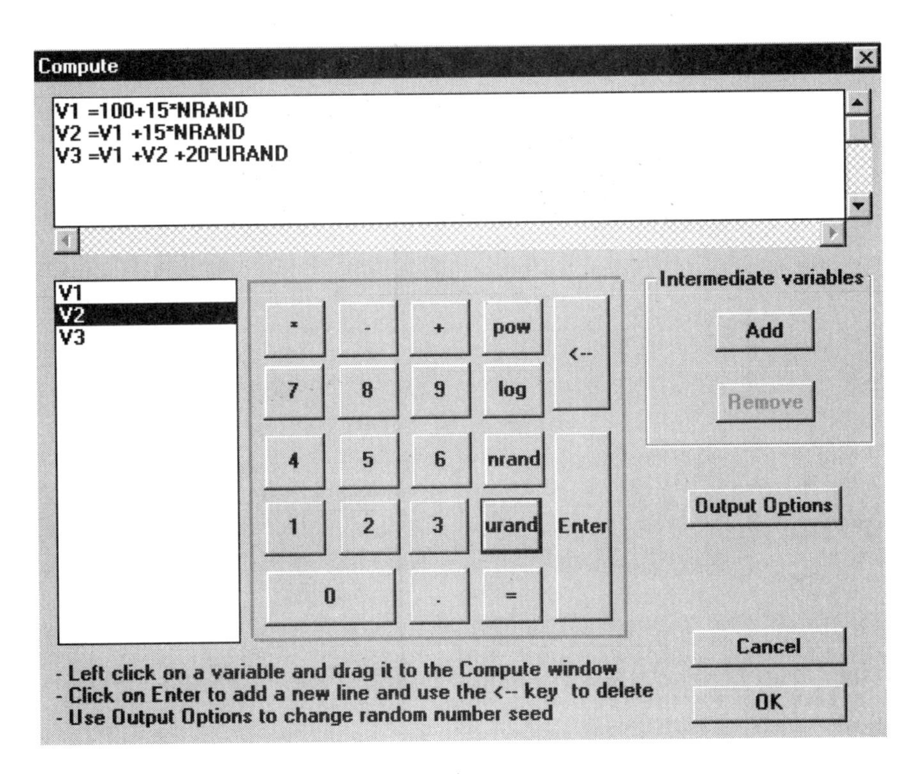

When done, click ***Output Options*** to obtain the ***Output*** dialog box. Select ***Covariances*** from the ***Moment Matrix*** drop-down list box. Check the ***Save to file*** check box directly below this and enter a filename, in this case **simul.cov**. Click ***OK*** to return to the ***Compute*** dialog box and ***OK*** again when done.

The PRELIS system file will be updated as shown below.

DATA1.PSF

	V1	V2	V3
1	92.687	90.355	188.528
2	79.718	90.591	188.970
3	116.803	136.488	262.218
4	124.160	132.843	264.053
5	132.124	119.102	269.195

3.10.5 Graphical Displays

To obtain a histogram of the distribution of *V1*-values, click the *V1* button on the spreadsheet to highlight this column and then click the ***Univariate Plot*** icon button to produce the graphical representation shown below.

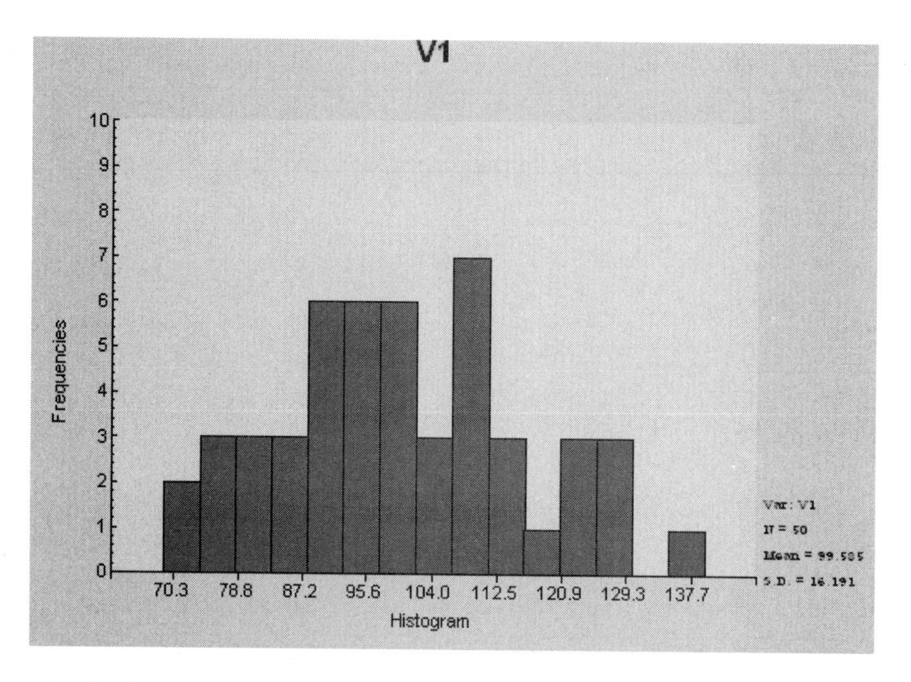

From the display, we see that the data follows a bell-shaped curve reasonably well. The sample mean and standard deviation values of 99.585 and 16.191 are close to the population values of 100 and 15 respectively.

Finally, a bivariate plot of the distribution of *V1* and *V3* may be obtained by selecting ***Bivariate*** from the ***Graphs*** menu. In the ***Bivariate Plot*** dialog box, select *V1* as the *Y*-variable and *V3* as the *X*-variable. Also select ***Scatter Plot*** and click ***Plot*** when done.

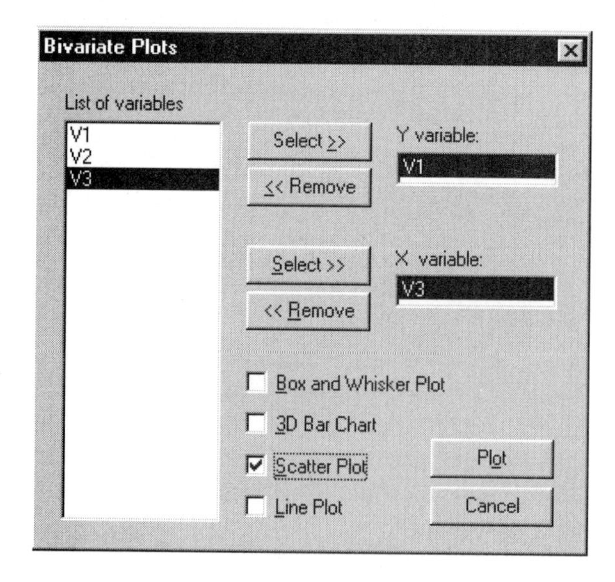

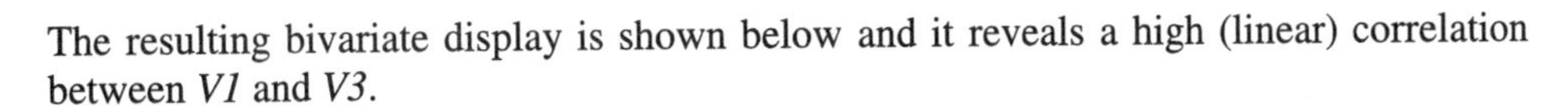

The resulting bivariate display is shown below and it reveals a high (linear) correlation between *V1* and *V3*.

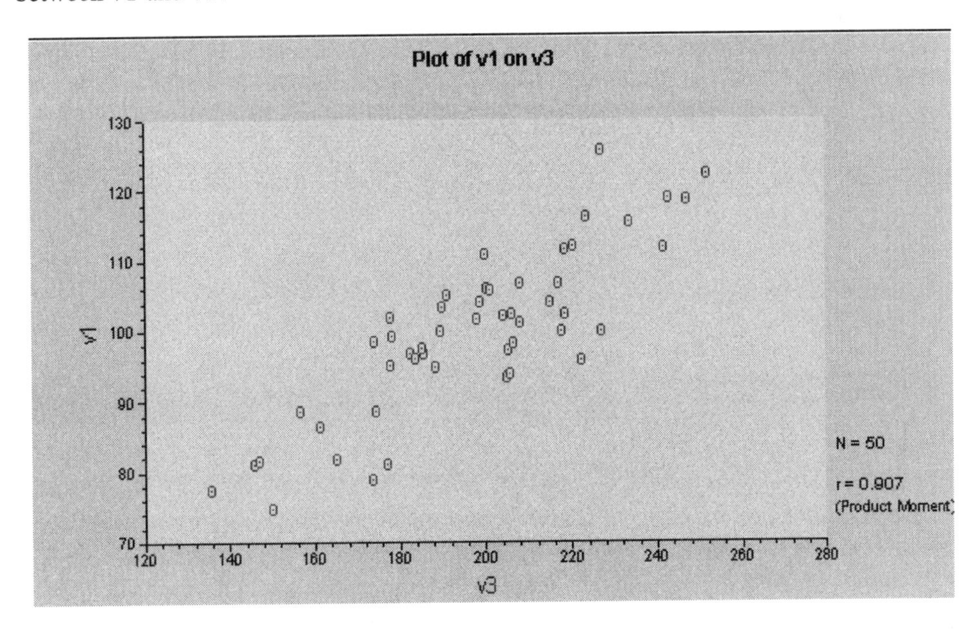

3.11 Multiple Imputation

Ryan and Joiner (1994, Table 9.1) report serum-cholesterol levels for n = 28 patients treated for heart attacks. Cholesterol levels were measured 2 days and 4 days after the attack. For 19 of the 28 patients, an additional measurement was taken 14 days after the attack.

Five rows of the data set, **chollev.dat**, are shown below.

chollev.dat - Notepad

File Edit Format Help

```
270.000   218.000    156.000
236.000   234.000    -9.000
210.000   214.000    242.000
142.000   116.000    -9.000
280.000   200.000    -9.000
```

Select the ***Import Data in Free Format*** option from the ***File*** menu. Select the **tutorial** folder in the ***Open Data File*** dialog box, select **chollev.dat** and click ***Open***.

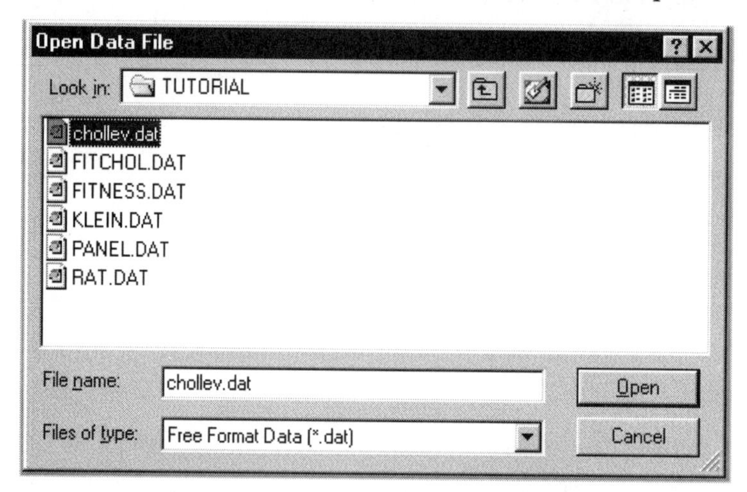

Enter the number of variables in the ***Number*** string field of the ***Enter Number of Variables*** dialog box.

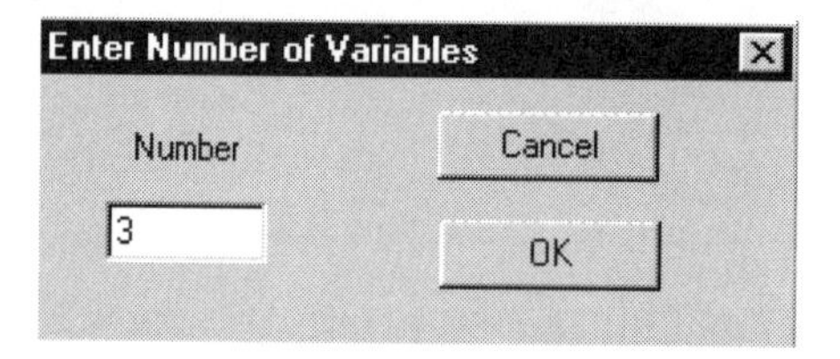

Click ***OK*** to create a PRELIS system file. From the main menu bar, select the ***Define Variables*** option from the ***Data*** menu to obtain the ***Define Variables*** dialog box.

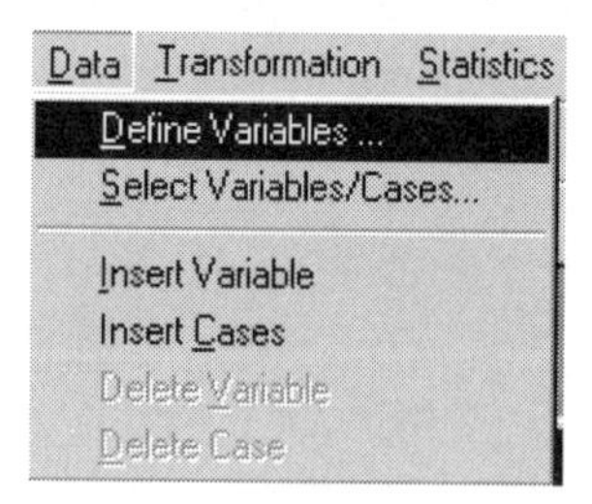

Use the dialog box to rename *VAR1*, *VAR2* and *VAR3* to *Chol_D2*, *Chol_D4* and *Chol_D14* respectively. Click ***Variable Type*** to define these variables as continuous and when done, assign the value of `-9.000` as the ***Global missing value***.

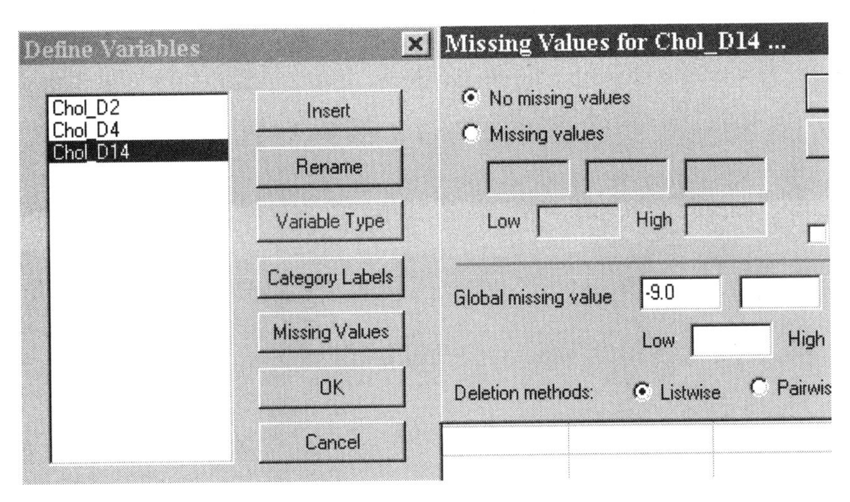

Select the ***Multiple Imputation*** option on the ***Statistics*** menu to obtain the ***Multiple Imputation*** dialog box. Select all the variables from the ***Variables List***.

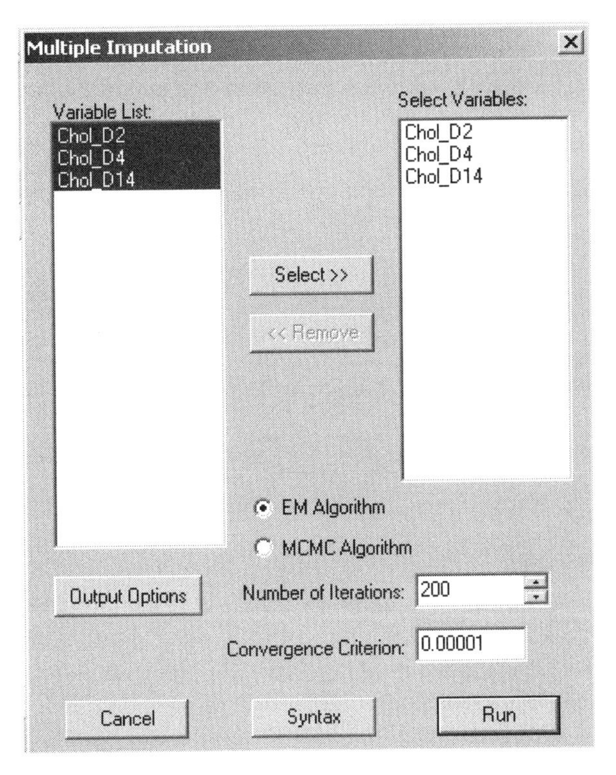

To impute data using the EM Algorithm, click the ***EM Algorithm*** radio button. To impute data using the MCMC algorithm, click the ***MCMC Algorithm*** radio button. Select the number of iterations and the convergence criterion.

To save the imputed data, means, and covariances to external files, click ***Output Options*** to obtain the ***Output*** dialog box. Enter the values shown below. Click ***OK*** when done to return to the ***Multiple Imputation*** dialog box.

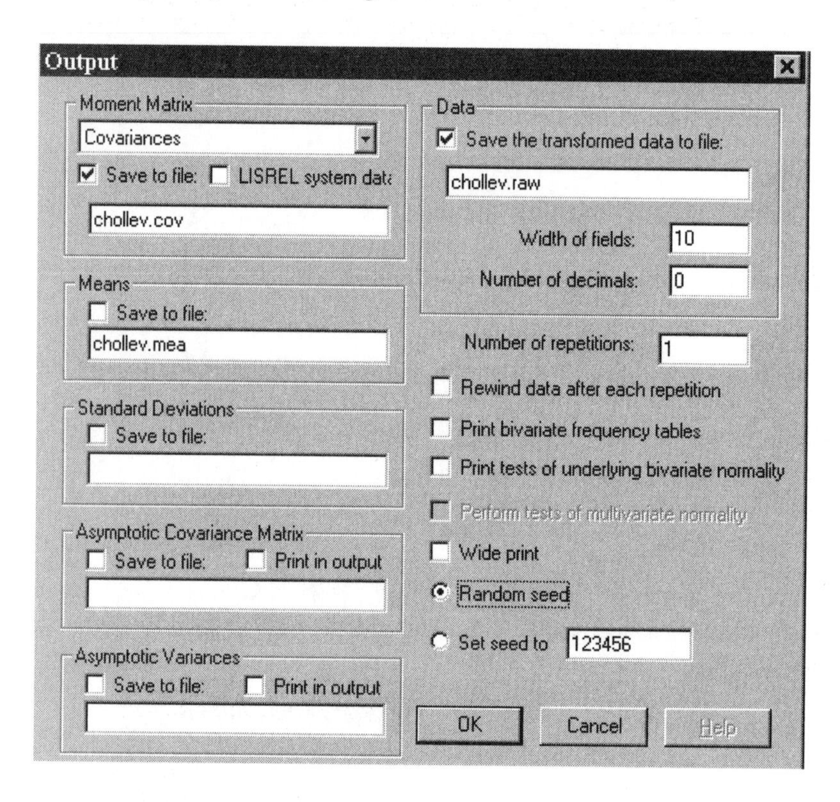

PRELIS syntax generated by clicking ***Syntax*** in the ***Multiple Imputation*** dialog box is shown below.

chollev.PR2

```
!PRELIS SYNTAX: Can be edited
SY=C:\lisrel850-student\TUTORIAL\chollev.PSF
SE 1 2 3
EM CC = 0.00001 IT = 200
OU MA=CM SM=chollev.cov RA=chollev.raw WI=10 ND=0 ME=chollev.mea XM XB XT
```

A portion of the PRELIS output is shown below.

```
                    Number of different missing-value patterns=         2
                    Convergence of EM-algorithm in      4 iterations
                    -2 Ln(L) =         753.83093

                    Percentage missing cases=   10.71

                    Estimated Means

Chol_D2                253.9286
Chol_D4                230.6429
Chol_D14               222.2372

                    Estimated Covariances

                       Chol_D2        Chol_D4        Chol_D14

Chol_D2                2276.2910
Chol_D4                1508.4921      2205.9418
Chol_D14                866.3386      1571.5957      2024.5375
```

The percentage of missing values is 10.71 and the EM procedure converged in 4 iterations.

The first 16 records of the imputed data set, written to a file named **imputed_data.psf**, are shown below.

imputed_data

	Chol_D2	Chol_D4	Chol_D14
1	270.000	218.000	156.000
2	236.000	234.000	228.000
3	210.000	214.000	242.000
4	142.000	116.000	146.000
5	280.000	200.000	192.000
6	272.000	276.000	256.000
7	160.000	146.000	142.000
8	220.000	182.000	216.000
9	226.000	238.000	248.000
10	242.000	288.000	271.000
11	186.000	190.000	168.000
12	266.000	236.000	236.000
13	206.000	244.000	241.000
14	318.000	258.000	200.000
15	294.000	240.000	264.000
16	282.000	294.000	269.000

3.11.1 Use of the MCMC Option to obtain more than one Imputed Data Set

In contrast to the EM algorithm, the MCMC algorithm is based on random draws from multivariate normal and inverse Wishart distributions. This fact enables the researcher to obtain more than one imputed data set. This may be accomplished by selecting the MCMC algorithm on the ***Multiple Imputation*** dialog box. Once this is done, click on ***Output Options*** to obtain the dialog box shown below.

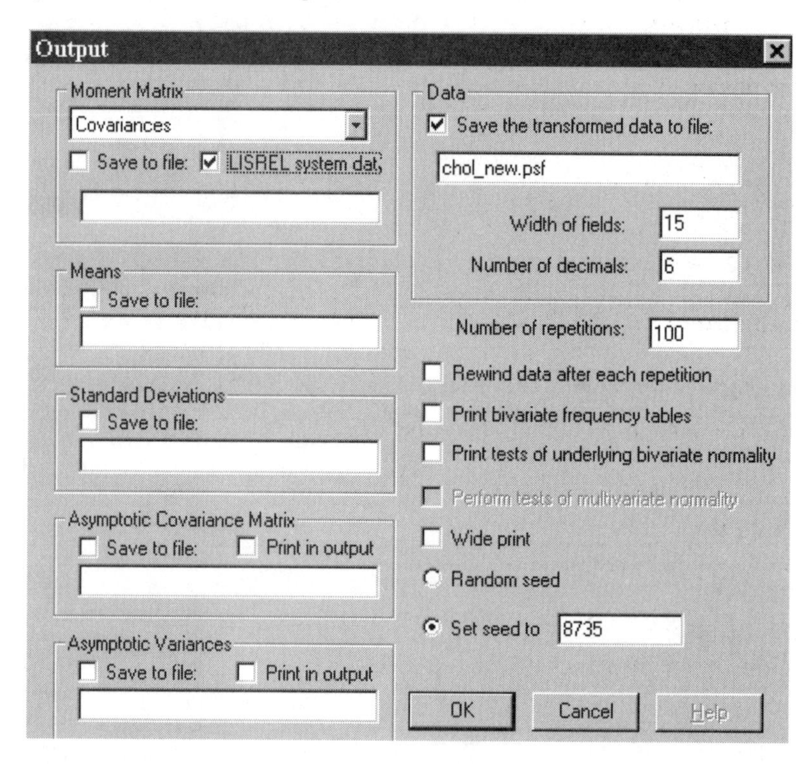

Enter the ***Number of repetitions*** and enter an integer value into the ***Set seed to*** field. For this example, the number of repetitions equals 100 and the seed is set to 8735. Note that the data set **chol_new.psf** will contain $28 \times 100 = 2800$ cases.

Click ***OK*** to return to the ***Multiple Imputation*** dialog box. If the ***Syntax*** button is clicked, the PRELIS syntax file shown below is generated.

```
!PRELIS SYNTAX: Can be edited
SY=C:\lisrel850\TUTORIAL\chollev.PSF
SE
EM CC = 0.00001 IT = 200
OU MA=CM RA=chol_new.psf IX=8735 RP=100 XM XB XT
```

4 LISREL Examples

4.1 Using Existing SIMPLIS Syntax Files

The examples given in the *LISREL 8: The SIMPLIS Command Language* guide are contained in the **splex** folder of LISREL. Select the ***Open*** option from the ***File*** menu to obtain the ***Open*** dialog box. Select **ex10a.spl** from the **splex** folder. Click ***Open*** when done. The window shown below contains the SIMPLIS syntax for example **ex10a.spl**.

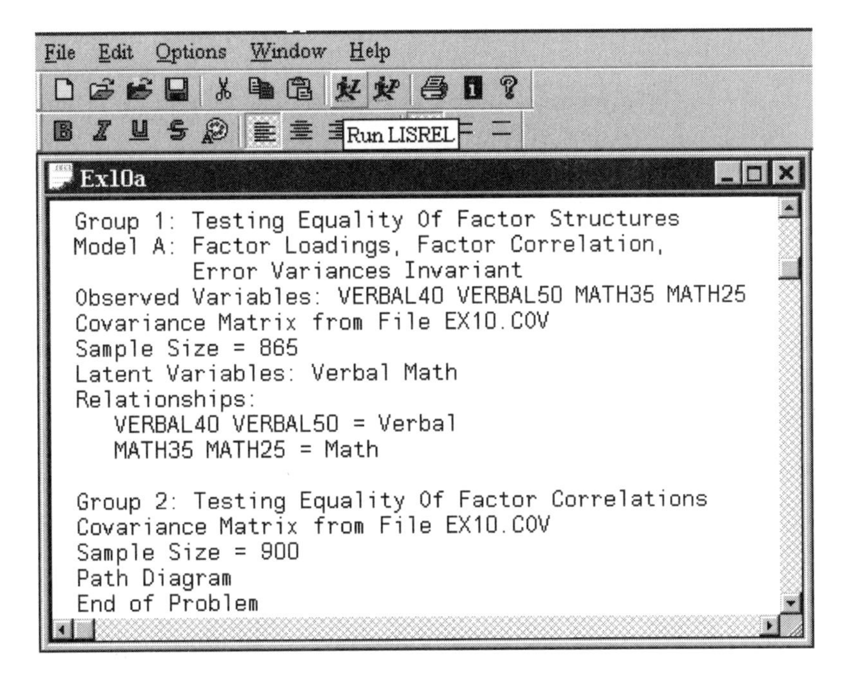

Click the ***Run LISREL*** icon button to run **ex10a.spl**.

Note that a path diagram is obtained by entering the command `Path Diagram` in the file **ex10a.spl**, just below the line `Sample Size = 900` as shown above. The path diagram is displayed in the path diagram window.

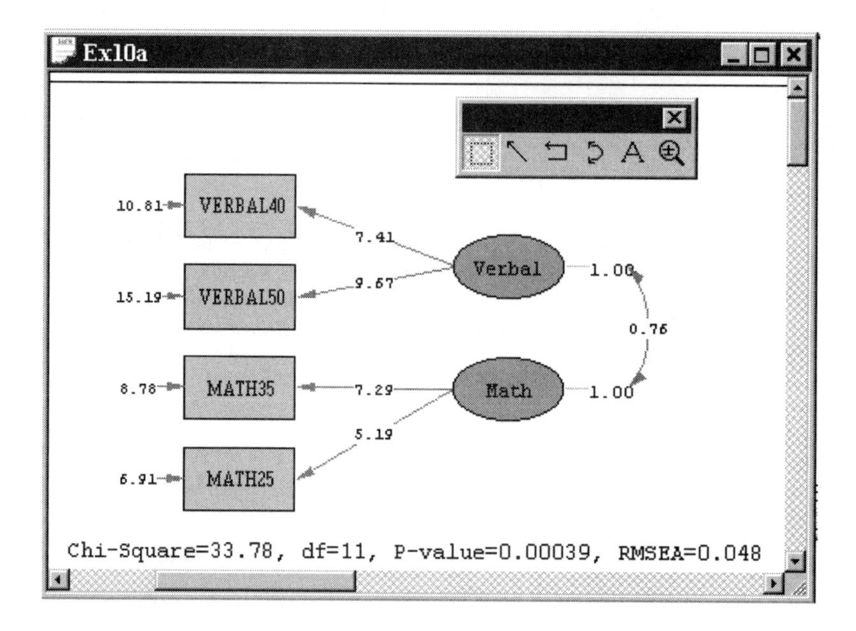

A text editor window is also opened. A selection of the output contained in **ex10a.out** is displayed below.

Ex10a.OUT

```
         Measurement Equations


 VERBAL40 = 7.41*Verbal, Errorvar.= 10.81, R² = 0.84
           (0.16)                  (0.85)
            47.41                   12.66

 VERBAL50 = 9.67*Verbal, Errorvar.= 15.19, R² = 0.86
           (0.20)                  (1.41)
            48.49                   10.75

   MATH35 = 7.29*Math, Errorvar.= 8.78 , R² = 0.86
           (0.15)                (0.86)
            47.95                 10.22

   MATH25 = 5.19*Math, Errorvar.= 6.91 , R² = 0.80
           (0.11)                (0.47)
            45.33                 14.67
```

Once a path diagram is produced, the LISREL menu bar will display a different selection of options, as shown below.

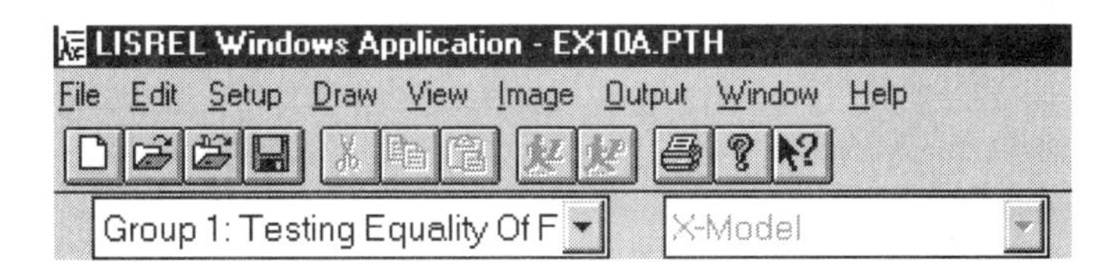

After a path diagram has been obtained, one may select the ***Draw*** menu to add paths, change text, etc. The items on this menu correspond to the ***Draw*** toolbox displayed next to it on the image shown below.

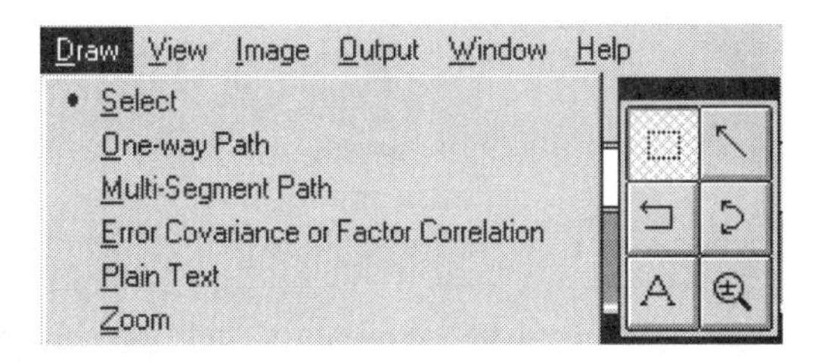

The default values shown on the path diagram are the parameter estimates. By selecting the ***Estimations*** option from the ***View*** menu, one may alternatively select the *t*-values, modification indices, etc.

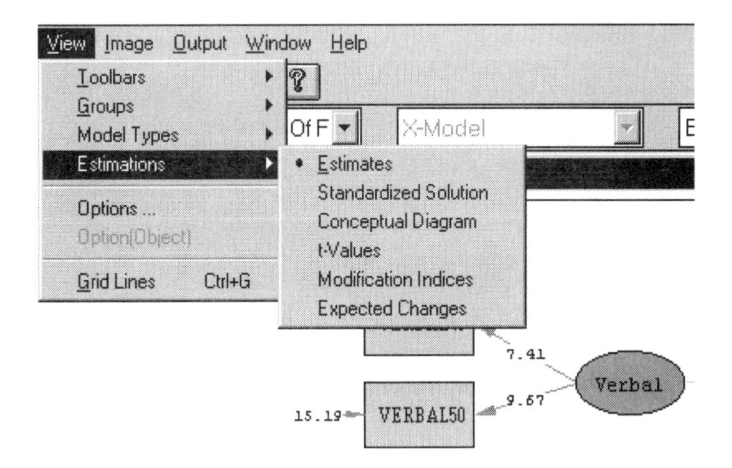

One may change the number of decimals displayed in the path diagram by selecting ***SIMPLIS outputs*** from the ***Output*** menu. On the dialog box displayed, change the number of decimals from the default value of 2.

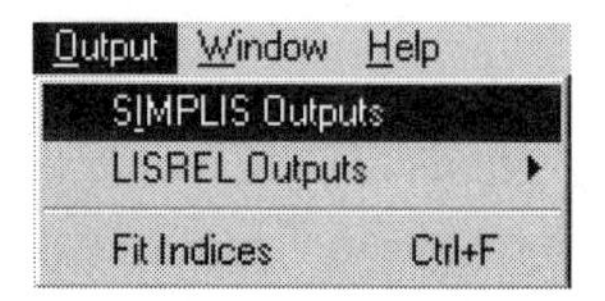

The ***SIMPLIS Outputs*** dialog box is shown below:

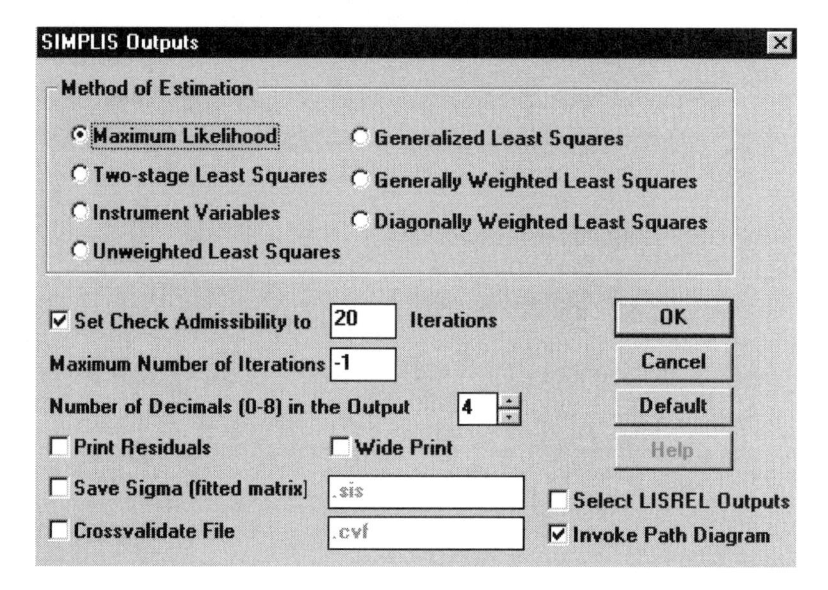

The LISREL syntax that corresponds to the SIMPLIS command language in **ex10a.spl** may be built from the path diagram. To do so, select the ***Setup, Build LISREL Syntax*** option as shown in the next image.

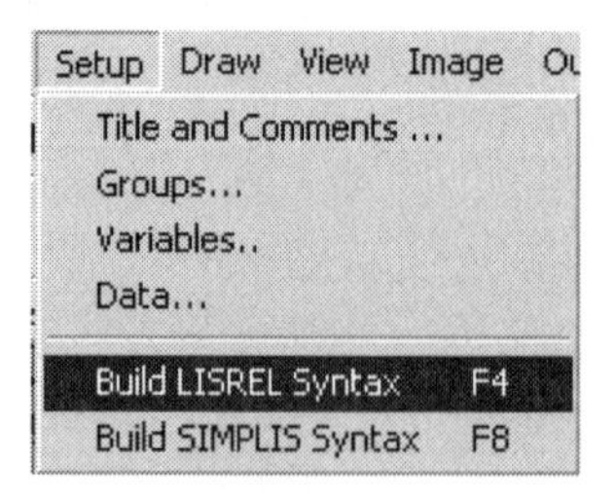

A LISREL project file (**ex10a.lpj**) is produced, containing the corresponding LISREL syntax, as shown below.

Ex10a

```
Group 1: Testing Equality Of Factor Structures
!DA NI=4 NO=865 NG=2 MA=CM
SY='C:\lisrel850\SPLEX\Ex10a.DSF' NG=2
SE
1 2 3 4 /
MO NX=4 NK=2 LX=FU,FI PH=SY,FR TD=DI,FR
LK
Verbal  Math
FI PH(1,1) PH(2,2)
FR LX(1,1) LX(2,1) LX(3,2) LX(4,2)
VA 1.00 PH(1,1) PH(2,2)
PD
OU ME=ML
Group 2: Testing Equality Of Factor Correlations
!DA NI=4 NO=900 NG=2 MA=CM
SY='C:\lisrel850\SPLEX\Ex10a.DSF' NG=2
```

4.1.1 Changing the Appearance of the Output

While the output file is opened, one may use the ***Edit*** menu to perform various operations such as finding a keyword (***Find***), replacing text (***Replace***), inserting files (***Insert File***), etc.

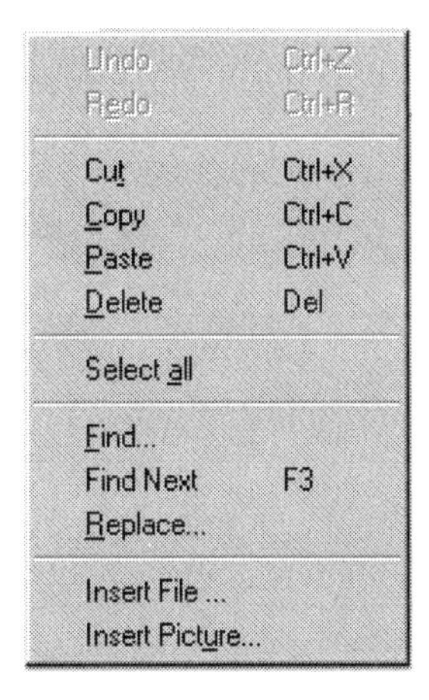

When output is displayed, one may also use the ***Options*** menu to change the ***Layout***, ***Font***, ***Format***, etc. of marked text.

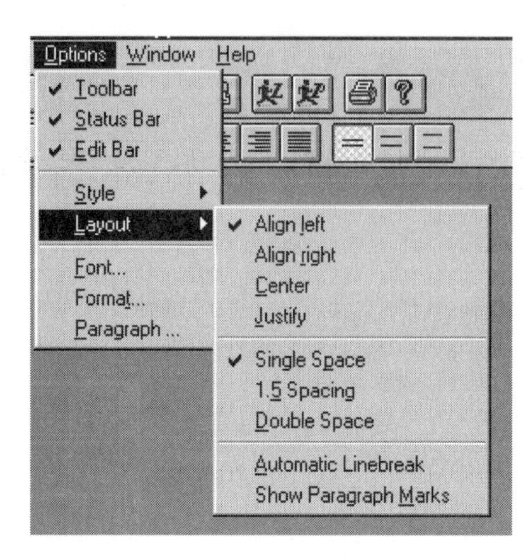

4.2 Using Existing LISREL Syntax Files

The examples given in the *LISREL 8: User's Reference Guide* are contained in the **ls8ex** folder. Select the ***Open*** option from the ***File*** menu to obtain the ***Open*** dialog box. Select the file **ex31a.ls8**. Click ***Open*** when done. The contents of the file **ex31a.ls8** is displayed.

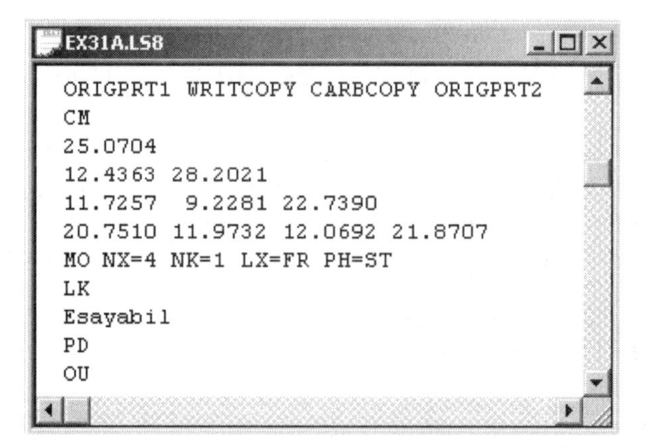


```
ORIGPRT1 WRITCOPY CARBCOPY ORIGPRT2
CM
25.0704
12.4363 28.2021
11.7257  9.2281 22.7390
20.7510 11.9732 12.0692 21.8707
MO NX=4 NK=1 LX=FR PH=ST
LK
Esayabil
PD
OU
```

To run LISREL, click the ***Run LISREL*** icon button. Make sure that the ***Run LISREL*** icon button is selected when running LISREL or SIMPLIS syntax and that the ***Run PRELIS*** icon button is selected when running PRELIS syntax.

To produce the path diagram shown below, the keyword PD (Path Diagram) was entered between the lines containing the words Esayabil and OU in **ex31a.ls8**.

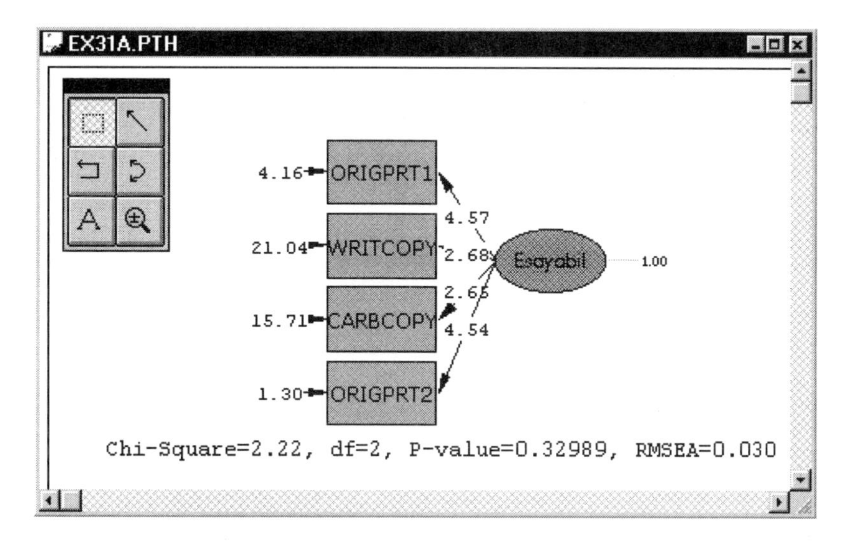

The file **ex31a.out** is also produced. This file may be viewed by either closing the path diagram window or by clicking on the output window, which is partially hidden behind the path diagram window.

4.3 Creating Syntax by Drawing Path Diagrams

LISREL for Windows allows the user to create LISREL or SIMPLIS syntax by drawing the appropriate path diagram on the screen and then building the corresponding syntax directly from the path diagram. This procedure will be demonstrated by fitting a CFA (Confirmatory Factor Analysis) model to the data. To obtain correct standard errors of the estimators of the parameters, the CFA model should be fitted to the observed covariance matrix. However, in order to illustrate how to build syntax from a path diagram, the model will be fitted to the available correlation matrix.

In confirmatory factor analysis, one builds a model assumed to describe or account for the empirical data in terms of relatively few parameters. The model is based on *a priori* information about the data structure in the form of a specified theory or hypothesis or knowledge from previous studies based on extensive data.

Holzinger & Swineford (1939) collected data on twenty-six psychological tests administered to 145 seventh- and eighth-grade children in the Grant-White school in Chicago. Six of these tests were selected and for this example it was hypothesized that these measure two common factors: visual perception and verbal ability such that the first three variables measure visual perception and the last three measure verbal ability. The six selected variables and their intercorrelations are given below:

VIS PERC	1.000					
CUBES	0.318	1.000				
LOZENGES	0.436	0.419	1.000			
PAR COMP	0.335	0.234	0.323	1.000		
SEN COMP	0.304	0.157	0.283	0.722	1.000	
WORDMEAN	0.326	0.195	0.350	0.714	0.685	1.000

The path diagram for this model is given below:

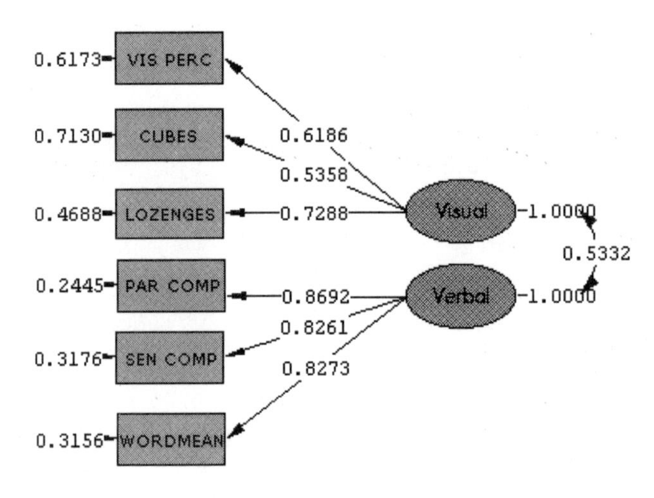

The corresponding SIMPLIS syntax file is:

```
Six Psychological Variables-A Confirmatory Factor Analysis
Observed variables
'VIS PERC' CUBES LOZENGES 'PAR COMP' 'SEN COMP' WORDMEAN
Correlation Matrix From File EX5.COR
Sample Size: 145
Latent Variables: Visual Verbal
Relationships:
'VIS PERC' - LOZENGES = Visual
'PAR COMP' - WORDMEAN = Verbal
Number of decimals = 4
Print Residuals
End of Problem
```

Some important aspects contained in the SIMPLIS syntax file given above are

- The correlation matrix is read from an external file
- There are two latent variables in the model: *Visual* and *Verbal*
- The statement `Number of decimals = 4` specifies that we wish to have the results in the output file given to an accuracy of four decimals. LISREL uses two decimals by default.

4.3.1 Title, Labels and Data

Select the ***File, New*** option and click on ***Path Diagram***. In the ***Save As*** dialog box, select a filename and a folder in which the path diagram should be saved. For the present example the folder chosen is **splex** and the file name selected is **cfa6.pth**. When done, click ***Save***.

From the ***Output*** menu select ***SIMPLIS Outputs***. The dialog box below will appear. Customize this dialog box according to your preferences. For example, change the ***Number of Decimals in the Output*** option to `4`. Make sure that the ***Invoke Path Diagram*** check box (the default) is checked. When done, click ***OK***.

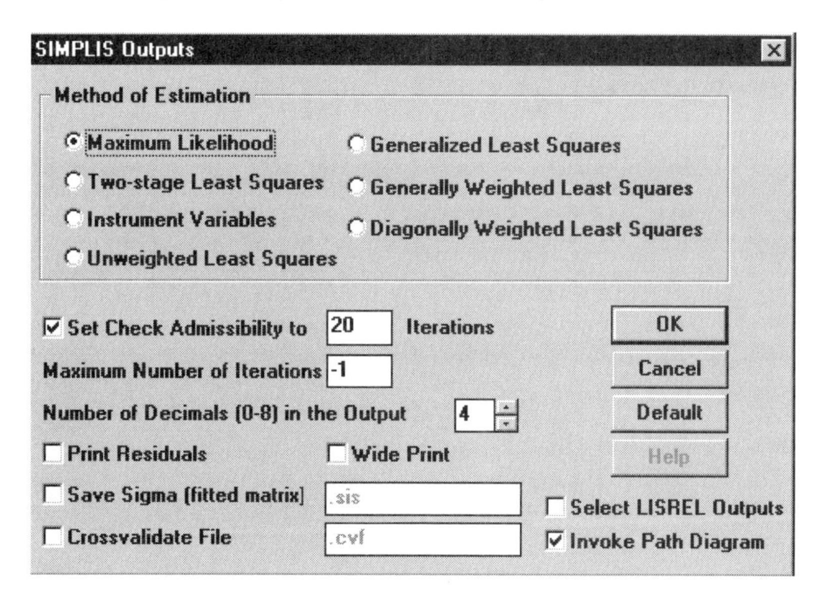

Before drawing the path diagram, select ***Toolbars*** from the ***View*** menu and ensure that the items shown below are selected. These are: ***Toolbar, Status Bar, Typebar, Variables, Drawing Bar***. Optionally, also check the ***Grid Lines*** option.

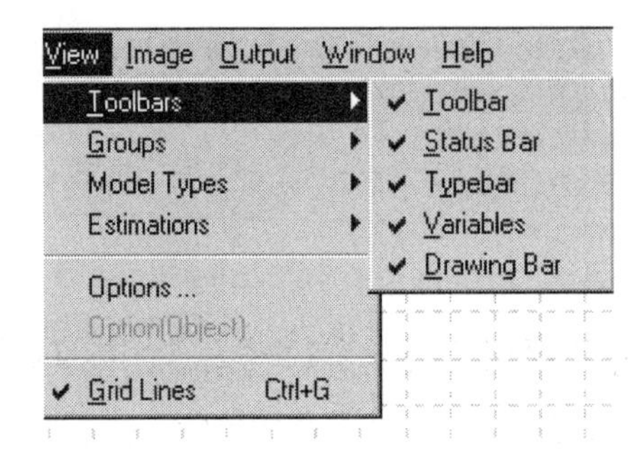

Select the ***Title and Comments*** option from the ***Setup*** menu to obtain the ***Title and Comments*** dialog box.

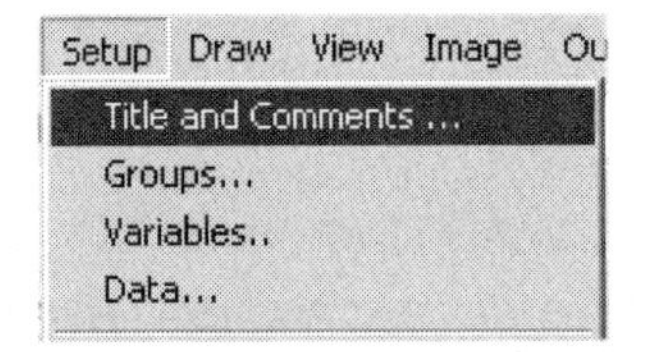

The first item on the ***Title and Comments*** dialog box is the ***Title*** for the particular analysis. Provision is also made for any additional ***Comments*** that the user may wish to enter. After typing in the title and (optionally) the comments, click on ***Next*** to go to the ***Group Names*** dialog box.

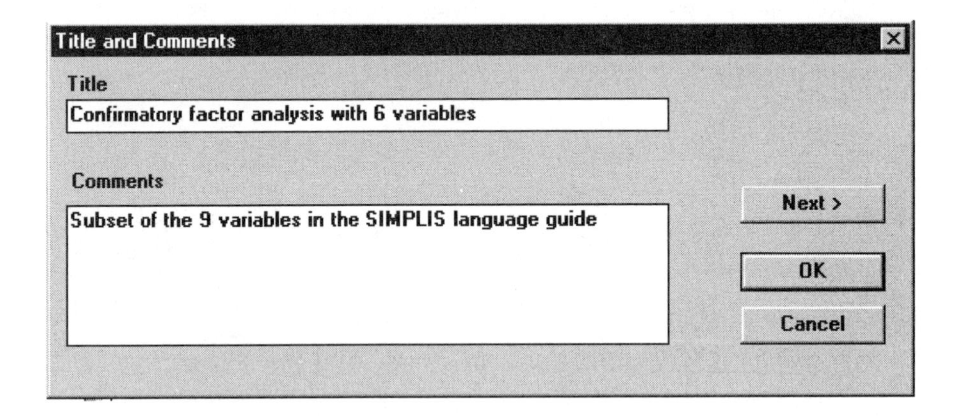

Since the present example is based on the analysis of one correlation matrix, nothing is entered in the space allowed for ***Group Labels*** and ***Next*** is clicked to go to the ***Labels*** dialog box. The default number of variables shown on the ***Labels*** dialog box is 2, these being *VAR 1* and *VAR 2*.

Since we have six observed variables, the ***Add/Read Variables*** option is selected. On the ***Add/Read Variables*** dialog box (of which the top half is shown below), check the ***Add list of variables*** radio button. Enter the labels *VIS PERC*, *CUBES*, …, *WORDMEAN* one by one in the ***Var List*** field. Note that a label name, which may include blanks, may not exceed 8 characters.

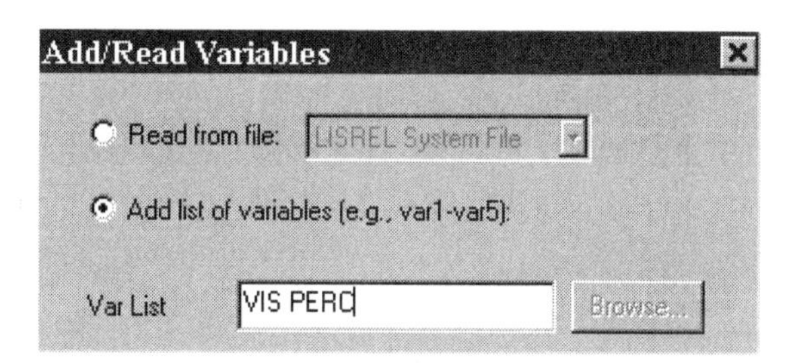

Since there are 2 latent variables, *Visual* and *Verbal*, the procedure outlined above for the observed variables is repeated by selecting ***Add List*** on the ***Latent Variables*** portion of the ***Labels*** dialog box. Add *var1-var2* and click ***OK***. Assign labels to these latent variables by clicking on the number of each variable. Finally, click ***Next*** to go to the ***Data*** dialog box.

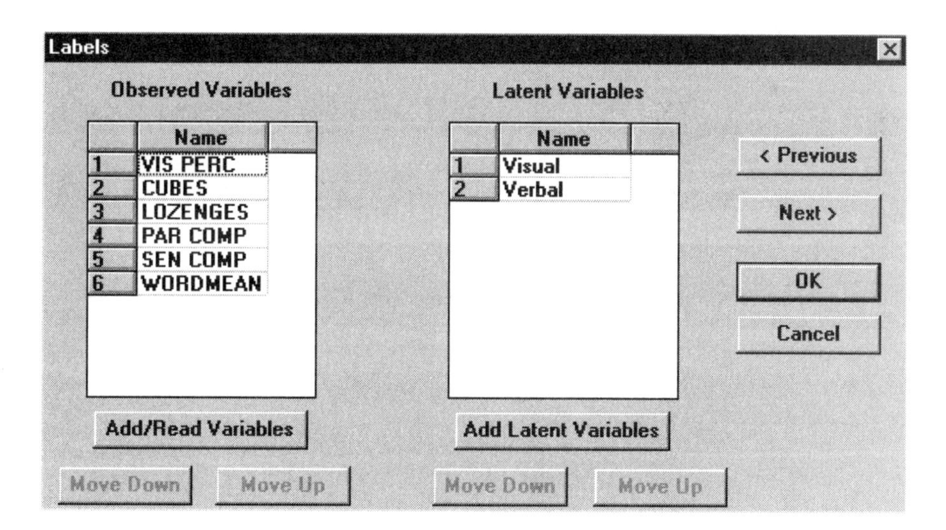

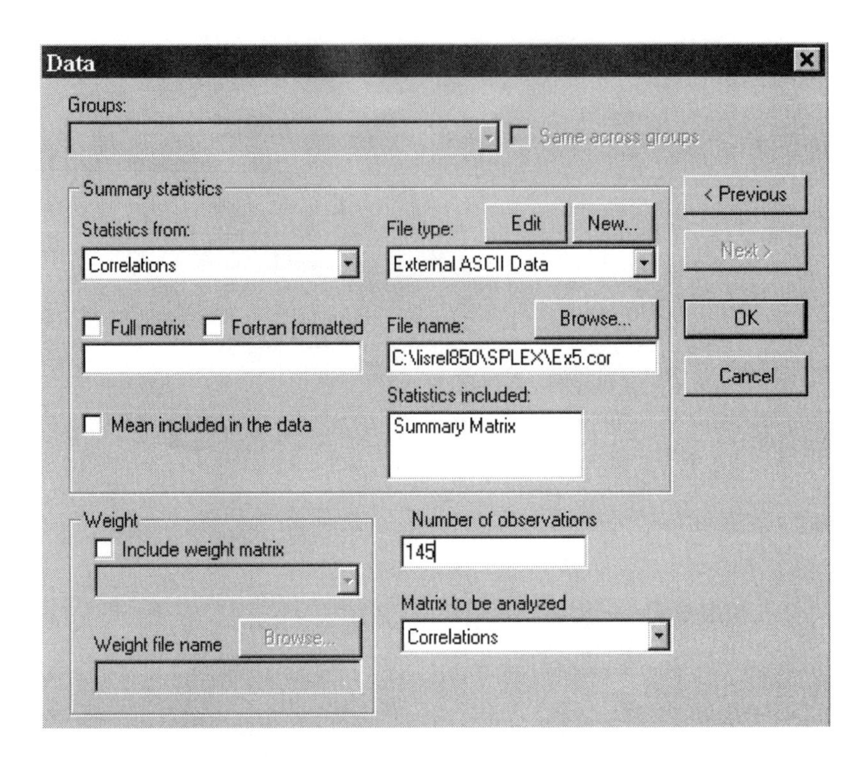

Select ***Correlations*** from the ***Statistics from*:** drop-down list box and also ***Correlations*** from the ***Matrix to be analyzed*** drop-down list box. In the ***Number of Observations*** string field type `145`. For ***File type***, select ***External ASCII Data*** and use the ***Browse*** button to locate **exs.cor** in the **splex** folder. Click ***OK*** when done.

4.3.2 Drawing the Path Diagram

We now proceed with the actual drawing of the path diagram. Start by clicking on the *VIS PERC* label under the ***Observed*** variables portion of the ***Labels*** window. Hold the mouse button down and "drag" the label to the draw area indicated by the grid lines. A rectangular-shaped object will appear on this part of the screen when the mouse button is released as shown below. Note also that

- The observed variables are assumed to be *X* (or independent) variables unless appropriate squares under the ***Y***-column are clicked.

- The latent variables are assumed to be *KSI* (or independent) variables unless appropriate squares under the ***Eta***-column are clicked.

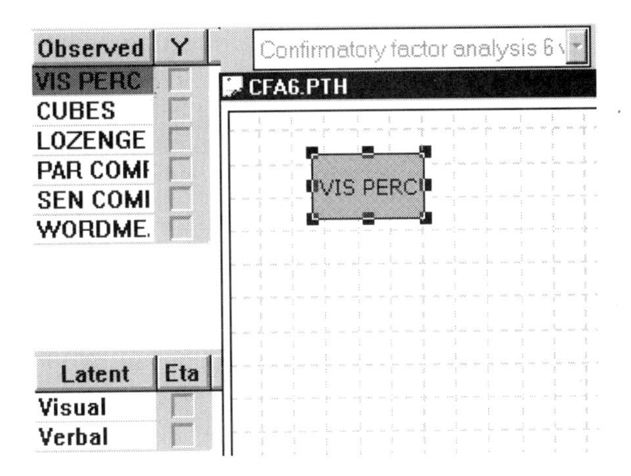

Repeat the same procedure for each of the remaining 5 variables by clicking on a label (left mouse button) and dragging the object to the draw area on the screen. The result of these operations should look similar to the image given below:

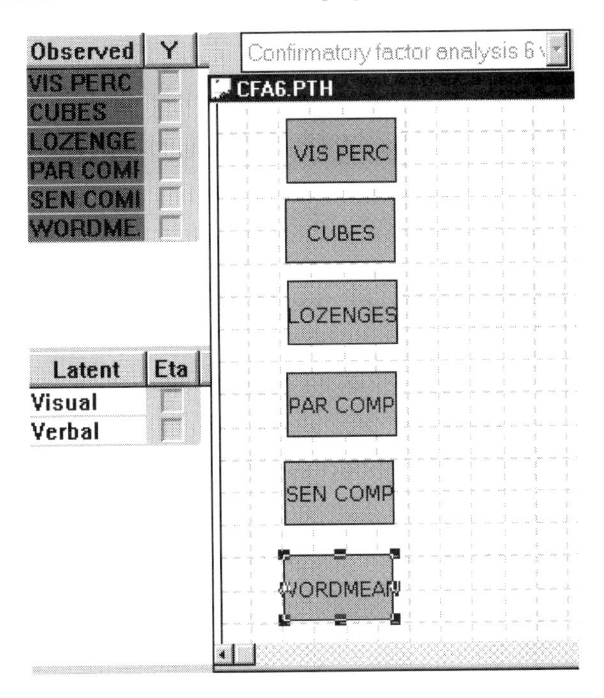

The rectangles representing the six observed variables can be properly aligned by going through the following steps:

- Choose ***Select all*** from the ***Edit*** menu, or draw a rectangle with the mouse pointer around all the objects to be included.

- Move the mouse pointer to the vicinity of the selected objects and click the right mouse button. The menu shown below will appear on the draw part of the screen.
- Select the ***Align>Left*** option. Once this is done, select the ***Even Space>Vertically*** option. One can, alternatively, use the ***Align*** option on the ***Image*** menu to achieve this.

Note that the alignment of objects is not really necessary. Once syntax is built from the path diagram and the program is run, a properly aligned path diagram is displayed.

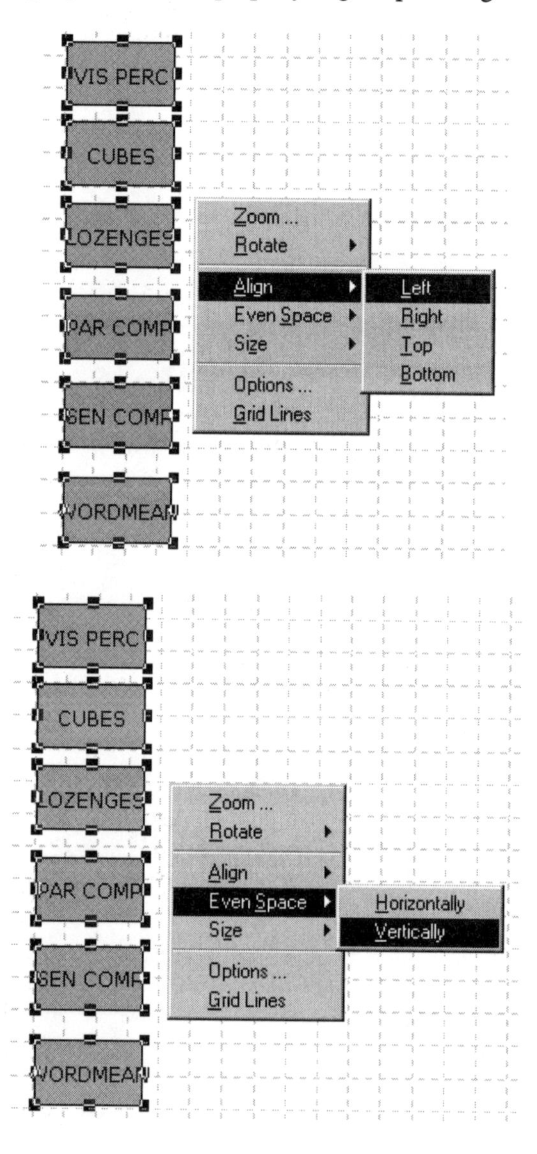

The rectangular objects will now be aligned and one can proceed by dragging the latent variables, *Visual* and *Verbal* to the path diagram window. Note that *latent* variables are represented by elliptically shaped objects.

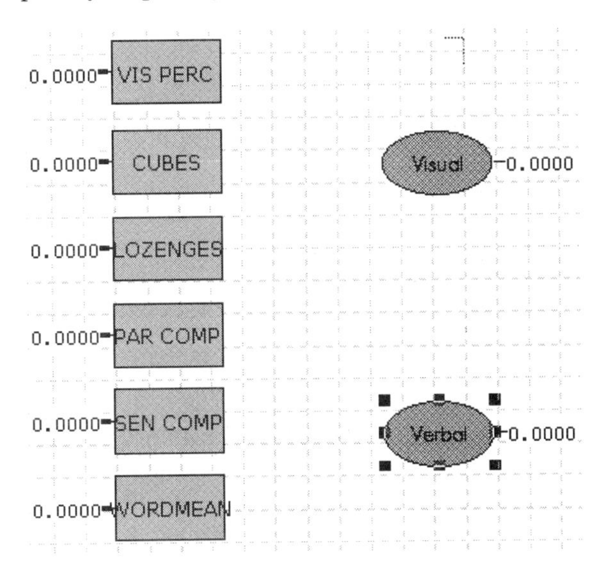

Finally, arrows can be drawn pointing from the latent variables to the observed variables. To accomplish this, click on the single-headed arrow of the ***Draw*** toolbar (seen on the right of the picture given below) and move the mouse pointer to within one of the elliptically shaped objects. With the left mouse button held down, drag the arrow to within a rectangular-shaped object. When the colors of both objects change (see diagram below) release the mouse button. Note that the ***Draw*** menu on the main menu bar can also be used when drawing path diagrams. Related draw tool options are contained in this menu.

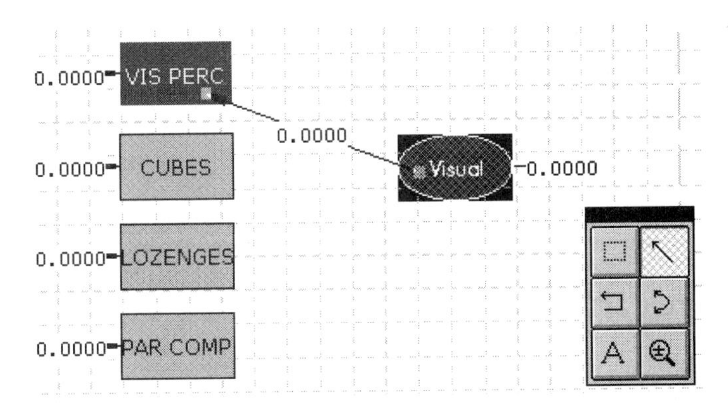

Proceed in a similar fashion to graphically display the relationships between the observed and latent variables as shown on the next image.

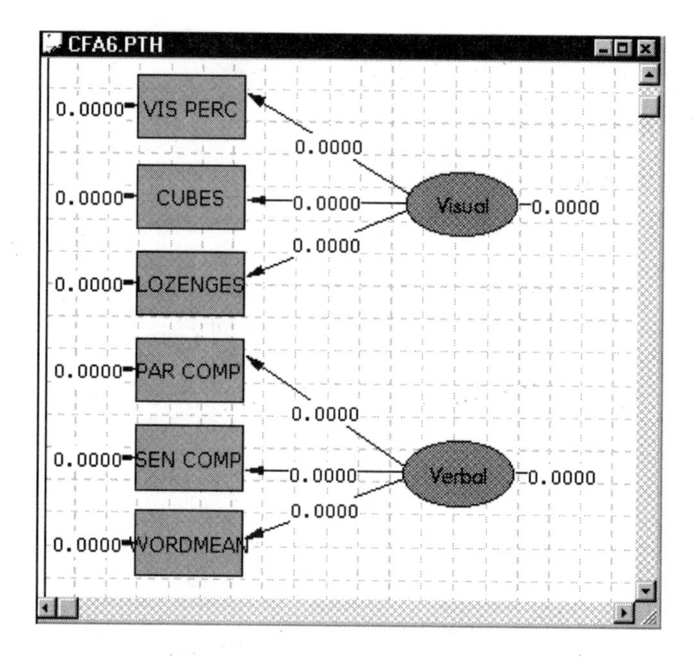

Select the ***Build SIMPLIS Syntax*** option from the ***Setup*** menu. SIMPLIS syntax (not shown below) will be built from the path diagram and is stored in a system file named **cfa6.spj**. Click the ***Run LISREL*** icon button to fit the CFA model to the data. Parameter estimates are shown on the path diagram given below.

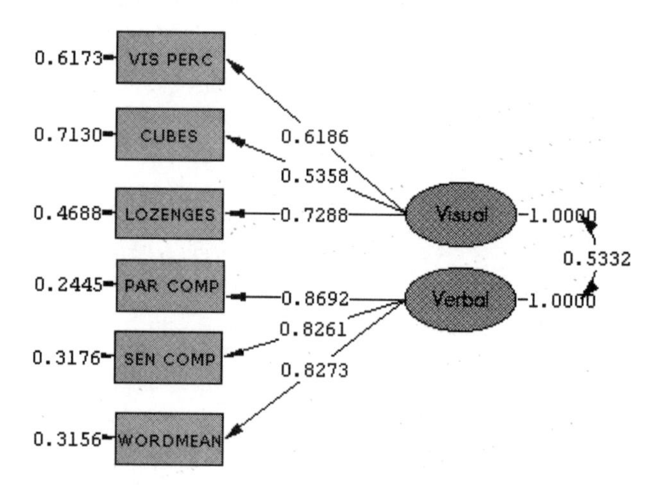

By altering the path diagram, model specifications may be changed. The covariance between *Visual* and *Verbal* may, for example, be fixed by clicking (right mouse button)

on the two-headed arrow representing this covariance. This will enable the pop-up menu shown below. Select the ***Fix*** option using the left mouse button.

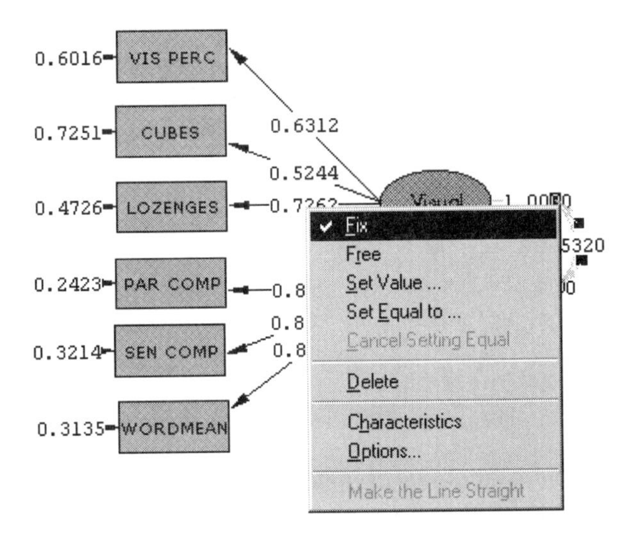

Click the ***Run LISREL*** icon button to fit the revised CFA model to the data. The path diagram, χ^2-fit statistic and RMSEA are shown below.

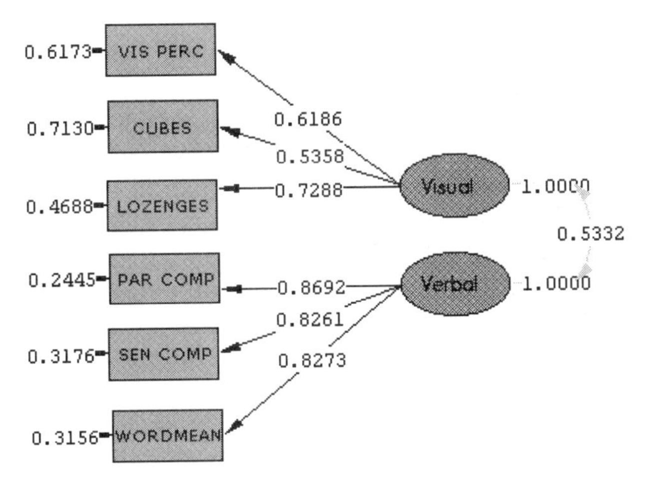

Chi-Square=4.37, df=9, P-value=0.88514, RMSEA=0.000

4.4 Multi-Sample Analyses Using Path Diagrams

In Section 4.3, we have given an example to show how LISREL models may be specified and estimated by drawing the appropriate path diagram from which the SIMPLIS or LISREL syntax is created. The example used was based on data from a single sample. The path diagram can, however, also be used to create syntax for models based on data from several samples simultaneously, according to a multiple-group LISREL model with some or all parameters constrained to be equal over groups. For examples of such simultaneous analyses, see the *LISREL 8: The SIMPLIS Command Language* guide.

Consider a set of *G* mutually exclusive groups of individuals. It is assumed that a number of variables have been measured on a number of individuals from each of these populations. This approach is particularly useful in comparing a number of treatment and control groups regardless of whether individuals have been assigned to the groups randomly or not.

Any LISREL model may be specified and fitted for each group of data. However, LISREL assumes by default that the models are identical over groups, *i.e.*, all relationships and all parameters are the same in each group. Thus, only differences between groups need to be specified.

4.4.1 Testing Equality of Factor Structures

Table 4.1 below gives sample covariance matrices for two samples ($N_1 = 865$, $N_2 = 900$, respectively) of candidates who took the Scholastic Aptitude Test in January 1971. The four measures are, in order, *VERBAL40*, a 40-item verbal aptitude section, *VERBAL50,* a separately timed 50-item verbal aptitude section, *MATH35,* a 35-item math aptitude section, and *MATH25*, a separately timed 25-item math aptitude section.

Table 4.1: Covariance Matrices for SAT Verbal and Math Sections

Covariance Matrix for Group 1

Tests	VERBAL40	VERBAL50	MATH35	MATH25
VERBAL40	63.382			
VERBAL50	70.984	110.237		
MATH35	41.710	52.747	60.584	
MATH25	30.218	37.489	36.392	32.295

Covariance Matrix for Group 2

Tests	VERBAL40	VERBAL50	MATH35	MATH25
VERBAL40	67.898			
VERBAL50	72.301	107.330		
MATH35	40.549	55.347	63.203	
MATH25	28.976	38.896	39.261	35.403

The data are used here to illustrate how one can test equality of factor loadings and factor correlations in a confirmatory factor analysis model, while allowing the error variances to be different.

We regard *VERBAL40* and *VERBAL50* as indicators of a latent variable *Verbal* and *MATH35* and *MATH25* as indicators of a latent variable *Math*. The model we consider is shown in the figure below.

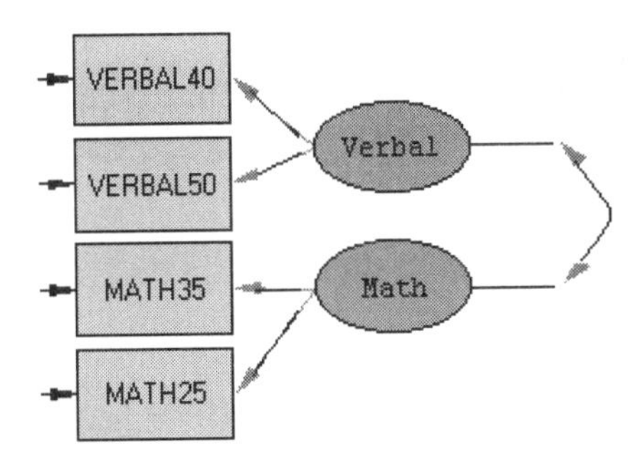

There are three sets of parameters in the model:

- the four factor loadings corresponding to the paths from *Verbal* and *Math* to the observed variables,
- the correlation between *Verbal* and *Math*, and
- the four error variances of the observed variables.

We begin by constructing a data file **pdex10.cov** containing the two covariance matrices. In free format, each covariance matrix can be written on one line. The file **pdex10.cov** (not part of the LISREL installation) is given below.

```
63.382  70.984  110.237  41.710  52.747  60.584  30.218
37.489  36.392  32.295
67.898  72.301  107.330  40.549  55.347  63.203  28.976
38.896  39.261  35.403
```

We assume that all parameters are the same in both groups. The syntax file **pdex10a.spl** is:

```
Group 1: Testing Equality Of Factor Structures Model A:
Factor Loadings, Factor Correlation, Error Variances
Invariant
Observed Variables: VERBAL40 VERBAL50 MATH35 MATH25
Covariance Matrix from File PDEX10.COV
Sample Size = 865
Latent Variables: Verbal Math
Relationships: VERBAL40 VERBAL50 = Verbal
MATH35 MATH25 = Math
Group 2: Testing Equality Of Factor Structures
Covariance Matrix from File PDEX10.COV
Sample Size = 900
Set the error variances of VERBAL40-MATH25 free
End of Problem
```

In the syntax above, the data in the covariance matrices and the sample sizes are different. The names of the variables, observed as well as latent, are the same, and the model is the same.

Select the ***New*** option from the ***File*** menu and select ***Path Diagram*** in the ***New*** dialog box. Click ***OK*** when done.

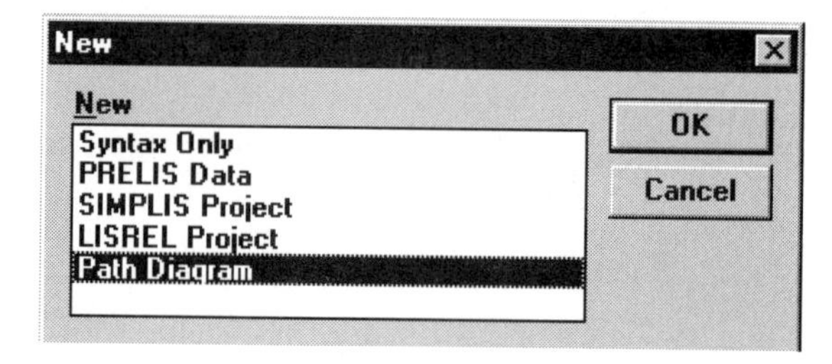

Select the **tutorial** folder and enter the name **pdex10.pth**. When done, click ***Save***.

It is possible to control specific SIMPLIS default features such as ***Maximum Number of Iterations*** or ***Method of Estimation***. From the ***Output*** menu, select ***SIMPLIS Outputs***.

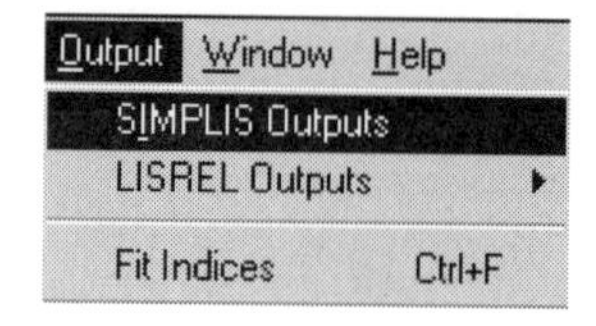

The ***SIMPLIS Outputs*** dialog box will appear. Customize this dialog box according to your preferences, for example, by changing the ***Number of Decimals in the Output*** option to 4 and by making sure that the ***Invoke path diagram*** option is selected. Click ***OK*** when done.

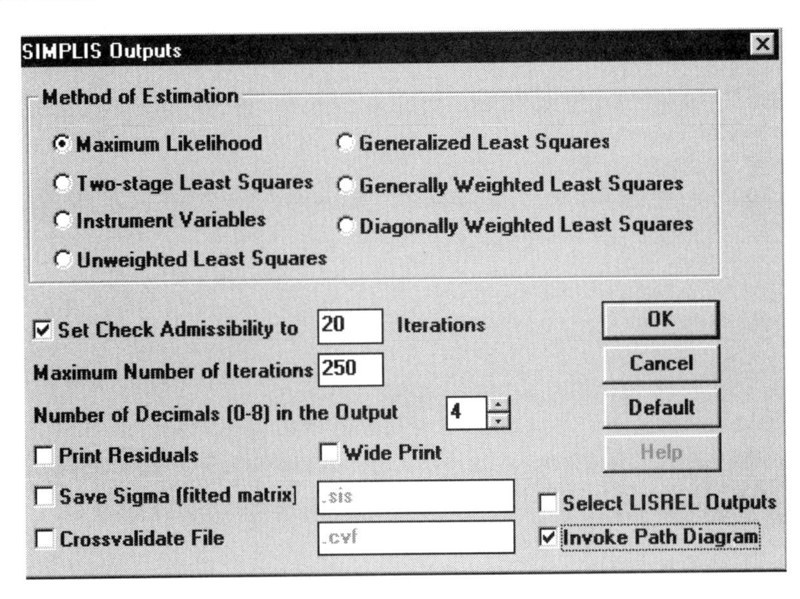

Before drawing the path diagram, select ***Toolbars*** from the ***View*** menu and ensure that the items shown below are selected. These are: ***Toolbar***, ***Status Bar***, ***Typebar***, ***Variables***, ***Drawing Bar***. Optionally, also check the ***Grid Lines*** option.

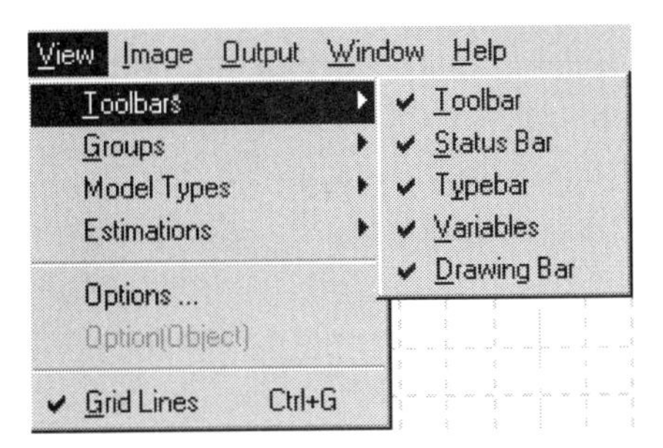

Select the ***Title and Comments*** option from the ***Setup*** menu to obtain the ***Titles and Comments*** dialog box.

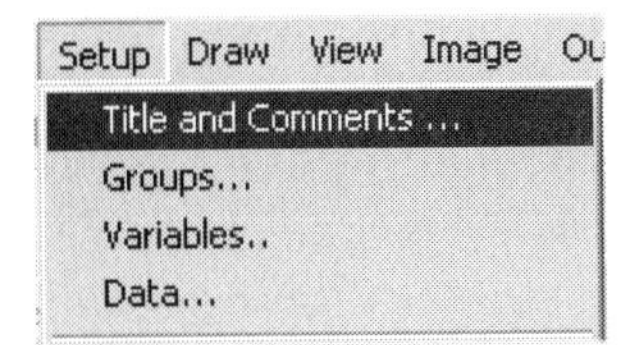

The first item on the ***Setup*** dialog box is the ***Title*** for the particular problem. Provision is also made for any additional ***Comments*** that the user may wish to enter. After typing in the title and (optionally) the comments, click ***Next*** to go to the ***Group Names*** dialog box.

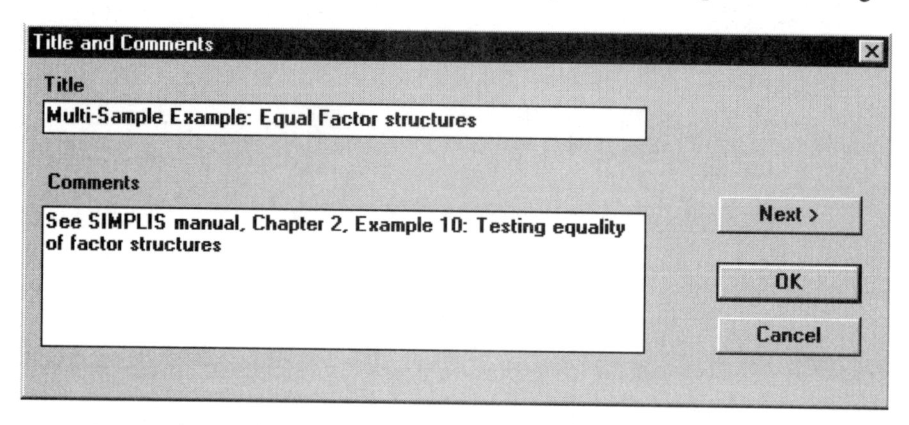

Next we enter a description for each group. Group labels can be inserted by clicking on the first string field, entering the label for the first group and then using the down arrow on the computer keyboard to create the next group's string field. When done, click ***Next*** to go to the ***Labels*** dialog box.

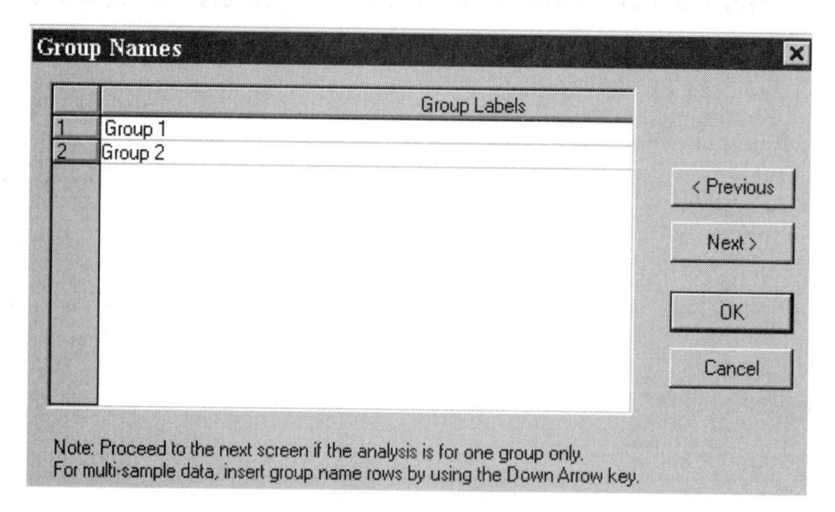

The default number of variables shown on the ***Labels*** dialog box is 2, these being *VAR 1* and *VAR 2*. Move the mouse pointer to the ***Observed Variables*** box and click in the string field of *Var 1*. Rename this variable to *VERBAL 40*. Press the "down arrow" on the keyboard to move to the second observed variable string field and enter the label *VERBAL 50*. Proceed in a similar way to enter the labels *MATH35*, *MATH 25* and the labels *Verbal* and *Math* for the latent variables. Click ***Next*** when done to go to the ***Data*** dialog box.

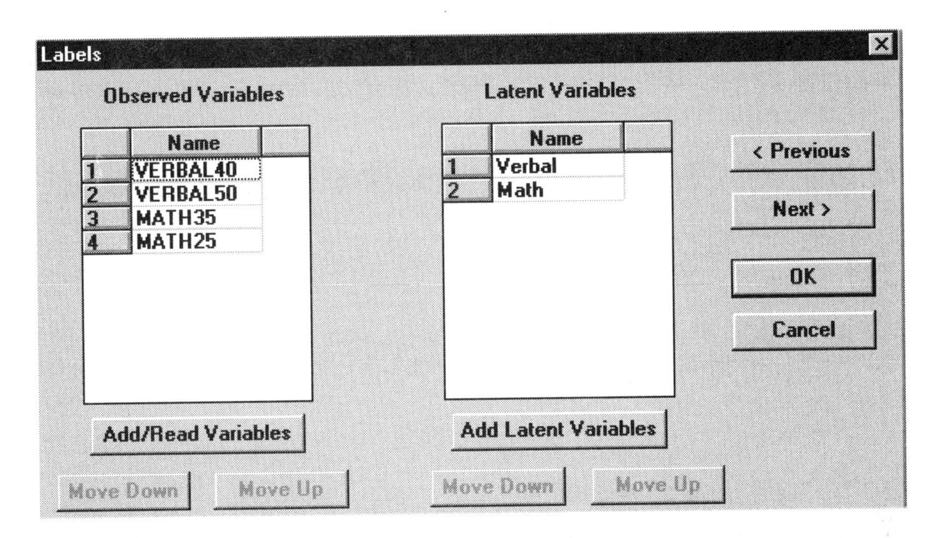

From the ***Data*** dialog box, select ***Covariances*** from the ***Statistics from:*** drop-down list box and also ***Covariances*** from the ***Matrix to be analyzed*** drop-down list box. Enter `865` in the ***Number of Observations*** string field. For ***File type*** select ***External ASCII Data***.

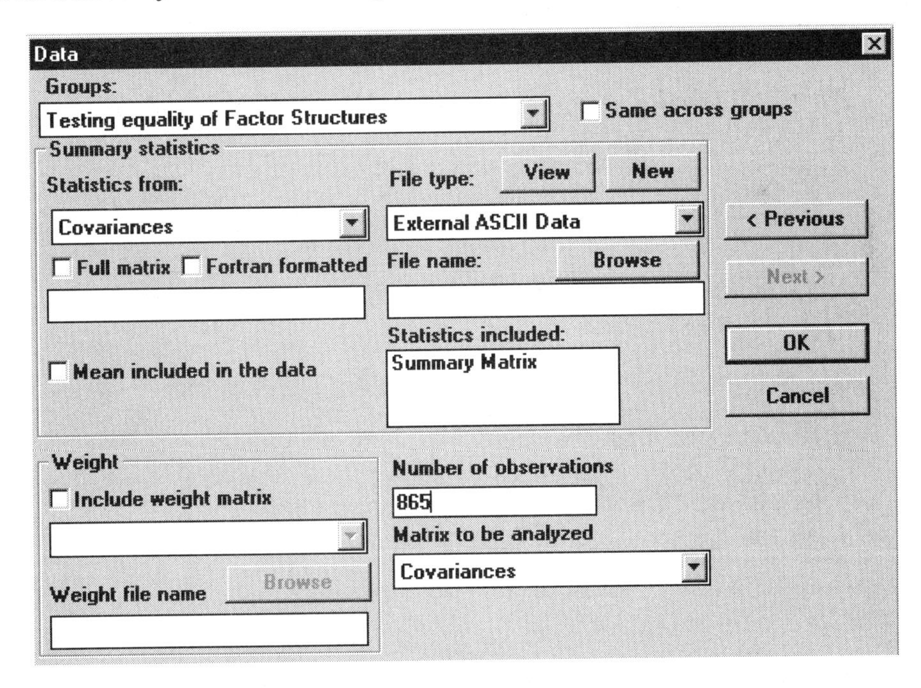

Click the ***Browse*** button to select the path and name of the file that contains the values of the covariance matrices of the 2 groups. For the present example, the folder is **splex** and the file is **pdex10.cov**, which was created earlier in this section. Click ***Open*** when done.

From the ***Data*** dialog box, select the second group from the ***Groups:*** drop-down list box. Proceed as outlined above to enter the number of observations (900), file type and matrix to be analyzed. The top part of the ***Data*** dialog box for the second group is shown below.

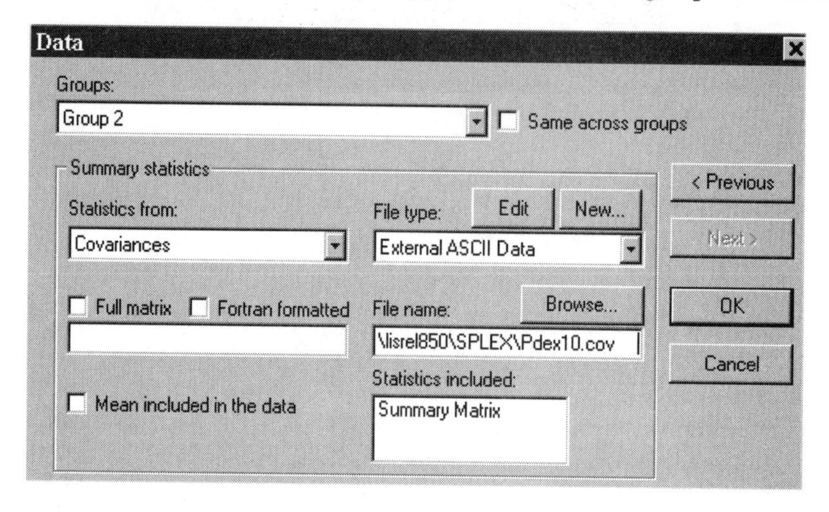

4.4.2 Drawing the Path Diagram

We now proceed with the actual drawing of the path diagram. Start by clicking on the *VERBAL40* label under the ***Observed*** variables portion of the ***Variables*** window. Hold the mouse button down and drag the label to the draw area indicated by the grid lines. A rectangular-shaped object will appear on this part of the screen when the mouse button is released as shown below. Repeat this procedure until all observed variables are dragged to the drawing area as shown in the next image.

Note also that:

- The observed variables are assumed to be *X* (or independent) variables unless appropriate check boxes under the ***Y***-column are checked.
- The latent variables are assumed to be *Ksi* (or independent) variables unless appropriate check boxes under the ***Eta***-column are checked.

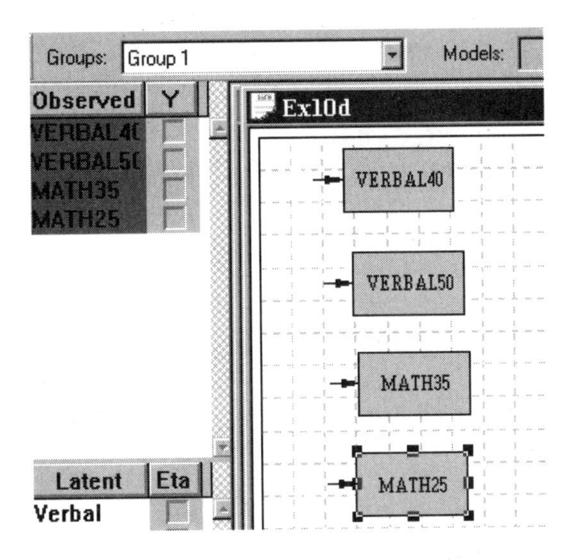

Proceed by dragging the latent variables, *Verbal* and *Math*, to the draw area of the path window. Note that the latent variables are represented by elliptically shaped objects.

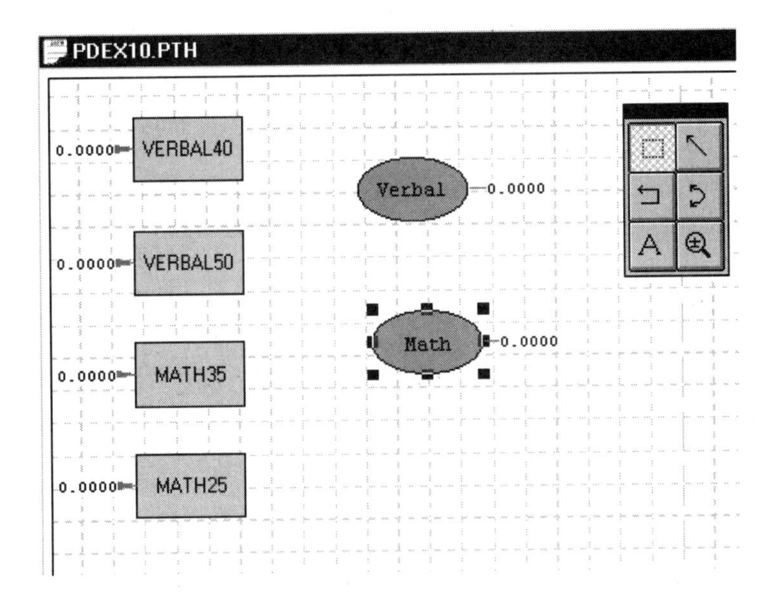

Finally, arrows can be drawn pointing from the latent variables to the observed variables. To accomplish this, click on the single-headed arrow of the ***Draw*** toolbar (seen on the right of the picture given below). As an alternative, use the ***Draw*** menu on the main menu bar. The related drawing tool options may be selected from this menu.

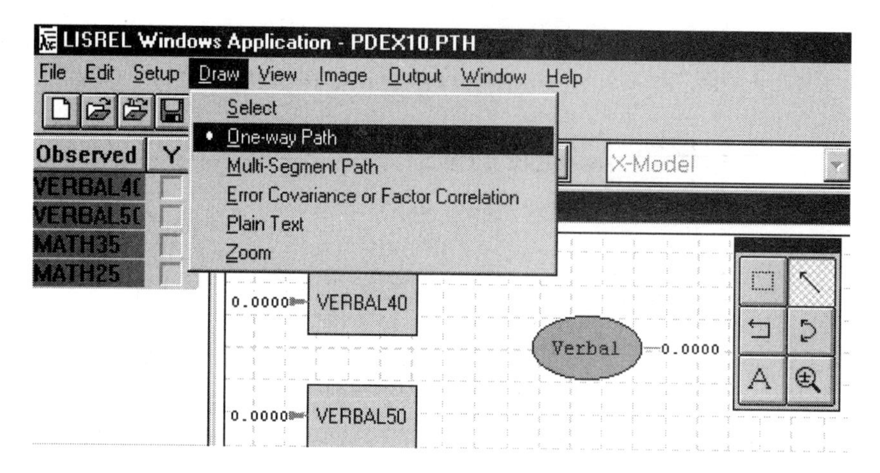

Once the single-headed arrow (or ***One-way Path*** option) has been selected, move the mouse pointer to within one of the elliptically shaped objects. With the left mouse button held down, "drag" the arrow to within a rectangular-shaped object. When the colors of both objects change (see diagram below) release the mouse button. Proceed in a similar way to graphically display the relationships between the observed and latent variables.

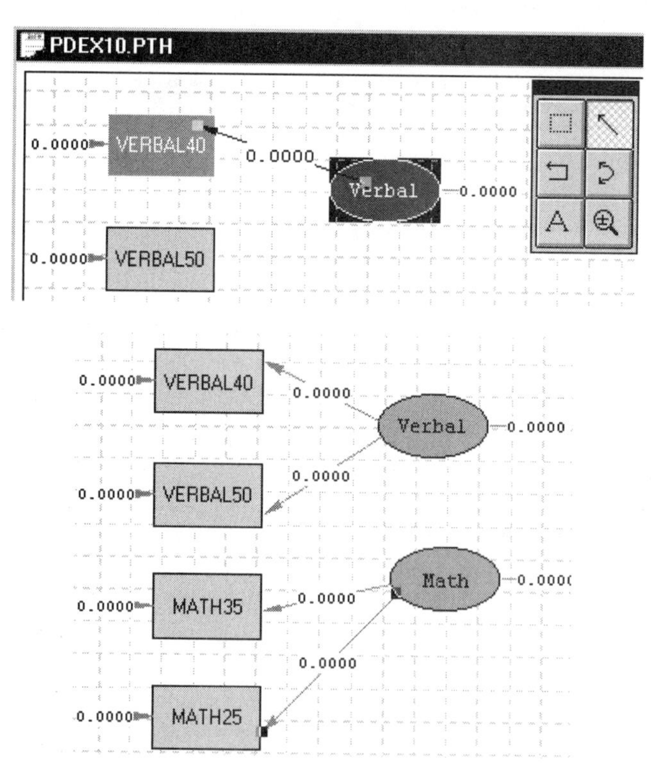

Select ***Build SIMPLIS Syntax*** from the ***Setup*** menu. SIMPLIS syntax will be built from the path diagram and is stored in a system file named **pdex10.spj** shown below.

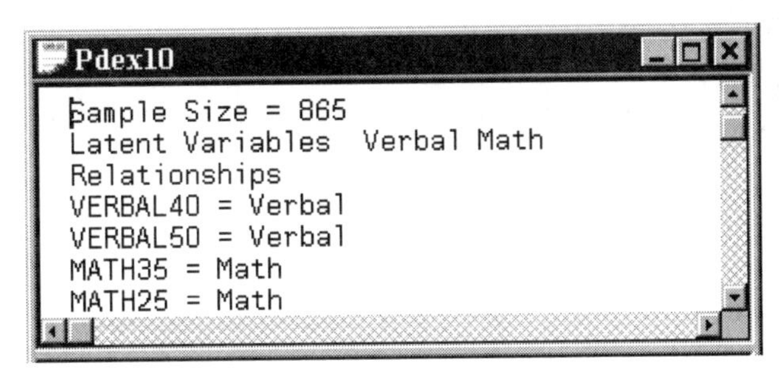

Pdex10

Sample Size = 865
Latent Variables Verbal Math
Relationships
VERBAL40 = Verbal
VERBAL50 = Verbal
MATH35 = Math
MATH25 = Math

Click the ***Run LISREL*** icon button to fit the multi-sample model to the data. The parameter estimates and χ^2 goodness-of-fit index are shown on the path diagram displayed below:

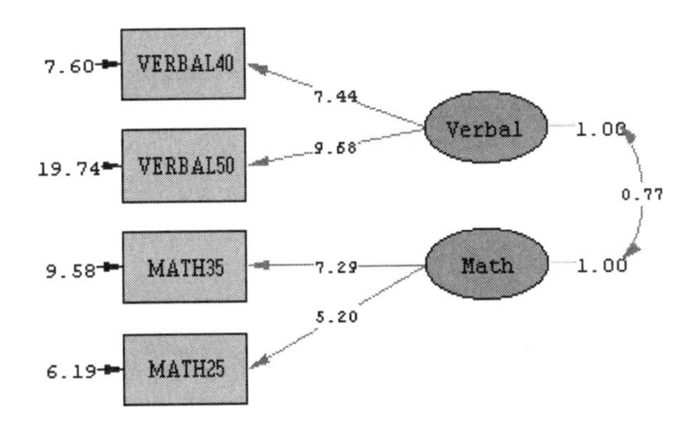

The values shown on the path diagram are the parameter estimates. This is the default selection. It is, however, also possible to view the *t*-values, modification indices or expected changes by selecting the ***View*** menu on the main menu bar. To view *t*-values, select the ***t-values*** option from the ***View, Estimations*** menu. The figure below shows how this is done and also indicates the *t*-values on the path diagram.

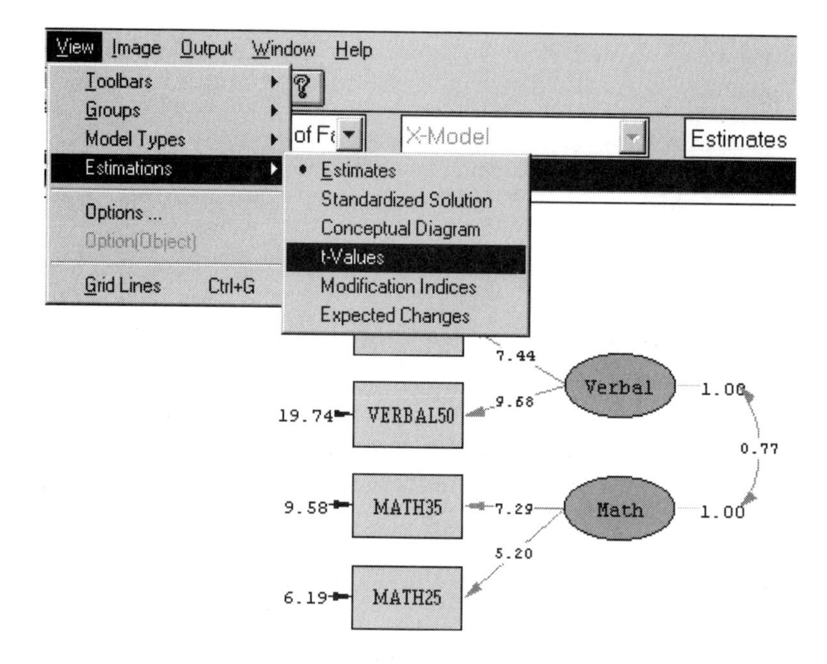

To view the corresponding *t*-values for the second group, select group 2 on the menu bar. The resultant changes in the path diagram are shown in the next image.

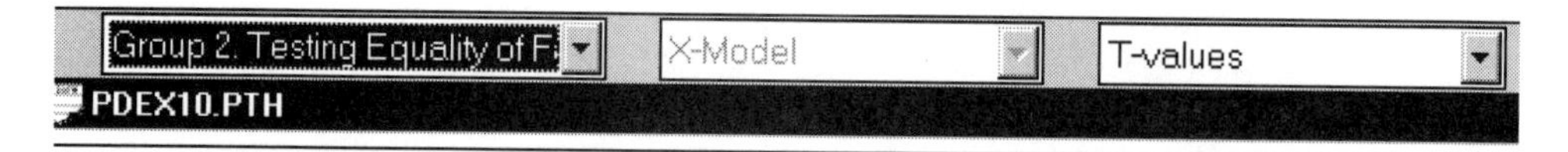

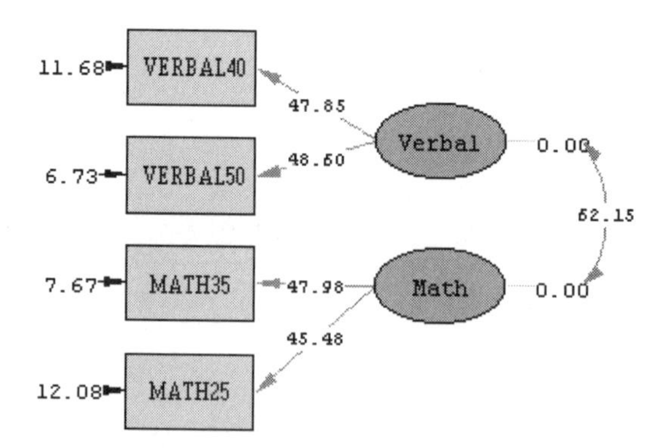

The output file reveals two solutions, one for each group. The value of χ^2 is reported only after the second group. In a multi-sample analysis, the χ^2 is a measure of fit of all models in all groups, and, in general, this χ^2 cannot be decomposed into a χ^2 for each group separately. For our example, χ^2 is 10.87 with 7 degrees of freedom ($p = 0.14$), so that the model does fit the data.

4.5 Analysis of Ordinal Data

Aish and Jöreskog (1990) analyzed data on political attitudes. Their data consist of six ordinal variables measured on the same people at two occasions (see *PRELIS 2: User's Reference Guide*, pages 174 to 178, and the *LISREL 8: User's Reference Guide*, pages 243 to 247). Labels assigned to the 12 variables are as follows: *NOSAY1*, *VOTING1*, *COMPLEX1*, *NOCARE1*, *TOUCH1*, *INTERES1* (first occasion) and *NOSAY2*, *VOTING2*, *COMPLEX2*, *NOCARE2*, *TOUCH2* and *INTERES2* (second occasion). These data are saved in the PRELIS system file **panelusa.psf** in the **tutorial** folder.

In this example, the following concepts are illustrated:

- calculation of polychoric correlations and asymptotic covariances
- drawing a path diagram for a given model
- creation of variable names for the path diagram
- creating LISREL and or SIMPLIS syntax from the path diagram
- the basic model and its components
- estimation types and the conceptual path diagram
- saving the path diagram in graphics format.

4.5.1 Calculation of Correlations and Asymptotic Covariance Matrix

Select the ***Open*** option from the ***File*** menu to obtain the ***Open*** dialog box. Select ***PRELIS Data*** (*.**psf**) from the ***Files of type:*** drop-down list box. Select **panelusa.psf** from the **tutorial** folder and click ***Open*** when done. The PRELIS system file (*.**psf**) is displayed in the form of a spreadsheet as shown below:

PANELUSA.PSF

	NOSAY1	VOTING1	COMPLEX1	NOCARE1	TOUCH1
1	2.000	2.000	1.000	1.000	1.000
2	2.000	3.000	3.000	3.000	2.000
3	3.000	2.000	2.000	3.000	3.000
4	2.000	2.000	1.000	1.000	2.000
5	3.000	2.000	2.000	3.000	3.000
6	2.000	2.000	2.000	2.000	1.000
7	3.000	1.000	2.000	2.000	2.000
8	2.000	1.000	2.000	2.000	1.000

Select ***Output Options*** from the ***Statistics*** menu to obtain the ***Output*** dialog box. The ***Output Options*** dialog box allows the user to save various matrices as well as the raw data as files. Select ***Correlations*** (to be saved as a LISREL Data System File with extension *.**dsf**) from the ***Moment Matrix*** drop-down list box. Check the ***LISREL System Data*** check box. Also check the ***Save to File*** check box below ***Asymptotic Covariance Matrix*** and enter the name **panelusa.acm** in the string field. With the required selections made, click ***OK*** to run PRELIS. Note that PRELIS computes polychoric correlations for all variables that are defined as ordinal, provided that the number of distinct values of the variables is less than 15.

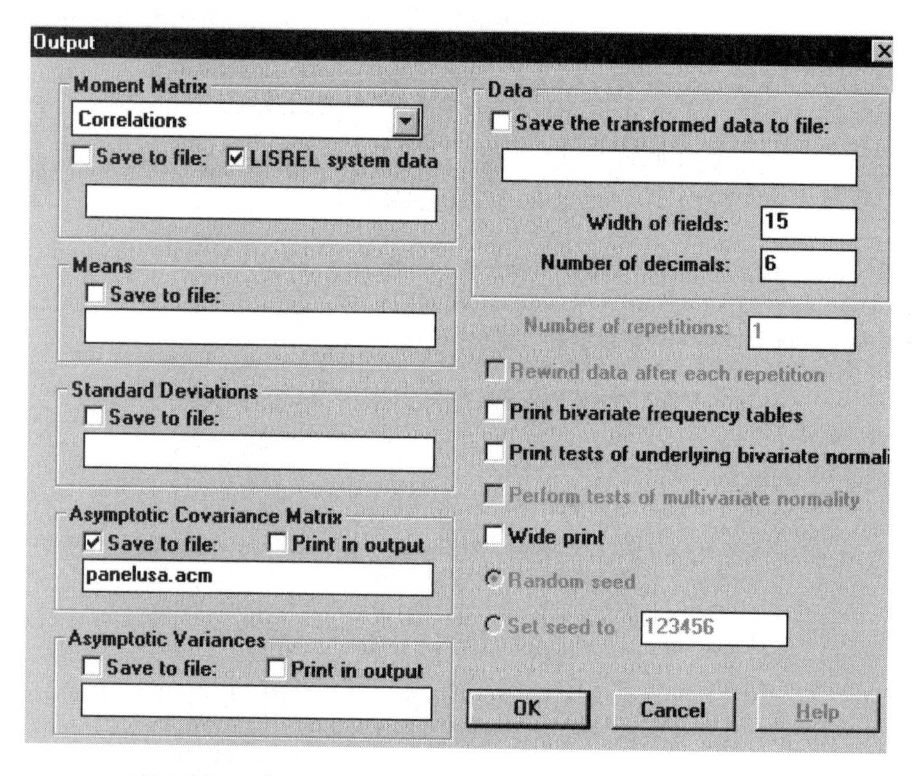

A portion of the PRELIS output is given below:

PANELUSA.OUT

```
Univariate Marginal Parameters

Variable       Mean St. Dev.    Thresholds
--------       ---- --------    ----------
  NOSAY1      0.000    1.000  -2.092   -1.265   -0.271    1.432
 VOTING1      0.000    1.000  -2.117   -0.923    0.155    1.742
```

4.5.2 Drawing a Path Diagram

We use PRELIS to compute the polychoric correlation matrix and corresponding asymptotic covariance matrix of these correlations. The next step is to create a path diagram from which the LISREL or SIMPLIS syntax may be built.

Select the ***New*** option the ***File*** menu to obtain the ***New*** dialog box. From the ***New*** dialog box, select ***Path Diagram*** and click ***OK*** when done.

The ***Save As*** dialog box will appear. Save the new path diagram as **panelusa.pth** in the **tutorial** folder.

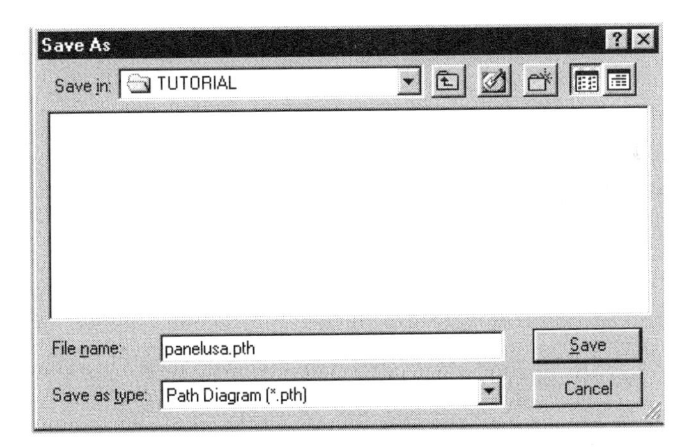

Before proceeding to the next step, ensure that the ***Toolbar***, ***Status Bar***, ***Type Bar***, ***Variables*** and ***Drawing Bar*** options are selected as explained in Sections 4.3 and 4.4.

Select the ***Title and Comments*** option from the ***Setup*** menu to obtain the ***Title and Comments*** dialog box. Use the ***Title and Comments*** dialog box to enter a title and to add additional comments as shown in the example below:

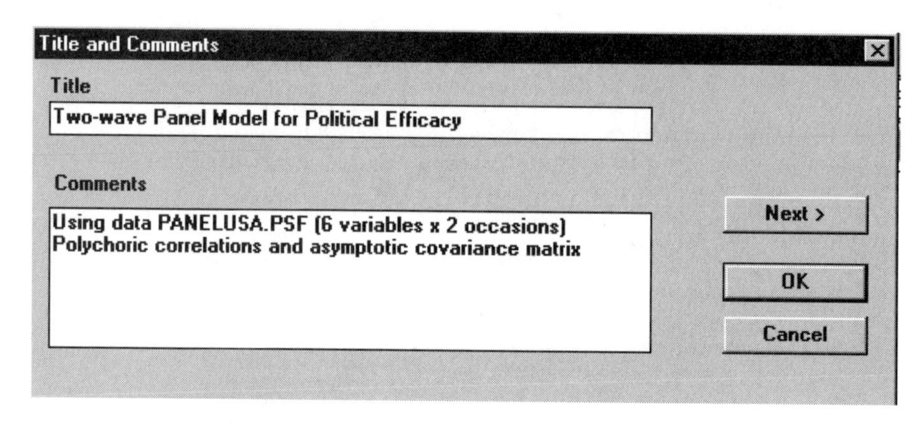

Click ***Next*** to go to the ***Group Names*** dialog box.

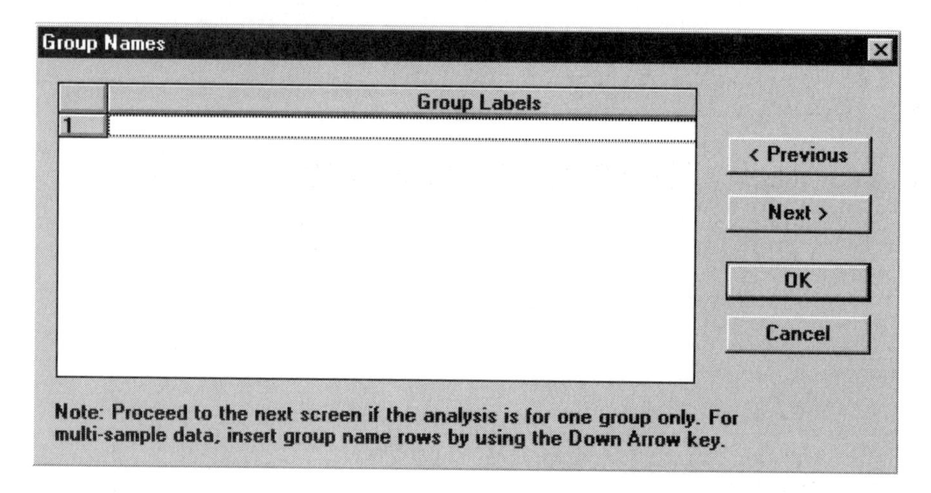

Since the present example is based on the analysis of a single group, we can leave this dialog box blank and go to the variable names (***Labels***) dialog box by clicking ***Next***.

4.5.3 Reading Variable Names from the DSF File

The data system file **panelusa.dsf**, which was created during the PRELIS analysis, contains the variable names, the location of the asymptotic covariance matrix, and the matrix of polychoric correlation coefficients. For more information on data system files, see Section 7.1. The variable names can be read in by clicking ***Add/Read Variables*** on the ***Labels*** dialog box. In doing so, the ***Add/Read Variables*** dialog box is obtained. Click the ***Read from File*** radio button and select ***LISREL System File*** from the drop-down list box.

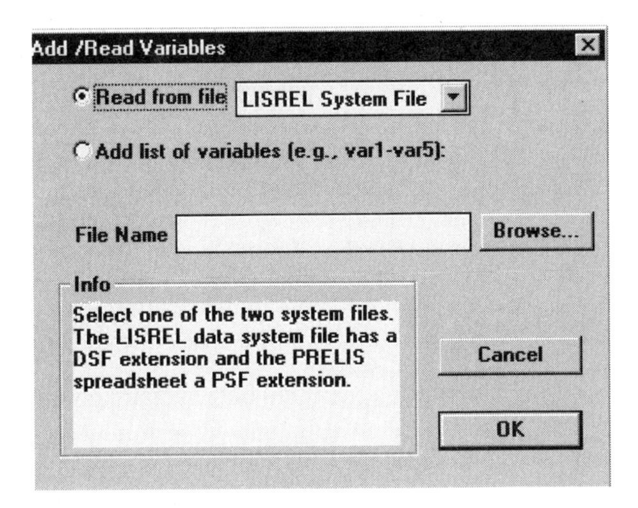

Click ***Browse*** to select the file **panelusa.dsf** from the **tutorial** folder. Click ***Open*** to return to the ***Add/Read Variables*** dialog box.

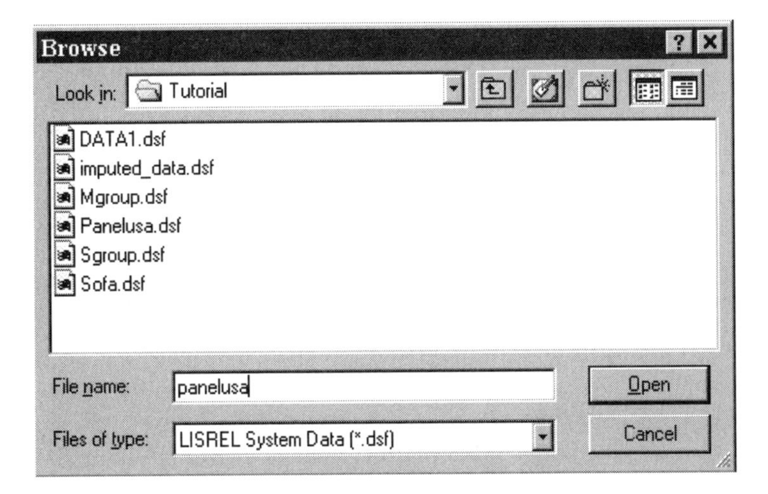

Click ***OK*** to obtain the list of twelve names shown in the ***Labels*** dialog box below. To complete the information in this dialog box, add the names of the two latent variables *Efficac1* and *Efficac2* in the ***Latent Variables*** window. See the notes at the bottom of the ***Labels*** dialog box. To start typing a name, click anywhere within the ***Latent Variables*** window and type in the name. Alternatively, use the ***Add Latent Variables*** button.

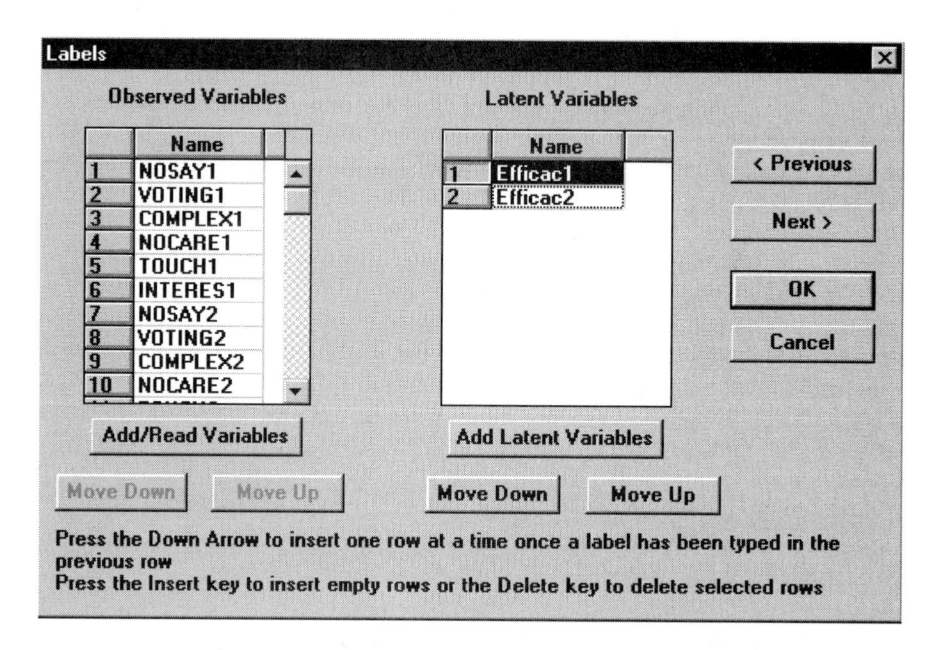

Once names for the observed and latent variables are entered, click ***Next*** to go to the ***Data*** dialog box. Select ***Correlations*** from the ***Statistics from*** drop-down list box.

Note:

Since the location of the asymptotic covariance matrix is stored in the ***.dsf** file, it is not necessary to enter the weight matrix details on the ***Data*** dialog box.

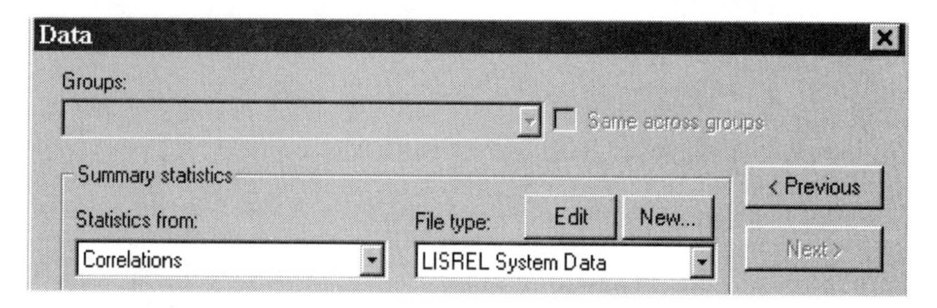

At this point, one may view the contents of the **panelusa.dsf** file by clicking the ***Edit*** button. A portion of the correlation matrix is shown below.

PANELUSA.DSF

	NOSAY1	VOTING1	COMPLEX1	NOCARE1
NOSAY1	1.000	0.000	0.000	0.000
VOTING1	0.442	1.000	0.000	0.000
COMPLEX1	0.403	0.293	1.000	0.000
NOCARE1	0.601	0.295	0.406	1.000
TOUCH1	0.468	0.292	0.322	0.619
INTERES1	0.513	0.302	0.369	0.681
NOSAY2	0.436	0.231	0.225	0.372

Close the spreadsheet to return to the ***Data*** dialog box and click ***OK*** when done. The list of variable names will appear on the left side on the computer screen. Define the variables *NOSAY1*, *VOTING1*, *COMPLEX1*, *NOSAY2*, *VOTING2* and *COMPLEX2* as *Y*-variables by checking the appropriate boxes under the heading ***Y***. Also check the ***Eta-variable*** boxes to define *Efficac1* and *Efficac2* as *Eta* latent variables.

The observed values to be used in the drawing of the path diagram can be dragged to the path diagram area by clicking on a variable name (mouse left button). By holding this button down, a variable can be dragged to a desired position.

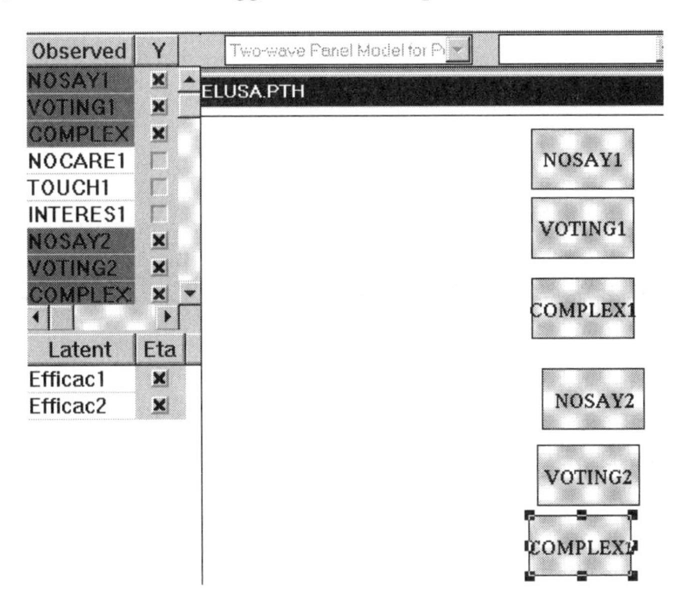

Drag the two latent variables to the path diagram window. To create paths between variables we proceed as follows: click on the single-headed arrow icon button on the ***Draw*** toolbar as shown. Move the mouse cursor to the middle of the *Efficac1* ellipse, and then click the left mouse button. Hold the button down while dragging the arrow to within the *NOSAY1* rectangle. The colors of both objects should change. Repeat the procedure to add the remaining paths as illustrated:

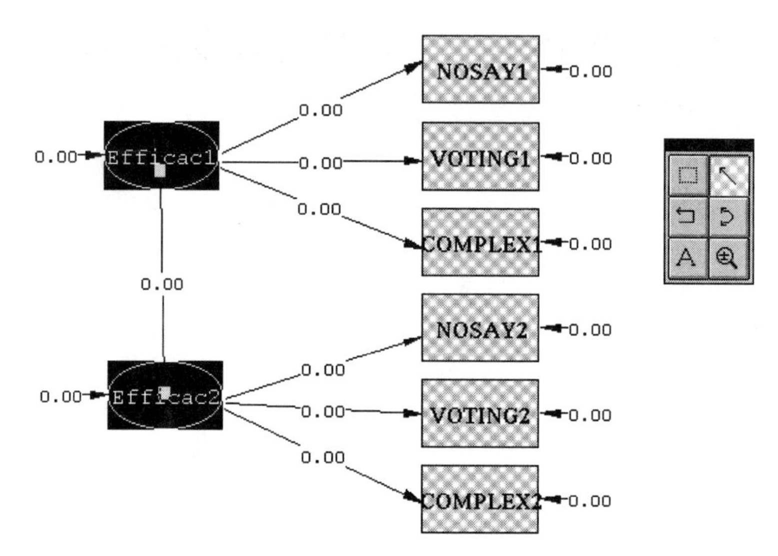

Creating LISREL or SIMPLIS Syntax from the Path Diagram

Prior to building the SIMPLIS or LISREL syntax, we adjust the default number of decimals from 2 to 3. This is accomplished by selecting the ***Output, LISREL outputs***, ***Selections*** option.

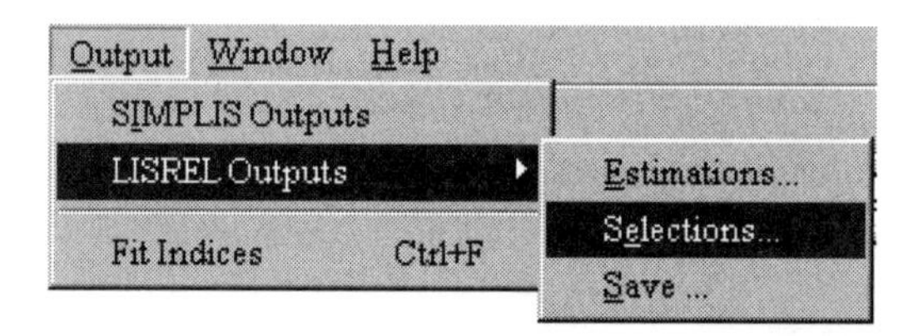

The method of estimation is also changed from ***Maximum Likelihood*** to ***Generally Weighted Least Squares*** by selecting the ***Output***, ***LISREL Outputs***, ***Estimations*** option.

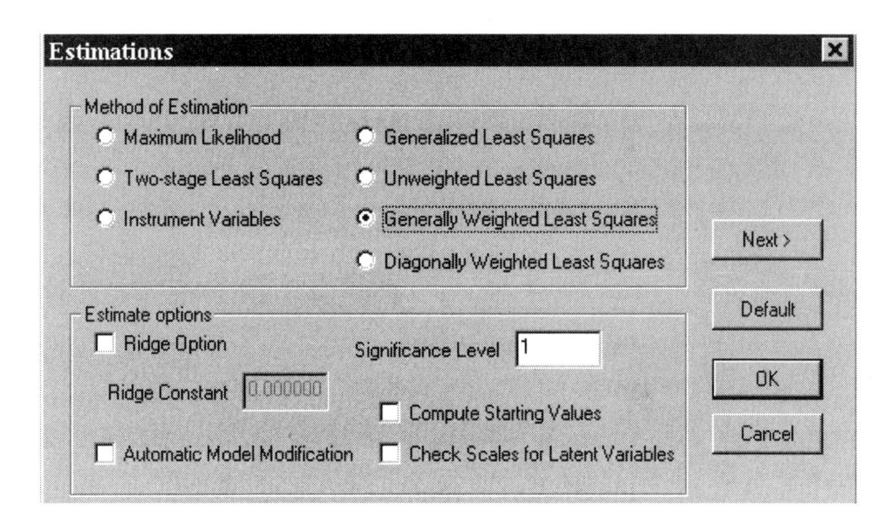

Select ***Build LISREL Syntax*** from the ***Setup*** menu to obtain the following LISREL project window which displays the contents of the LISREL syntax file **panelusa.lpj**.

panelusa

```
TI Two-wave Panel Model for Political Efficacy
!DA NI=12 NO=0 NG=1 MA=CM
SY='C:\lisrel850\TUTORIAL\Panelusa.dsf' NG=1
!AC FI=C:\lisrel850\TUTORIAL\panelusa.acm
SE
1 2 3 7 8 9 /
MO NY=6 NE=2 LY=FU,FI BE=FU,FI PS=DI,FR TE=DI,FR
LE
Efficac1 Efficac2
FR LY(1,1) LY(2,1) LY(3,1) LY(4,2) LY(5,2) LY(6,2) BE(2,1)
PD
OU ME=WL ND=3 IT=250
```

Click the ***Run LISREL*** icon button to start the analysis. The path diagram shows the estimated parameter values and χ^2-value.

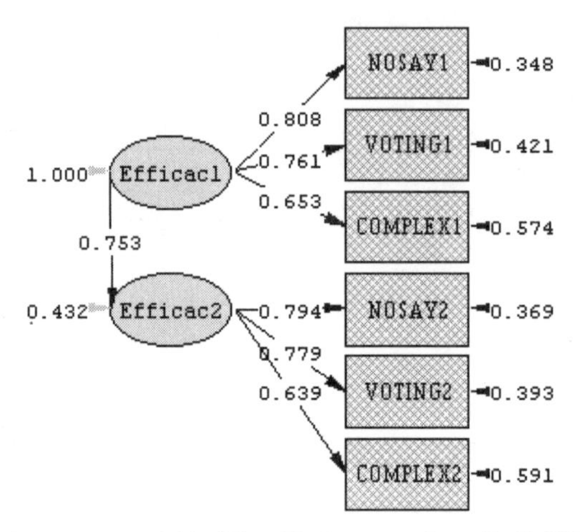

Chi-Square=213.95, df=8, P-value=0.00000

The corresponding SIMPLIS input syntax can be built from the existing path diagram. This is accomplished by selecting the ***Build SIMPLIS Syntax*** option from the ***Setup*** menu.

A selection of the SIMPLIS syntax (**panelusa.spj**) is displayed in the SIMPLIS project window as shown in the next image.

panelusa

```
Relationships
NOSAY1 = Efficac1
VOTING1 = Efficac1
COMPLEX1 = Efficac1
NOSAY2 = Efficac2
VOTING2 = Efficac2
COMPLEX2 = Efficac2
Efficac2 = Efficac1
Path Diagram
Number of Decimals = 4
Iterations = 250
Method of Estimation: Weighted Least Squares
End of Problem
```

This syntax produces a model that consists of *X*-variables and *Y*-variables, a *KSI* latent variable and an *Eta* latent variable. Click the ***Run LISREL*** icon button to run the analysis. It is interesting to note that the *Y*-only model and the model shown below are equivalent.

4.5.4 The Basic Model and its Components

The model shown above is called the ***Basic Model***, which can be divided into the ***X-Model***, the ***Y-Model*** and the ***Structural Model***. Each of these models may be viewed separately by clicking on the ***Basic Model*** drop-down list box to obtain the list of models. By selecting the ***X-Model*** model type, this portion of the path diagram is displayed. The ***X-Model*** selection and ***Y-Model*** selection are shown below.

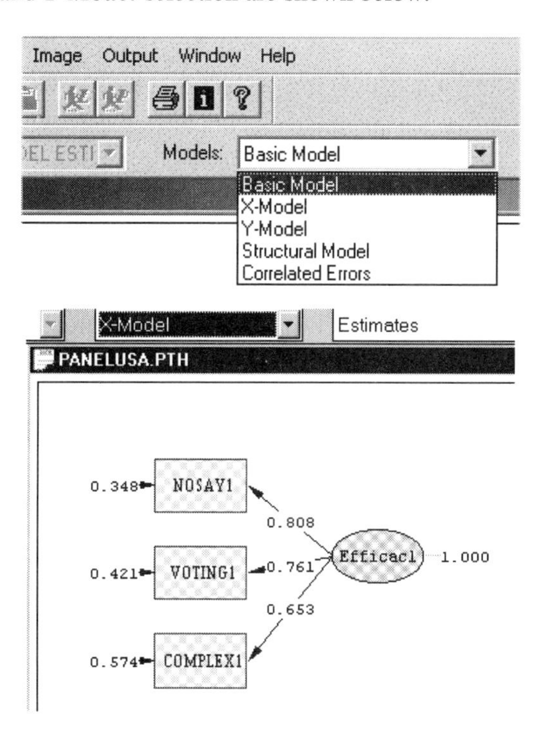

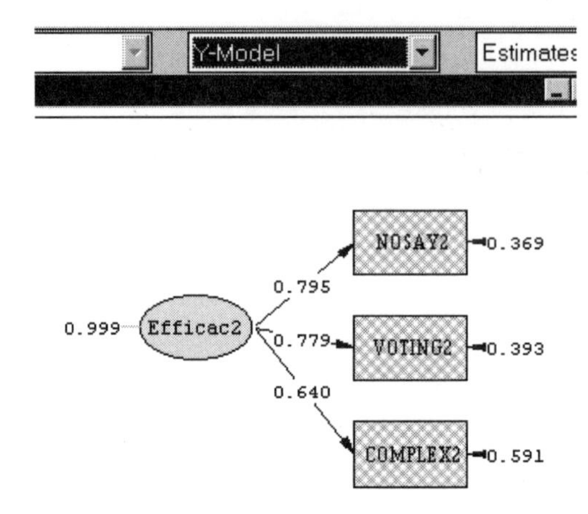

The selection of the ***Structural Model*** option produces that part of the path diagram:

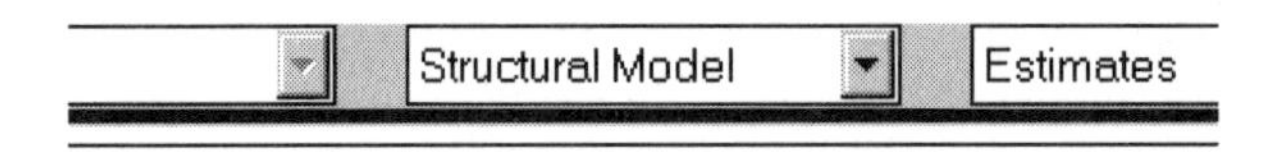

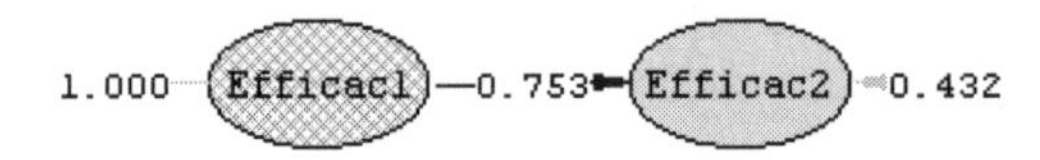

Finally, the selection of the ***Correlated errors*** option produces the following diagram.

Note:

Use was made of the two-headed arrow (corresponding to the ***Error Covariance or Factor Correlation*** option on the ***Draw*** menu) to add the error covariance paths to the diagram.

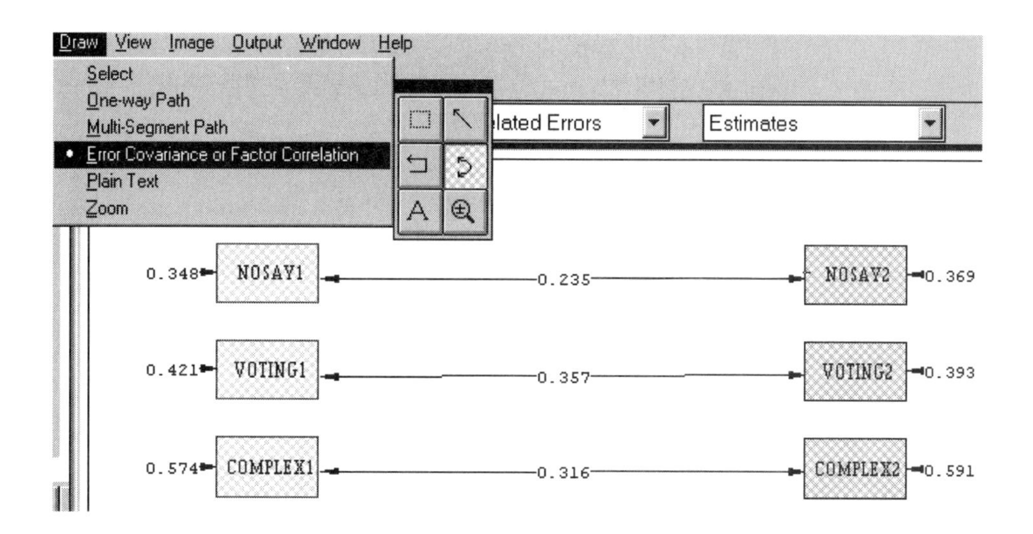

When the ***Run LISREL*** icon button is clicked, the following path diagram is obtained. From this display, it is seen that the χ^2-value has significantly decreased as a result of the addition of the error covariances.

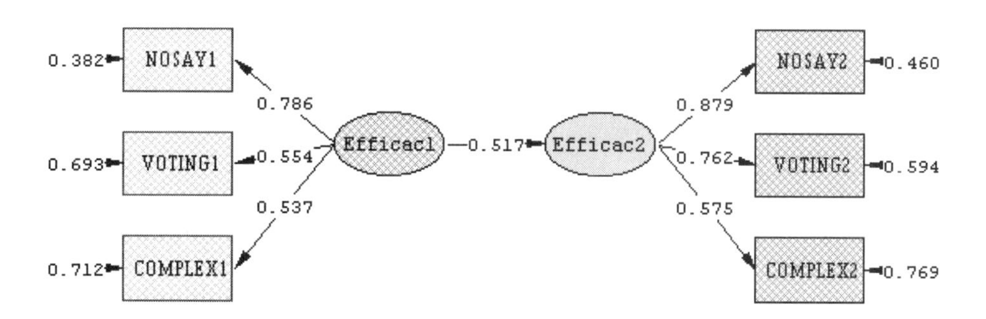

4.5.5 Estimation Types and the Conceptual Path Diagram

LISREL for Windows also allows the user to view the parameter ***Estimates***, ***Completely Standardized Solution***, ***Conceptual Diagram***, ***T-values***, etc. by clicking on ***View*** (main menu bar) or by clicking on the ***Estimates*** drop-down list box as shown below:

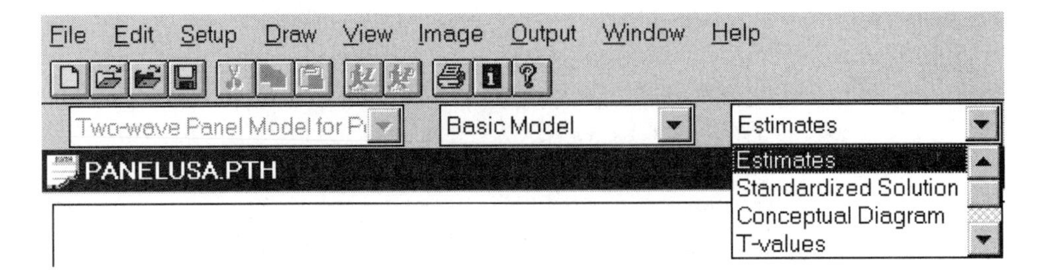

The selection of the ***Conceptual Diagram*** option produces a path diagram that contains only variable names and paths between variables.

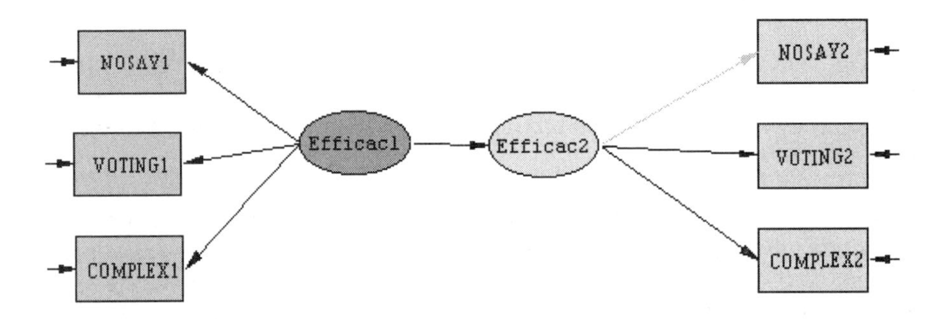

4.5.6 Saving the Path Diagram in Graphics Format

The conceptual path diagram, or one with an alternative selection from the ***Estimations*** bar, can be saved as a Windows Metafile (*.**wmf**) or a Graphic Interchange Format (*.**gif**) file, which may be included as a graphic in, for example, Microsoft Word or Powerpoint. To save as a *.**wmf** file, select the ***Export as Metafile*** option from the ***File*** menu.

4.6 Using SIMPLIS Project Files

LISREL allows the user to create SIMPLIS syntax by making use of a ***Relationships*** keypad. This keypad, see illustration below, is operated by clicking (left mouse button) on the appropriate symbol. Note that the **<==** symbol represents the ***backspace*** key while the symbol **<--|** denotes the ***enter*** key. The path diagram produced by running the program can subsequently be modified to change the model specifications.

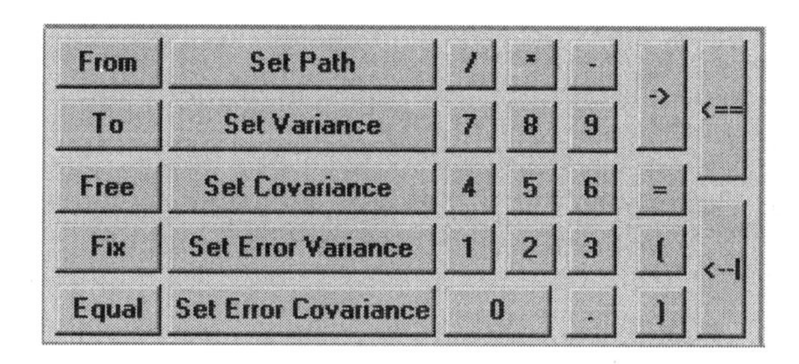

As an illustration, we consider path analysis for latent variables, which, in its most general form, is a structural equation system for a set of latent variables classified as dependent or independent. In the application described below, the system is recursive (see the *LISREL 8: SIMPLIS command language* guide.)

Recursive models are particularly useful for analyzing data from longitudinal studies in psychology, education and sociology. The characteristic feature of a longitudinal research design is that the same measurements are used on the same people at two or more occasions.

Wheaton, *et al.* (1977) reported on a study concerned with the stability over time of attitudes such as alienation, and the relation to background variables such as education and occupation. Data on attitude scales were collected from 932 persons in two rural regions in Illinois at three points in time: 1966, 1967, and 1971. The variables used for the present example are the Anomia subscale and the Powerlessness subscale, taken to be indicators of *Alienation*. This example uses data from 1967 and 1971 only.

The background variables are the respondent's education (years of schooling completed) and Duncan's Socio-economic Index (*SEI*). These are taken to be indicators of the respondent's socio-economic status (*Ses*). The sample covariance matrix of the six observed variables is given below:

	y1	y2	y3	y4	x1	x2
ANOMIA67	11.834					
POWERL67	6.947	9.364				
ANOMIA71	6.819	5.091	12.532			
POWERL71	4.783	5.028	7.495	9.986		
EDUC	-3.839	-3.889	-3.841	-3.625	9.610	
SEI*	-2.190	-1.883	-2.175	-1.878	3.552	4.503

*The variable *SEI* has been scaled down by a factor of 10.

The path diagram for this model is shown below:

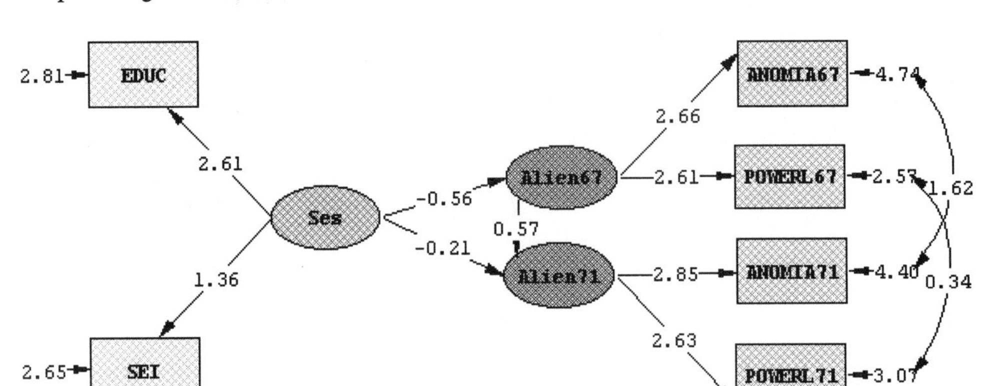

Notes:

- The error terms of *ANOMIA67* and *ANOMIA71* and *POWERL67* and *POWERL71* are specified to be correlated over time.
- The four one-way arrows to the right of the diagram represent the measurement errors in *ANOMIA67*, *POWERL67*, *ANOMIA71* and *POWERL71* respectively.
- The two-way arrows on the right indicate that some of the measurement errors are correlated. The covariance between the two error terms for each variable can be interpreted as a specific error variance. For other models for the same data, see Jöreskog & Sörbom (1989, pp. 213-223).

To set up this model for SIMPLIS is straightforward as shown in the following syntax file:

```
Stability of Alienation
Observed Variables
ANOMIA67 POWERL67 ANOMIA71 POWERL71 EDUC SEI
Covariance matrix
 11.834
  6.947    9.364
  6.819    5.091   12.532
  4.783    5.028    7.495    9.986
 -3.839   -3.889   -3.841   -3.625    9.610
 -2.190   -1.883   -2.175   -1.878    3.552    4.503
Sample Size 932
Latent Variables Alien67 Alien71 Ses
```

```
Relationships
ANOMIA67 POWERL67 = Alien67
ANOMIA71 POWERL71 = Alien71
EDUC SEI = Ses
Alien67 = Ses
Alien71 = Alien67 Ses
Set Error Covariance of ANOMIA67 and ANOMIA71 Free
Set Error Covariance of POWERL67 and POWERL71 Free
Path Diagram
End of Problem
```

Notes:

- The model is specified in terms of relationships. The first three lines under `Relationships` specify the relationships between the observed and the latent variables. For example, `ANOMIA71 POWERL71 = Alien71` means that the observed variables *ANOMIA71* and *POWERL71* depend on the latent variable *Alien71*.
- The Line `Alien71 = Alien67 Ses` means that the latent variable *Alien71* depends on the two latent variables *Alien67* and *Ses*. This is one of the structural relationships.

4.6.1 Building the SIMPLIS Syntax

The files **ex6a.spl** and **ex6b.spl** in the **splex** folder both contain the numeric values of the covariance matrix of the observed variables. From the ***File*** menu, select ***Open*** and locate **ex6a.spl** in the **splex** folder. Use the mouse cursor (left button down) to mark the covariances as shown below. Use the ***Edit*** menu and click ***Copy*** or, alternatively, press the ***Ctrl - C*** keys on the keyboard.

From ***File***, ***New*** select ***Syntax Only*** and click ***OK***. Use the ***Edit***, ***Paste*** function to copy the contents of the clipboard to the syntax window or, alternatively, use the ***Ctrl - V*** keys on the keyboard. Use the ***File***, ***Save As*** option to save the covariances to the file **wheaton.cov**.

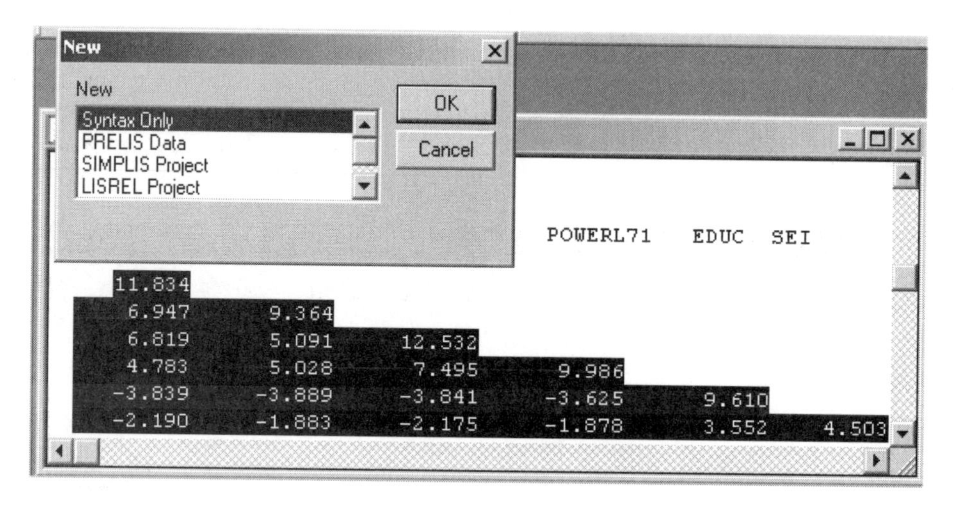

Select the ***New*** option from the ***File*** menu, and select ***SIMPLIS Project*** in the ***New*** dialog box. Click ***OK*** to open the ***Save As*** dialog box.

Select a filename and a folder in which the SIMPLIS project should be saved. For the present example the folder chosen is **splex** and the filename selected is **simplis6.spj**. When done, click ***Save*** to go to the SIMPLIS project (SPJ) window.

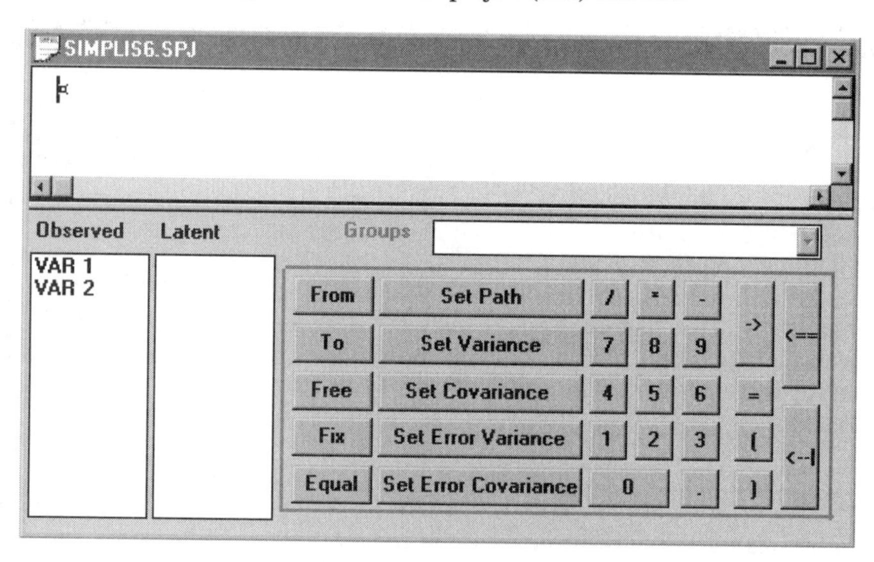

Before generating syntax, you may optionally select ***SIMPLIS Outputs*** from the ***Output*** menu. Customize this dialog box according to your preferences; for example, change the ***Number of Decimals in the Output*** option to 4. When done, click ***OK***.

The following step is to provide information regarding ***Title and Comments, Groups, Variables***, and ***Data***. Select the ***Titles and Comments*** option from the ***Setup*** menu. The first item on the ***Title and Comments*** dialog box is the title for the particular problem. After typing in the title and (optionally) the comments, click on ***Next*** to go on with the setup.

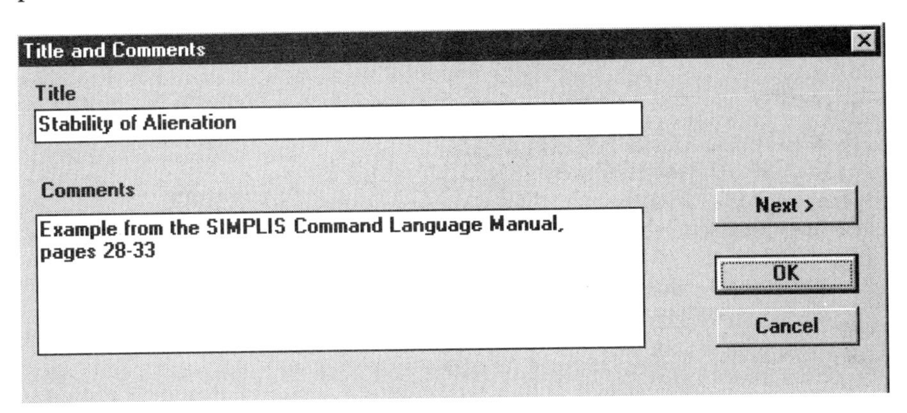

Since the present example is based on the one group only, nothing is entered in the space allowed for ***Group Labels***, and ***Next*** is clicked to go to the ***Labels*** dialog box.

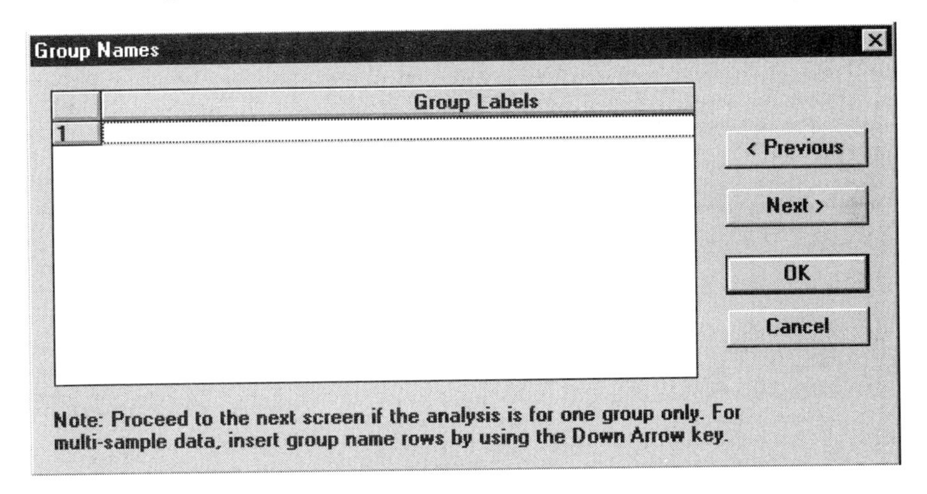

Click ***Add/Read Variables*** to obtain the ***Add/Read Variables*** dialog box. Click the ***Add list of variables*** radio button. Enter the labels *ANOMIA67*, ..., *SEI* one at a time. Click ***Add Latent Variables*** to obtain the ***Add Variables*** dialog box. Enter the labels *Alien67*, *Alien71* and *SES* one at a time to obtain the following dialog box. Finally, click ***Next*** to go to the ***Data*** dialog box.

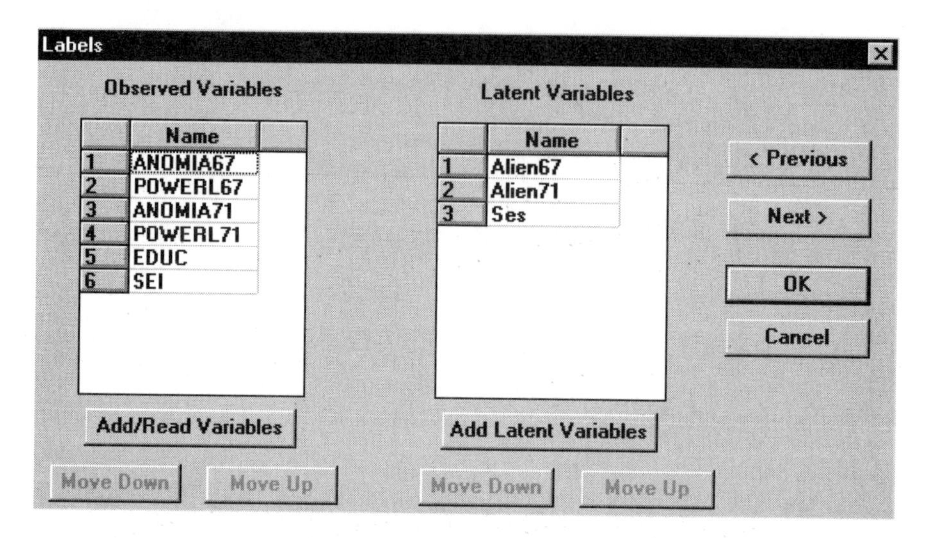

On the ***Data*** dialog box select ***Covariances*** from the ***Statistics from*:** drop-down list box. Type `932` in the ***Number of Observations*** string field. For ***File type***, select ***External ASCII data*** and use the ***Browse*** button to select **wheaton.cov**.

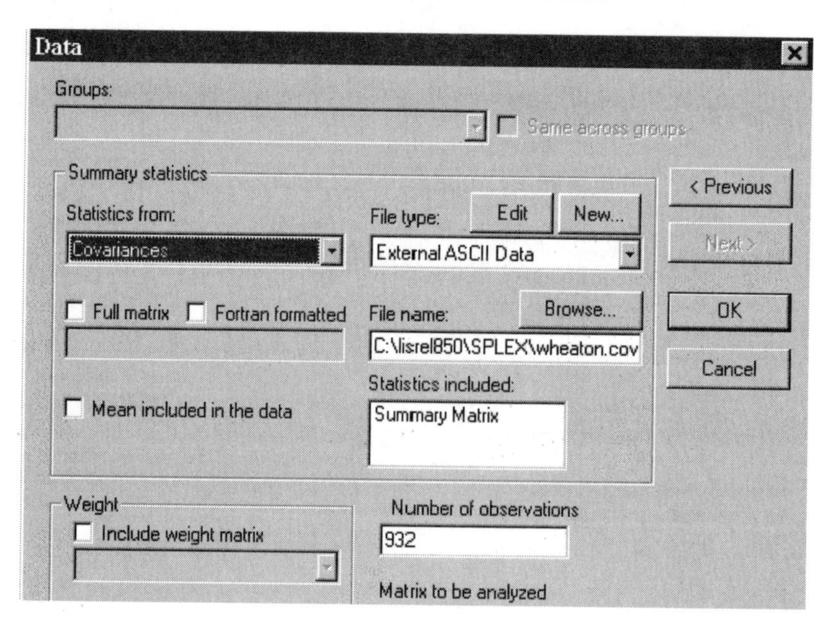

By clicking the ***Edit*** button on the ***Data*** dialog box, one may view the contents of **wheaton.cov**.

simplis6.DSF

	ANOMIA67	POWERL67	ANOMIA71
ANOMIA67	11.834	0.000	0.000
POWERL67	6.947	9.364	0.000
ANOMIA71	6.819	5.091	12.532
POWERL71	4.783	5.028	7.495
EDUC	-3.839	-3.889	-3.841

The number of decimals displayed in the spreadsheet can be changed as follows: from the main LISREL menu bar, click on the ***Edit*** menu and select the ***Format*** option.

Change the number of decimals in the ***Data Format*** dialog box to the number that should be displayed, for example, to `2`. Click ***OK***.

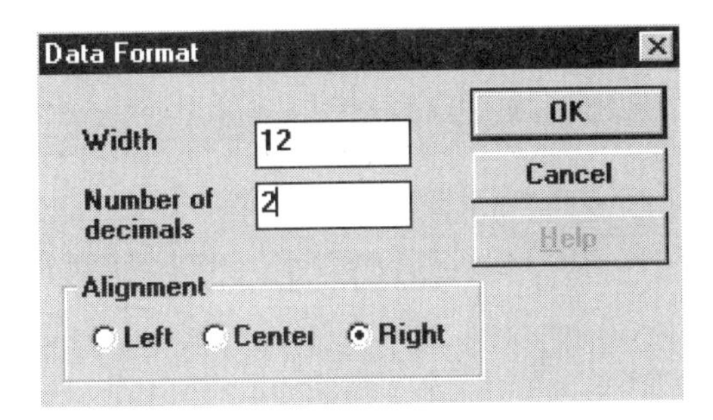

The revised spreadsheet will display the data values to the number of decimal places requested. Note that if the actual data were in a format with more decimal places, these values would be retained in the analysis, regardless of what was seen in the screen display.

Select ***Build SIMPLIS Syntax*** from the ***Setup*** menu to create the basic SIMPLIS syntax.

4.6.2 Completing the SIMPLIS Syntax

The ***Build SIMPLIS Syntax*** option on the ***Setup*** menu provides the user with a skeleton of SIMPLIS commands. There are a few commands that must be included to complete the SIMPLIS syntax. The first of these commands is:

```
ANOMIA67 POWERL67=Alien67
```

To add a line in the syntax window, move the mouse cursor to the end of the keyword `Relationships` and click once. Now move the cursor to the ***Enter*** key (**<--|**) on the keypad and click to insert a blank line below the `Relationships` line.

To enter the line

```
ANOMIA67 POWERL67 = Alien67
```

proceed as follows:

- Click on the observed variable *ANOMIA67* (first variable under the heading ***Observed***) and with the mouse button held down drag the variable name to the syntax window.
- Repeat by dragging the variable name *POWERL67* next to *ANOMIA67*. To continue, click on the "=" key and then drag the latent variable name *Alien67* (first variable under the heading ***Latent***).

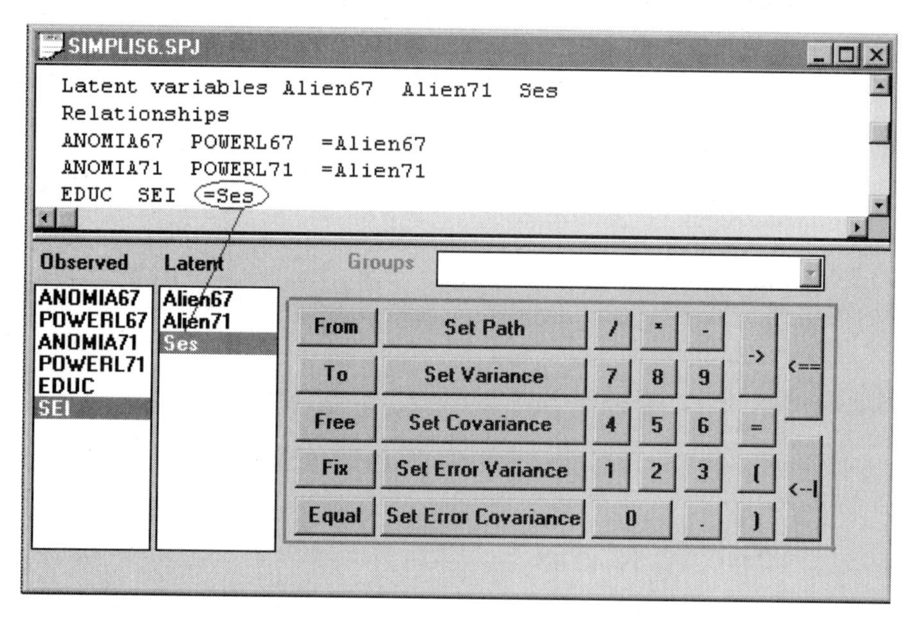

To complete, move the mouse cursor to the end of the line

```
ANOMIA67 POWERL67 = Alien67
```

and click. Then click on the ***Enter*** key and repeat the steps outlined above to insert the lines:

```
ANOMIA71 POWERL71 = Alien71
EDUC SEI = Ses
Alien67 = Ses
Alien71 = Alien67 Ses
```

Finally, enter the commands:

```
Set Error Covariance of ANOMIA67 ANOMIA71 free
```

and

```
Set Error Covariance of POWERL67 POWERL71 free
```

by using the keypad and by dragging the variable names to the appropriate positions. The completed syntax should correspond to the lines shown in the SPJ window below:

```
SIMPLIS6.SPJ
Latent Variables  Alien67 Alien71 Ses
Relationships
ANOMIA67  POWERL67  = Alien67
ANOMIA71  POWERL71   =Alien71
EDUC  SEI  =Ses
Alien67  =Ses
Alien71  =Alien67  Ses
Set Covariance of ANOMIA67 and ANOMIA71  Free
Set Covariance of POWERL67 and POWERL71  Free
```

To run SIMPLIS, click the ***Run LISREL*** icon button on the main menu bar.

A path diagram as shown below is produced, as well as the usual output file.

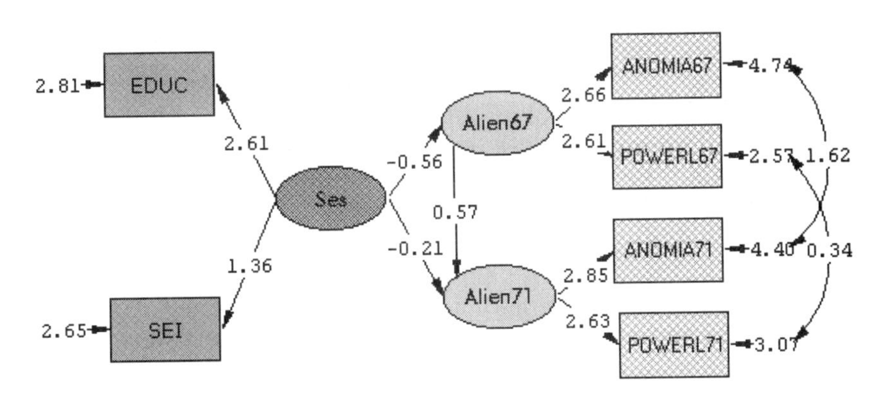

Chi-Square=4.72, df=4, P-value=0.31731, RMSEA=0.014

4.7 Using LISREL Project Files

LISREL allows the user to create LISREL syntax interactively by making use of a set of menus. As an illustration, we consider a second-order factor analysis model. This model is described in detail in the *LISREL 8: User's Reference Guide* and therefore only a brief description is given here.

Second-order factor analysis

The following equation

$$\mathbf{y} = \boldsymbol{\Lambda}_y \boldsymbol{\eta} + \boldsymbol{\varepsilon}$$

is in the form of a factor analysis model for $\mathbf{y}$ with first-order factors $\boldsymbol{\eta}$ and measurement errors $\boldsymbol{\varepsilon}$. Now suppose that the variables $\boldsymbol{\eta}$ in turn can be accounted for by a set of so called second-order factors $\boldsymbol{\xi}$, so that

$$\boldsymbol{\eta} = \boldsymbol{\Gamma}\boldsymbol{\xi} + \boldsymbol{\varsigma}$$

where $\boldsymbol{\Gamma}$ is a matrix of second-order factor loadings and $\boldsymbol{\varsigma}$ is a vector of unique components for $\boldsymbol{\eta}$.

To illustrate the model, we use data on some cognitive ability tests. The standard deviations and correlations of two forms of each of five tests are given in the table below. The sample size (*N*) is 267.

Table 4.2: Correlations and Standard Deviations for Some Cognitive Tests

GESCOM-A	2.42	1									
GESCOM-B	2.80	0.74	1								
CONWORD-A	3.40	0.33	0.42	1							
CONWORD-B	3.19	0.34	0.39	0.65	1						
HIDPAT-A	1.94	0.26	0.21	0.15	0.18	1					
HIDPAT-B	1.79	0.23	0.24	0.22	0.21	0.77	1				
THIROUND	5.63	0.15	0.12	0.14	0.11	0.17	0.20	1			
THIBLUE	3.10	0.14	0.14	0.14	0.15	0.06	0.09	0.42	1		
VOCABU_A	3.05	-.04	0.03	0.09	0.16	0.06	0.09	0.19	0.21	1	
VOCABU_B	2.25	0.02	0.02	0.10	0.23	0.024	0.07	0.09	0.21	0.72	1

We shall examine the hypothesis that the two forms of each test are tau-equivalent, except for the two-word fluency tests *Things Round* and *Things Blue* that are only assumed to be congeneric. The five true scores are postulated to depend on two factors, the first, *Speed Closure*, being measured by the first three tests and the second, *Vocabulary*, being measured by the last two tests. The model specification is:

$$\Lambda_y = \begin{pmatrix} 1 & 0 & 0 & 0 & 0 \\ 1 & 0 & 0 & 0 & 0 \\ 0 & 1 & 0 & 0 & 0 \\ 0 & 1 & 0 & 0 & 0 \\ 0 & 0 & 1 & 0 & 0 \\ 0 & 0 & 1 & 0 & 0 \\ 0 & 0 & 0 & 1 & 0 \\ 0 & 0 & 0 & * & 0 \\ 0 & 0 & 0 & 0 & 1 \\ 0 & 0 & 0 & 0 & 1 \end{pmatrix}, \quad \Gamma = \begin{pmatrix} * & 0 \\ * & 0 \\ * & 0 \\ 0 & * \\ 0 & * \end{pmatrix}, \quad \Phi = \begin{pmatrix} 1 & \\ * & 1 \end{pmatrix}$$

Here * denotes parameters to be estimated and "0" and "1" are fixed parameters. The LISREL syntax file for this model is as follows:

```
Second order factor analysis
DA NI=10 NO=267
LA FI=EX62.DAT
KM FI=EX62.DAT
SD FI=EX62.DAT
MO NY=10 NE=5 NK=2 GA=FI PH=ST PS=DI
LE
GESCOM CONWOR HIDPAT THINGS VOCABU
LK
SPEEDCLO VOCABUL
VA 1 LY 1 1 LY 2 1 LY 3 2 LY 4 2 LY 5 3 LY 6 3 LY 7 4 LY 9
5 LY 10 5
FR LY 8 4 GA 1 1 GA 2 1 GA 3 1 GA 4 2 GA 5 2
ST 1 ALL
OU SE TV SS NS
```

The goodness-of-fit statistic χ^2 is 53.06 with 33 degrees of freedom.

4.7.1 Building the LISREL Syntax

From the ***File*** menu, select ***New*** to obtain the ***New*** dialog box. Select ***LISREL Project*** and click ***OK*** to obtain the ***Save As*** dialog box.

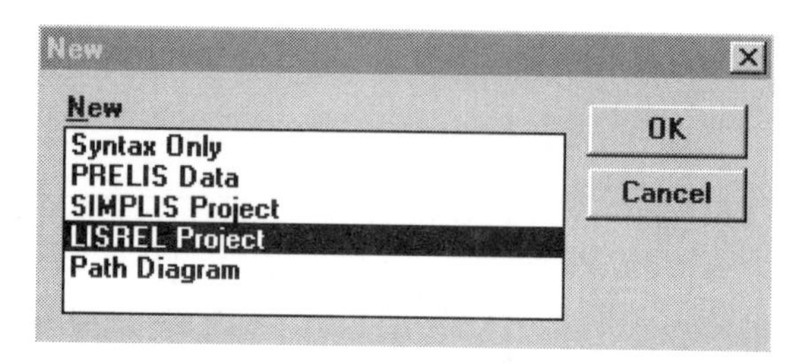

Select the **tutorial** folder and filename **sofa.lpj**. When done, click ***Save***.

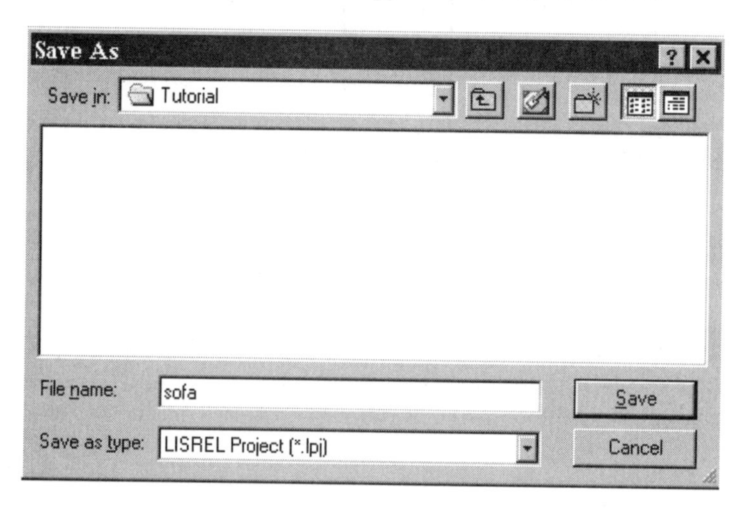

The following step is to provide information regarding the title, additional comments to be included in the syntax file, the data file and the variables in it. This is accomplished by selecting the ***Title and Comments*** option from the ***Setup*** menu. The ***Title and Comments*** dialog box appears. Enter a title and (optional) comments in the ***Title*** string field and the ***Comments*** text box.

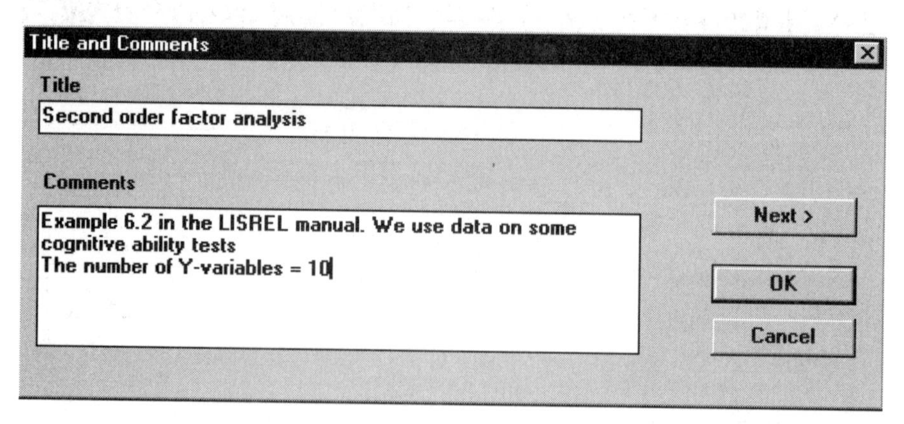

Since the present example is based on the analysis of a single group, nothing is entered in the space allowed for ***Group Labels***, and ***Next*** is clicked to go to the ***Labels*** dialog box.

The default number of variables on the ***Labels*** dialog box is 2, these being *VAR 1* and *VAR 2*. Since the present analysis calls for 10 variables, the mouse cursor is moved to the ***Observed Variables*** box. By pressing the ***Insert*** key on the computer keyboard eight times, provision is made for eight additional observed variables, and therefore the total number becomes ten. Alternatively, the ***Add List*** option may be used and *var3-var10* entered (or, for example, *X3-X10*). If the ***Add List*** option was used, click ***OK*** to return to the ***Labels*** dialog box. Variable labels can alternatively be read in from an existing ***.psf** or ***.dsf** file. See the section on LISREL system files in Chapter 7 for more information on this option.

A label may be assigned to each of the 10 variables by clicking on the corresponding button (numbered 1,2,3, ... under the ***Observed Variables*** column). The background color of the space allowed for entering the variable name will change and one can go ahead and type the appropriate label. For the present example, the labels are *GESCOM A*, *GESCOM B*, *CONWOR A*, *CONWOR B*, *HIDPAT A*, *HIDPAT B*, *THIROUND*, *THIBLUE*, *VOCABU A* and *VOCABU B* respectively. Note that a label name, which may include blanks, may not exceed 8 characters. Note also that the labels for the variables may instead be entered by, for example, clicking on *VAR 1*. Type in *GESCOM A* and use the down arrow on the keyboard to create the next space. Type *GESCOM B* and continue in a similar fashion until all the labels are typed in.

Since we have 7 latent variables, *Gescom*, *Conwor*, *Hidpat*, *Things*, *Vocabu*, *Speedclo* and *Vocabul*, the procedure outlined above for the observed variables is repeated by selecting ***Add List*** on the ***Latent Variables*** portion of the ***Labels*** dialog box. Add *var1-var7* and click on ***OK*** (alternatively the keyboard down arrow may be used). Assign labels to these latent variables by clicking on the number of each variable. Finally, click ***Next*** to go to the ***Data*** dialog box.

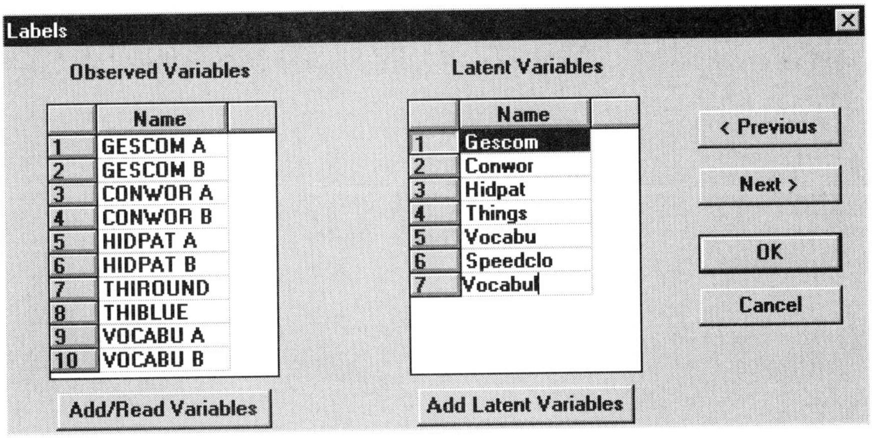

Labels
Observed Variables
Name
1 GESCOM A
2 GESCOM B
3 CONWOR A
4 CONWOR B
5 HIDPAT A
6 HIDPAT B
7 THIROUND
8 THIBLUE
9 VOCABU A
10 VOCABU B
Add/Read Variables
Latent Variables
Name
1 Gescom
2 Conwor
3 Hidpat
4 Things
5 Vocabu
6 Speedclo
7 Vocabul
Add Latent Variables
< Previous
Next >
OK
Cancel

From the ***Data*** dialog box, select ***Covariances*** from the ***Statistics from:*** drop-down list box and also ***Covariances*** from the ***Matrix to be analyzed*** drop-down list box. Type 267 in the ***Number of Observations*** string field. For ***File type***, select ***External ASCII Data*** and then click ***Browse*** to go to the ***Browse*** dialog box.

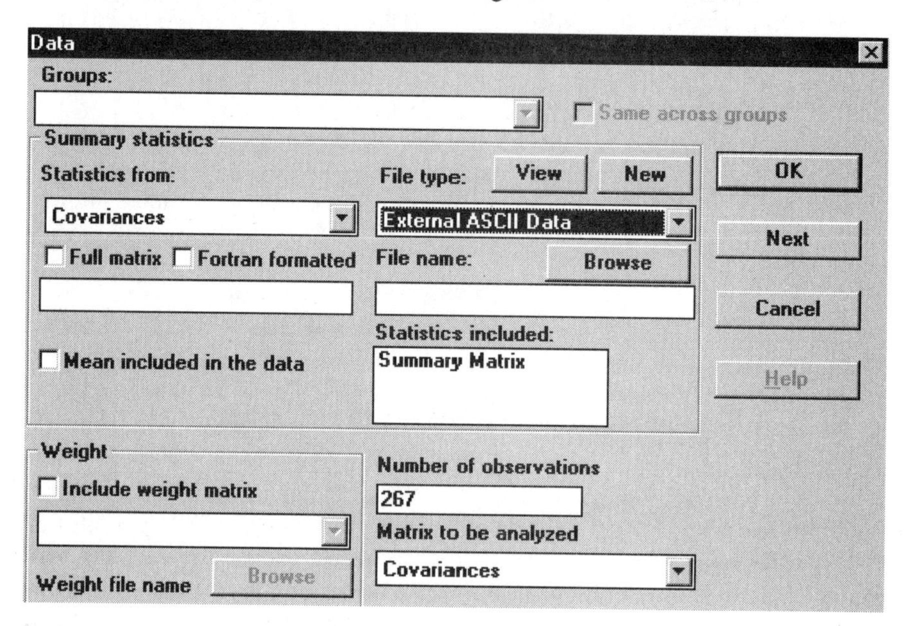

Select the ASCII file **sofa.cov** from the **tutorial** folder. Click ***Open*** to return to the ***Data*** dialog box, and then select ***Next*** to go to the ***Define Observed Variables*** dialog box.

The program will now produce the ***Define Observed Variables*** dialog box, on which the user may select ***Y-variables*** and ***X-variables*** from the list of ***Variable Names***. Since the present application calls for *Y*-variables only, all of the variable names are selected as *Y*-variables. This is achieved by sequentially clicking on each variable number box and then on the ***Select as Y*** option. Alternatively, with the left mouse button down, drag from variable number 1 to number 10, release the mouse button and click on the ***Select as Y*** button.

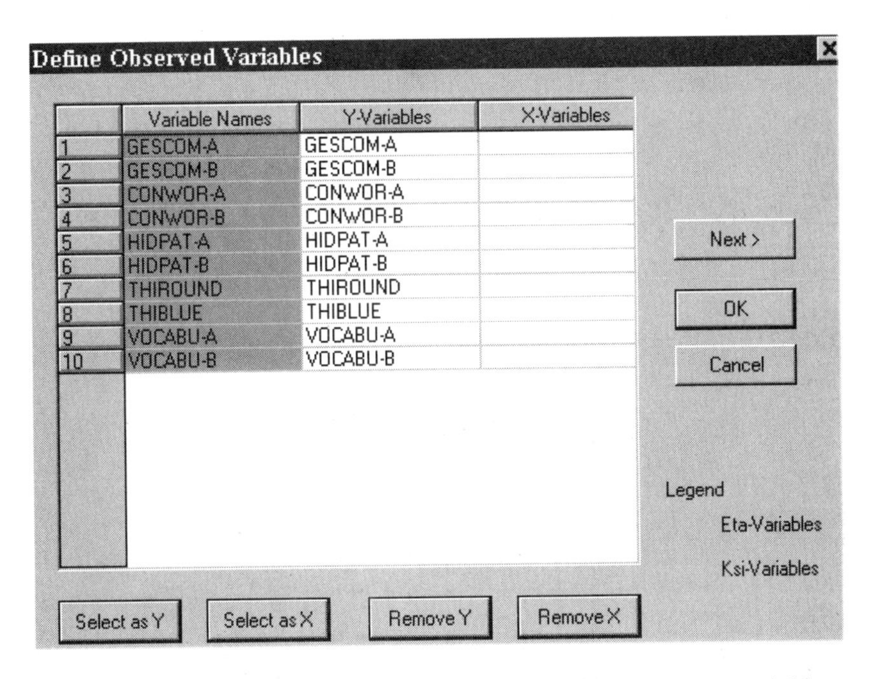

Click ***Next*** to go to the ***Define Latent Variables*** dialog box. The latent variables consist of 5 *Eta*-variables and 2 *Ksi*-variables. Click on the appropriate square under ***Variable Names*** and then choose ***Select Eta*** for the *Eta*-variables (the first 5) or ***Select Ksi*** for the *Ksi*-variables (the last 2 variables). Once this is done, click ***Next*** to go to the ***Model Parameters*** dialog box.

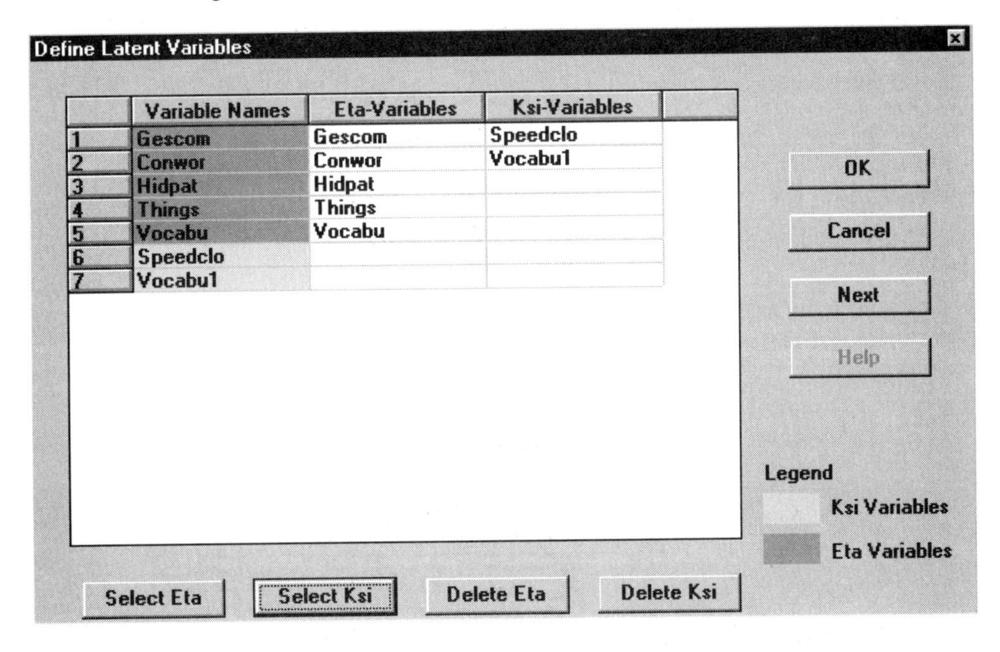

The ***Model Parameters*** dialog box shown below allows the user to:

- specify the form and mode of the matrices listed,
- specify which elements are fixed and which are free, and
- assign values to the elements of each of the matrices.

Since some changes need to be made to the default specifications for the *Lambda-Y* matrix, highlight the ***Lambda-Y Full Matrix, Fixed*** option by clicking on it. This action will cause the words `Full Matrix` and `Fixed` to appear in the ***Matrix Form*** and ***Matrix Mode*** fields respectively. Proceed by clicking ***Specify***. The default values for these fields are given in *LISREL 8: User's Reference Guide*, Table 2.1 on page 67.

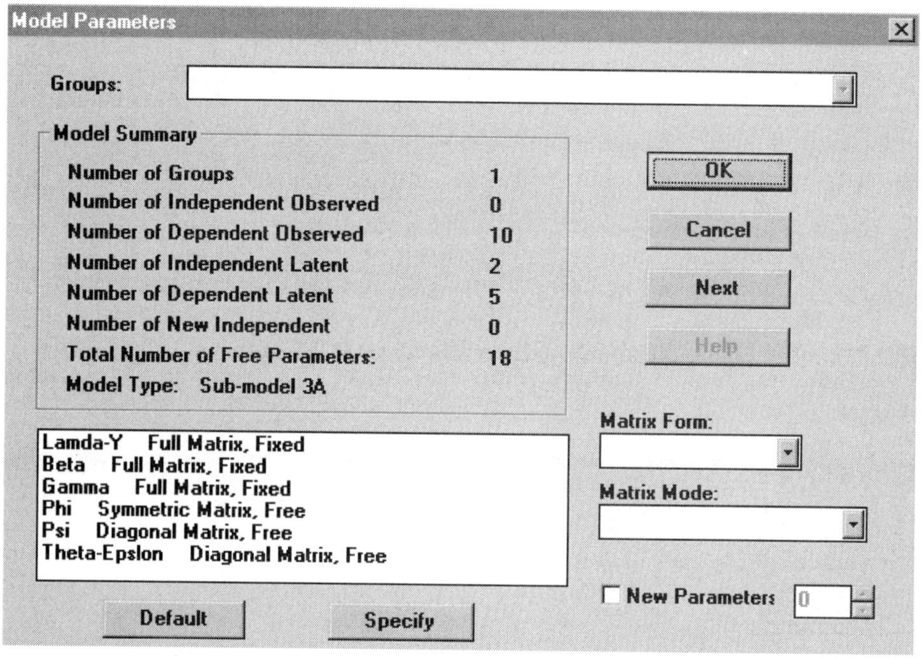

Model Parameters

Groups:

Model Summary

Number of Groups	1
Number of Independent Observed	0
Number of Dependent Observed	10
Number of Independent Latent	2
Number of Dependent Latent	5
Number of New Independent	0
Total Number of Free Parameters:	18
Model Type: Sub-model 3A	

OK

Cancel

Next

Help

Lamda-Y Full Matrix, Fixed
Beta Full Matrix, Fixed
Gamma Full Matrix, Fixed
Phi Symmetric Matrix, Free
Psi Diagonal Matrix, Free
Theta-Epslon Diagonal Matrix, Free

Matrix Form:

Matrix Mode:

New Parameters 0

Default Specify

Lamda-Y Full Matrix, Fixed
Beta Full Matrix, Fixed
Gamma Full Matrix, Fixed
Phi Symmetric Matrix, Free
Psi Diagonal Matrix, Free
Theta-Epslon Diagonal Matrix, Free

Matrix Form: Full Matrix

Matrix Mode: Fixed

New Parameters 0

Default Specify

The initial dialog box will indicate that all the elements of *Lambda-Y* are fixed and equal to zero. To obtain the pattern shown in the dialog box below, which corresponds with the statements

```
VA 1.0 LY(1,1) LY(2,1) LY(3,2) LY(4,2) LY(5,2) LY(6,3)
LY(7,4) LY(9,5) LY(10,5)
FR LY(8,4)
```

proceed as follows:

- Click on the first rectangle next to the variable description button (that is under the heading *Gescom*) and use the computer keyboard to type in the value of `1`.
- Proceed in a similar fashion to enter `1` in the second row, and finally a `1` in the rectangle corresponding to the row *VOCABU B* and column *Vocabu.*
- By clicking on the rectangle in the row *THIBLUE* and column *Things* and then on the ***Free*** button, the color of the rectangle will change to indicate that LY(8,4) is free.

Note that one may move the mouse pointer to a vertical divider line on the dialog box below. Column widths can subsequently be adjusted by clicking the left button and dragging the vertical line to the left or the right of its present position.

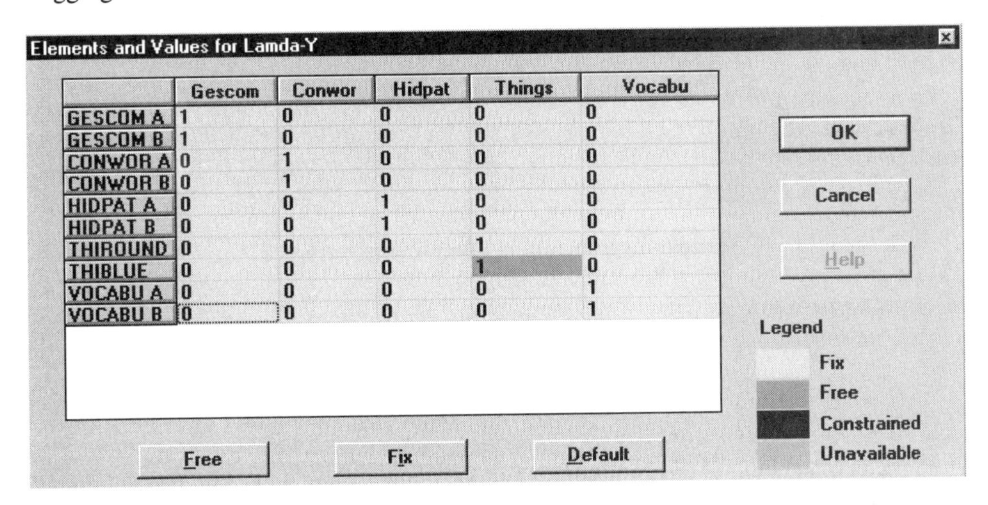

The input specifications require *Beta* to be a full matrix with all elements fixed at a value of 0. Since this is the default, one can proceed to specify the pattern and values for the *Gamma* matrix.

Lamda-Y Full Matrix, Fixed
Beta Full Matrix, Fixed
Gamma Full Matrix, Fixed
Phi Symmetric Matrix, Free
Psi Diagonal Matrix, Free
Theta-Epslon Diagonal Matrix, Free

Full Matrix
Matrix Mode:
Fixed
New Parameters 0
Default Specify

Highlight the ***Gamma Full Matrix, Fixed*** option and click on ***Specify*** to specify the following statements

```
FR GA(1,1) GA(2,1) GA(3,1) GA(4,2) GA(5,2)
VA 1.0 GA(1,1) GA(2,1) GA(3,1) GA(4,2) GA(5,2)
```

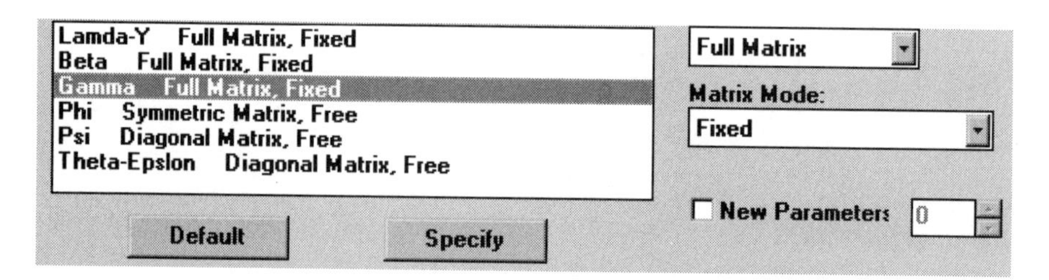

This is accomplished by clicking on the rectangle that corresponds to the *Eta*-variable *Gescom* and the *Ksi*-variable *Speedclo*. Use the computer keyboard to type in 1 and click again to activate this rectangle. Then select the ***Free*** button to change the status of the GA(1,1) parameter to free. Continue in a similar fashion to obtain the pattern shown below.

Elements and Values for Gamma

	Speedclo	Vocabul
Gescom	1	0
Conwor	1	0
Hidpat	1	0
Things	0	1
Vocabu	0	1

Lamda-Y Full Matrix, Fixed
Beta Full Matrix, Fixed
Gamma Full Matrix, Fixed
Phi Std-Symmetric
Psi Diagonal Matrix, Free
Theta-Epslon Diagonal Matrix, Free

Std-Symmetric
Matrix Mode:
New Parameters 0
Default Specify

Since the default settings for the parameter matrices *Psi* and *Theta-Epsilon* are correctly set at the default values (see dialog box above), click on ***Next*** to go to the ***Constraints*** dialog box.

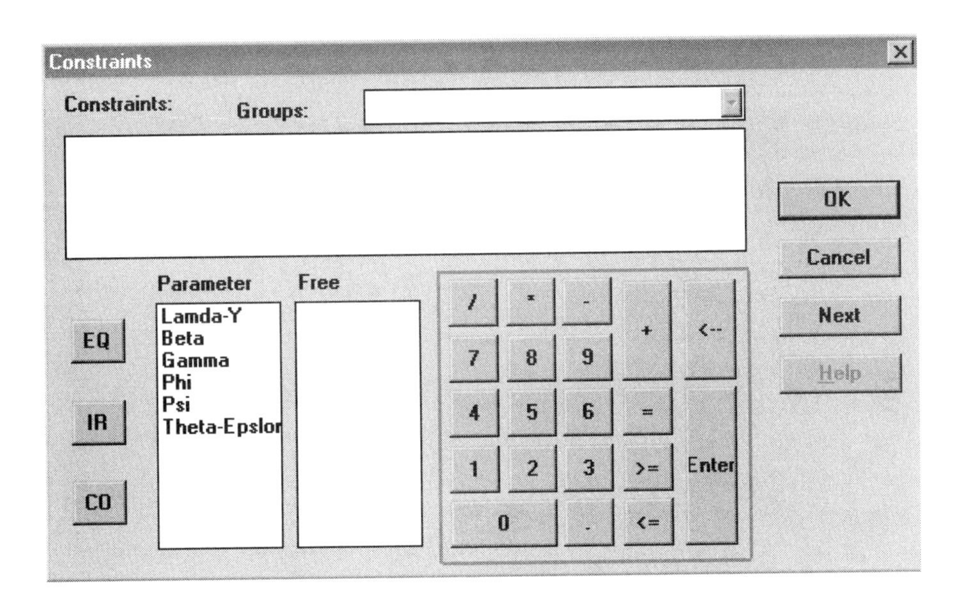

In the present example, no constraints are imposed on the elements of the parameter matrices and therefore ***Next*** is clicked to go to the ***Output Selections*** dialog box.

The ***Output Selections*** dialog box enables the user to create an output file that contains selected parts of the LISREL printout. Make the appropriate selections by clicking on the boxes at the left of the different options. Once this is done, click ***Next*** to go to the final dialog box, which is the ***Save Matrices*** dialog box.

Any of the matrices or statistics shown on the dialog box illustration below may be selected by clicking on the boxes. As an example, we selected ***Lambda-Y***, ***Gamma*** and ***t-values***. These files will have the same filename as the LISREL project file but with different file extensions, these being *.**lys**, *.**gas**, and *.**tvs** respectively.

Selections

Selected Printout

☑ Correlation Matrix of Parameter Estimators
☑ Residuals,Standardized Residuals, Q-plot and Fitted Covariance
☐ Total Effects and Indirect Effects
☐ Factor-scores Regression
☑ Standardized Solution
☐ Completely Standardized Solution
☐ Technical Output
☐ Miscellaneous Results (see Sec 1.11)
☑ Excluding Modification Indices
☐ Print All

Number of Decimals (0-8) in the Printed Output 4

☑ Invoke Path Diagram ☐ Wide Print

OK
Cancel
Next
Help

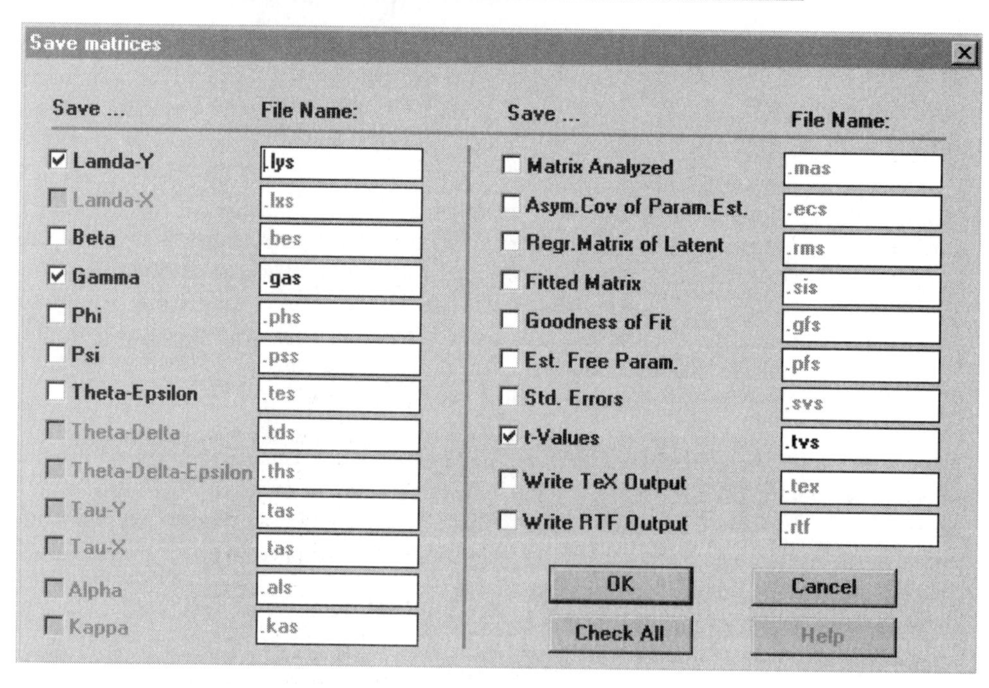

Click ***OK*** to return to the LISREL project (LPJ) window, and select ***Build LISREL syntax*** from the ***Setup*** menu to generate the syntax file.

The LISREL syntax, saved in the file **sofa.lpj**, is shown below:

```
Example 6.2 in the LISREL manual
We use data on some cognitive ability tests
The number of Y-variables = 10
Second order factor analysis
DA NI=10 NO=267 NG=1 MA=KM
LA
'GESCOM A' 'GESCOM B' 'CONWOR A' 'CONWOR B' 'HIDPAT A'
THIROUND THIBLUE 'VOCABU A' 'VOCABU B'
CN FI=SOFA.COV
SE
1 2 3 4 5 6 7 8 9 10/
MO NY=10 NK=2 NE=5 LY=FU,FI BE=FU,FI GA=FU,FI PH=SY,FR
PS=DI,FR
TE=DI,FR
LE
Gescom Conwor Hidpat Things Vocabu
LK
Speedclo Vocabul
FR LY(8,4) GA(1,1) GA(2,1) GA(3,1) GA(4,2) GA(5,2)
VA 1.00 LY(1,1) LY(2,1) LY(3,2) LY(4,2) LY(5,3) LY(6,3)
PD
OU ME=ML PC RS SS XM IT=250 LY=.lys GA=.gas TV=.tvs
```

Click the ***Run LISREL*** icon button on the main menu bar to run the problem. The path diagram for this example is shown below and the output is saved to the file **sofa.out**.

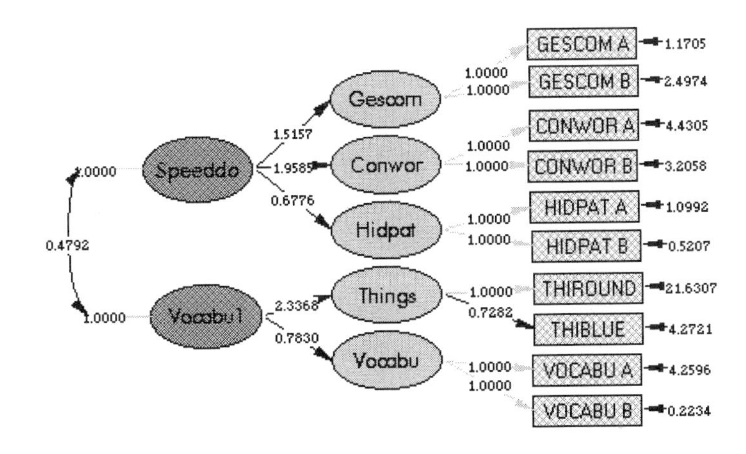

Chi-Square=53.67, df=33, P-value=0.01294, RMSEA=0.049

The output can be converted to rich text format (*.**rtf**), a *.**tex** file or a web document (*.**htm**) by selecting one of these options from the ***File*** menu. One may also modify the path diagram by deleting, fixing or adding paths.

4.8 Analysis with Missing Data

4.8.1 Single-Group Analysis with Missing Data using LISREL Syntax

Data on attitude scales were collected from 932 persons in two rural regions in Illinois at three points in time: 1966, 1967 and 1971. The variables used for the present example are the Anomia subscale and the Powerlessness subscale, taken to be indicators of Alienation. This example uses data from 1967 and 1971 only.

The background variables are the respondent's education (years of schooling completed) and Duncan's Socioeconomic Index (*SEI*). These are taken to be indicators of the respondent's socioeconomic status (*SES*). The sample covariance matrix of the six observed variables is given in Table 6.5 of the *LISREL 8: User's Reference Guide* (see pages 213-223).

Observed Variables:

Anomia 67
Powerlessness 67
Anomia 71
Powerlessness 71
Education
SEI

Latent Variables:

SES
Alienation 67
Alienation 71

The model shown in the path diagram below was fitted by running **ex64d.ls8** (in the **lis8ex** folder).

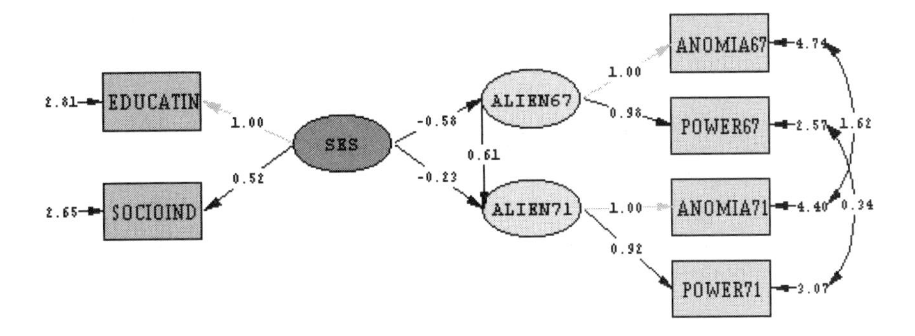

By adding `SI=filename` to the `OU` line of the syntax file, the fitted covariance matrix may be saved to an external file.

Subsequently, a data set of size 1500, with 15% of the values missing at random, was simulated by regarding the fitted covariance matrix as the true population covariance matrix. This data set is stored as the text file **wmas.dat** in the **missingex** folder. A corresponding *.**psf** file, containing the variable names, missing value code, number of records and the data is stored as **wmas.psf** in the **tutorial** folder. Save the path diagram in the **tutorial** folder. In the dialog box below, we used the name **sgroup.pth**.

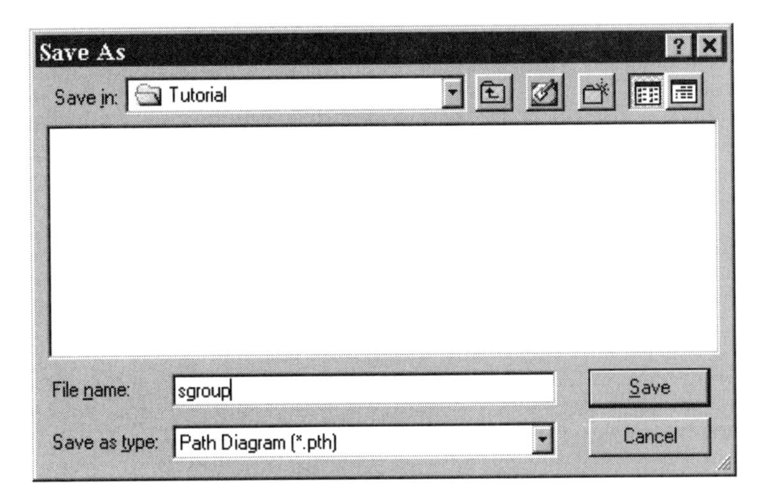

Click ***Save*** when done and select the ***Title and Options*** option from the ***Setup*** menu. Enter a title and optional comments. Click ***Next*** to proceed to the ***Group Names*** dialog box and again click ***Next*** to proceed to the ***Labels*** dialog box.

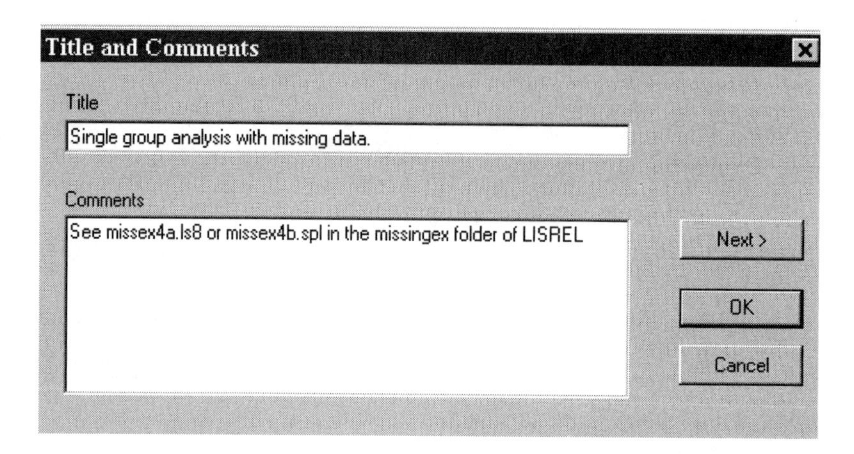

Click the ***Add/Read Variables*** button and on the ***Add/Read Variables*** dialog box select `Prelis System file` from the ***Read from file*** drop-down list. Use the ***Browse*** button to locate **wmas.psf** in the **tutorial** folder.

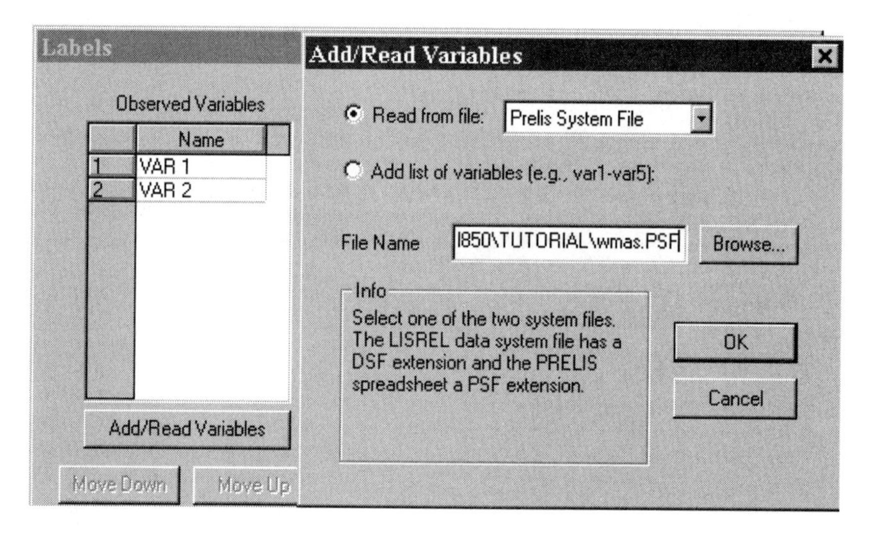

Click ***OK*** when done and add the latent variable names as shown below. By default, LISREL 8.50 for Windows uses the file selected for reading in variable names as the data file. It is therefore not necessary to proceed to the ***Data*** menu. Click ***OK*** to return to the path diagram window.

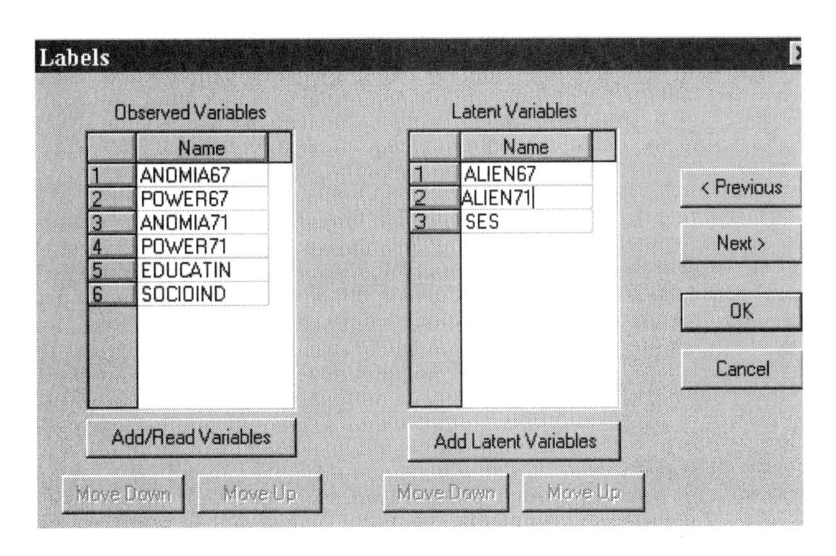

Mark the first four variables as *Y*-variables and the first two latent variables as *Eta*-variables. (Left) click on each variable name, and with the mouse button down, drag the variable to the path diagram (PTH) window.

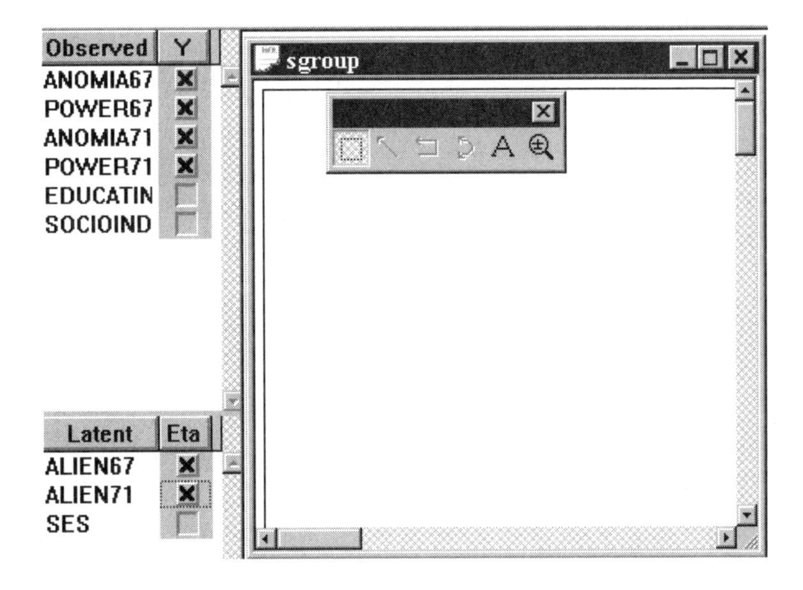

Start with the *Y*-variables first, then continue with the *Eta*-variables, then the *Ksi*-variables and finally the *X*-variables. In other words, start with *ANOMIA67* and end with *SOCIOIND.*

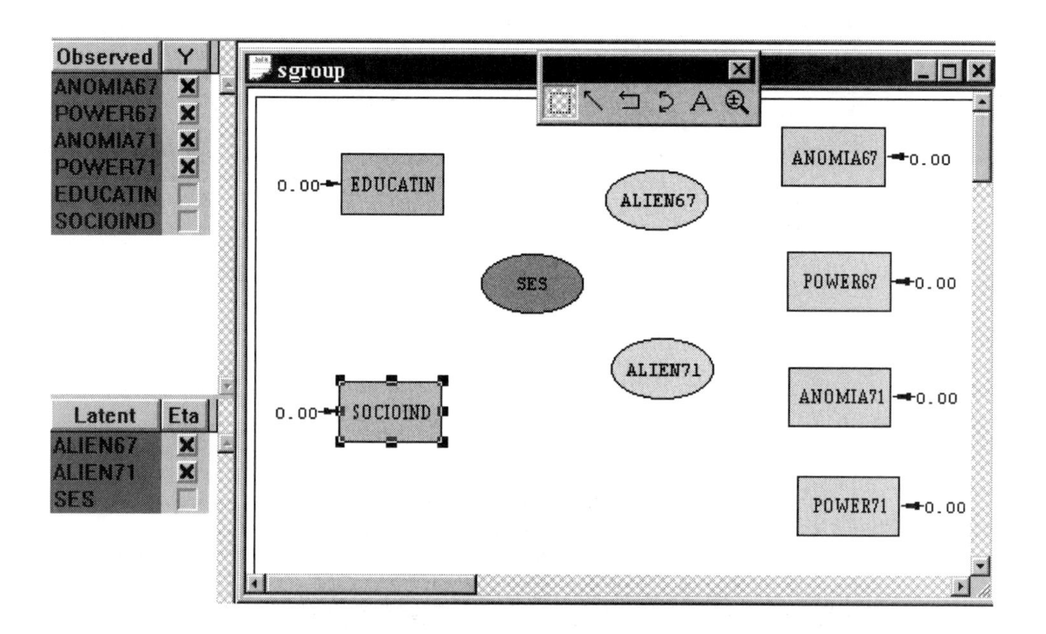

Click on the single-headed arrow icon button on the ***Draw*** toolbar and add all the paths as shown below. Start by moving the mouse cursor to the center of *ALIEN67*, and, with the mouse button down, drag the arrow to the center of *ANOMIA67* and release the button.

Once all the single paths have been drawn, click on the double-headed arrow icon button. Move the cursor to the center of the error variance arrow of *ANOMIA67*, then, with the left mouse button held down, drag to the center of the error variance arrow of *ANOMIA71*. Repeat this procedure to allow for a covariance between *POWER67* and *POWER71*.

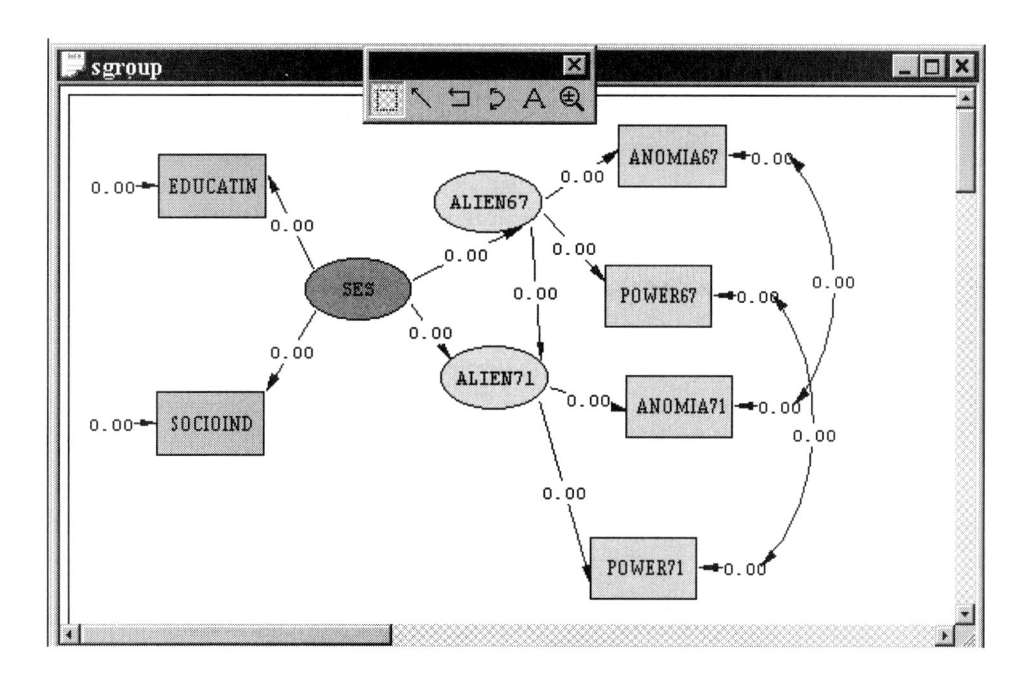

Select ***Build SIMPLIS syntax*** from the ***Setup*** menu to produce the **sgroup.spj** file. This file is displayed in the SIMPLIS project (SPJ) window shown below.

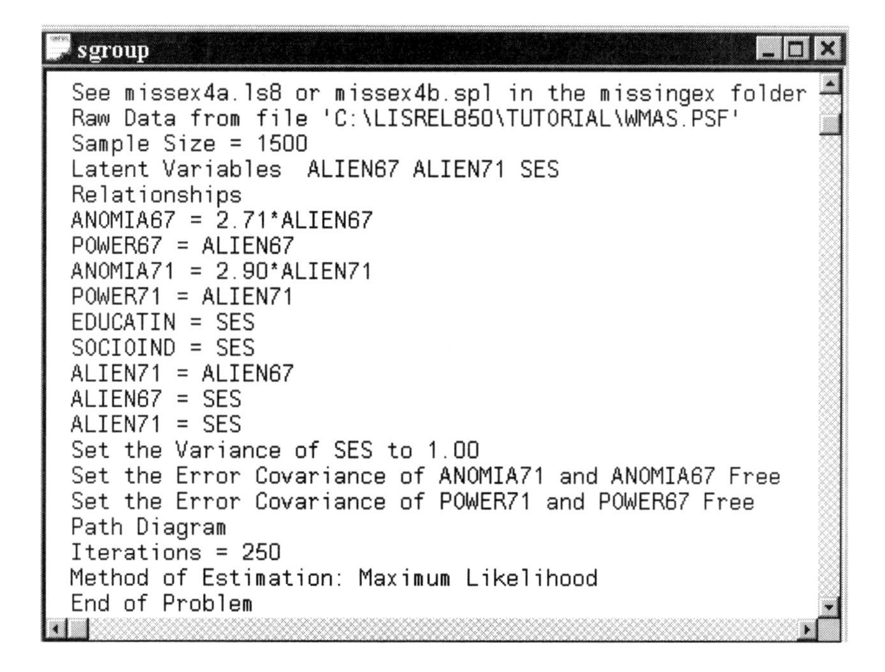

```
sgroup
See missex4a.ls8 or missex4b.spl in the missingex folder
Raw Data from file 'C:\LISREL850\TUTORIAL\WMAS.PSF'
Sample Size = 1500
Latent Variables  ALIEN67 ALIEN71 SES
Relationships
ANOMIA67 = 2.71*ALIEN67
POWER67 = ALIEN67
ANOMIA71 = 2.90*ALIEN71
POWER71 = ALIEN71
EDUCATIN = SES
SOCIOIND = SES
ALIEN71 = ALIEN67
ALIEN67 = SES
ALIEN71 = SES
Set the Variance of SES to 1.00
Set the Error Covariance of ANOMIA71 and ANOMIA67 Free
Set the Error Covariance of POWER71 and POWER67 Free
Path Diagram
Iterations = 250
Method of Estimation: Maximum Likelihood
End of Problem
```

The syntax files **missex4a.ls8** and **missex4b.spl** in the **missingex** folder contain the corresponding LISREL and SIMPLIS syntax for fitting the model. A FIML analysis is carried out if the `DA` line in the LISREL syntax contains the keyword `MI=<value>`. Missing values for this analysis are indicated by a value of `-9`.

```
Stability of Alienation, Model D (Correlated Errors for
ANOMIA67 and ANOMIA71 and for POWERL67 and POWERL71)
DA NI=6 NO=1500 MI=-9
LA
ANOMIA67 POWER67 ANOMIA71 POWER71 EDUCATIN SOCIOIND
RA FI=WMAS.DAT
MO NY=4 NX=2 NE=2 NK=1 BE=SD PS=DI TE=SY
LE
ALIEN67 ALIEN71
LK
SES
FR LY(2,1) LY(4,2) LX(2,1) TE(3,1) TE(4,2)
VA 1 LY(1,1) LY(3,2) LX(1,1)
PD
OU SE TV MI ND=3 PT
```

Click the ***Run LISREL*** icon button to obtain a path diagram and output file. Once the data are read, estimates are obtained of the means and covariances and a $-2\ln L$ value is printed. This value is the $-2\ln L$ value for the unrestricted model.

Portions of the output file are shown below:

(i) –2*Log(likelihood) and percentage values missing

The EM-algorithm is used to obtain estimates of the population means and covariances. LISREL uses these values to obtain starting values for the maximum likelihood procedure. Convergence is attained in 8 iterations, and at convergence $-2\ln L$ equals 35822.65. This value is also referred to as the $-2\ln L$ value for the saturated model.

```
              --------------------------------
               EM Algorithm for missing Data:
              --------------------------------

          Number of different missing-value patterns = 47
          Convergence of EM-algorithm in     8 iterations
          -2 Ln(L) =     35822.64668
          Percentage missing values=  15.32

          Sample Size =  1500
```

(ii) Estimated covariance matrix (EM algorithm)

	ANOMIA67	POWER67	ANOMIA71	POWER71	EDUCATIN	SOCIOIND
ANOMIA67	11.84					
POWER67	6.75	9.51				
ANOMIA71	6.57	4.94	11.99			
POWER71	4.37	5.07	7.29	10.05		
EDUCATIN	-3.79	-3.74	-3.40	-3.47	8.96	
SOCIOIND	-2.02	-1.83	-1.59	-1.66	3.21	4.20

(iii) LISREL Estimates (Maximum Likelihood)

```
 Measurement Equations

  ANOMIA67 = 2.71*ALIEN67, Errorvar.= 5.22  , R**2 = 0.56
                                      (0.42)
                                      12.54

  POWER67 = 2.77*ALIEN67, Errorvar.= 2.63  , R**2 = 0.72
            (0.17)                   (0.39)
            16.54                    6.67

  ANOMIA71 = 2.90*ALIEN71, Errorvar.= 4.43  , R**2 = 0.63
                                      (0.48)
                                      9.21

  POWER71 = 2.80*ALIEN71, Errorvar.= 2.98  , R**2 = 0.70
            (0.18)                   (0.44)
            15.72                    6.79

  EDUCATIN = 2.55*SES, Errorvar.= 2.44  , R**2 = 0.73
          (0.11)                (0.47)
          22.76                 5.21

  SOCIOIND = 1.26*SES, Errorvar.= 2.62  , R**2 = 0.38
          (0.068)               (0.15)
          18.57                 17.18
```

(iv) Global Goodness of Fit Statistics, Missing Data Case

```
     -2ln(L) for the saturated model =        35822.647
     -2ln(L) for the fitted model    =        35828.671

                      Degrees of Freedom = 4
     Full Information ML Chi-Square  = 6.02 (P = 0.20)
 Root Mean Square Error of Approximation (RMSEA) = 0.018
```

4.8.2 Multiple-groups with Missing Data

Sörbom (1981) reanalyzed some data from the Head Start summer program previously analyzed by Magidson (1977). Sörbom used data on 303 white children consisting of a Head Start sample ($N = 148$) and a matched Control sample ($N = 155$). The correlations, standard deviations and means are given in Table 2.9 of the *LISREL 8: The SIMPLIS Command Language* (pages 79-84). The children were matched on gender and kindergarten attendance, but no attempt had been made to match on social status variables.

The variables used in Sörbom's re-analysis were:

X_1 = Mother's education (*MOTHEDUC*)
X_2 = Father's education (*FATHEDUC*)
X_3 = Father's occupation (*FATHOCCU*)
X_4 = Family income (*FAMILINC*)
Y_1 = Score on the Metropolitan Readiness Test (*MRT*)
Y_2 = Score on the Illinois Test of Psycholinguistic Abilities (*ITPA*)

The following issues were examined:

- Test whether X_1, X_2, X_3 and X_4 can be regarded as indicators of a single construct *Ses* (socioeconomic status) for both groups. Is the measurement model the same for both groups? Is there a difference in the mean of *Ses* between groups?
- Assuming that Y_1 and Y_2 can be used as indicators of another construct *Ability* (cognitive ability), test whether the same measurement model applies to both groups. Test the hypothesis of no difference in the mean of *Ability* between groups.
- Estimate the structural equation:

$$Ability = \alpha + \gamma Ses + z$$

- Is γ the same for the two groups? Test the hypothesis $\alpha = 0$ and interpret the results.

A conceptual path diagram for the model fitted to the data is shown below.

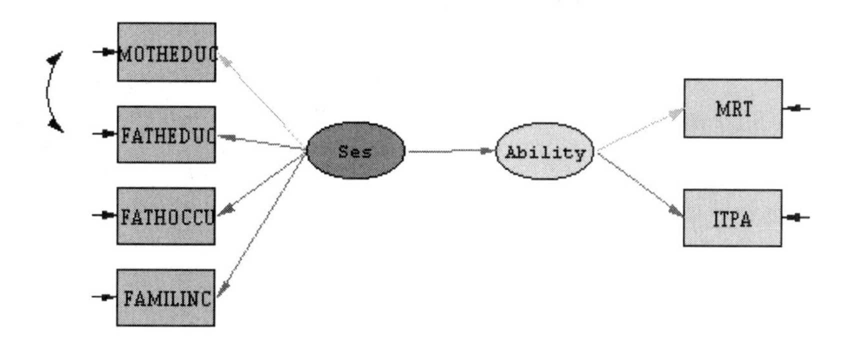

The fitted covariance matrices obtained from **ex16d.spl** (in the **splex** folder) were used to simulate a control group data set (sample size 550 and percentage missing 15%) and to simulate a Head Start data set (sample size of 600 and percentage missing 10%).

Note:

To invoke the FIML procedure for the analysis of missing data, the following three statements must be given in the SIMPLIS syntax in the order shown below.

```
Missing Value Code <value>
Sample Size: <nsize>
Raw data from file <filename>
```

The complete SIMPLIS syntax file is shown below.

```
FIML: Example 1: SIMPLIS syntax
Group = Control
Observed Variables: MOTHEDUC FATHEDUC FATHOCCU FAMILINC MRT ITPA
Missing Value Code -9
Sample Size: 550
Raw Data from File CONTROL.DAT
Latent Variables: Ses Ability
Relationships:
   MOTHEDUC            = CONST + 1*Ses
   FATHEDUC - FAMILINC = CONST +   Ses
   MRT                 = CONST + 1*Ability
   ITPA                = CONST +   Ability
   Ability = Ses
Let the Errors of MOTHEDUC and FATHEDUC correlate

Group = Head Start
Missing Value Code -9
Sample Size: 600
Raw Data from File EXPERIM.DAT
```

```
Relationships:
   Ses     = CONST
   Ability = CONST + Ses
Set the Error Variances of MOTHEDUC - ITPA free
Set the Variance of Ses free
Set the Error Variance of Ability free
Let the Errors of MOTHEDUC and FATHEDUC correlate
LISREL Output: ND=3
Path Diagram
End of Problem
```

If the data are stored in a PRELIS system file (*.**psf**), the `Observed Variables`, `Missing Value Code`, `Sample Size` and `Raw data from file control.dat` can be replaced with the statement

```
Raw data from file control.psf
```

If a *.**psf** file contains missing data, the user should ensure that a global missing value code is assigned. This can be done by using the ***Data***, ***Define Variables*** option.

Subsequently, we illustrate how to build the SIMPLIS (or LISREL) syntax by drawing a path diagram.

Select the ***New*** option from the ***File*** menu. On the ***New*** dialog box, select ***Path Diagram*** and save the path diagram in the **tutorial** folder as **mgroup.pth**. These steps are described in detail in earlier sections of this chapter. From the ***Setup*** menu, select ***Title and Comments*** and provide a title and any optional comments. When done, click ***Next*** to go to the ***Group Names*** dialog box.

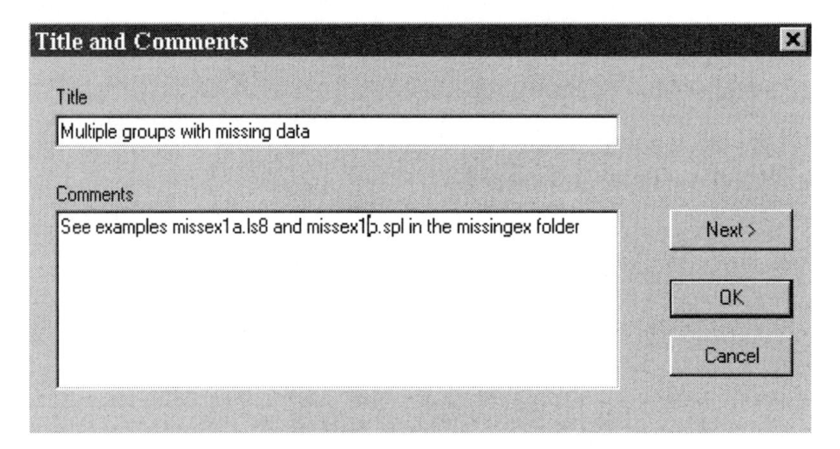

Use the instructions on the bottom of the ***Group Names*** dialog box to enter group names, then click ***Next*** to go to the ***Labels*** dialog box.

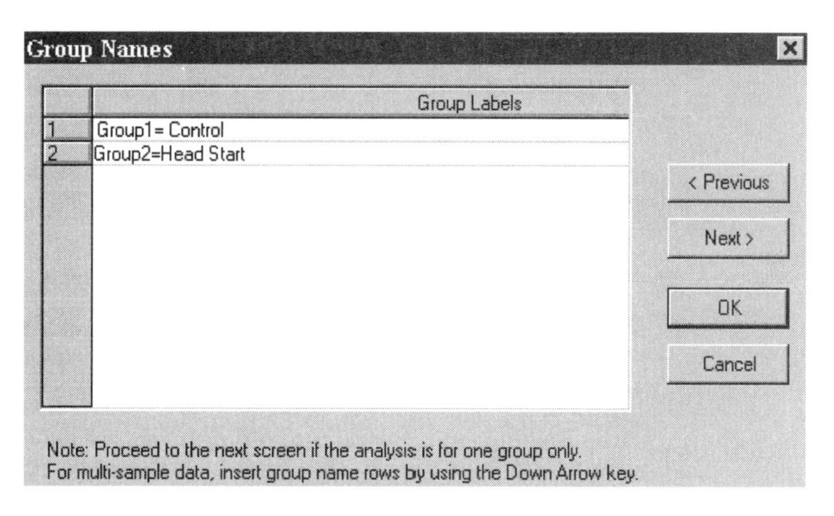

The files **control.psf** and **experim.psf** containing the Group1 and Group2 data respectively, are stored in the **tutorial** folder. Click the ***Add/Read Variables*** button, select ***Prelis System File*** from the ***Add/Read Variables*** dialog box and use the ***Browse*** button to locate the file **control.psf** in the **tutorial** folder. Click ***OK*** when done to return to the ***Labels*** dialog box. Use the ***Add Latent Variables*** button to insert the names *Ses* and *Ability*.

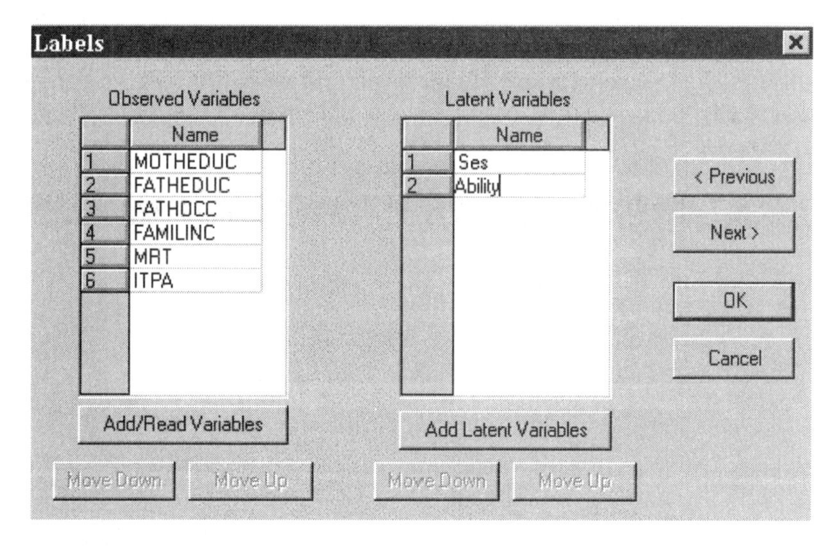

Click the ***Next*** button to go to the ***Data*** dialog box. Since a ***.psf** file was selected in the previous step, the ***Data*** dialog box shows the ***File type:*** as `PRELIS System Data` and the filename as **control.psf**. Use the ***Groups:*** drop-down list box to select `Group2=Head Start`. Once this is done, use the ***Browse*** button to locate the file **experim.psf**. Click ***Open*** to return to the ***Data*** dialog box.

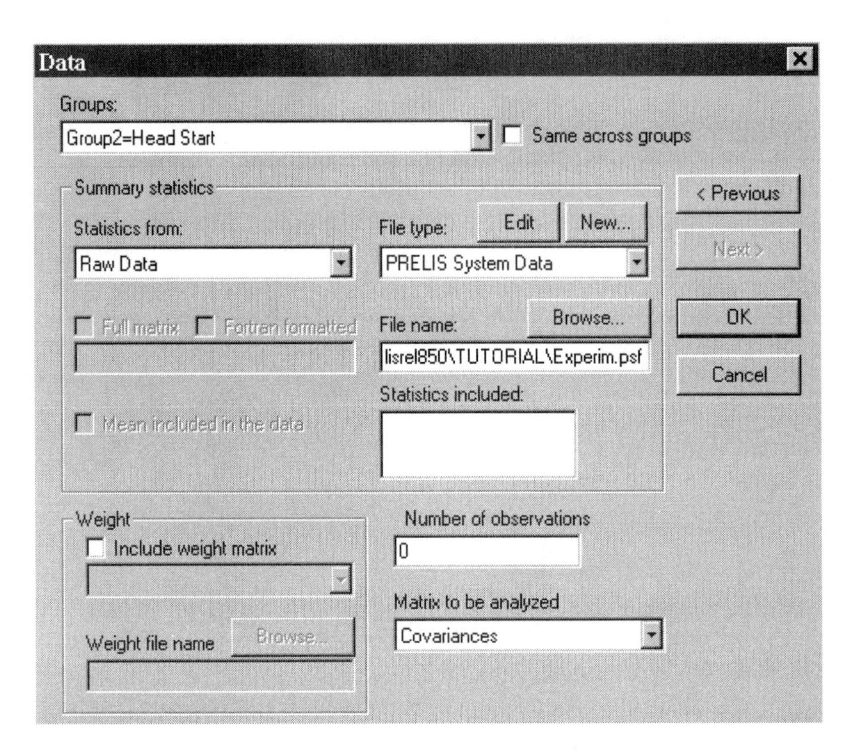

Click ***OK*** when done. Select the variables *MRT* and *ITPA* as *Y*-variables and *Ability* as an *Eta*-variable. Drag variable names to the path diagram screen in the order *MRT*, *ITPA*, *Ability*, *Ses*, *MOTHEDUC*, *FATHEDUC*, *FATHOCC*, and *FAMILINC*. Use the ***Draw*** toolbar to draw the paths as shown below.

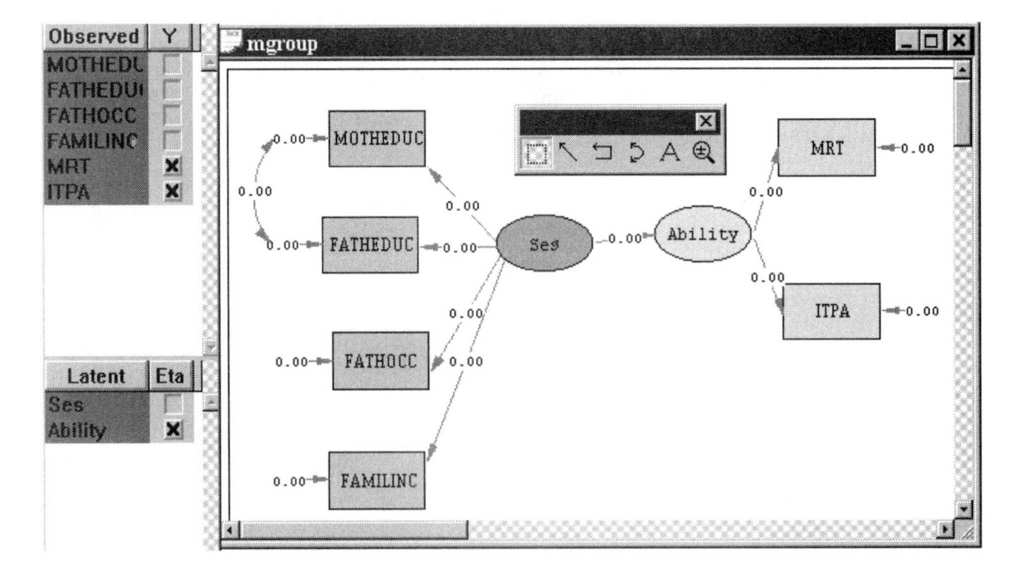

A more detailed description of this process is given in preceding sections of this chapter.

When the path diagram is completed, use the ***Build SIMPLIS syntax*** option from the ***Setup*** menu to create the SIMPLIS syntax file which is displayed in the SIMPLIS project window, as shown below.

```
mgroup
Set the Variance of Ses to 1.00
Set the Error Covariance of FATHEDUC and MOTHEDUC Free
Path Diagram
Iterations = 250
Method of Estimation: Maximum Likelihood

Group2=Head Start
Raw Data from file 'C:\LISREL850\TUTORIAL\EXPERIM.PSF'
Sample Size = 600
Latent Variables  Ability Ses
Relationships
End of Problem
```

Note that no relationships are given for Group2 under the `Relationships` keyword. This implies that all parameters are constrained to be equal across groups.

Click the ***Run LISREL*** icon to create an output file and the path diagram shown below.

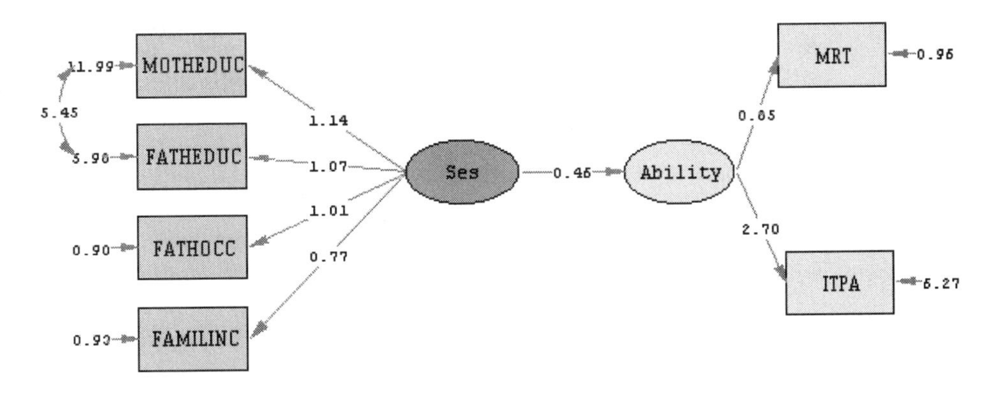

Chi-Square=72.69, df=27, P-value=0.00000, RMSEA=0.054

Since it is not realistic to assume equal error variances across groups, we can set each error variance free by moving the mouse to the appropriate arrow (*e.g. MRT* ← 0.96). Right click to obtain the menu below and select the ***Free*** option.

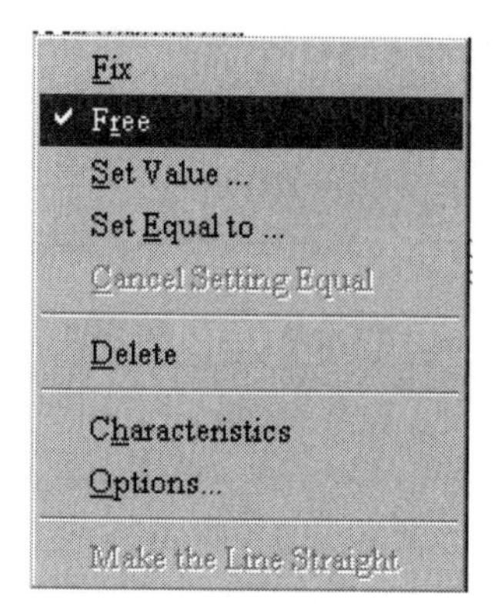

Proceed in a similar way to free the error variances of *ITPA*, *MOTHEDUC*, *FATHEDUC*, *FATHOCC* and *FAMILINC* as well as the covariance (5.45) between *MOTHEDUC* and *FATHEDUC*. Also right click on the arrow between *Ses* and *Ability* to free that path. When this is done, rebuild the SIMPLIS syntax using the ***Build SIMPLIS syntax*** option on the ***File*** menu.

A portion of the SIMPLIS project file showing the resultant relationships for group 2 is shown below.

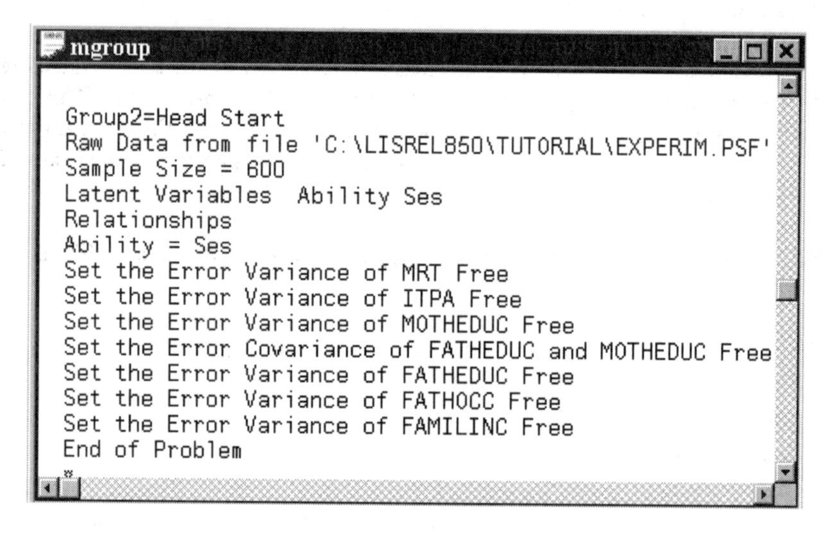

Click the ***Run LISREL*** icon button to produce the path diagram shown below.

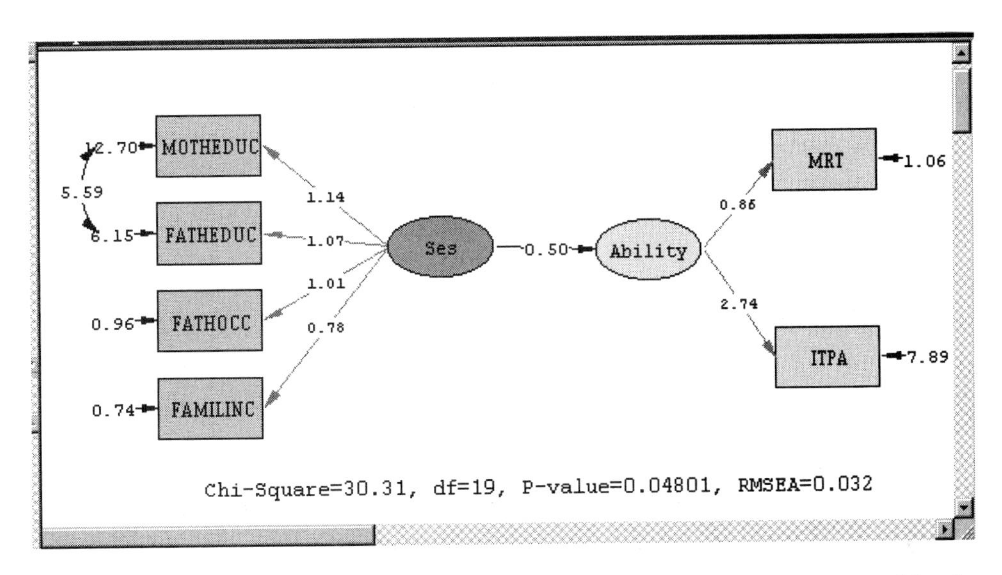

When the missing value code, sample size and raw data file information are read in, the EM algorithm for estimating the means and covariances under the unrestricted model is started. From this portion of the output a percentage of 14.61% missing values is reported for group 1 and 9.92% for group 2.

Note that the estimated means and covariances are used to obtain starting values for the FIML procedure. In addition, a $-2\ln L$ value is reported for each group. This value is minus two times the log likelihood value obtained when no restrictions are imposed on means and covariance matrices. From the output it follows that the sum of the $-2\ln L$ values for the groups equals 11358.595 + 12716.065 = 24074.660.

The FIML procedure converged in 6 iterations. Portions of the output file are given below.

(i) LISREL Estimates (Maximum Likelihood) for the control group

```
Number of Iterations =  6

 Measurement Equations


 MRT = 0.86*Ability, Errorvar.= 1.06  , R**2 = 0.20
      (0.11)                   (0.084)
       7.78                     12.58


 ITPA = 2.74*Ability, Errorvar.= 7.89 , R**2 = 0.26
                                (0.69)
                                 11.41
```

```
MOTHEDUC = 1.14*Ses, Errorvar.= 12.70, R**2 = 0.093
     (0.14)               (0.85)
     8.42                 14.92


FATHEDUC = 1.07*Ses, Errorvar.= 6.15 , R**2 = 0.16
     (0.098)              (0.43)
     10.90                14.26


FATHOCC = 1.01*Ses, Errorvar.= 0.96 , R**2 = 0.51
     (0.055)              (0.11)
     18.35                8.39


FAMILINC = 0.78*Ses, Errorvar.= 0.74   , R**2 = 0.45
           (0.047)              (0.076)
           16.65                9.81
```

(ii) Global Goodness of Fit Statistics, Missing Data Case

```
-2ln(L) for the saturated model =       24074.660
-2ln(L) for the fitted model    =       24104.968

Degrees of Freedom = 19
Full Information ML Chi-Square  = 30.31 (P = 0.048)
Root Mean Square Error of Approximation (RMSEA) = 0.032
```

The FIML χ^2 is obtained as the difference between -2 ln L (24105.18) for the fitted model and -2 ln L for the unrestricted model (11358.60 + 12716.07) and equals 30.3.

4.9 Latent Variable Scores

4.9.1 Calculation of Latent Variable Scores

To obtain scores for the latent variables of a structural equation model, the following three steps are required.

Step 1

If the data are not available as a PRELIS system file (***.psf**) yet, use the ***Import Data in Free Format*** or ***Import External Data in Other Formats*** options from the ***File*** menu to create a ***.psf** file.

Step 2

Use this *.**psf** file to compute the desired sample covariance or correlation matrix to which the structural equation model should be fitted. This may be done by running the appropriate PRELIS syntax file or by using the ***Output Options*** option from the ***Statistics*** menu. Either action will result in the creation a Data System File (*.**dsf**). A *.**dsf** file contains all the data information that LISREL requires to fit the structural equation model to the data.

Step 3

In SIMPLIS syntax, use the line

```
System File from file <filename>.DSF
```

to specify the data to be analyzed and insert the command line

```
PSFfile <filename>.PSF
```

after the `Relationships` paragraph where <filename> denotes the path and name of the *.**psf** file. The `System File` command replaces the `Observed Variables` paragraph and the `Sample Size` line.

In LISREL syntax, use the command line

```
SY = <filename>.DSF
```

to specify the *.**dsf** file and the

```
PS = <filename>.PSF
```

command line to specify the *.**psf** file. The `SY` command replaces the `LA` and the `DA` commands.

When done, run the syntax file by clicking the ***Run LISREL*** icon button. When the output file is displayed, open the relevant *.**psf** file using the ***Open*** option from the ***File*** menu. The latent variable scores appear as the last set of columns of this file.

4.9.2 An Illustrative Example

Chow (2000) used five indicators of self esteem (*Selfest*) and four indicators of depression (*Depress*) to formulate a measurement model for the latent variables *Self Esteem* and *Depression*. A path diagram of this measurement model is shown below.

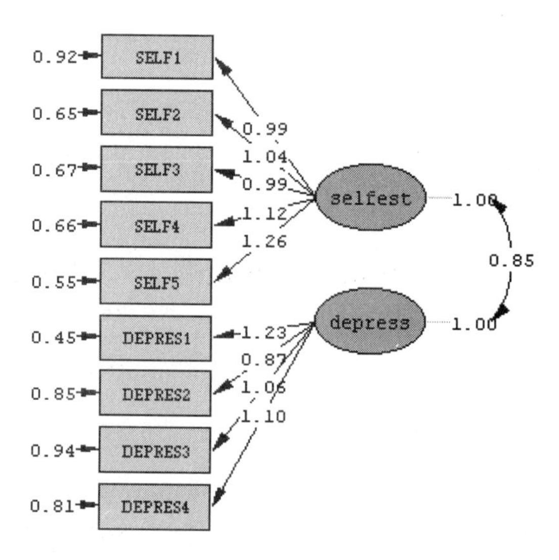

The nine indicators were observed for a random sample size of 230 college graduates. The resulting raw data are listed in the PRELIS System File **depression.psf**, located in the **tutorial** folder. A portion of the spreadsheet is shown below.

Depression

	SELF1	SELF2	SELF3	SELF4	SELF5	DEPRES1	DEPRES2
1	3.00	2.00	3.00	4.00	4.00	4.00	2.00
2	2.00	1.00	2.00	3.00	2.00	3.00	0.00
3	2.00	1.00	4.00	2.00	2.00	2.00	0.00
4	1.00	1.00	2.00	2.00	4.00	4.00	3.00
5	2.00	0.00	1.00	2.00	3.00	2.00	1.00
6	4.00	3.00	3.00	2.00	4.00	2.00	1.00
7	0.00	0.00	1.00	2.00	1.00	2.00	0.00
8	4.00	2.00	2.00	2.00	2.00	2.00	0.00
9	3.00	3.00	2.00	2.00	3.00	4.00	2.00
10	0.00	3.00	3.00	3.00	1.00	3.00	1.00

The PRELIS syntax file **depression.pr2** contains the required syntax to compute the sample covariance and to create the desired Data System File **depression.dsf**. This file is generated by selecting the ***Output Options*** option from the ***Statistics*** menu.

A portion of the ***Output*** dialog box is shown below. Select ***Covariances*** and check the ***LISREL system data*** box. Click ***OK*** to run PRELIS.

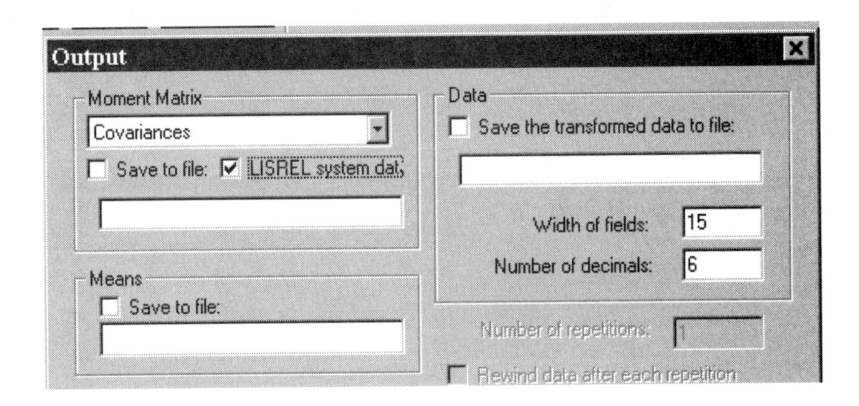

The contents of **depression.pr2** are listed below.

```
SY= Depress.psf
OU MA=CM XM XB XT
```

The SIMPLIS syntax file **depress.spl** in the **tutorial** folder is used to compute the latent variable scores for the latent variables *Self Esteem* and *Depression*. The contents of this file are listed below.

```
System File from File DEPRESS.DSF
Latent Variables: selfest depress
Relationships
SELF1-SELF5 = selfest
DEPRES1-DEPRES4 = depress
PSFFile DEPRESSION.PSF
Lisrel Output: ND=3
Path Diagram
End of Problem
```

If LISREL is executed, the PRELIS system file **depression.psf** is updated with the latent variable scores as shown below.

DEPRES1	DEPRES2	DEPRES3	DEPRES4	selfest	depress
4.000	2.000	0.000	4.000	1.148	0.978
3.000	0.000	0.000	1.000	0.010	-0.235
2.000	0.000	4.000	4.000	0.220	0.482
4.000	3.000	4.000	4.000	0.399	1.512
2.000	1.000	4.000	4.000	-0.127	0.549
2.000	1.000	3.000	4.000	1.033	0.594
2.000	0.000	2.000	1.000	-0.888	-0.429
2.000	0.000	4.000	4.000	0.294	0.493
4.000	2.000	4.000	4.000	0.676	1.434
3.000	1.000	4.000	4.000	0.171	0.916

4.10 Multilevel Confirmatory Factor Analysis

4.10.1 Description of the Data

This example illustrates

- importing of external data such as SPSS ***.sav** files
- drawing a multiple-group path diagram
- building SIMPLIS/LISREL syntax from a path diagram
- fitting a multilevel structural equation model to the data.

The data set used in this example forms part of the data library of the Multilevel Project at the University of London, and comes from the Junior School Project (Mortimore *et al*, 1988). Mathematics and language tests were administered in three consecutive years to more than 1000 students from 50 primary schools, which were randomly selected from primary schools maintained by the Inner London Education Authority. The data set is stored in the **tutorial** folder as an SPSS for Windows file named **jsp2.sav**.

In LISREL 8.50 for Windows it is convenient to work with PRELIS system files (***.psf**) since LISREL can read the variable names, the number of variables, the number of records and other relevant information directly from these files. It is therefore advantageous to convert files from other software packages to the ***.psf** format.

4.10.2 Import of External Data

Use the ***Import External Data in Other Formats*** option from the ***File*** menu to obtain the ***Input Database*** dialog box. Select SPSS for Windows (***.sav**) from the ***Files of type*** drop-down menu list and from the **tutorial** folder select **jsp2.sav**. Click ***Open*** to obtain the ***Save As*** dialog box. Enter the name **jsp2.psf** in the ***File name*** string field. By clicking the ***Save*** button a PRELIS system file is created and displayed in the form of a data spreadsheet.

Although no missing value codes were assigned in **jsp2.sav**, data values of -9 indicate missing values and can be defined as such by selecting the ***Define Variables*** option from the ***Data*** menu.

Click on *MATH1* (or any other variable) to select it. Click ***Missing Values*** to obtain the ***Missing Values*** dialog box. Enter the value -9.0 in the ***Global missing value*** field and change the method of deletion from listwise to pairwise. The pairwise option ensures that recoding and computing of variables do not change the number of cases.

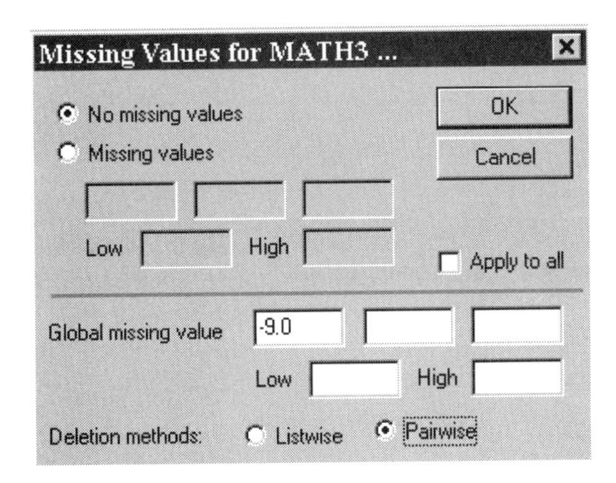

4.10.3 Draw the Path Diagram

Click the ***OK*** buttons of the ***Missing Values*** and ***Define Variables*** dialog boxes to return to the main window. Use the ***Save*** option from the ***File*** menu to save the changes to **jsp2.psf**. Use the ***New*** option from the ***File*** menu to open the ***New*** dialog box and select the ***Path Diagram*** option from the ***New*** dialog box and assign a file name, for example, **msemex.pth**. By clicking the ***Save*** button, the path diagram window appears.

Select the ***Title and Comments*** option from the ***Setup*** menu to open the ***Title and Comments*** dialog box. Enter the title `Multilevel CFA model for Numeric and Verbal Ability` in the ***Title*** string field to produce the following ***Title and Comments*** dialog box.

Title and Comments

Title

Multilevel CFA model for Numeric and Verbal Ability

Comments

Next >

OK

Cancel

Click ***Next*** to open the ***Groups*** dialog box. Enter the label `Group1: Between Schools` in the first string field. Use the down arrow key to create a string field for the second group. Enter the label `Group2: Within Schools` in the second string field to produce the following ***Groups*** dialog box.

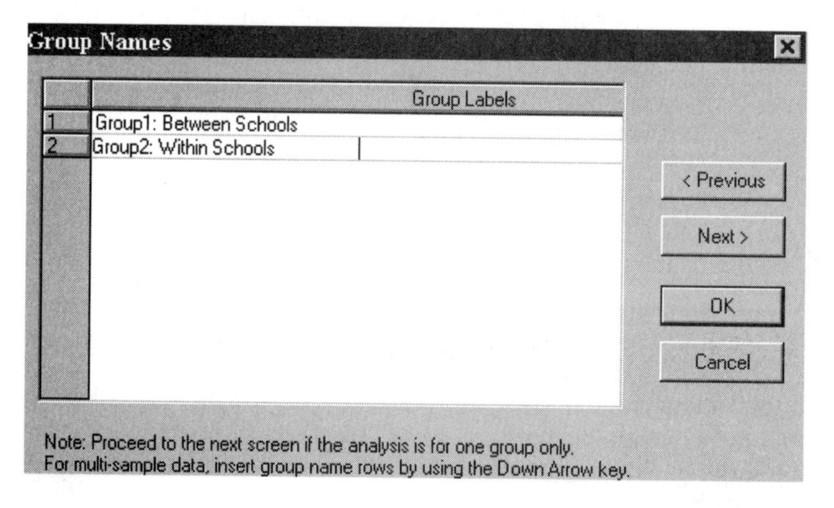
Group Names

	Group Labels
1	Group1: Between Schools
2	Group2: Within Schools

< Previous

Next >

OK

Cancel

Note: Proceed to the next screen if the analysis is for one group only.
For multi-sample data, insert group name rows by using the Down Arrow key.

Click ***Next*** to proceed to the ***Labels*** dialog box. Since the observed variables are stored in the file **jsp2.psf**, it is convenient to read the variable names from this file. Click ***Add/Read Variables*** to open the ***Add/Read Variables*** dialog box. Select the ***Prelis System File*** option from the ***Read from file*** drop-down list box.

Use the ***Browse*** button to select the file **jsp2.psf** in the **tutorial** folder to produce the following ***Add/Read Variables*** dialog box. Click ***OK*** to return to the ***Labels*** dialog box.

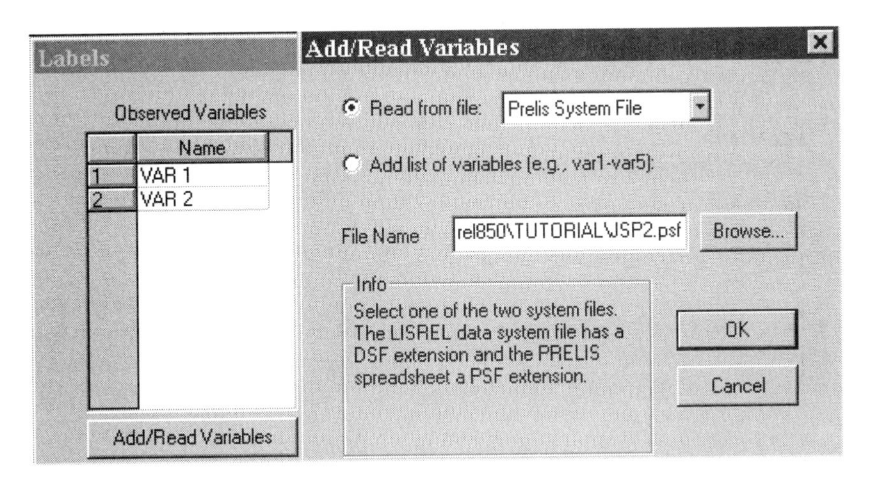

Click ***Add Latent Variables*** to open the ***Latent Variables*** dialog box. Enter the label *NABILITY* for Numeric Ability to produce the following dialog box.

Add Variables
Add one or list of variables here (e.g., var1 - var5):
NABILTY
OK
Cancel

Click ***OK***. Repeat the process to enter the label *VABILITY* for Verbal Ability. Click ***OK*** to produce the following ***Labels*** dialog box.

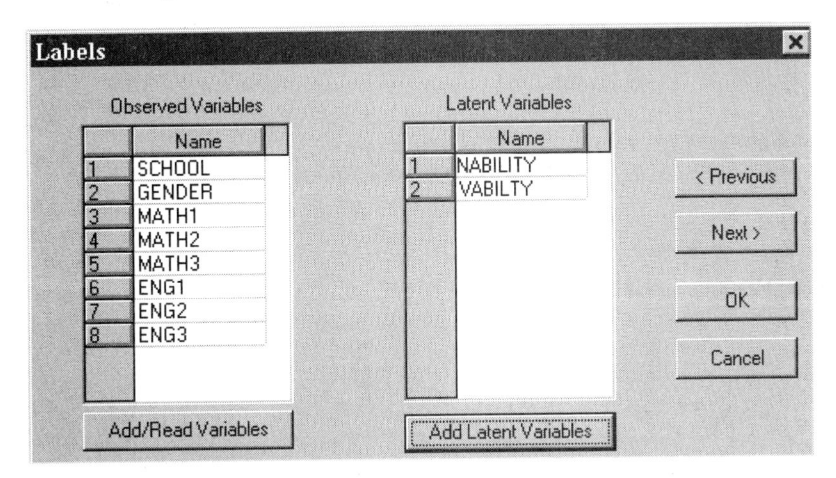

Click ***OK*** to return to the path diagram window. Click and drag the observed variables *MATH1*, *MATH2*, *MATH3*, *ENG1*, *ENG2* and ENG3 one by one into the path diagram window. Also click and drag the latent variables *NABILITY* and *VABILITY* one by one into the path diagram window.

Use the one-directional arrow icon button on the ***Draw*** toolbar to insert paths from *NABILITY* to *MATH1*, *MATH2* and *MATH3* and also to insert paths from *VABILITY* to *ENG1*, *ENG2* and *ENG3* as shown below.

Note:

Once this arrow is selected, move the mouse to within an ellipse representing a latent variable. With the left mouse button down, drag the arrow to an observed variable. Release the mouse button when the arrow is in the rectangle representing the observed variable.

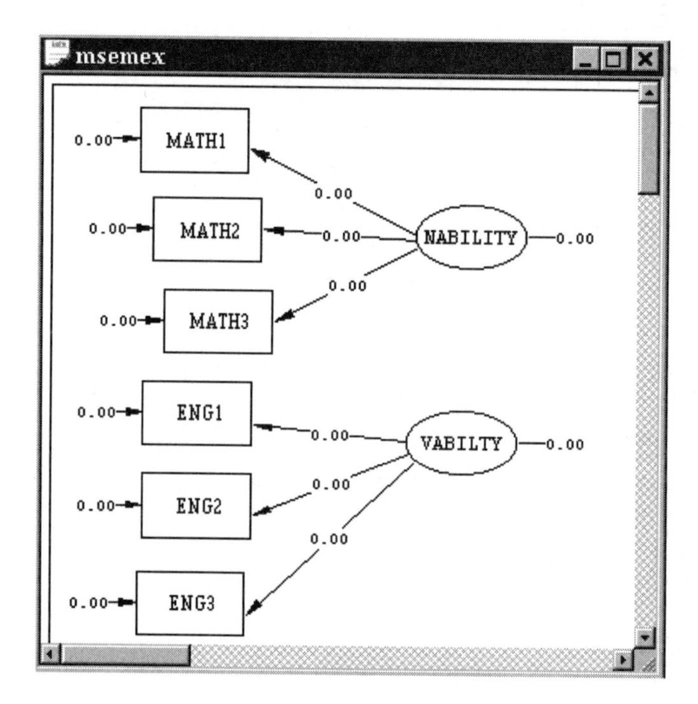

4.10.4 Build SIMPLIS/LISREL Syntax

Select the ***Build SIMPLIS Syntax*** option from the ***Setup*** menu to produce the text editor window shown overleaf. Insert the command line `$CLUSTER SCHOOL` after the `Raw Data from File` command. Change the command line `MATH1=NABILITY` to `MATH1=1*NABILITY` to set the scale of *NABILITY*. Change the command line `ENG1=VABILITY` to `ENG1=1*VABILITY` to set the scale of *VABILITY*. Copy and paste the relationships of the Between Schools group to the `Relationships` paragraph of the Within Schools group. In the `Relationships` paragraph of the Within Schools group, insert the following SIMPLIS commands

```
Set the Variance of NABILITY Free
Set the Variance of VABILITY Free
Set the Covariance between NABILITY and VABILITY Free
```

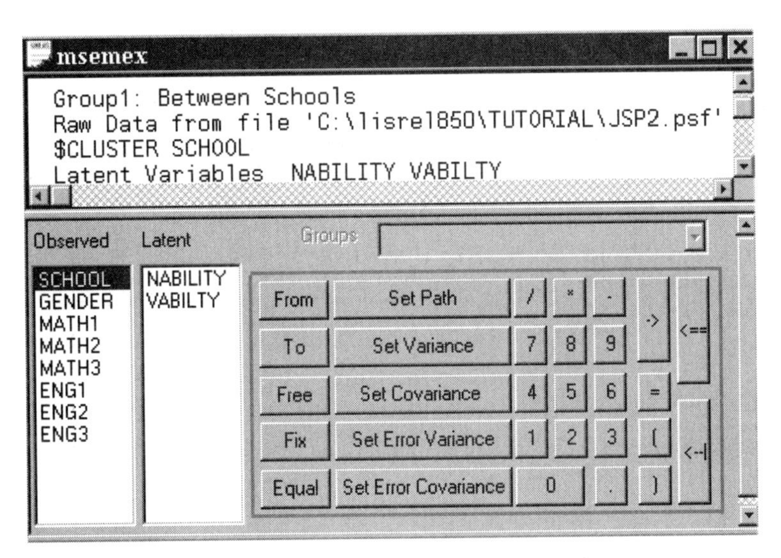

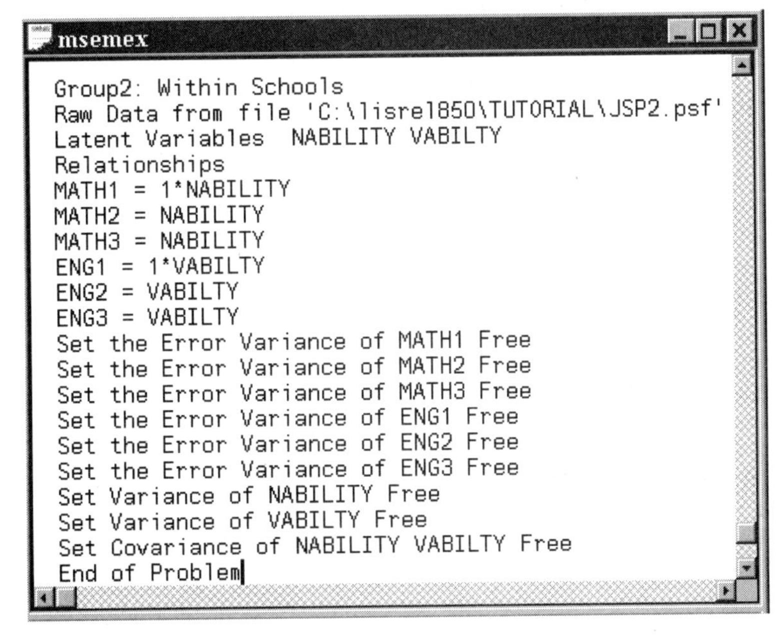

4.10.5 Fit the Model to the Data

Click the ***Run LISREL*** icon button on the main toolbar to produce the path diagram shown below:

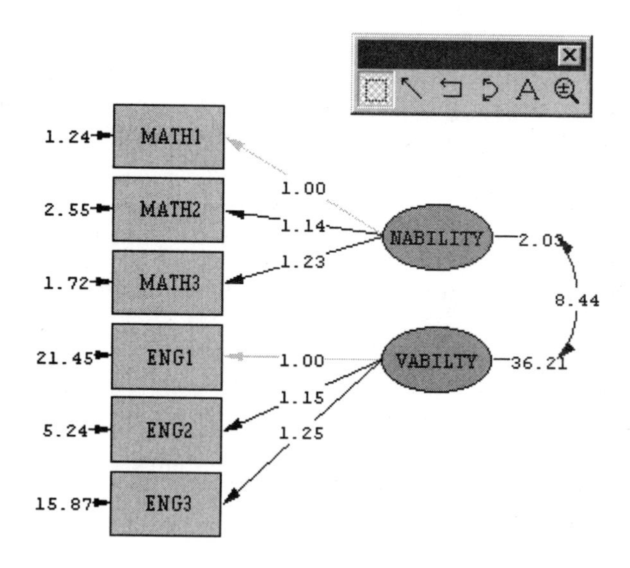

LISREL produces an output file, **msemex.out** which contains the results of the analysis.

5 Multilevel Examples

5.1 Analysis of Multilevel Models

Multilevel models deal with the analysis of data where observations are nested within groups. Social, behavioral and even economic data often have a hierarchical structure. A frequently cited example is in education, where students are grouped in classes. Classes are grouped in schools, schools in education departments and so on. We thus have variables describing individuals, but the individuals may be grouped into larger or higher order units. Multilevel data structures are pervasive. As Kreft and de Leeuw (1998) note: "Once you know that hierarchies exist, you see them everywhere". It was not until the 1980s and 1990s that techniques to analyze multilevel data properly became widely available.

These techniques use maximum likelihood procedures to estimate random coefficients and are often referred to as "multilevel random coefficient models" (MRCM) in contrast to "hierarchical linear models" (HLM). For a more comprehensive discussion of this technique and the syntax needed to run such models with LISREL, the reader is referred to the *LISREL 8: New Statistical Features Guide*.

In the following section, latent growth curve models will be fitted to a small data set as first examples of the use of the multilevel procedure available within the LISREL program.

5.2 A 2-level Growth Curve Model

5.2.1 Repeated Measures Data

In this example, it will be illustrated how multilevel modeling may be used to explicitly recognize the hierarchical structure of repeated measurement data. Five models will be fitted and discussed, these being:

- A variance decomposition model
- Modeling linear growth
- Modeling non-linear growth
- Introducing a covariate when modeling non-linear growth
- A model with complex variation at level-1 of the hierarchy

The sections describing these models are best viewed in sequence. A description of the data is given here.

The data set used contain repeated measurements on 82 striped mice and was obtained from the Department of Zoology at the University of Pretoria, South Africa (see du Toit, 1979). A number of male and female mice were released in an outdoor enclosure with nest boxes and sufficient food and water. They were allowed to multiply freely. Occurrence of birth was recorded daily and newborn mice were weighed weekly, from the end of the second week after birth until physical maturity was reached. The data set consists of the weights of 42 male and 40 female mice. For male mice, 9 repeated weight measurements are available and for the female mice 8 repeated measurements.

The first 11 observations from this data set, contained in **mouse.psf**, and the variable names to be used are shown below:

MOUSE.PSF

	iden2	iden1	weight	constant	time	timesq	gender
1	1.00	1.00	15.00	1.00	1.00	1.00	1.00
2	1.00	2.00	17.00	1.00	2.00	4.00	1.00
3	1.00	3.00	23.00	1.00	3.00	9.00	1.00
4	1.00	4.00	24.00	1.00	4.00	16.00	1.00
5	1.00	5.00	26.00	1.00	5.00	25.00	1.00
6	1.00	6.00	31.00	1.00	6.00	36.00	1.00
7	1.00	7.00	37.00	1.00	7.00	49.00	1.00
8	1.00	8.00	42.00	1.00	8.00	64.00	1.00
9	1.00	9.00	46.00	1.00	9.00	81.00	1.00
10	2.00	1.00	11.00	1.00	1.00	1.00	1.00

The response variable *weight* contains the weight measurements (in gram) for all mice at the different times of measurement. The explanatory variables which may be used are the time points at which measurements were made (*time*), the squared values of these time points (*timesq*) and the gender of the mice (*gender*). It is also assumed that the growth of the mice during this period can be adequately described by a parabolic function.

A hierarchical level-2 structure is incorporated where the individual mice are the level-2 units. Unique numbers identifying the mice are contained in the variable *iden2*, which will be used as the level-2 ID for the analysis. The variable *iden1* identifies the occasions on which measurements for a particular mouse were made and will be used as the level-1 ID. From the description of the data set as given above, it follows that there are 82 level-2 units, with either 8 or 9 measurements nested within each level-2 unit.

The variable *constant* consists of a column of 1's and will be used to estimate the intercept term in the model. It may also be used to estimate the amount to which the intercept varies over the different levels of the hierarchy, as illustrated in the following section.

5.2.2 A Variance Decomposition Model

The simplest multilevel model is equivalent to a one-way ANOVA with random effects. Although this model is not interesting in itself, it is useful as a preliminary step in a multilevel analysis as it provides important information about the outcome variability at each of the levels of the hierarchy. It may also function as a baseline with which more sophisticated models may be compared.

Let the subscript i denote the i-th level 2 unit, in this case the i-th mouse. The subscript j refers to the j-th weight measurement for the i-th mouse. Using this notation, the one-way ANOVA model can be written as

$$y_{ij} = \beta_0 + u_{0i} + e_{ij}$$

where u_{0i} represents the random variation in intercepts on level-2 of the model. It is assumed that u_{0i} has an expected value of 0 and a variance of $\Phi_{(2)}$. The variance $\Phi_{(2)}$ may be interpreted as the 'between-group' variability. Likewise, it is assumed that e_{ij} is $N(0, \Phi_{(1)})$ distributed. Thus $\Phi_{(1)}$ may be interpreted as the 'within-group' variability.

In matrix notation, this model can be written as

$$y_{ij} = [constant][\beta_0] + [constant][u_{0i}] + [constant]e_{ij},$$

where *constant* represents the intercept. This model is also known as a fully unconditional model (Bryk and Raudenbush, 1992) as no predictors are specified at either level of the hierarchy.

To fit this model to the data in **mouse.psf** we proceed as follows. Select the ***Title and Options*** option from the ***Multilevel, Linear Model*** menu to activate the ***Title and Options*** dialog box.

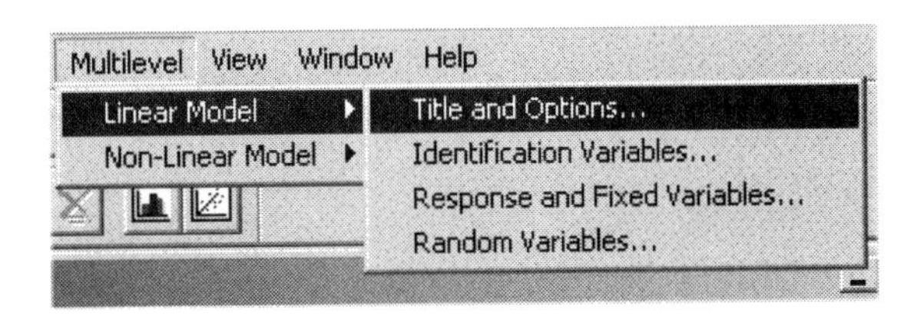

Type a title command for this analysis. Note that the dialog box shown below shows all program defaults for the `OPTIONS` command. Click ***Next*** to go to the ***Identification Variables*** dialog box.

Title and Options

Title (Maximum 70 characters):
Variance decomposition model for mouse data

Maximum Number of Iterations: 10

Convergence Criterion: 0.001

Missing Data Value: -999999

Missing Dep Value: -999999

Use OLS for Starting Values

Additional Output

Residuals

Empirical Bayes Estimates

Next >> Cancel OK

To build Syntax, proceed to the Random Variables screen and click the Finish Button

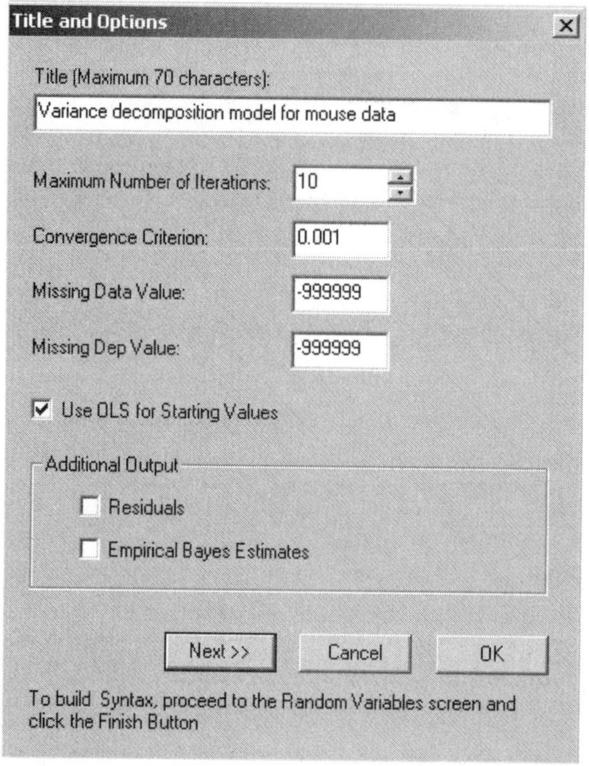

On the ***Identification Variables*** dialog box, select the variable *iden2* and use the ***Add*** button to enter this variable in the level-2 identification field. Enter *iden1* in the level-1 identification field in a similar manner.

Note:

Although we used the names iden2 and iden1 for the identification variables, any other names up to 8 characters in length may be used. For example, *mousenum* and *occasion* are valid alternative names for the identification variables.

Click ***Next*** to go to the ***Select Response and Fixed Variables*** dialog box.

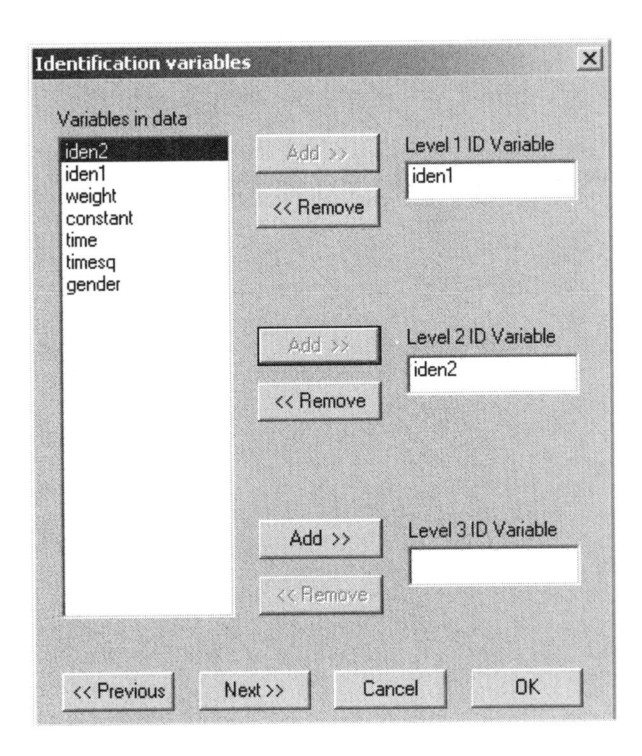

Select the variable *weight* as response variable and use the ***Add*** button to enter it into the response variable field. Enter the variable *constant* in the fixed variable field. Click ***Next*** to go to the ***Random Variables*** dialog box.

Note that response variables are also called level-1 outcome variables. Likewise, fixed variables are also known as predictors.

On the ***Random Variables*** dialog box, we select *constant*, the variable representing the intercept term, and add it to both the level-1 and level-2 random variable fields, thus associating it with the error terms at these levels. To generate the syntax file for this analysis, click ***Finish***.

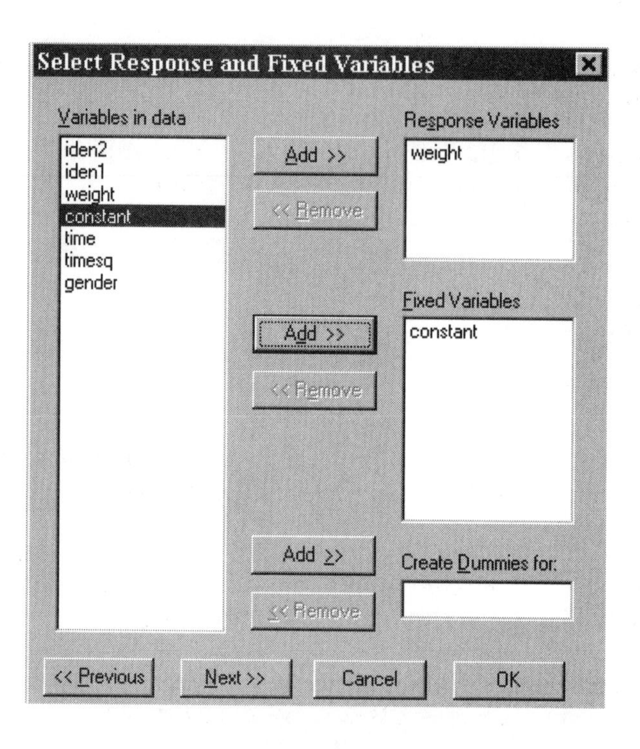
Select Response and Fixed Variables
Variables in data
iden2
iden1
weight
constant
time
timesq
gender
Add >>
<< Remove
Response Variables
weight
Add >>
<< Remove
Fixed Variables
constant
Add >>
<< Remove
Create Dummies for:
<< Previous
Next >>
Cancel
OK

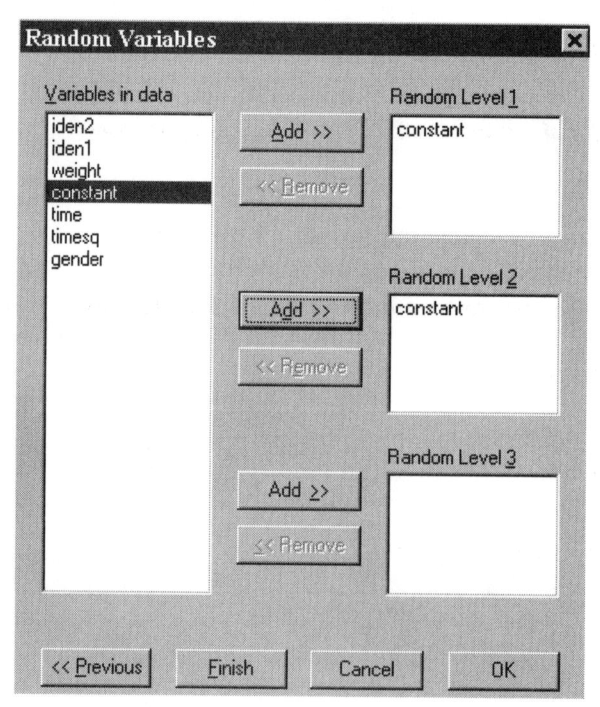
Random Variables
Variables in data
iden2
iden1
weight
constant
time
timesq
gender
Add >>
<< Remove
Random Level 1
constant
Add >>
<< Remove
Random Level 2
constant
Add >>
<< Remove
Random Level 3
<< Previous
Finish
Cancel
OK

The generated syntax file **mouse.pr2** (see **mouse1.pr2** in the **mlevelex** folder of LISREL) is shown below:

```
OPTIONS OLS=YES CONVERGE=0.001000 MAXITER=10 OUTPUT=STANDARD ;
TITLE=Variance decomposition for mouse data;
SY=MOUSE.PSF;
ID1=iden1;
ID2=iden2;
RESPONSE=weight;
FIXED=constant ;
RANDOM1=constant;
RANDOM2=constant;
```

With the exception of the use of the optional `TITLE` command, this syntax file is the most basic one that can be used for the analysis of a level-2 model.

Note that in the `OPTIONS` command the default values of the options `MAXITER`, `CONVERGE` and `OUTPUT` are used (see the section on multilevel syntax in the *LISREL 8: New Statistical Features Guide* for more detailed information).

Click the ***Run PRELIS*** icon button to run the analysis. The output obtained for this analysis is discussed in detail in the *LISREL 8: New Statistical Features Guide.*

5.2.3 Modeling Linear Growth

The variance decomposition model is now extended by including the variable *time* as a fixed effect. The model thus becomes

$$y_{ij} = \beta_0 + (time)_{ij}\beta_1 + u_{0i} + e_{ij}$$

with the variable *time* used as a predictor of the response measurements.

In matrix notation, the model can now be written as

$$y_{ij} = \left[constant \quad time\right]\begin{bmatrix}\beta_0 \\ \beta_1\end{bmatrix} + \left[constant\right]\left[u_{0i}\right] + \left[constant\right]e_{ij}\,.$$

An equivalent formulation for this model is

$$y_{ij} = b_{0i} + (time)_{ij}b_{1i} + e_{ij}$$

where

$$b_{0i} = \beta_0 + u_{0i}$$
$$b_{1i} = \beta_1.$$

Note that b_{0i} and b_{1i} are often referred to as level-2 outcome variables.

The only change to the syntax file created in the previous section (alternatively, open **mouse1.pr2** in the **mlevelex** folder) is in the `FIXED` command, which now becomes:

```
FIXED=constant time;
```

Alternatively, this change can be made in the ***Select Response and Fixed Variables*** dialog box of the ***Multilevel, Linear Model*** menu as shown below. Proceed to the ***Random Variables*** dialog box and click ***Finish*** to generate a revised syntax file (**mouse2.pr2** in the **mlevelex** folder).

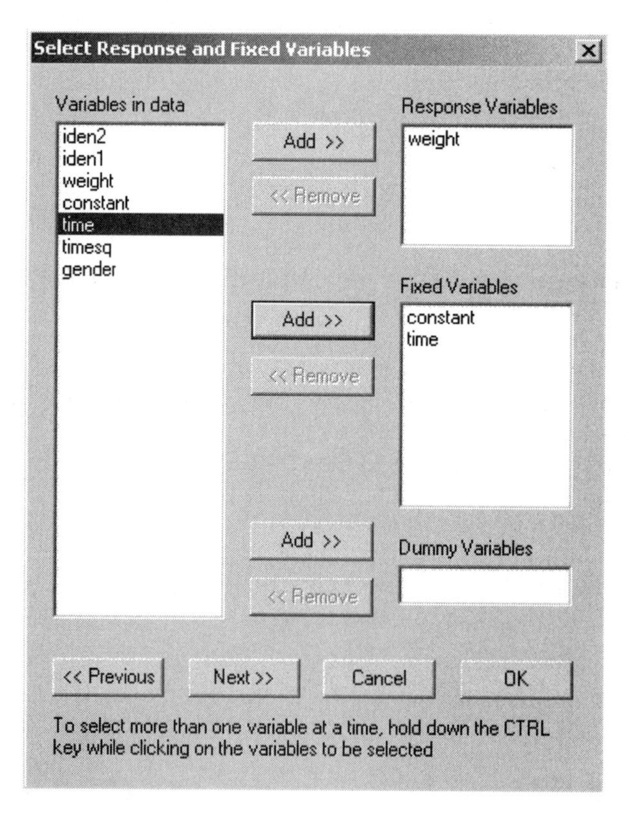

As can be seen from the syntax file given below, the TITLE command was changed to reflect the addition of a predictor to the model.

```
OPTIONS OLS=YES CONVERGE=0.001000 MAXITER=10 OUTPUT=STANDARD ;
TITLE=Mouse data: linear growth model one;
SY=MOUSE.PSF;
ID1=iden1;
ID2=iden2;
RESPONSE=weight;
FIXED=constant time;
RANDOM1=constant;
RANDOM2=constant;
```

The output obtained for this analysis is discussed in detail in the *LISREL 8: New Statistical Features Guide.*

It is expected that the linear growth rate may vary from mouse to mouse around its mean value, rather than be fixed. The random component on level-2 of the hierarchy is thus extended to

```
RANDOM2 = constant time ;
```

which implies a revised model

$$y_{ij} = \left[constant \quad time \right] \begin{bmatrix} \beta_0 \\ \beta_1 \end{bmatrix} + \left[constant \quad time \right] \begin{bmatrix} u_{0i} \\ u_{1i} \end{bmatrix} + \left[constant \right] e_{ij}.$$

This change can be made to the previous syntax file (**mouse2.pr2** in the **mlevelex** folder) by simply editing this file in the text editor window of LISREL. Alternatively, the variable *time* may also be added to the level-2 random variable list by using the ***Random Variables*** dialog box.

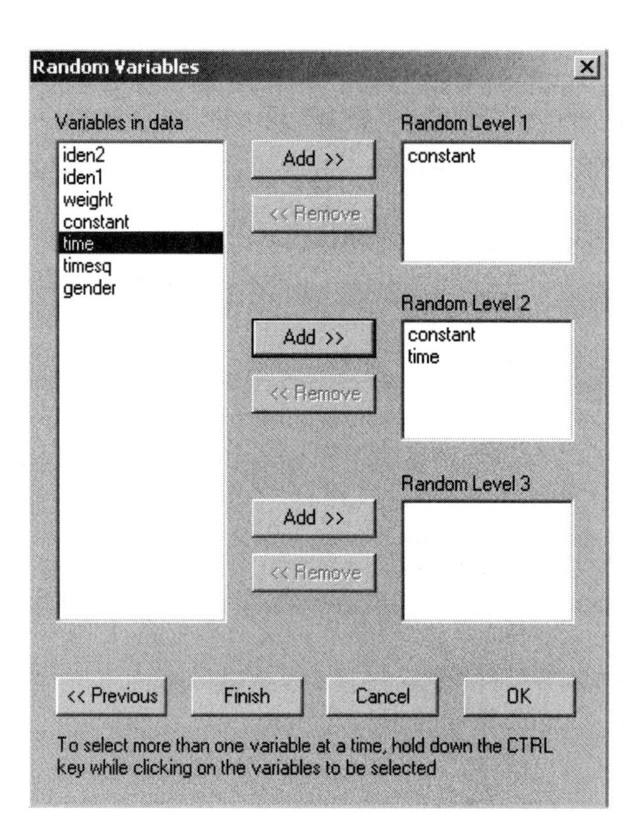

Clicking ***Finish*** on this dialog box will generate a new syntax file (available as **mouse3.pr2** in the **mlevelex** folder). As before, the `TITLE` command has been used to reflect changes made to the model.

The syntax file thus obtained is:

```
OPTIONS OLS=YES CONVERGE=0.001000 MAXITER=10 OUTPUT=STANDARD ;
TITLE=Mouse data: linear growth model two;
SY=MOUSE.PSF;
ID1=iden1;
ID2=iden2;
RESPONSE=weight;
FIXED=constant time;
RANDOM1=constant;
RANDOM2=constant time;
```

Note that the inclusion of the variable names imply that the researcher assumes that the regression coefficients of these variables are random. It is further assumed that the coefficients follow a multivariate normal distribution with covariance matrix $\mathbf{\Phi}_{(n)}$, where n denotes the level.

The total variation on a particular level of the hierarchy may also be calculated. In this case, the total variance on level-2 is the variance of the sum of the two random coefficients associated with *constant* and *time* and may be written as

$$\begin{aligned} &Var(\text{constant} \times u_0 + \text{time} \times u_1) \\ &= Var(u_0) + 2 \times \text{time} \times Cov(u_0, u_1) + \text{time} \times Var(u_1) \\ &= \Phi_{(2)1,1} + 2 \times \text{time} \times \Phi_{(2)2,1} + \text{time} \times \text{time} \times \Phi_{(2)2,2} \end{aligned}$$

We can thus write the total variation on level-2 as a quadratic function of the variable *time*.

The output obtained for this analysis is discussed in detail in the *LISREL 8: New Statistical Features Guide.*

5.2.4 Modeling Non-Linear Growth

In data of this nature, it is unlikely that the increase in weight measurement will be linear for all mice over the time period concerned. A non-linear component may be introduced in the model discussed in the previous section by adding a quadratic term (the variable *timesq*) to the model.

The model previously given is thus extended to

$$y_{ij} = \beta_0 + (time)_{ij}\beta_1 + (timesq)_{ij}\beta_2 + u_{0i} + (time)_{ij}u_{1i} + (timesq)_{ij}u_{2i} + e_{ij}$$

or, in matrix notation,

$$y_{ij} = \begin{bmatrix} constant & (time)_{ij} & (timesq)_{ij} \end{bmatrix} \begin{bmatrix} \beta_0 \\ \beta_1 \\ \beta_2 \end{bmatrix} + \begin{bmatrix} constant & (time)_{ij} & (timesq)_{ij} \end{bmatrix} \begin{bmatrix} u_{0i} \\ u_{1i} \\ u_{2i} \end{bmatrix} + \begin{bmatrix} constant \end{bmatrix} \begin{bmatrix} e_{ij} \end{bmatrix}$$

The addition of the variable *timesq* in this case leads to the following changes in the FIXED and RANDOM2 commands contained in the syntax file.

```
FIXED = constant time timesq ;
RANDOM2 = constant time timesq ;
```

In order to obtain the empirical Bayes residuals for the level-2 models and the fitted values for each observation, the option OUTPUT=BAYES is added to the OPTIONS command by checking the radio button next to the ***Empirical Bayes Estimates*** option on the ***Title and Options*** dialog box.

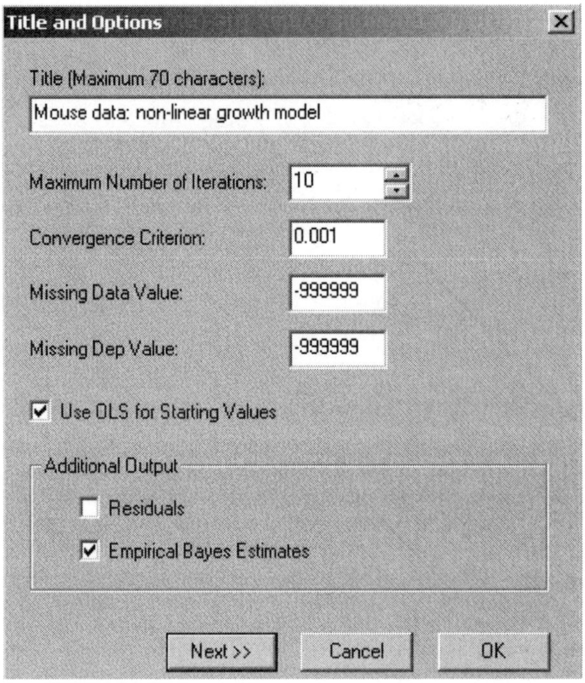

To make the changes to the `FIXED` and `RANDOM2` commands, add the variables *time* and *timesq* to the ***Fixed variables*** field on the ***Select Response and Fixed Variables*** dialog box of the ***Multilevel, Linear Model*** menu. Click ***Next*** to make the required changes to the ***Random Variables*** dialog box.

Add *timesq* to the ***Random Level 2*** field and click ***Finish*** to generate the new syntax file.

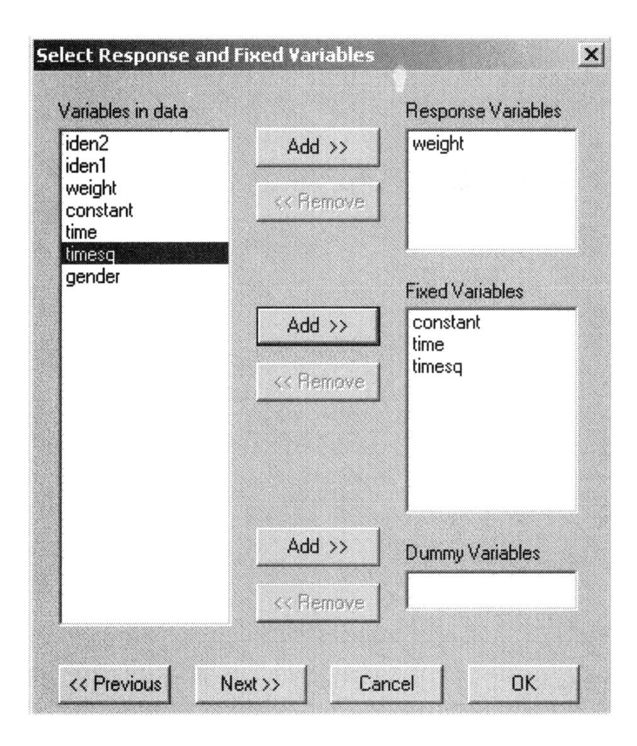
Select Response and Fixed Variables
Variables in data
iden2
iden1
weight
constant
time
timesq
gender
Add >>
<< Remove
Response Variables
weight
Fixed Variables
constant
time
timesq
Add >>
<< Remove
Add >>
<< Remove
Dummy Variables
<< Previous
Next >>
Cancel
OK

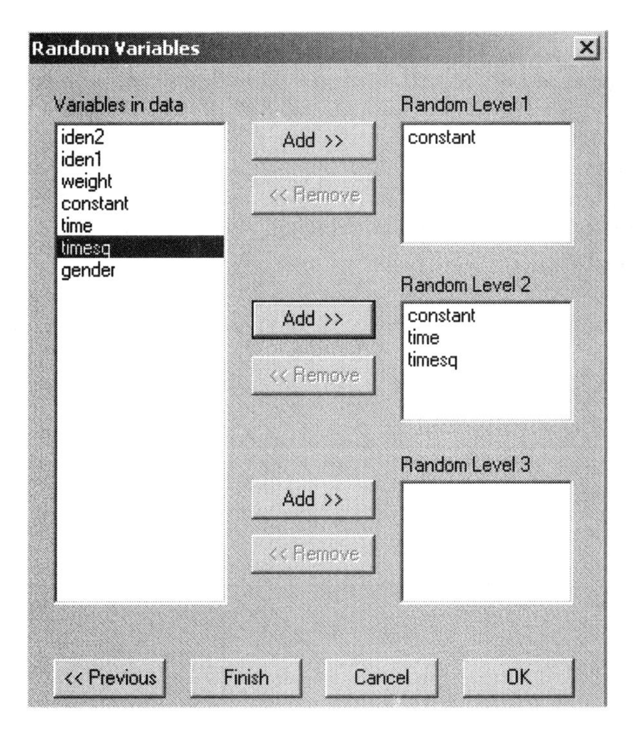
Random Variables
Variables in data
iden2
iden1
weight
constant
time
timesq
gender
Add >>
<< Remove
Random Level 1
constant
Random Level 2
constant
time
timesq
Add >>
<< Remove
Random Level 3
Add >>
<< Remove
<< Previous
Finish
Cancel
OK

The syntax generated is shown below (see **mouse4.pr2** in the **mlevelex** folder).

```
OPTIONS OLS=YES CONVERGE=0.001000 MAXITER=10 OUTPUT=BAYES ;
TITLE=Mouse data: non-linear growth model;
SY=MOUSE.PSF;
ID1=iden1;
ID2=iden2;
RESPONSE=weight;
FIXED=constant time timesq;
RANDOM1=constant;
RANDOM2=constant time timesq;
```

Click the ***Run PRELIS*** icon button to run the analysis. Convergence of the iterative procedure was reached in five iterations. The output obtained for this analysis is discussed in detail in the *LISREL 8: New Statistical Features Guide.*

5.2.5 Introducing a Covariate while Modeling Non-Linear Growth

In this example, we want to determine whether there is a significant difference between the growth pattern of the male and female mice, as modeled in the non-linear growth model discussed previously. This can be determined by adding the gender of the mice as covariate to the model fitted in the previous section. The variable *gender* is introduced as covariate by modifying the `FIXED` command to the following:

```
FIXED = constant time timesq gender gender*time gender*timesq;
```

This is equivalent to changing the model to:

$$\begin{aligned} y_{ij} &= \beta_0 + (time)_{ij}\beta_1 + (timesq)_{ij}\beta_2 \\ &\quad +(gender)_{ij}\beta_3 + (gender \times time)_{ij}\beta_4 + (gender \times timesq)_{ij}\beta_5 \\ &\quad +u_{0i} + (time)_{ij}u_{1i} + (timesq)_{ij}u_{2i} + e_{ij} \end{aligned}$$

In matrix notation, this implies extending the fixed part of the model to:

$$\begin{bmatrix} constant \\ (time)_{ij} \\ (timesq)_{ij} \\ (gender)_{ij} \\ (gender)_{ij} \times (time)_{ij} \\ (gender)_{ij} \times (timesq)_{ij} \end{bmatrix}' \begin{bmatrix} \beta_0 \\ \beta_1 \\ \beta_2 \\ \beta_3 \\ \beta_4 \\ \beta_5 \end{bmatrix}$$

The fact that the variable *gender* is a level-2 variable, thus having the same value for all level-1 observations nested within each level-2 unit, is perhaps best illustrated by rewriting the model as:

$$y_{ij} = b_{0i} + (time)_{ij} b_{1i} + (timesq)_{ij} b_{2i} + e_{ij}$$

$$b_{0i} = \beta_0 + (gender)_{ij} \beta_3 + u_{0i}$$
$$b_{1i} = \beta_1 + (gender)_{ij} \beta_4 + u_{1i}$$
$$b_{2i} = \beta_2 + (gender)_{ij} \beta_5 + u_{2i}.$$

To add the covariate *gender* to the model, edit the file **mouse4.pr2** directly. The variable *gender* may be added to the syntax file through use of the ***Response and Fixed Variable*** dialog box, but the terms *gender*time* and *gender*timesq* cannot. Alternatively, one may add the variables *gentim* (*gender*time*) and *gentimsq* (*gender*timesq*) by using the ***Compute*** option of the ***Transformations*** menu. Change the optional `TITLE` command to reflect the changes to the model.

Click the ***Run PRELIS*** icon button to fit the model to the data.

The revised syntax file (see **mouse5.pr2** in the **mlevelex** folder) is given below.

```
OPTIONS OLS=YES CONVERGE=0.001000 MAXITER=10 OUTPUT=BAYES ;
TITLE=Mouse data: non-linear growth model with a covariate;
SY=MOUSE.PSF;
ID1=iden1;
ID2=iden2;
RESPONSE=weight;
FIXED=constant time timesq gender gender*time gender*timesq;
RANDOM1=constant;
RANDOM2=constant time timesq;
```

Convergence is reached after 4 iterations and the output obtained for this analysis is discussed in detail in the *LISREL 8: New Statistical Features Guide.*

5.2.6 Complex Variation at Level-1 of the Model

In the final example of a level-2 model, the linear growth model fitted previously in **mouse3.pr2** (see Section 5.3.2) is extended to include complex variation on both levels of the hierarchy. The term "complex variation" refers to the existence of two or more random variables at the same level of the hierarchy. We include the variable *time* in this model to illustrate such a model.

We modify the RANDOM1 command previously used to

```
RANDOM1 = constant time;
```

The model considered here is thus

$$y_{ij} = \begin{bmatrix} constant & (time)_{ij} & (timesq)_{ij} \end{bmatrix} \begin{bmatrix} \beta_0 \\ \beta_1 \\ \beta_2 \end{bmatrix} + \begin{bmatrix} 1 & (time)_{ij} & (timesq)_{ij} \end{bmatrix} \begin{bmatrix} u_{0i} \\ u_{1i} \\ u_{2i} \end{bmatrix} + \begin{bmatrix} constant & (time)_{ij} \end{bmatrix} \begin{bmatrix} e_{1ij} \\ e_{2ij} \end{bmatrix}$$

This change in the level-1 covariance structure implies that the total variation at this level of the model can now be written as:

$$\begin{aligned} & Var(constant \times e_{1ij} + time \times e_{2ij}) \\ & = \Phi_{(1)constant,constant} + 2 \times time \times \Phi_{(1)time,constant} + time \times time \times \Phi_{(1)time,time} \end{aligned}$$

To include the variable *time* in the RANDOM1 command, enter *time* in the ***Random Level 1*** field on the ***Random Variables*** dialog box of the ***Multilevel, Linear Model*** menu. Click ***Finish*** to generate a revised syntax file. Alternatively, open the syntax file **mouse3.pr2** and add the variable *time* to the RANDOM1 command.

```
OPTIONS OLS=YES CONVERGE=0.001000 MAXITER=25 OUTPUT=STANDARD ;
TITLE=Mouse data: complex variation ;
SY=MOUSE.PSF;
ID1=iden1;
ID2=iden2;
RESPONSE=weight;
FIXED=constant time;
RANDOM1=constant time;
RANDOM2=constant time;
```

The model is fitted to the data when the ***Run PRELIS*** icon button is clicked. The output obtained for this model is discussed in detail in the *LISREL 8: New Statistical Features Guide.* For this model, 18 iterations were needed before convergence was reached. The MAXITER option of the OPTIONS command was used to increase the number of iterations from the default of 10 to 25.

5.3 Analysis of Air Traffic Control Data

5.3.1 Description of the Air Traffic Control Data

The data used in this example are described by Kanfer and Ackerman (1989). The data consist of information for 141 U.S. Air Force enlisted personnel. The personnel carried out a computerized air traffic controller task developed by Kanfer and Ackerman.

Note:

The subjects were instructed to accept planes into their hold pattern and land them safely and efficiently on one of four runways, varying in length and compass directions, according to rules governing plane movements and landing requirements. For each subject, the success of a series of between three and six 10-minute trials was recorded. The measurement employed was the number of correct landings per trial. The Armed Services Vocational Aptitude Battery (ASVAB) was also administered to each subject. A global measure of cognitive ability, obtained from the sum of scores on 10 subscales, is included in the data.

The data for this example can be found in **kanfer.psf**. The first few records of this data file are shown below:

KANFER.PSF

	control	time	measure	ability	constant	timesq
1	1.00	1.00	24.00	142.16	1.00	1.00
2	1.00	2.00	27.00	142.16	1.00	4.00
3	1.00	3.00	30.00	142.16	1.00	9.00
4	1.00	4.00	32.00	142.16	1.00	16.00
5	1.00	5.00	38.00	142.16	1.00	25.00
6	1.00	6.00	41.00	142.16	1.00	36.00
7	2.00	1.00	2.00	-7.63	1.00	1.00
8	2.00	2.00	3.00	-7.63	1.00	4.00

The variables in the data set are:

control	: The identifying number of the air traffic controller
time	: The number of the trial (between 1 and 6)
measure	: The number of successful landings for the trial
ability	: The cognitive ability score (combined ASVB score)
constant	: The intercept term, with value 1 throughout
timesq	: time × time a quadratic term.

Three models will be fitted to these data. The first model will investigate the variation in the number of correct landings over subjects and also over measurements for each subject. In the next model, a non-linear growth model will be considered. Finally, the cognitive ability measure will be introduced into the non-linear growth model.

5.3.2 Variance Decomposition Model for Air Force Data

In order to fit a fully unconditional model similar to the model discussed in Section 5.2.2

$$y_{ij} = \beta_0 + u_{0i} + e_{ij}$$

where u_{0i} represents the random variation in intercepts on level-2 of the model, the data spreadsheet, as contained in the file **kanfer.psf** of the **mlevelex** folder, is opened. Select the ***Title and Options*** option from the ***Multilevel, Linear Model*** menu on the main tool bar, and provide an optional title for the analysis.

For the moment, we use the default settings for the amount of output requested, maximum number of iterations and convergence criterion. Click ***Next*** to go to the ***Identification Variables*** dialog box.

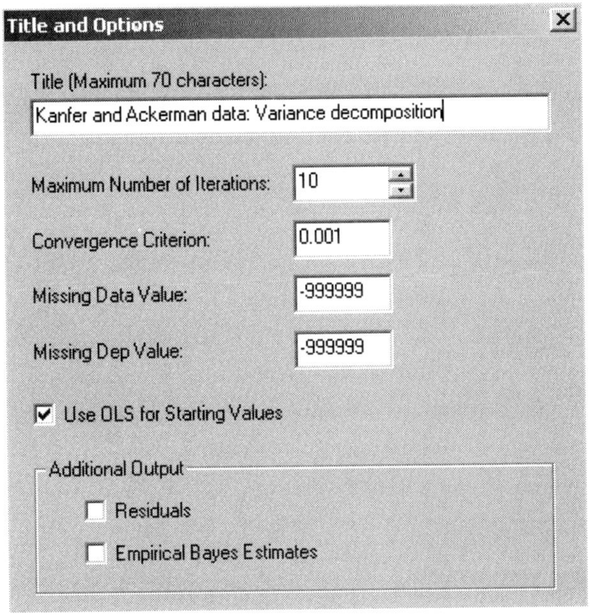

Title and Options

Title (Maximum 70 characters):
Kanfer and Ackerman data: Variance decomposition

Maximum Number of Iterations: 10

Convergence Criterion: 0.001

Missing Data Value: -999999

Missing Dep Value: -999999

☑ Use OLS for Starting Values

Additional Output

☐ Residuals

☐ Empirical Bayes Estimates

The variable *control*, identifying the air traffic controller, is used as level-2 identification. The variable *time*, indicating the number of the trial, is used as level-1 identification. Click ***Next*** to go to the ***Select Response and Fixed Variables*** dialog box, where the response variable and fixed effects, if any, can be selected.

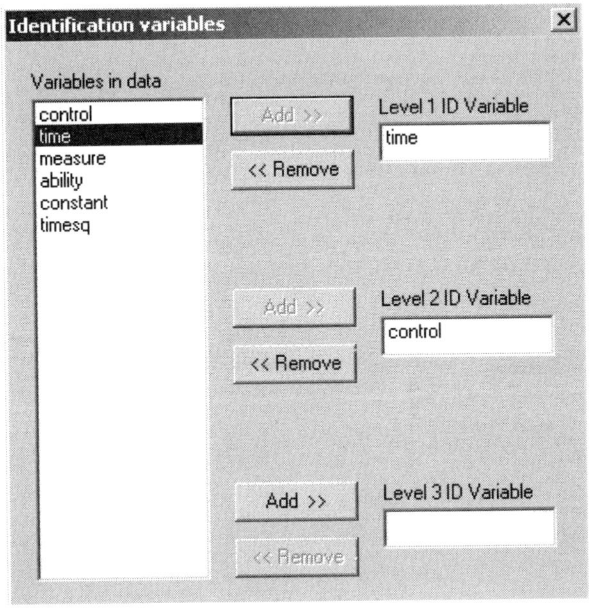

Identification variables

Variables in data
control
time
measure
ability
constant
timesq

Add >> | << Remove | Level 1 ID Variable: time

Add >> | << Remove | Level 2 ID Variable: control

Add >> | << Remove | Level 3 ID Variable:

The number of successful landings per trial is represented by the variable *measure*, which is selected from the list of variables on the left by clicking on it. Clicking ***Add*** next to the response variable list box identifies *measure* as the response variable for this particular model. We include the variable *constant*, representing the intercept term, as fixed effect in a similar way as shown below.

Clicking ***Next*** activates the ***Random Variables*** dialog box, where the variables whose coefficients are allowed to vary randomly over the levels of the hierarchy may be selected.

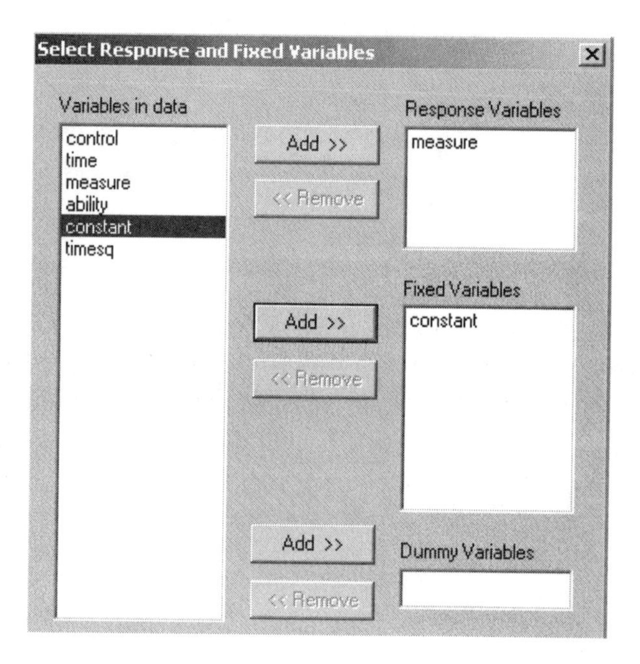

On the ***Random Variables*** dialog box, the intercept term, as represented by the variable *constant* is selected and entered in both the ***Random Level 1*** and ***Random Level 2*** fields.

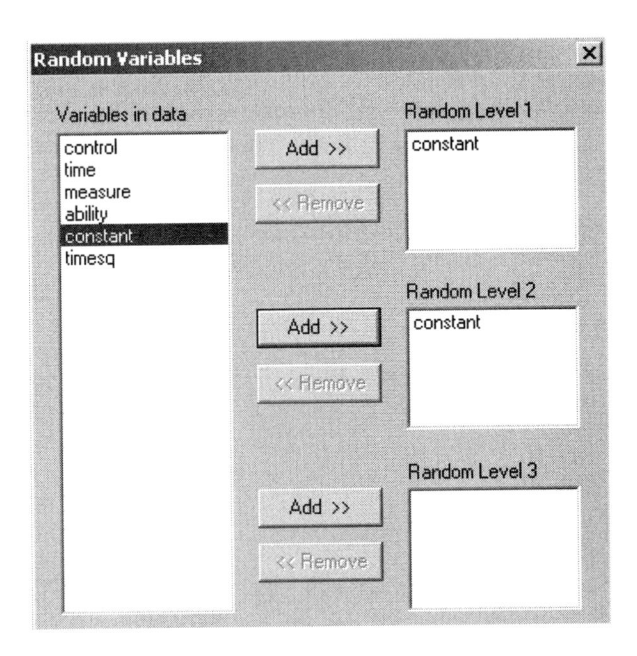

Once ***Finish*** is clicked, the syntax file for this analysis is automatically generated and displayed (see **kanfer1.pr2** in the **mlevelex** folder). Note that the command indicating the data file to be used is automatically generated and included in the syntax file.

```
OPTIONS OLS=YES CONVERGE=0.001000 MAXITER=10 OUTPUT=STANDARD ;
TITLE=Kanfer and Ackerman data: Variance decomposition;
SY='D:\lisrel850\MLEVELEX\KANFER.PSF';
ID1=time;
ID2=control;
RESPONSE=measure;
FIXED=constant;
RANDOM1=constant;
RANDOM2=constant;
```

When the ***Run PRELIS*** icon button is clicked, the analysis is performed. The output obtained for this model is discussed in detail in the *LISREL 8: New Statistical Features Guide.*

5.3.3 Non-Linear Model for Air Traffic Data

In the previous example, a baseline model for the analysis of the air traffic data of Kanfer and Ackerman was considered.

In data of this nature, it can be expected that the number of successful landings per trial may be influenced by previous experience. In order to take this into account, we now introduce the order of the measurements into the model. The variable *time* indicates the number of the trial, and *timesq* the quadratic term equal to *time* $\times$*time*. The model previously given (see **kanfer2.pr2** in the **mlevelex** folder) is thus extended to

$$y_{ij} = \beta_0 + (time)_{ij}\beta_1 + (timesq)_{ij}\beta_2 + u_{0i} + (time)_{ij}u_{1i} + (timesq)_{ij}u_{2i} + e_{ij}$$

or, in matrix notation,

$$y_{ij} = \begin{bmatrix} constant & (time)_{ij} & (timesq)_{ij} \end{bmatrix} \begin{bmatrix} \beta_0 \\ \beta_1 \\ \beta_2 \end{bmatrix} + \begin{bmatrix} constant & (time)_{ij} & (timesq)_{ij} \end{bmatrix} \begin{bmatrix} u_{0i} \\ u_{1i} \\ u_{2i} \end{bmatrix}$$

$$+ \begin{bmatrix} constant \end{bmatrix} \begin{bmatrix} e_{ij} \end{bmatrix}$$

To fit such a model, the `FIXED` and `RANDOM2` commands used previously has to be changed to:

```
FIXED = constant time timesq;
RANDOM2 = constant time timesq;
```

These changes can be made by opening **kanfer1.pr2** in the **mlevelex** folder and making the required changes. Alternatively, if the syntax file **kanfer.pr2** file generated in the previous section is still open, one can directly edit the syntax file obtained in the previous analysis. With the ***.psf** file **kanfer.psf** open, one can also add the required variables using the ***Select Response and Fixed Variables*** and ***Random Variables*** dialog box of the ***Multilevel, Linear Model*** menu.

In the image below, changes to the syntax file **kanfer1.pr2** have been made to reflect the changes to the model (see **kanfer2.pr2** in the **mlevelex** folder).

KANFER2.PR2

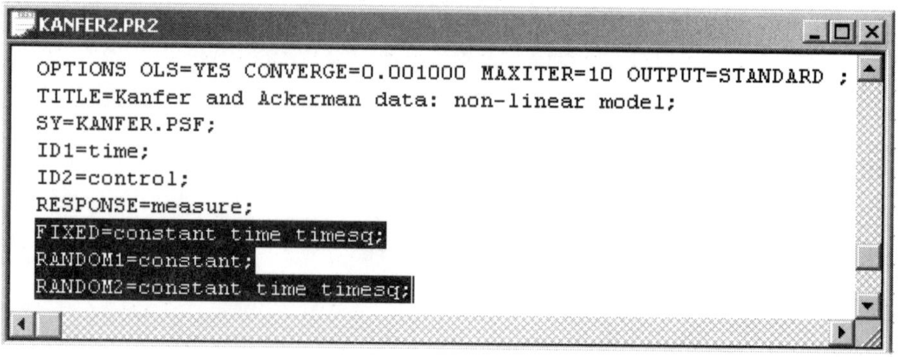

```
OPTIONS OLS=YES CONVERGE=0.001000 MAXITER=10 OUTPUT=STANDARD ;
TITLE=Kanfer and Ackerman data: non-linear model;
SY=KANFER.PSF;
ID1=time;
ID2=control;
RESPONSE=measure;
FIXED=constant time timesq;
RANDOM1=constant;
RANDOM2=constant time timesq;
```

Click the ***Run PRELIS*** icon button to run this model. The output obtained for this model is discussed in detail in the *LISREL 8: New Statistical Features Guide.* In the two models considered thus far, the measure of cognitive ability available for each controller has not been considered. In the third and final analysis of this data, we will include this measure in the model.

5.3.4 Including Additional Variables in the Air Traffic Data Analysis

Recall that the data set also includes a measure of cognitive ability (composite ASVAB score) for each of the 141 air traffic controllers, denoted by *ability* in **kanfer.psf**. We now include this variable as a fixed effect (covariate) in the model.

To add *ability* to the model, the syntax file used in the previous example may be edited. Alternatively, the variable may be added to the model in the ***Select Response and Fixed Variables*** dialog box of the ***Multilevel, Linear Model*** menu as shown below. After it has been added to the fixed variables field, proceed to the ***Random Variables*** dialog box by clicking ***Next***. Click ***Finish*** to generate the syntax file for this analysis.

This is equivalent to changing the model to:

$$\begin{aligned} y_{ij} &= \beta_0 + (time)_{ij}\beta_1 + (timesq)_{ij}\beta_2 + (ability)_{ij}\beta_3 \\ &\quad + u_{0i} + (time)_{ij}u_{1i} + (timesq)_{ij}u_{2i} + e_{ij} \end{aligned}$$

In matrix notation, this implies extending the fixed part of the model to:

$$\begin{bmatrix} constant & (time)_{ij} & (timesq)_{ij} & (ability)_{ij} \end{bmatrix} \begin{bmatrix} \beta_0 \\ \beta_1 \\ \beta_2 \\ \beta_3 \end{bmatrix}$$

Alternatively, the model for the different levels of the hiearchy can be expressed as:

$$y_{ij} = b_{0i} + (time)_{ij}b_{1i} + (timesq)_{ij}b_{2i} + (ability)_{ij}b_{3i} + e_{ij}$$

$$\begin{aligned} b_{0i} &= \beta_0 + +u_{0i} \\ b_{1i} &= \beta_1 + u_{1i} \\ b_{2i} &= \beta_2 + u_{2i} \\ b_3 &= \beta_3. \end{aligned}$$

The syntax file (**kanfer3.pr2** in the **mlevelex** folder) for this analysis is shown below:

```
OPTIONS OLS=YES CONVERGE=0.001000 MAXITER=10 OUTPUT=STANDARD ;
TITLE=Kanfer and Ackerman data: non-linear model plus ability;
SY=KANFER.PSF;
ID1=time;
ID2=control;
RESPONSE=measure;
FIXED=constant time timesq ability;
RANDOM1=constant;
RANDOM2=constant time timesq;
```

Click the ***Run PRELIS*** icon button to run this model. The output obtained for this model is discussed in detail in the *LISREL 8: New Statistical Features Guide.*

5.4 Multivariate Analysis of Educational Data

5.4.1 Description of Data and Multivariate Model

In practice, data sets often contain more than one dependent variable. One typical example is data from a longitudinal study, such as weight measurements of children from ages 3 to 14 in 30 countries. In this context, the countries may be regarded as the level-3 units, the children as the level-2 units and the weight measurements as the level-1 units. In this example, the variables *Weight3*, *Weight4* to *Weight14* are the dependent (response) variables. Possible predictors of weight are gender, ethnic group, average calorie intake and the weight of parents. Another example is data from a cross-sectional study where the response variables are aptitude test scores *Test1*, *Test2* to *Test10* of students from 30 different schools.

The multivariate model for 2 response variables y_1 and y_2 and a single predictor x can be written as

$$y_{ijk} = D_1(b_{01j} + b_{11j}x) + D_2(b_{02j} + b_{12j}x)$$

where $k = 1$ or 2 and D_1 and D_2 are dummy variables coded as follows:

$$D_1 = 1 \text{ if } k = 1 \text{ and 0 otherwise}$$
$$D_2 = 1 \text{ if } k = 2 \text{ and 0 otherwise.}$$

The level-2 random coefficients may be written as

$$
\begin{aligned}
b_{01j} &= c_{01} + u_{01j} \\
b_{02j} &= c_{02} + u_{02j} \\
b_{11j} &= c_{11} + u_{11j} \\
b_{12j} &= c_{12} + u_{12j},
\end{aligned}
$$

where $(u_{01j}, u_{02j}, u_{11j}, u_{12j})$ has covariance matrix $\mathbf{\Phi}_{(2)}$. In turn, the level-3 random coefficients can be written as

$$
\begin{aligned}
c_{01} &= \beta_{01} + v_{01} \\
c_{02} &= \beta_{02} + v_{02} \\
c_{11} &= \beta_{11} + v_{11} \\
c_{12} &= \beta_{12} + v_{12},
\end{aligned}
$$

where $(v_{01}, v_{02}, v_{11}, v_{12})$ has covariance matrix $\mathbf{\Phi}_{(3)}$.

The model described above consists of the fixed part $\beta_{01}, \beta_{02}, \beta_{11}$ and β_{12} and a random part described by the within covariance matrix $\mathbf{\Phi}_{(2)}$ and between covariance matrix $\mathbf{\Phi}_{(3)}$. The number of dummy variables depend on the number of response variables and, in general, $D_k = 1$ for the *k*-th response and 0 otherwise. There is no random component on level-1.

The data set used in this section forms part of the data library of the Multilevel Project at the University of London, and come from the Junior School Project (Mortimore *et al*, 1988). Mathematics and language tests were administered in three consecutive years to more than 1000 students from 50 primary schools, which were randomly selected from primary schools maintained by the Inner London Education Authority.

The following variables are available in the data file **jsp.psf**:

school	School code (1 to 50)
class	Class code (1 to 4)
student	Student ID (1 to 1402)
gender	Gender (boy = 1; girl = 0)
ravens	Ravens test score in year 1(score 4 - 36)
math1	Score on mathematics test in year 1 (score 1 - 40)
math2	Score on mathematics test in year 2 (score 1 - 40)
math3	Score on mathematics test in year 3 (score 1 - 40)

eng1	Score on language test in year 1 (score 0 - 98)
eng2	Score on language test in year 2 (score 0 - 98)
eng3	Score on language test in year 3 (score 0 - 98)
constant	Intercept, value = 1 throughout

The school number (*school*) is used as the level-3 identification variable, with Student ID (*student*) as the level-2 identification variable. The level-1 units are the scores obtained by a student for the mathematics and language tests, as represented by *math1* to *math3* and *eng1* to *eng3*.

The aim of this analysis is to examine the variation in test scores over students. It would also be interesting to determine the extent to which schools vary with respect to the response variable(s). One of the main benefits of analyzing different responses simultaneously in one multivariate analysis is that the way in which measurements relate to the explanatory variables can be directly explored. The gender, Ravens test score and an intercept term (*gender*, *ravens* and *constant* respectively) will be used as explanatory variables in the second of the models described in this section.

As the residual covariance matrices for the second and third levels of the hierarchy are also obtained from this analysis, differences between coefficients of explanatory variables for different responses can be studied.

Finally, each respondent does not have to be measured on each response, as is the case for the data set we consider here, where scores for all three years are not available for all students. This type of model is one in which the `MISSING_DEP` and `MISSING_DAT` commands can be used to good advantage, as will be shown. In the case of missing responses, a multilevel multivariate analysis of the responses that are available for respondents can be used to provide information in the estimation of those that are missing.

The first nine observations in the **jsp.psf** file are shown below:

JSP.PSF

	school	student	gender	ravens	math1	math2	math3	eng1
1	1.00	1.00	0.00	23.00	23.00	24.00	23.00	72.00
2	1.00	2.00	1.00	15.00	14.00	11.00	-9.00	7.00
3	1.00	3.00	1.00	22.00	36.00	32.00	39.00	88.00
4	1.00	4.00	1.00	14.00	24.00	26.00	32.00	12.00
5	1.00	5.00	0.00	19.00	22.00	23.00	-9.00	67.00
6	1.00	6.00	0.00	16.00	19.00	23.00	11.00	52.00
7	1.00	7.00	1.00	17.00	22.00	22.00	26.00	37.00
8	1.00	8.00	0.00	21.00	18.00	29.00	28.00	57.00
9	1.00	9.00	1.00	30.00	30.00	31.00	-9.00	42.00

One line of information is given for each student. Note that, for the second and fourth students, no mathematics score was available in the third year of the study (*math3*). A missing data code of `-9` was assigned to all missing values on both explanatory and response variables in the data set.

In order to perform a multilevel analysis, we need one line of information for each level-1 unit, in this case each of the six test scores. The data manipulation required in creating this revised data file format will be performed automatically by the program in the case of a multivariate model. Six dummy variables are created for each of the explanatory variables used. For the explanatory variable *gender*, for example, the dummy variables *gender1*, *gender2*,... *gender6* will denote the gender effect for each of the response variables.

The construction of two syntax files and a discussion of results obtained are given.

5.4.2 A Variance Decomposition Model

As a first step, we open the PRELIS system file **jsp.psf** located in the **mlevelex** folder and select the ***Title and Options*** option from the ***Multilevel, Linear Model*** menu, and provide a title for the analysis. No changes are made to the other keywords on the `OPTIONS` command, and thus the default convergence criterion and maximum number of iterations will be used. Only the default output file will be produced.

As a final step, the optional `MISSING_DEP` command is used to identify –9 as the missing data code for all response variables. Add the value `-9` to the ***Missing Dep Value*** field. We click ***Next*** to go to the ***Identification Variables*** dialog box.

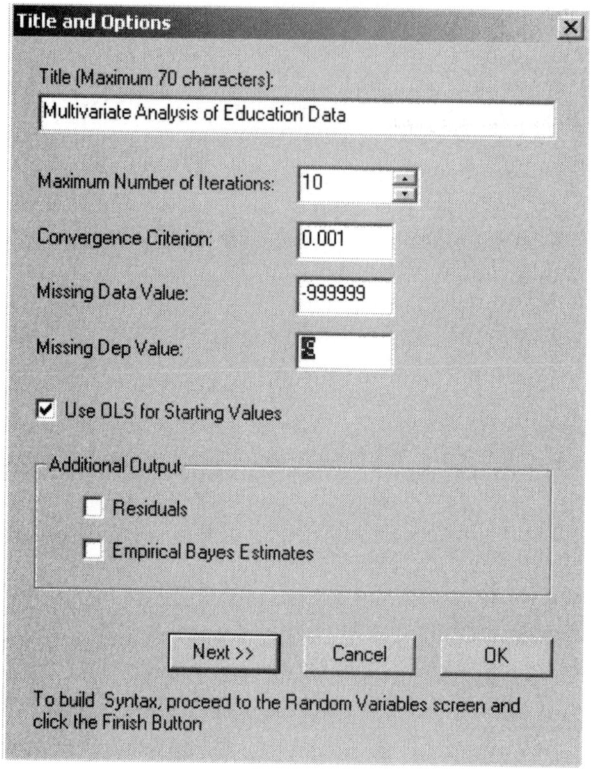

Title and Options

Title (Maximum 70 characters):
Multivariate Analysis of Education Data

Maximum Number of Iterations: 10

Convergence Criterion: 0.001

Missing Data Value: -999999

Missing Dep Value: 8

Use OLS for Starting Values

Additional Output

Residuals

Empirical Bayes Estimates

Next >> Cancel OK

To build Syntax, proceed to the Random Variables screen and click the Finish Button

In the case of a multivariate model, only levels 2 and 3 of the hierarchy have to be identified, as the actual measurements for each level-2 unit serve as level-1 units. In this case, we select *school* as the level-3 identification variable on the list of variables and click the ***Add*** button next to the ***Level 3 ID Variable*** field. The Student ID (*student*) is selected and entered as the ***Level 2 ID variable***. Click ***Next*** to move on to the ***Select Response and Fixed Variables*** dialog box.

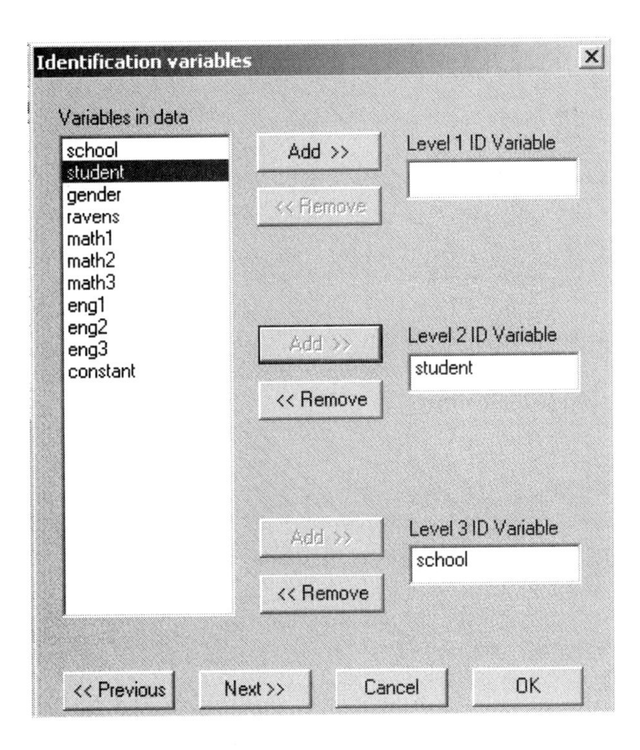

By holding down the CTRL key, the six variables *math1* to *math3* and *eng1* to *eng3* are selected and added to the ***Response Variables*** field by clicking the ***Add*** button as shown below. Clicking ***Next*** brings us to the last dialog box, where the random coefficients are to be selected.

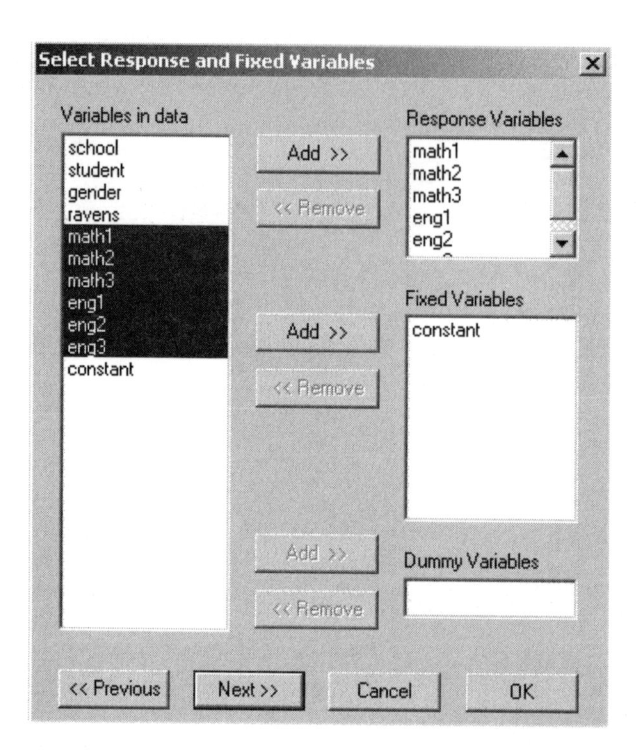

In this model, we wish to start our analysis of the data by studying the differences in intercepts over students and schools. Note that, in this analysis, no RANDOM commands have been included. These commands are only required when fitting a multivariate model and the coefficients of additional predictors, other than the intercept coefficients are believed to vary randomly on the school or student level. If no RANDOM commands are included, the inclusion of an intercept term on higher levels is automatically assumed by the program, and dummy variables for the explanatory predictors given in the FIXED command are generated. The ***Random Variables*** dialog box is thus left empty and the ***Finish*** button is clicked, after which the syntax file for this analysis is automatically generated by PRELIS.

The syntax file (**jsp1.pr2** in the **mlevelex** folder) is shown below. Click the ***Run PRELIS*** icon button on the main menu bar when done to start the analysis.

```
OPTIONS OLS=YES CONVERGE=0.001000 MAXITER=10 OUTPUT=STANDARD ;
TITLE=Multivariate Analysis of Education Data;
MISSING_DEP =-9.000000 ;
SY='D:\lisrel850\MLEVELEX\JSP.PSF';
ID2=student;
ID3=school;
RESPONSE=math1 math2 math3 eng1 eng2 eng3;
FIXED=constant;
MISSING_DEP=-9;
```

The output for this model is written to the default output file **jsp.out**, and is displayed after completion of the analysis. The output obtained for this model is discussed in detail in the *LISREL 8: New Statistical Features Guide.*

5.4.3 Adding Explanatory Variables to the Model

Using the model discussed in the previous section as point of departure, we now proceed to add fixed effects to the model. The variables *gender* and *ravens*, indicating the gender of a student and the student's score on the Ravens test in the first year of the study respectively, are added to the FIXED command.

This implies use of the FIXED command:

```
FIXED = constant gender ravens;
```

To make this change, the syntax file previously used (**jsp.pr2**) can be edited directly, or the required change can be made through the interface by selecting the ***Select Response and Fixed Variables*** option from the ***Multilevel, Linear Model*** menu and adding the variables *gender* and *ravens* to the ***Fixed Variables*** field while **jsp.psf** is still open. Clicking ***Finish*** on the ***Random Variables*** dialog box will automatically update the syntax file.

The revised syntax file (**jsp2.pr2** in the **mlevelex** folder) is given below.

```
OPTIONS OLS=YES CONVERGE=0.001000 MAXITER=10 OUTPUT=STANDARD ;
TITLE=Multivariate Analysis: added fixed effects;
SY=JSP.PSF;
ID2=student;
ID3=school;
RESPONSE=math1 math2 math3 eng1 eng2 eng3;
FIXED=constant gender ravens;
MISSING_DEP=-9;
```

In analysis of the type described above, one can use the values of either the level-3 covariance matrix (the schools) or the level-2 covariance matrix (students) as input to LISREL to fit an appropriate structural equation model to the data.

The output obtained for this model is discussed in detail in the *LISREL 8: New Statistical Features Guide.*

5.5 3-level model for CPC Survey Data

5.5.1 Description of CPC Survey Data

In multilevel modeling, regression coefficients are allowed to vary randomly at any level of the hierarchy. As an example, suppose that verbal ability (y) is to be predicted from an aptitude test score (x). Suppose also that students are nested within schools and that schools are nested within education departments.

A possible level-3 model is

$$y_{ijk} = b_{0j} + b_{1j} x_{ijk} + e_{ijk},$$

where i denotes education departments, j schools and k students within schools. b_{0j} and b_{1j} become outcome (dependent) variables on level 2 (students), with

$$b_{0j} = c_0 + u_{0j}$$
$$b_{1j} = c_1 + u_{1j}.$$

Similarly, c_0 and c_1 become outcome variables on level-3 (education departments), where $c_0 = \beta_0 + v_0$ and $c_1 = \beta_1 + v_1$.

The coefficients c_0 and c_1 vary randomly over education departments while b_{0j} and b_{1j} vary randomly over schools. Substitution of c_0 and c_1 into the multilevel random coefficient model gives

$$y_{ijk} = \beta_0 + \beta_1 x_{ijk} + v_0 + u_{0j} + (v_1 + u_{1j}) x_{ijk} + e_{ijk}$$

or, in matrix notation

$$y_{ijk} = \begin{bmatrix} 1 & x \end{bmatrix} \begin{bmatrix} \beta_0 \\ \beta_1 \end{bmatrix} + \begin{bmatrix} 1 & x \end{bmatrix} \begin{bmatrix} v_0 \\ v_1 \end{bmatrix} + \begin{bmatrix} 1 & x \end{bmatrix} \begin{bmatrix} u_{0j} \\ u_{1j} \end{bmatrix} + e_{ijk}.$$

It is usually assumed that (v_0, v_1) has a multivariate normal distribution with zero means and covariance $\mathbf{\Phi}_{(3)}$, independently distributed of (u_{0j}, u_{1j}) which has a multivariate normal distribution with zero means and covariance matrix $\mathbf{\Phi}_{(2)}$.

β_0 and β_1 are the fixed parameters while the elements of $\mathbf{\Phi}_{(1)}$, $\mathbf{\Phi}_{(2)}$, and $\mathbf{\Phi}_{(3)}$ describe the distribution of the random part of the model.

In this example, data from the March 1995 Current Population Survey are used. The data set is a subset of data obtained from the Data Library at the Department of Statistics at UCLA. A small number of demographic variables for two occupation groups were extracted, and all analyses are based on unweighted data.

Only respondents between the ages of twenty-one and sixty-five, who held full-time positions in 1994 and had an annual income of at least $1 were considered. The two groups we will focus on here are defined as follows:

Educational sector : Respondents with professional specialty in the educational sector
Construction sector : Operators, fabricators and laborers in the construction sector.

The variable *group* in the data spreadsheet **income.psf** represents the groups, with $group = 0$ for the respondents in the construction sector and $group = 1$ for respondents in the educational sector.

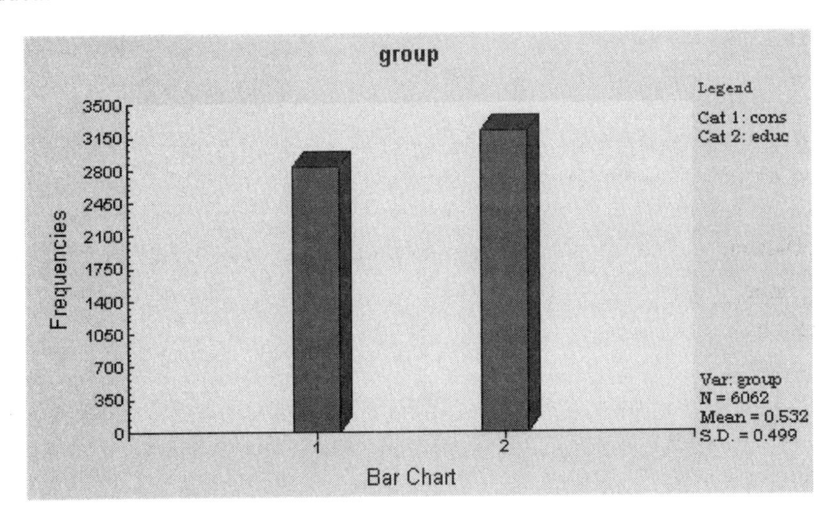

Other demographic variables and their codes are:

gender	0 = female; 1 = male
age	Age in single years
marital	1 = married; 0 = other
hours	Hours worked during last week prior to survey at all jobs
citizen	1 for native Americans, 0 for all foreign born respondents
income	The natural logarithm of the personal income during 1994
degree	1 for respondents with master's degrees, professional school degree or doctoral degree; 0 otherwise

Respondents were from 9 divisions of the USA, and the state of residence was also given. The variables *region* and *state* represent this information. For the full description of the regions and states within regions, see the *LISREL 8: New Statistical Features Guide*. On the respondent level, the variable *person* is a respondent's identity number.

Finally, the variable *constant* denotes the intercept term and has a value of 1 for all respondents. The variable *income* will be used as response variable.

5.5.2 3-level Model for CPC Survey Data

In the model to be fitted to the CPC data, the predictors *age*, *gender*, *marital*, *hours*, *citizen*, *constant*, *degree* and *group* are used to predict *income*. In addition, it is assumed that the intercept varies randomly over regions and states.

The model to be fitted can be written as

$$y_{ijk} = \beta_{0j} + \beta_{1j}(age)_{ij} + \beta_{2j}(gender)_{ij} + \beta_{3j}(marital)_{ij} + \beta_{4j}(hours)_{ij} + \beta_{5j}(citizen)_{ij} + \beta_{6j}(degree)_{ij} + \beta_{7j}(group)_{ij} + v_0 + u_{0j} + e_{ijk},$$

or, in matrix notation, as

$$y_{ijk} = \begin{bmatrix} constant & age_{ij} & gender_{ij} & marital_{ij} & hours_{ij} & citizen_{ij} & degree_{ij} & group_{ij} \end{bmatrix} \begin{bmatrix} \beta_{0j} \\ \beta_{1j} \\ \beta_{2j} \\ \beta_{3j} \\ \beta_{4j} \\ \beta_{5j} \\ \beta_{6j} \\ \beta_{7j} \end{bmatrix}$$

$$+[constant][v_0]+[constant]\left[u_{0j}\right]+[constant]e_{ijk}$$

The data set used, as described in the previous section, is contained in the data spreadsheet **income.psf**. The first six records of this file are shown below:

INCOME.PSF

	region	state	age	gender	marital	hours	citizen
1	1.00	11.00	59.00	0.00	0.00	40.00	1.00
2	1.00	11.00	56.00	1.00	1.00	40.00	1.00
3	1.00	11.00	64.00	0.00	1.00	12.00	1.00
4	1.00	11.00	30.00	1.00	1.00	40.00	1.00
5	1.00	11.00	27.00	0.00	1.00	40.00	1.00
6	1.00	11.00	49.00	0.00	1.00	40.00	1.00

We begin setting up the model by selecting the ***Title and Options*** option from the ***Multilevel, Linear Model*** menu on the main menu bar. In this dialog box, a title for the analysis (optional) is provided. When done, click ***Next*** to go to the ***Identification Variables*** dialog box. Doing so implies that the default settings for the maximum number of iterations, convergence criterion and output are used.

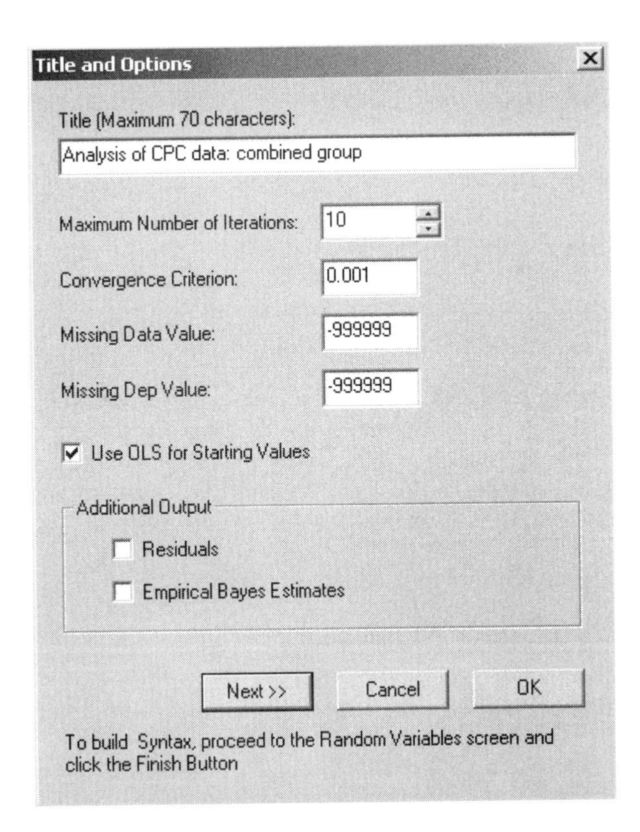

As all respondents are nested within state of residence, and states are in turn nested within the nine divisions, we select *region* and use the ***Add*** button to add this variable to the level-3 identification variable field. The variables *state* and *person* are selected in similar fashion as level-2 and level-1 identification variables respectively. Clicking ***Next*** leads to the display of the ***Select Response and Fixed Variables*** dialog box.

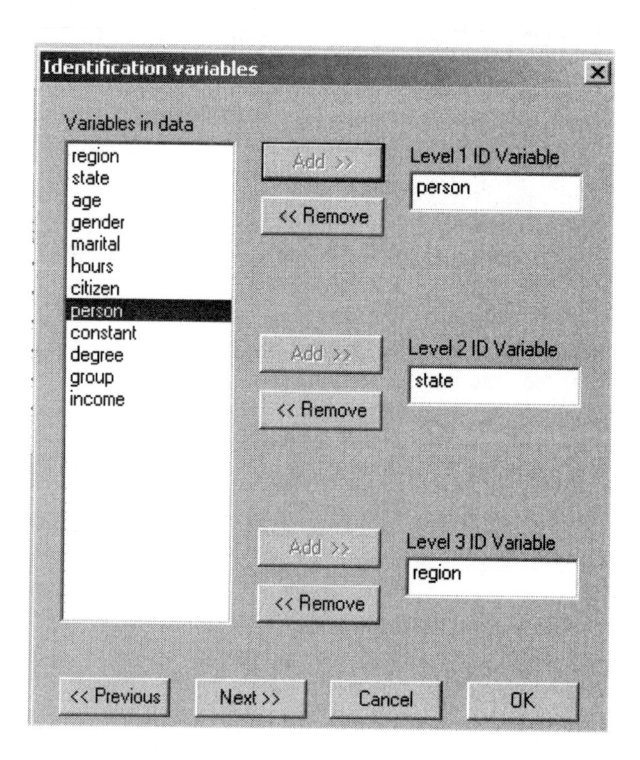

In the ***Select Response and Fixed Variables*** dialog box, we select *income*, representing the natural logarithm of personal income, as the response variable (level-1 outcome) for this analysis. By holding down the CTRL key while clicking on the variables *age*, *gender*, *marital*, *hours*, *citizen*, *constant*, *degree* and *group*, and clicking ***Add*** next to the ***Fixed Variables*** field, we enter all these variables into the model as fixed effects.

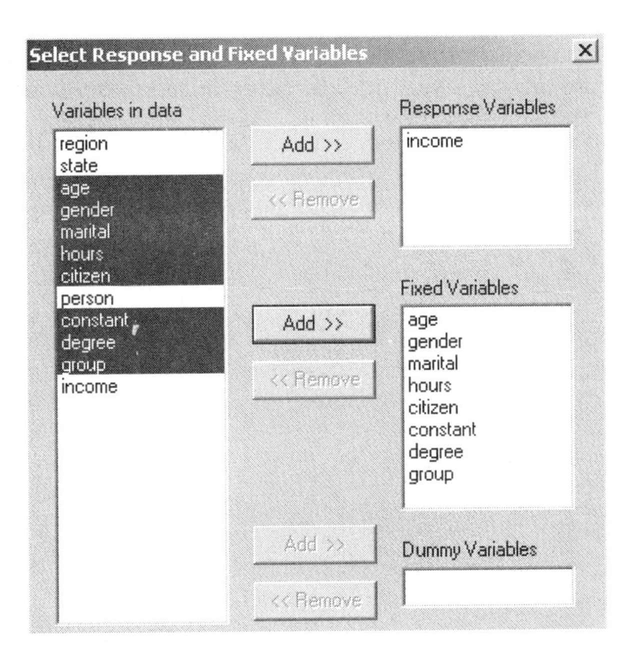

Clicking ***Next*** leads to the display of the final dialog box, in which the random effects are identified.

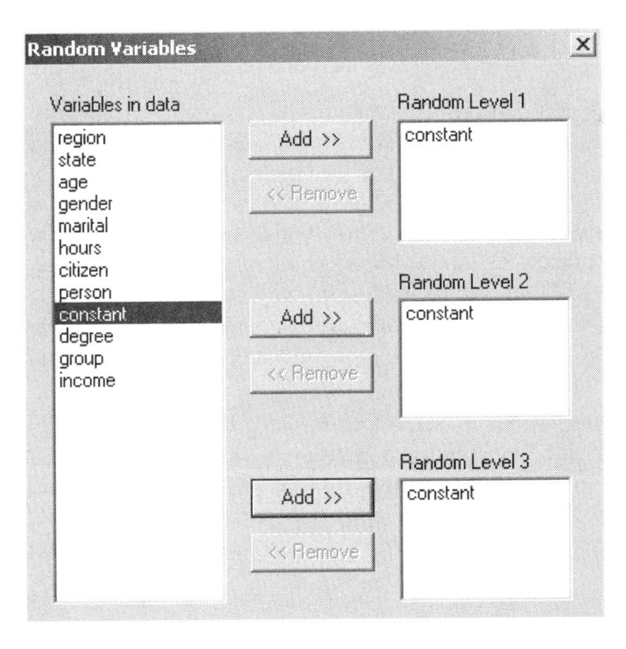

When the ***Finish*** button is clicked after the variable *constant*, representing the intercept term, is included as a random effect on all levels of the hierarchy, the syntax file shown below is automatically generated:

```
OPTIONS OLS=YES CONVERGE=0.001000 MAXITER=10 OUTPUT=STANDARD ;
TITLE=Analysis of CPC data: combined group;
SY='D:\lisrel850\MLEVELEX\INCOME.PSF';
ID1=person;
ID2=state;
ID3=region;
RESPONSE=income;
FIXED=age gender marital hours citizen constant degree group;
RANDOM1=constant;
RANDOM2=constant;
RANDOM3=constant;
```

Click the ***Run PRELIS*** icon button to start the analysis. The output obtained for this model, and similar models for the two groups (education and construction) are discussed in detail in the *LISREL 8: New Statistical Features Guide.*

5.6 Latent Growth Curve Models

5.6.1 Description of the Data

The data set discussed in this section contains repeated measurements made on 9 rats and a covariate value for each rat. The first value in each row denotes the rat number, and the next 8 values are reaction time over age and age-related changes in bio-markers, measured on 8 equally spaced occasions. The last value, which is the covariate, is a measure of the amount of the damage to a specific area of the brain prior to the onset of the experiment (Wiesen, C.A., University of North Carolina at Chapel Hill). The complete data set is given in Table 5.1.

In this section, two models are fitted to the rat data. In the first example, an intercepts-and-slopes model is fitted. The second example illustrates a model that is extended by assuming that the response function is a second degree polynomial and by including brain damage as a covariate. The empirical Bayes estimates and residuals for this model are discussed. Finally it is shown how this type of growth curve model can also be fitted using SIMPLIS syntax.

Table 5.1: Rat data

Rat	Measures of reaction time	Damage							
1	397.4	462.0	417.0	427.8	462.6	453.0	434.2	434.8	-0.543
2	480.6	450.0	455.6	472.4	477.8	468.0	452.4	447.0	-0.450
3	406.6	439.6	445.0	528.0	530.6	549.8	571.8	530.8	-0.399
4	424.6	428.4	467.4	470.0	469.4	481.8	478.8	480.0	-0.153
5	372.0	379.8	373.2	367.4	399.4	412.8	407.6	606.0	1.013
6	358.6	377.0	386.8	394.8	395.4	426.8	403.8	416.4	1.019
7	403.0	383.2	392.4	389.4	449.2	431.8	437.0	478.8	1.331
8	378.6	372.2	359.8	368.2	364.8	386.4	368.0	382.2	1.477
9	380.4	412.2	398.8	389.6	396.6	411.2	441.6	473.2	2.030

5.6.2 A 2-level Intercept-and-Slopes Model

In this example, a 2-level MRCM is fitted to the rat data set. It is assumed that an intercepts-and-slopes model will provide an adequate description of the within- and between-rats variation of the reaction time measurements. The data set, **rat.psf**, can be found in the **tutorial** folder. The first few lines of the ***.psf** file are shown below.

RAT.PSF

	ratnumb	reaction	intercep	time	timesq	damage
1	1.00	397.40	1.00	1.00	1.00	-0.54
2	1.00	462.00	1.00	2.00	4.00	-0.54
3	1.00	417.00	1.00	3.00	9.00	-0.54
4	1.00	427.80	1.00	4.00	16.00	-0.54
5	1.00	462.60	1.00	5.00	25.00	-0.54
6	1.00	453.00	1.00	6.00	36.00	-0.54
7	1.00	434.20	1.00	7.00	49.00	-0.54
8	1.00	434.80	1.00	8.00	64.00	-0.54
9	2.00	480.60	1.00	1.00	1.00	-0.45
10	2.00	450.00	1.00	2.00	4.00	-0.45

The variables available are:

ratnumb:	The identification number of the rat (level-2 ID variable)
reaction:	The reaction time of the rat to the stimulus
intercep:	A column of 1's to be used to represent the intercept in the model
time:	The occasion of measurement
timesq:	$time \times time$
damage:	The amount of brain damage sustained by the rat

In the model to be fitted, it is assumed that the time of measurement is related to the reaction time of the rat to the stimulus. Additionally, both intercept and time are assumed to vary randomly over the rats (the level-2 units). The outcome variable is *response*, and *time* denotes the occasion of measurement.

The model to be fitted is:

$$y_{ij} = b_{0i} + b_{1i}(time)_{ij} + e_{ij}$$

where

$$b_{0i} = \beta_0 + u_{0i}$$
$$b_{1i} = \beta_1 + u_{1i}.$$

The combined model can be written as

$$y_{ij} = \beta_0 + \beta_1(time)_{ij} + u_{0i} + u_{1i}(time)_{ij} + e_{ij}$$

or, in matrix notation, as

$$y_{ij} = \begin{bmatrix} constant & (time)_{ij} \end{bmatrix} \begin{bmatrix} \beta_0 \\ \beta_1 \end{bmatrix} + \begin{bmatrix} constant & (time)_{ij} \end{bmatrix} \begin{bmatrix} u_{0i} \\ u_{1i} \end{bmatrix} + e_{ij}$$

The first step in creating syntax for this model is to open the data file **rat.psf** in the **tutorial** folder. Select the ***Title and Options*** option from the ***Multilevel, Linear Model*** menu on the main menu bar and provide a title for the analysis, for example `Reaction time of 9 rats: slopes and intercepts`. When done, click ***Next*** to go to the ***Identification Variables*** dialog box. As the repeated measurements of the rats' responses are nested within the individual rats, select *ratnumb* and use the ***Add*** button to add this variable to the level-2 identification variable field. Use the variable *time* as level-1 identification in order to distinguish between the multiple measurements available for each rat. Clicking ***Next*** leads to the display of the ***Select Response and Fixed Variables*** dialog box.

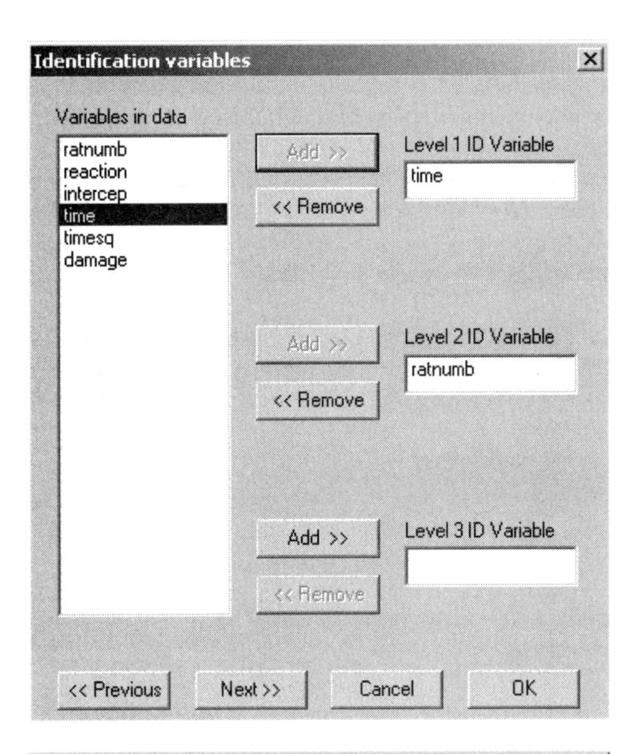
Identification variables
Variables in data
ratnumb
reaction
intercep
time
timesq
damage
Add >>
<< Remove
Level 1 ID Variable
time
Add >>
<< Remove
Level 2 ID Variable
ratnumb
Add >>
<< Remove
Level 3 ID Variable
<< Previous
Next >>
Cancel
OK

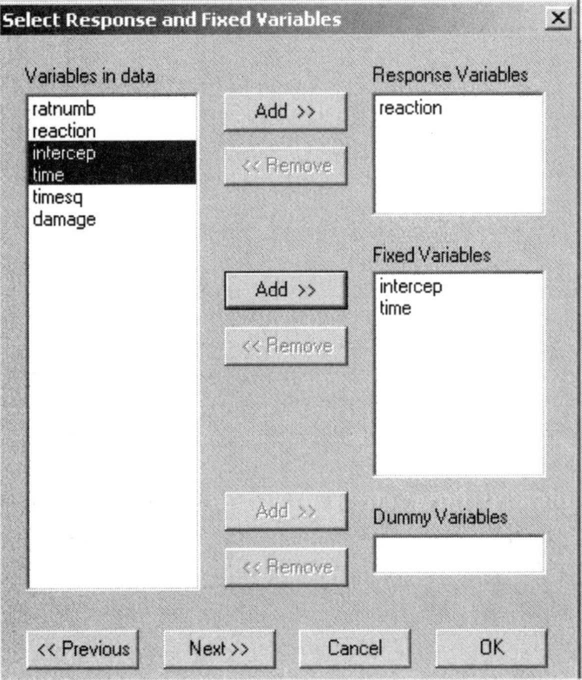
Select Response and Fixed Variables
Variables in data
ratnumb
reaction
intercep
time
timesq
damage
Add >>
<< Remove
Response Variables
reaction
Fixed Variables
intercep
time
Add >>
<< Remove
Add >>
<< Remove
Dummy Variables
<< Previous
Next >>
Cancel
OK

In the ***Select Response and Fixed Variables*** dialog box, select *reaction* as the response variable for this analysis. Select the variables *intercep* and *time* and click ***Add*** next to the ***Fixed Variables*** field to enter these variables into the model as fixed effects.

Clicking ***Next*** leads to the display of the final dialog box, in which the random effects are identified. Select the variable *intercep*, representing the intercept term, and use the ***Add*** button to include this variable in the ***Random Level 1*** field. Select and add *intercep* and *time* in a similar way to the ***Random Level 2*** field.

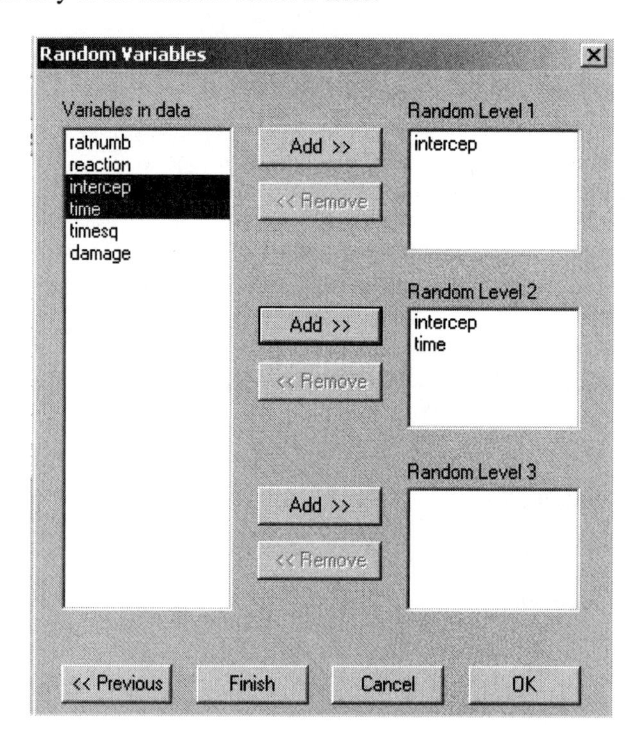

Click ***Finish*** to generate the syntax file (**rat1.pr2** in the **tutorial** folder) shown below.

```
OPTIONS OLS=YES CONVERGE=0.001000 MAXITER=10 OUTPUT=STANDARD ;
TITLE=Reaction time of 9 rats: slopes and intercepts;
SY='D:\lisrel850\TUTORIAL\RAT.PSF';
ID1=time;
ID2=ratnumb;
RESPONSE=reaction;
FIXED=intercep time;
RANDOM1=intercep;
RANDOM2=intercep time;
```

Click the ***Run PRELIS*** icon button to start the analysis. Partial output is given below.

```
DATA SUMMARY


NUMBER OF LEVEL 2 UNITS : 9
NUMBER OF LEVEL 1 UNITS : 72

N2 : 1 2 3 4 5 6 7 8
N1 : 8 8 8 8 8 8 8 8


N2 : 9
N1 : 8
```

The data summary section contains information on the number of units at the different levels of the hierarchy. It can be concluded that there were eight measurements (level-1 units) for each level-2 unit, the level-2 units being the 9 rats.

This part of the output is followed by details of the fixed part of the model, the -2 log likelihood function value and the estimates of the random components (the variances and covariances of the random coefficients).

```
Reaction time of 9 rats: slopes and intercepts model


ITERATION NUMBER 2


FIXED PART OF MODEL


COEFFICIENTS     BETA-HAT   STD.ERR.   Z-VALUE    PR > |Z|
intercep        388.58889  14.81303   26.23292   0.00000
time              9.39259   2.71065    3.46507   0.00053
```

From the fixed part of the model it follows that both the intercept and slope coefficients (*intercep* and *time*) are highly significant.

```
-2 LOG-LIKELIHOOD = 712.997140197850
```

A -2 log likelihood value (the so-called deviance statistic) of 712.997 is obtained.

```
RANDOM PART OF MODEL


LEVEL 2                  TAU-HAT  STD.ERR.   Z-VALUE    PR > |Z|
intercep/intercep     1545.41875 934.60543   1.65355    0.09822
time /intercep        -141.09366 141.24094  -0.99896    0.31782
time /time              49.28907  31.34142   1.57265    0.11580
```

```
LEVEL 1                 TAU-HAT    STD.ERR.   Z-VALUE    PR > |Z|
intercep/intercep      707.26842  136.11387  5.19615    0.00000


LEVEL 2 COVARIANCE MATRIX


           intercep        time
intercep   1545.41875
time       -141.09366      49.289


LEVEL 2 CORRELATION MATRIX


           intercep   time
intercep    1.0000
time       -0.5112    1.0000


LEVEL 1 COVARIANCE MATRIX


           intercep
intercep   707.26842


LEVEL 1 CORRELATION MATRIX


           intercep
intercep   1.0000
```

All *p*-values for the Tau-parameters at level-2 of the model are larger than 0.05. According to these values, there is no significant variation in the slopes and intercepts over rats. This is a result that should be interpreted with caution, however, since standard errors of the variances and covariances tend to be inaccurate when dealing with small samples.

In the next section, a non-linear model with a covariate will be fitted.

5.6.3 A Non-Linear Model with Covariate

In this example, a quadratic term is included in the model, as well as the brain damage measurement as a covariate (level-2 predictor).

The MRCM model for this case is

$$y_{ij} = b_{0i} + b_{1i}(time)_{ij} + b_{2i}(timesq)_{ij} + e_{ij}$$

where

$$\begin{aligned} b_{0i} &= \beta_0 + \gamma_0(damage)_{ij} + u_{0i} \\ b_{1i} &= \beta_1 + \gamma_1(damage)_{ij} + u_{1i} \\ b_{2i} &= \beta_2 + \gamma_2(damage)_{ij} + u_{2i} \end{aligned}$$

or, in matrix notation, as

$$y_{ij} = \begin{bmatrix} constant & (time)_{ij} & (timesq)_{ij} \end{bmatrix} \begin{bmatrix} b_{0i} \\ b_{1i} \\ b_{2i} \end{bmatrix} + \begin{bmatrix} constant & (time)_{ij} & (timesq)_{ij} \end{bmatrix} \begin{bmatrix} u_{0i} \\ u_{1i} \\ u_{2i} \end{bmatrix} + e_{ij},$$

and where we assume that (u_{0i}, u_{1i}, u_{2i}) are i.i.d. multivariate normal with zero means and covariance matrix $\mathbf{\Phi}_{(2)}$. Note that the subscript *i* denotes rat number *i* while *j* denotes the *j*-th measurement.

Substitution of the expressions for the random coefficients b_{0i}, b_{1i}, and b_{2i} into the MRCM model gives

$$\begin{aligned} y_{ij} = \beta_0 + \gamma_0(damage)_{ij} + \beta_1(time)_{ij} + \gamma_1(time)_{ij}(damage)_{ij} + \\ \beta_2(timesq)_{ij} + \gamma_2(timesq)_{ij}(damage)_{ij} + \\ u_{0i} + u_{1i}(time)_{ij} + u_{2i}(timesq)_{ij} + e_{ij}. \end{aligned}$$

The fixed part of the model consists of 6 parameters $(\beta_0, \beta_1, \beta_2, \gamma_0, \gamma_1, \gamma_2)$ while the random part has 7 parameters, these being the 6 non-duplicated elements of $\mathbf{\Phi}_{(2)}$ and the random error variance $\mathbf{\Phi}_{(1)}$. The total number of parameters to be estimated in the current model are therefore 13, as opposed to 6 in the slopes-and-intercepts model described in the previous section.

To add the covariate *damage* to the model, edit the syntax file obtained in the previous section directly (see **rat2.pr2** in the **tutorial** folder). The variable *damage* may be added to the syntax file through use of the ***Response and Fixed Variables*** dialog box, but the terms *damage*time* and *damage*timesq* can not. Alternatively, one may create new variables *damtim* (*damage*time*) and *damtimsq* (*damage*timesq*) in the *.**psf** file by using the ***Compute*** option of the ***Transformation*** menu. Change the optional TITLE command to reflect the changes to the model.

The syntax to run this example (**rat2.pr2** in the **tutorial** folder) is as follows:

```
OPTIONS OLS=YES CONVERGE=0.001000 MAXITER=10 OUTPUT=ALL ;
TITLE=Reaction time: nonlin response function and covariate;
SY=RAT.PSF;
ID1=time ;
ID2=ratnumb ;
RESPONSE=reaction ;
FIXED=intercep  time timesq damage damage*time damage*timesq;
RANDOM1=intercep ;
RANDOM2=intercep  time timesq;
```

Note that the OUTPUT = ALL option is used here, in order to obtain both Bayes and residual files. The contents of these additional output files are discussed in the next section.

Click the ***Run PRELIS*** icon button to start the analysis. At convergence, the output shown below is obtained.

```
FIXED PART OF MODEL


COEFFICIENTS      BETA-HAT    STD.ERR.    Z-VALUE     PR > |Z|
intercep         396.13190   14.89758    26.59035    0.00000
time              16.93480    9.22519     1.83571    0.06640
timesq            -0.90394    1.09096    -0.82857    0.40735
damage            -0.39462   13.58795    -0.02904    0.97683
time *damage     -20.15988    8.41421    -2.39593    0.01658
timesq *damage     2.35140    0.99506     2.36308    0.01812


-2 LOG-LIKELIHOOD = 681.809481715634


RANDOM PART OF MODEL


LEVEL 2                 TAU-HAT    STD.ERR.    Z-VALUE     PR > |Z|
intercep/intercep      583.90036  690.03911    0.84618    0.39745
time /intercep        -352.59671  391.49755   -0.90064    0.36778
time /time             326.64996  259.94228    1.25662    0.20889
timesq /intercep        36.27301   43.29603    0.83779    0.40215
timesq /time           -37.80417   29.84338   -1.26675    0.20524
timesq /timesq           5.04842    3.61898    1.39498    0.16302
```

```
LEVEL 1                  TAU-HAT    STD.ERR.   Z-VALUE    PR > |Z|
intercep/intercep       427.36419  90.09628   4.74342    0.00000


LEVEL 2 COVARIANCE MATRIX


           intercep          time         timesq
intercep    583.90036
time       -352.59671        326.64996
timesq       36.27301        -37.80417    5.04842


LEVEL 2 CORRELATION MATRIX


           intercep          time         timesq
intercep    1.0000
time       -0.8074            1.0000
timesq      0.6681           -0.9309      1.0000


LEVEL 1 COVARIANCE MATRIX


           intercep
intercep   427.36419


LEVEL 1 CORRELATION MATRIX


           intercep
intercep   1.0000
```

Note that the number of parameters estimated for the fixed part of the model is 6, while the number of coefficients for the random part of the model (Tau) is 7.

Hypothesis test:

To test the hypothesis that the non-linear model with covariate fits the data better than the intercept-and-slopes model of Section 5.6.2, we calculate the difference between the $-2\ln L$ value obtained for the model in Section 5.6.2 and the $-2\ln L$ value obtained here. It can be shown that this difference

712.997 - 681.809 = 31.18,

has a χ^2 distribution with 13 - 6 = 7 degrees of freedom. The degrees of freedom is the difference in the number of parameters estimated in the two examples. Since the p-value for this test is smaller than 0.01, it is concluded that the model described in this section provides a better description of the data than the model in the previous section. In the next section, empirical Bayes estimates and residuals are discussed.

5.6.4 Empirical Bayes Estimates and Residuals

In this section, the Empirical Bayes estimates and residuals obtained in the previous example are discussed. Estimates of the fixed coefficients are given in the output file in the previous section. Table 5.2 contains the estimates of these coefficients.

Table 5.2: Estimated parameters for non-linear model fitted to rat data

Coefficient	Estimate
β_0	396.132
β_1	16.935
β_2	-0.904
γ_0	-0.395
γ_1	-20.160
γ_2	2.351

By specifying `PRINT = ALL` in the `OPTIONS` command of the syntax file (see **rat2.pr2** in the **tutorial** folder), the program produces a file **rat2.ba2** containing the empirical Bayes residuals (the estimated values of u_0, u_1 and u_2 respectively) and their variances as well as a file **rat2.res** which contains the differences between the observed and predicted reaction times of the rats.

The empirical Bayes residuals and variances for rats numbers 2 and 8 are given in Table 5.3 below.

Table 5.3: Empirical Bayes residuals and variances for selected rats

Rat Number	Residual	Variance	Predictor
2	23.665 (u_0)	327.40	*intcept*
2	-4.994 (u_1)	107.46	*time*
2	-0.26 (u_2)	1.43	*time_sq*
8	-2.982 (u_0)	327.40	*intcept*
8	0.407 (u_1)	107.46	*time*
8	-1.048 (u_2)	1.43	*time_sq*

Using the results of Tables 5.2 and 5.3 we obtain for rat 2:

$$y = 396.13 - 0.395\,damage + 23.665 + (16.935 - 20.160\,damage - 4.994)\times time$$
$$+ (-0.904 + 2.351\,damage - 0.216)\times time\_sq$$

Similarly, for rat 8:

$$y = 396.13 - 0.395\,damage - 2.982 + (16.935 - 20.160\,damage + 0.407)\times time$$
$$+ (-0.904 + 2.351\,damage - 1.048)\times time\_sq$$

Figure 5.1 below shows the predicted values for an "average rat" (estimated values of u_0, u_1 and u_2 all equal to zero) and brain damage values of -2, -1, 0, 1 and 2 respectively.

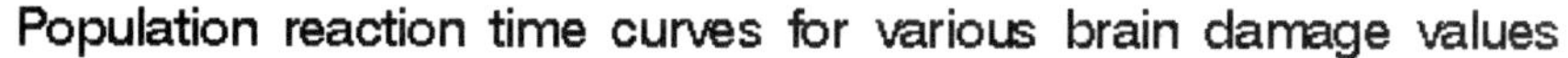

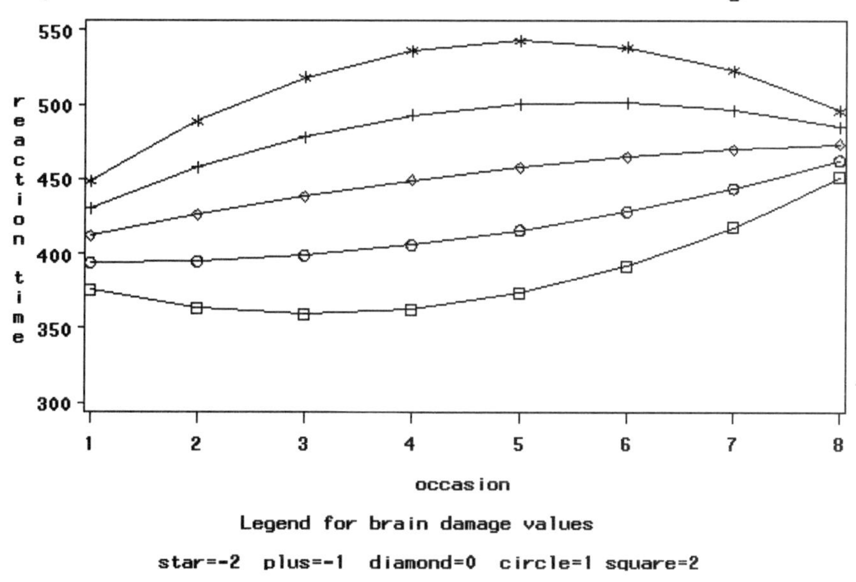

Figure 5.1: Predicted values for an average rat and various brain damage values

Shown below are the observed values, predicted values and residuals (e_i) for rat number 8 contained in **rat2.res**. The first column is the observation number, followed by the rat number and the level-1 ID. Columns 4, 5 and 6 are the observed, predicted and residual values respectively.

```
57  8   1   378.60   385.28   -6.6768
58  8   2   372.20   380.14   -7.9427
59  8   3   359.80   380.15   -20.347
```

```
60  8   4   368.20   385.29   -17.089
61  8   5   364.80   395.57   -30.769
62  8   6   386.40   410.99   -24.588
63  8   7   368.00   431.54   -63.544
64  8   8   382.20   457.24   -75.039
```

In Figure 5.2, the residuals for all rats are plotted by observation number.

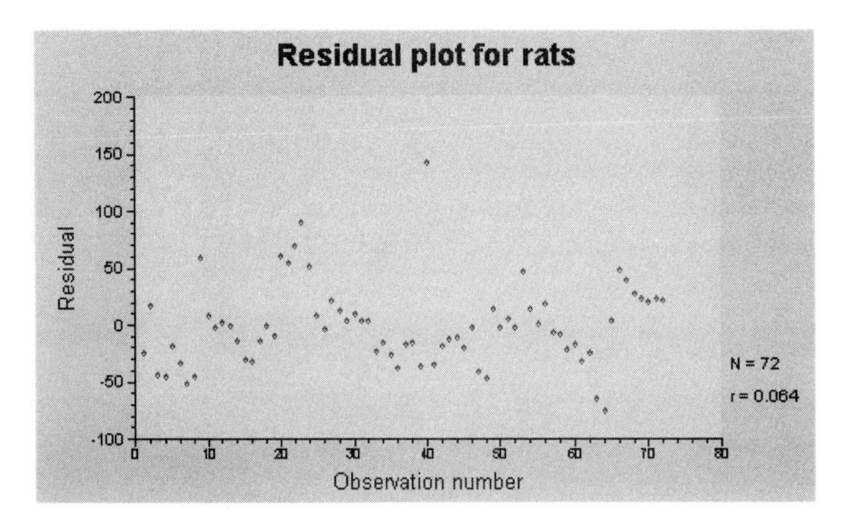

Figure 5.2: Plots of residuals by observation number

5.6.5 Latent Growth Curve Model using SIMPLIS Syntax

It is possible to fit a random coefficient growth curve model (second degree polynomial) to the rat data using SIMPLIS syntax. The data given in Table 5.1 in the previous section are used to calculate the sample covariance matrix and means of the eight repeated measurements made on each of the nine rats.

The elements of the covariance matrix and the mean vector are stored in the files **rat.cov** and **rat.mea**, respectively. The observed variables are the eight reaction time measurements and the latent variables the *intercept*, *time* and *time_sq* predictors, respectively.

To obtain an estimate of the level-1 error variance, the error variances of *React1* to *React8* are constrained to be equal. The SIMPLIS syntax is given below.

```
SIMPLIS Syntax for growth curve model
Reaction times of 9 rats, Second degree polynomial
```

```
Observed Variables
React1 React2 React3 React4 React5 React6 React7 React8

Covariance Matrix
1312.60
828.89 1206.10
1066.40 1061.50 1436.50
1283.60 1463.20 1884.40 3103.90
1222.80 1391.40 1672.20 2706.00 2781.30
869.83 1144.50 1504.20 2624.30 2440.60 2377.30
842.27 1216.50 1627.90 2882.50 2715.30 2665.10 3310.30
-108.15 -5.73 231.19 488.29 947.68 926.36 1509.90 4336.10
Means
400.20 411.60 410.67 423.07 438.42 446.84 443.91 472.13
Sample Size = 9
Latent Variables intcept time time_sq

Relationships
React1 = 1*intcept 1*time 1*time_sq
React2 = 1*intcept 2*time 4*time_sq
React3 = 1*intcept 3*time 9*time_sq
React4 = 1*intcept 4*time 16*time_sq
React5 = 1*intcept 5*time 25*time_sq
React6 = 1*intcept 6*time 36*time_sq
React7 = 1*intcept 7*time 49*time_sq
React8 = 1*intcept 8*time 64*time_sq
intcept time time_sq = CONST
Equal error variances: React1 - React8
Path Diagram
Iterations = 250
Method of Estimation: Maximum Likelihood
End of Problem
```

Extracts from the output file is given below. From the LISREL estimates part of the output it follows that the estimated error variance (level-1 variance $\Phi_{(1)}$) equals 475.25.

```
LISREL Estimates (Maximum Likelihood)

React1  =  1.00*intcept  +  1.00*time  +  1.00*time_sq
Errorvar.=  475.25  ,  R2  =  0.54
          (100.19)
             4.74


React2  =  1.00*intcept  +  2.00*time  +  4.00*time_sq,
Errorvar.=  475.25  ,  R2  =  0.72
          (100.19)
             4.74
```

```
React3    =  1.00*intcept  +  3.00*time   +  9.00*time_sq,
Errorvar.=  475.25   ,  R2   =  0.81
          (100.19)
              4.74

React4    =  1.00*intcept  +  4.00*time   +  16.00*time_sq,
Errorvar.=  475.25   ,  R2   =  0.84
          (100.19)
              4.74

React5    =  1.00*intcept  +  5.00*time   +  25.00*time_sq,
Errorvar.=  475.25   ,  R2   =  0.84
          (100.19)
              4.74

React6    =  1.00*intcept  +  6.00*time   +  36.00*time_sq,
Errorvar.=  475.25   ,  R2   =  0.83
          (100.19)
              4.74

React7    =  1.00*intcept  +  7.00*time   +  49.00*time_sq,
Errorvar.=  475.25   ,  R2   =  0.82
          (100.19)
              4.74

React8    =  1.00*intcept  +  8.00*time   +  64.00*time_sq,
Errorvar.=  475.25   ,  R2   =  0.86
          (100.19)
              4.74
```

The estimated elements of the level-2 covariance matrix is shown below. Values in parentheses indicate standard errors and values just below that, the *t*-values.

```
Covariance Matrix of Independent Variables
            intcept      time time_sq
intcept       667.90
             (775.83)
                0.86

time         -394.12     759.84
             (511.23)   (474.26)
               -0.77       1.60
time_sq        40.42     -88.26      11.01
              (57.38)    (54.81)     (6.55)
               0.70       -1.61       1.68
```

The estimated fixed coefficients (β_0, β_1 and β_2) of the *intercept*, *time* and *time_sq* predictors are given below, together with the standard errors in parentheses and *t*-values.

```
Mean Vector of Independent Variables

intcept     time   time_sq
 395.90     5.01      0.49
 (13.30)  (10.54)   (1.24)
  29.76     0.47      0.39
```

Figure 5.3 is a conceptual path diagram representation of the growth model. All path coefficients corresponding to the paths from observed variables to latent variables are fixed. For illustrative purposes, only the coefficients of *React8* are shown.

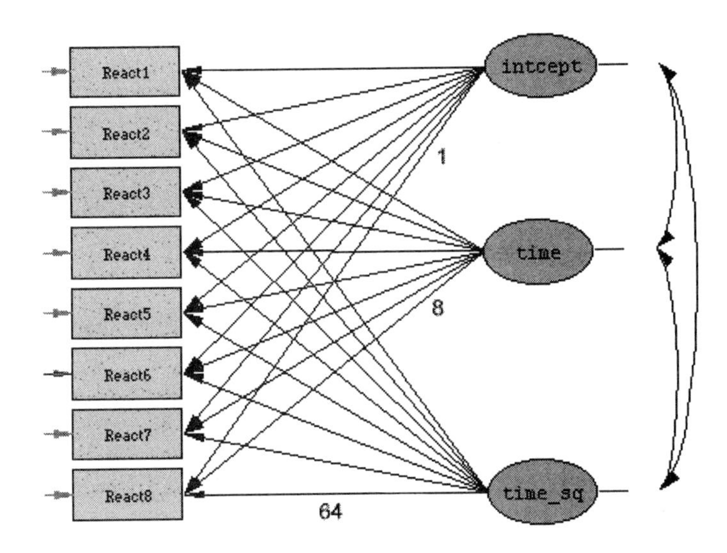

Figure 5.3: Conceptual path diagram

A representation of the mean model can also be obtained and is shown in Figure 5.4 below.

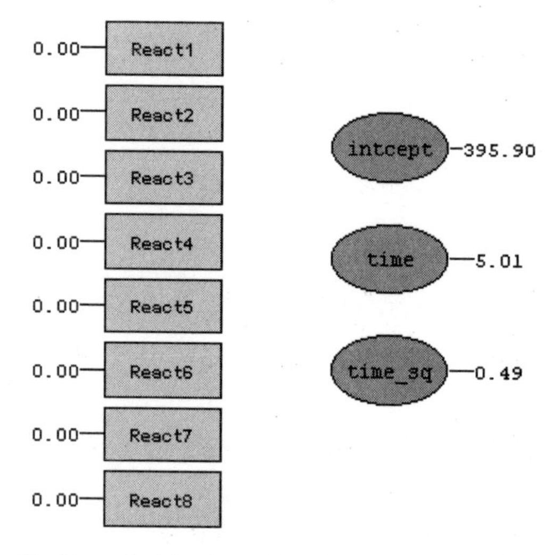

Figure 5.4: Mean model

5.7 Testing of Contrasts

5.7.1 Introduction

The construction of contrasts or linear functions of the parameters is a useful statistical analysis tool and enables the researcher to perform hypothesis testing concerning the equality of subsets of parameters. Suppose, for example, that the fixed part of the model has 6 parameters $\beta_1, \beta_2, \ldots, \beta_6$. We may want to test the following 3 hypotheses:

$$
\begin{aligned}
H_{01} &: \beta_1 = \beta_2 \\
H_{02} &: \beta_1 = \beta_3 \\
H_{03} &: \beta_1 = \beta_4.
\end{aligned}
$$

Each of these hypotheses can alternatively be written as

$$
\begin{aligned}
H_{01} &: 1\beta_1 - 1\beta_2 + 0\beta_3 + 0\beta_4 + 0\beta_5 + 0\beta_6 = 0 \\
H_{02} &: 1\beta_1 - 0\beta_2 - 1\beta_3 + 0\beta_4 + 0\beta_5 + 0\beta_6 = 0 \\
H_{03} &: 1\beta_1 + 0\beta_2 + 0\beta_3 - 1\beta_4 + 0\beta_5 + 0\beta_6 = 0
\end{aligned}
$$

In order to test these contrasts, a file is created which has the following structure

```
1
1 -1 0 0 0 0
1
1 0 -1 0 0 0
1
1 0 0 -1 0 0
```

The value of 1 in line 1 indicates that a single contrast is to be tested. The contrast is specified in the next line. Note that the sum of the values in this line equals zero and that there should be as many values as predictors included in the fixed part of the model.

5.7.2 Weight of Chicks on 4 Protein Diets

The PRELIS system file **chicks.psf** in the **tutorial** folder contains the weights (in grams) of chicks on four different protein diets. Measurements were taken on alternate days (Crowder and Hand, 1990, Example 5.3). The first few lines of the *.**psf** file is shown below.

CHICKS.PSF

	Chick_IC	Weight	Intcept	Diet1	Diet2	Diet3	Diet4	Day	Day_Sq
1	1.0	42.0	1.0	1.0	0.0	0.0	0.0	0.0	0.0
2	1.0	51.0	1.0	1.0	0.0	0.0	0.0	2.0	4.0
3	1.0	59.0	1.0	1.0	0.0	0.0	0.0	4.0	16.0
4	1.0	64.0	1.0	1.0	0.0	0.0	0.0	6.0	36.0
5	1.0	76.0	1.0	1.0	0.0	0.0	0.0	8.0	64.0
6	1.0	93.0	1.0	1.0	0.0	0.0	0.0	10.0	100.0
7	1.0	106.0	1.0	1.0	0.0	0.0	0.0	12.0	144.0
8	1.0	125.0	1.0	1.0	0.0	0.0	0.0	14.0	196.0

The first variable is the chick ID (*Chick_ID*), variable 2 is the *weight* and variable 3 assumes a value of one throughout and is labeled *Intcept*. Variables 4 to 7 are labeled *Diet1*, *Diet2*, *Diet3* and *Diet* 4 respectively. These variables are so-called dummy variables, coded as given in Table 5.4.

Table 5.4: Coding of dummy variables

	Diet1	*Diet2*	*Diet3*	*Diet4*
Protein diet 1	1	0	0	0
Protein diet 2	0	1	0	0
Protein diet 3	0	0	1	0
Protein diet 4	0	0	0	1

The last 2 variables of the data set (columns 8 and 9) are *Day* and *Day_sq*, where *Day* represents the actual days on which measurements were taken and *Day_sq* = *Day* × *Day*.

This data set contains a number of missing observations, denoted by `-9`. The model to be fitted to these data is

$$y_{ij} = \beta_{1j}(Diet1)_{ij} + \beta_{2j}(Diet2)_{ij} + \beta_{3j}(Diet3)_{ij} + \beta_{4j}(Diet4)_{ij} + \\ b_{1i} + b_{2i}(Day)_{ij} + b_{3i}(Day\_sq)_{ij} + (Day)_{ij} e_{ij}$$

where

$$b_{1i} = u_{1i}$$
$$b_{2i} = \beta_5 + u_{2i}$$
$$b_{3i} = \beta_6 + u_{3i}.$$

An intercept term is not included in the fixed part of the model, since its inclusion together with the 4 dummy variables will cause a singular design matrix, and thus one of the coefficients will be non-estimable.

To generate this syntax file interactively, open the PRELIS system file **chicks.psf** in the **tutorial** folder. Select the ***Title and Options*** option from the ***Multilevel, Linear Model*** menu on the main menu bar and provide a title for the analysis, for example

```
Weight of 50 Chicks on 4 Protein Diets
```

Enter the value `-9` in the ***Missing Data Value*** field. When done, click ***Next*** to proceed to the ***Identification Variables*** dialog box.

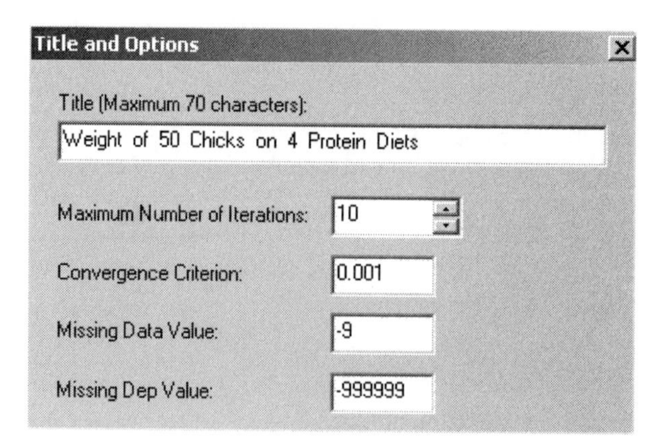

Select *Chick_ID* and use the ***Add*** button to add this variable to the level-2 identification variable field. Use the variable *Day* as level-1 identification. Clicking ***Next*** leads to the display of the ***Select Response and Fixed Variables*** dialog box.

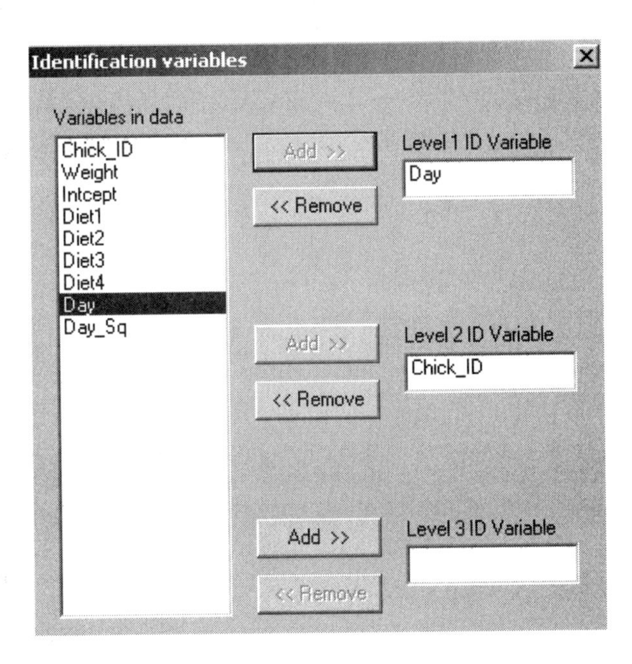

In the ***Select Response and Fixed Variables*** dialog box, the variable *Weight* is selected as the response variable for this analysis. Select the variables *Diet1* to *Diet4*, *Day*, and *Day_sq* and click ***Add*** next to the ***Fixed Variable*** field to enter these variables into the model as fixed effects.

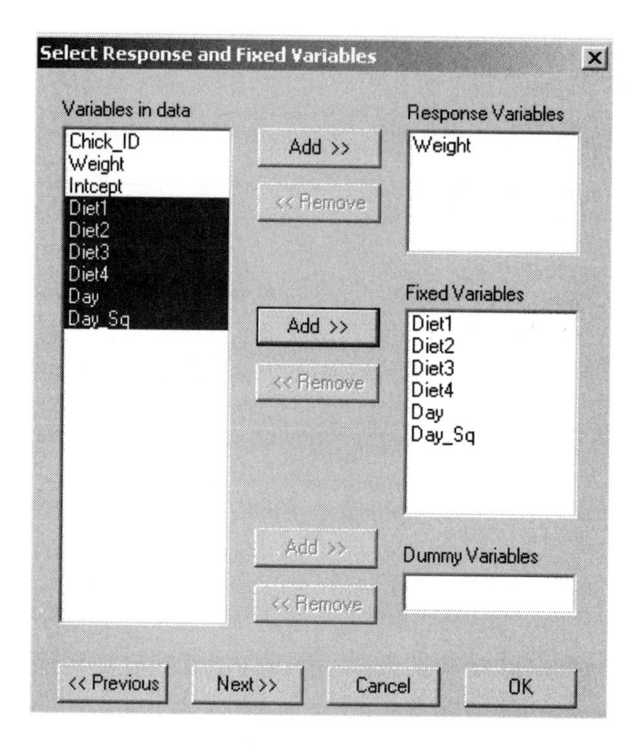

Clicking ***Next*** leads to the display of the final dialog box, in which the random effects are identified. Select the variable *Day* and use the ***Add*** button to include this variable in the ***Random Level 1*** field. Select and add *Intcept*, *Day*, and *Day_sq* in a similar way to the ***Random Level 2*** field.

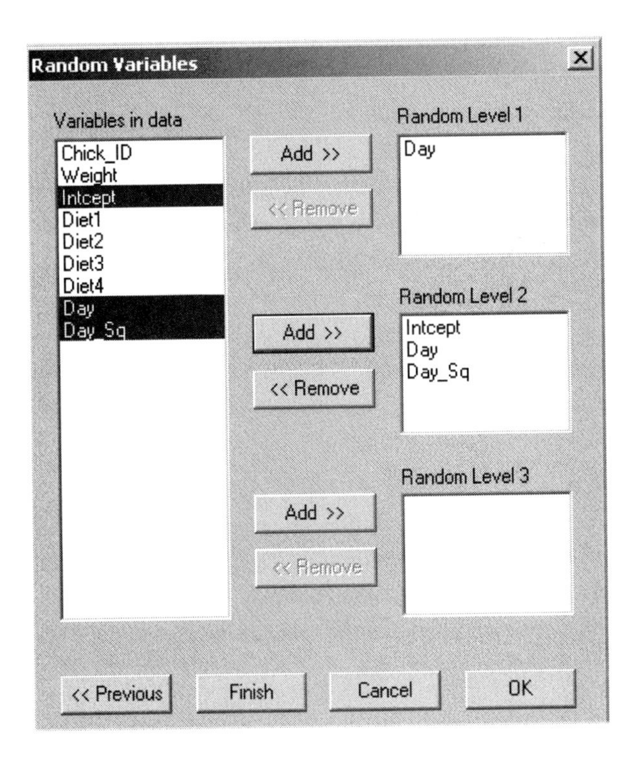

Click ***Finish*** to generate the syntax file (**chicks1.pr2** in the **tutorial** folder) shown below.

```
OPTIONS OLS=YES CONVERGE=0.001000 MAXITER=10 OUTPUT=STANDARD ;
TITLE=Weight of 50 Chicks on 4 Protein Diets;
MISSING_DAT =-9.000000 ;
SY='D:\lisrel850\TUTORIAL\CHICKS.PSF';
ID1=Day;
ID2=Chick_ID;
RESPONSE=Weight;
FIXED=Diet1 Diet2 Diet3 Diet4 Day Day_Sq;
RANDOM1=Day;
RANDOM2=Intcept Day Day_Sq;
```

However, the `CONTRAST` command

```
contrast=chicks.ctr;
```

cannot be generated through the dialog boxes and has to be added by editing this syntax file. Add this line of syntax to obtain the final syntax file (see **chicks.pr2** in the **tutorial** folder). Click the ***Run PRELIS*** icon button to start the analysis.

Remarks

The file **chicks.ctr** contains the following lines of input:

```
1
1 -1 0 0 0 0
1
1 0 -1 0 0 0
1
1 0 0 -1 0 0
1
0 1 -1 0 0 0
1
0 1 0 -1 0 0
1
0 0 1 -1 0 0
```

Since it may be expected that the variation in body weight measurements may increase over time (see Figure 5.5), the level-1 error variance is expressed as $Day \times e$, where e is the within-chicks variance component.

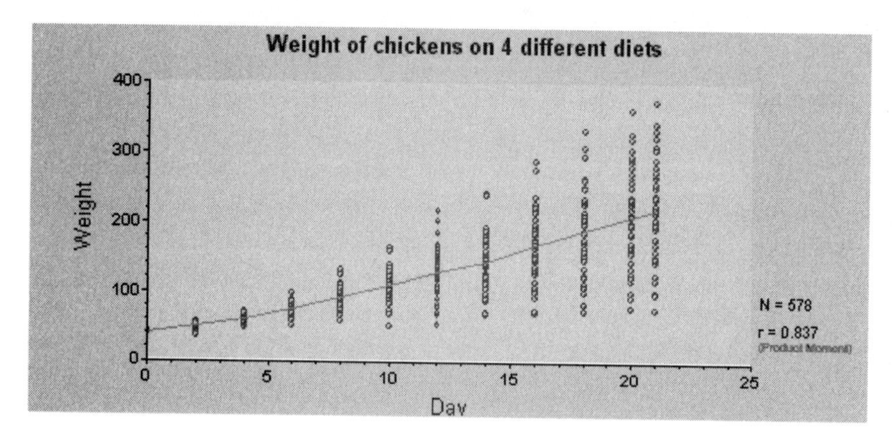

Figure 5.5: Weight of chicken on four different diets

Convergence was obtained in 18 iterations and sections of the output are given below.

```
Data Summary

NUMBER OF LEVEL 2 UNITS : 50
NUMBER OF LEVEL 1 UNITS : 578


N2 : 1   2  3  4  5  6  7  8
N1 : 12 12 12 12 12 12 12 11
```

```
N2 : 9  10 11 12 13 14 15 16
N1 : 12 12 12 12 12 12  8  7

...

N2 : 49 50
N1 : 12 12


Fixed part of the model


COEFFICIENTS      BETA-HAT    STD.ERR.    Z-VALUE     PR > |Z|
Diet1             34.76201    1.11539     31.16584    0.00000
Diet2             37.47334    1.29961     28.83427    0.00000
Diet3             39.09494    1.29961     30.08202    0.00000
Diet4             40.29282    1.29991     30.99664    0.00000
Day                5.49497    0.42191     13.02409    0.00000
Day_Sq             0.14663    0.03172      4.62272    0.00000


-2 LOG-LIKELIHOOD = 3963.22086775959


Random part of the model


LEVEL 2                    TAU-HAT   STD.ERR.   Z-VALUE    PR > |Z|
Intcept /Intcept          41.94030   9.78976    4.28410    0.00002
Day /Intcept             -15.70024   3.90165   -4.02400    0.00006
Day /Day                   7.77127   1.76391    4.40571    0.00001
Day_Sq /Intcept            0.32470   0.22965    1.41389    0.15739
Day_Sq /Day               -0.31089   0.10712   -2.90227    0.00370
Day_Sq /Day_Sq             0.04603   0.00995    4.62417    0.00000


LEVEL 1     TAU-HAT    STD.ERR.    Z-VALUE     PR > |Z|
Day /Day    0.41172    0.02986     13.78701    0.00000


LEVEL 2 COVARIANCE MATRIX


            Intcept     Day         Day_Sq
Intcept     41.94030
Day        -15.70024    7.7700
Day_Sq       0.32470   -0.31089     0.04603
```

```
LEVEL 2 CORRELATION MATRIX


             Intcept     Day         Day_Sq
Intcept       1.0000
Day          -0.8697     1.0000
Day_Sq        0.2337    -0.5198     1.0000


LEVEL 1 COVARIANCE MATRIX


          Day
Day   0.41172


LEVEL 1 CORRELATION MATRIX


         Day
Day   1.0000
```

Estimates of the variances and covariances are significant on both levels.

```
Testing of contrasts: output


CHISQ  STATISTIC                                   :        5.546
  DEGREES  OF  FREEDOM                             :            1
  EXCEEDENCE  PROBABILITY                          :        0.019


CHISQ  STATISTIC                                   :       14.163
  DEGREES  OF  FREEDOM                             :            1
  EXCEEDENCE  PROBABILITY                          :        0.000


CHISQ  STATISTIC                                   :       23.073
  DEGREES  OF  FREEDOM                             :            1
  EXCEEDENCE  PROBABILITY                          :        0.000


CHISQ  STATISTIC                                   :        1.492
  DEGREES  OF  FREEDOM                             :            1
  EXCEEDENCE  PROBABILITY                          :        0.222
```

```
CHISQ  STATISTIC                              :      4.509
  DEGREES   OF   FREEDOM                      :          1
  EXCEEDENCE  PROBABILITY                     :      0.034


CHISQ  STATISTIC                              :      0.814
  DEGREES   OF   FREEDOM                      :          1
  EXCEEDENCE  PROBABILITY                     :      0.367
```

As could be expected from the Fixed Part results, the most significant contrast is number 3,

```
1  0  0  -1  0  0
```

which corresponds to testing the hypothesis

$$H_{03} : \beta_1 = \beta_4.$$

This hypothesis is rejected ($p < 0.000$) and we conclude that there is a highly significant difference between diets 1 and 4. The MRCM model considered in this example accounts for both between chicks variation and heteroscedastic level-1 error variances.

5.8 Multilevel Models for Categorical Outcomes

5.8.1 Introduction

Survey data usually consist of a mixture of biographical, geographical and response variables and frequently have a hierarchical structure. Quite often, these response variables are categorical in nature. In the last few decades, a wide variety of methods for the analysis of categorical data have been proposed. Many of these are generalizations of continuous data analysis methods (see, for example, Agresti 1990). In order to accommodate the structure of hierarchical data, a multilevel modeling approach is used.

The basic idea is to express the dependent variable y as a function of the cell frequencies in a two-way contingency table, where the columns of the table correspond to the categories of the response variable and the rows to S mutually exclusive subpopulations.

It is assumed that the response variables are the natural logarithms of the ratio of cell frequencies in the two-way table as illustrated in the next example.

Residents of a specific city were asked whether they were satisfied with the public transportation system in that city. The responses for subpopulations of residents are given in Table 5.5.

Table 5.5: Results of Transportation study

Subpopulation	Response		Total
	Yes	No	
Male, low income	59	47	106
Male, high income	43	83	126
Female, low income	70	75	145
Female, high income	47	64	111

For each subpopulation, a y-value is obtained as follows.

$$\begin{aligned} y_1 &= \ln(59/47) \\ y_2 &= \ln(43/83) \\ y_3 &= \ln(70/75) \\ y_4 &= \ln(47/64), \end{aligned}$$

or in general,

$$y_j = \ln(f_{j1}/f_{j2}).$$

Suppose that the following model is fitted to the data:

$$y_j = \beta_0 + \beta_1 GENDER_j + \beta_2 INCOME_j + e_j.$$

It can be shown that the level-1 covariance matrix of e_1, e_2, e_3 and e_4 has the following form

$$\mathbf{\Phi}_{(1)} = \begin{bmatrix} \psi_{11}/106 & 0 & 0 & 0 \\ 0 & \psi_{11}/126 & 0 & 0 \\ 0 & 0 & \psi_{11}/145 & 0 \\ 0 & 0 & 0 & \psi_{11}/111 \end{bmatrix}$$

or, in general

$$\mathbf{\Phi}_{(1)ij} = \psi_{11}/f_{j\cdot}.$$

where $f_{j.} = f_{j1} + f_{j2}$, the row total for subpopulation j.

Remark:

It follows from

$$y_j = \ln(f_{j1} / f_{j2})$$

that

$$y_j = \ln(f_{j1} / f_{j.})/(f_{j2} / f_{j.}) = \ln(p_{j1} / p_{j2}).$$

Applying rules for logarithms, it follows that

$$\frac{p_{j1}}{(1-p_{j1})} = \exp(y_j)$$

or

$$p_{j1} = \frac{1}{1+\exp(-y_j)}.$$

The predicted value of y_j can thus be used to estimate p_{j1} and hence p_{j2} for each of the subpopulations.

5.8.2 Generalizations to Higher Level Models

Suppose that the contingency table shown in Section 5.8.1 was obtained for city i and that we have similar data for an additional 19 cities.

In this context, cities may be regarded as the level-2 units and since it can be expected that the regression coefficients may vary randomly over cities, the equivalent multilevel model is written as

$$y_{ij} = b_{0i} + b_{1i}GENDER_j + b_{2i}INCOME_j + e_{ij},$$

where, for example,

$$b_{0i} = \beta_0 + u_{0i}$$
$$b_{1i} = \beta_1 + u_{1i}$$
$$b_{2i} = \beta_2 + u_{2i}$$

or

$$b_{0i} = \beta_0 + u_{0i}$$
$$b_{1i} = \beta_1$$
$$b_{2i} = \beta_2.$$

The first set of equations implies that the *intercept*, *gender* and *income* coefficients are allowed to vary randomly on level-2. In contrast, the second set of equations implies that only the *intercept* coefficient is allowed to vary randomly on level-2 while the remaining coefficients are fixed parameters.

In the following section, a 2-level model for social mobility is discussed. Finally, Empirical Bayes estimates and predicted probabilities for the social mobility example are given.

5.8.3 A 2-level Model for Social Mobility

The PRELIS data set **socmob1.psf** in the **tutorial** folder used in this example is based on a 30% random sample of the data discussed in Biblarz and Raftery (1993). The data set contains six variables, *F_Career*, *Son_ID*, *S_Choice*, *Famstruc*, *Race* and *Intcept*. The first variable, *F_Career*, defines seventeen level-2 units, each being the father's occupation. *Son_ID* is used to number the sons within each level-2 unit. *S_Choice* is a dichotomous outcome variable that is defined as follows

S_Choice = 1: Son's current occupation is different from his father's occupation
S_Choice = 2: Son's current occupation is the same as his father's occupation

The variable *Famstruc* is family structure, coded 1 in the case of an intact family background, and 2 if the family background was not intact. The fifth variable, *Race*, is coded 1 in the case of white respondents and 2 in the case of black respondents. The father's occupation, *F_Career*, is coded as given in Table 5.6.

Table 5.6: Coding for occupation of father

Code	Description
17	Professional, Self-Employed
16	Professional, Salaried
15	Manager
14	Salesman-Non retail
13	Proprietor
12	Clerk
11	Salesman-Retail

Table 5.6: Coding for occupation of father (continued)

10	Craftsman-Manufacturing
9	Craftsman-Other
8	Craftsman-Construction
7	Service Worker
6	Operative-Non manufacturing
5	Operative-Manufacturing
4	Laborer-Manufacturing
3	Laborer-Non manufacturing
2	Farmer/Farm Manager
1	Farm Laborer

Figure 5.6 is a bar chart representation of the frequency distribution of *S_Choice*. A study of this figure reveals that the majority of sons do not have the same occupation than their fathers.

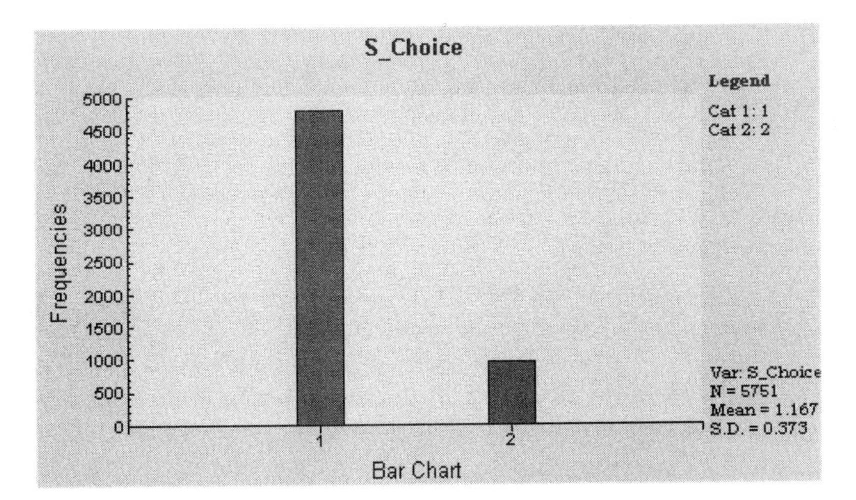

Figure 5.6: Bar chart of S_Choice

We would like to investigate the following:

- Is the ratio of different to same occupation consistent over the 17 occupations
- Is this ratio influenced by race and/or the intactness of the family background.

The level-2 model considered here is:

$$y_{ij} = \beta_0 + (\beta_1 + u_{1i})Famstruc_{ij} + (\beta_2 + u_{2i})Race_{ij} + e_{ij},$$

where y_{ij} is the natural logarithm of the ratio of "no" to "yes" for a specific combination of the categories of *Famstruc* and *Race*. The "no" response corresponds to a value of 1 for the variable *S_Choice* while "yes" corresponds to a value of 2 (same occupation as the father).

The multilevel syntax used in an attempt to answer these questions is (see **socmob1.pr2** in the **tutorial** folder) as follows:

```
OPTIONS OLS=YES CONVERGE=0.001000 add=0.1 MAXITER=20 OUTPUT=ALL ;
TITLE=17 Careers of Fathers and sons careers (1=diff 2=same);
SY=SOCMOB1.PSF;
ID2=F_Career ;
RESPONSE=S_Choice ;
FIXED=Intcept  Famstruc  Race ;
RANDOM2=Intcept Famstruc  Race ;
SUBPOP= Famstruc  Race ;
```

Remarks

Most of the syntax can be generated using the dialog boxes accessed through the ***Multilevel***, ***Linear Model*** menu. As this syntax file is very similar to previous syntax files discussed in detail in previous sections, the process of creating this syntax file is not discussed here. Rather, the interpretation of the output files obtained will be focused on. There are, however, two items added to the syntax file that are specific to the analysis of categorical outcome variables.

- `ADD = 0.1`, forms part of the `OPTIONS` command and indicate that 0.1 will be added to each frequency in the 2-way tables. This is done to ensure that $y = \ln(f_1 / f_2)$ is always defined.
- `SUBPOP = Famstruc Race;` defines the subpopulations (rows of the two-way tables).

It is important to note that, except for the intercept term (*intcept*), only predictors used to form subpopulations may be included in the fixed part of the model. One can use either `PRINT = ALL` or `PRINT = TABLES` to obtain a printout of tables for each level-2 unit. These frequency tables are contained in a file with extension ***.frq**. Partial contents of the additional output file **socmob1.frq** are shown below.

```
Frequency Table For: ID2= 1
--------------------------------
Famstruc    Race        Choice
1             1        118  67
1             2        40   18
2             1        21   13
2             2        18    9
```

```
Frequency Table For: ID2= 17
--------------------------------
Famstruc    Race      Choice
1           1          61 17
1           2           3  0
2           1           6  0
2           2           0  0
```

```
Aggregated Frequency Table
--------------------------
Famstruc Race          Choice
1        1        3775    770    4545
1        2         323     59     382
2        1         559    100     659
2        2         134     31     165
```

Convergence was attained in 10 iterations. Portions of the output file are given below.

```
DATA SUMMARY


NUMBER OF LEVEL 2 UNITS : 17
NUMBER OF LEVEL 1 UNITS : 66


N2 : 01 02 03 04 05 06 07 08
N1 : 04 04 04 04 04 04 04 04


N2 : 09 10 11 12 13 14 15 16
N1 : 04 04 04 04 04 03 04 04


N2 : 17
N1 : 03
```

From the output above we see that, except for occupations 14 and 17, there are 4 observations (the four subpopulations formed from the categories of *Famstruc* and *Race*) for all the remaining subpopulations.

FIXED PART OF MODEL

COEFFICIENTS	BETA-HAT	STD.ERR.	Z-VALUE	PR > |Z|
Intc1(Y1)1_2	2.20435	0.23517	9.37334	0.00000
FamS1(Y1)1_2	-0.25340	0.10073	-2.51559	0.01188
Race1(Y1)1_2	-0.15686	0.11515	-1.36218	0.17314

From the fixed part, it follows that there is a highly significant average effect (*Intcept*) and a significant family structure effect (*Famstruc*). The race effect is not significant.

RANDOM PART OF MODEL

LEVEL 2	TAU-HAT	STD.ERR.	Z-VALUE	PR > |Z|
Intc1(Y1)1_2/Intc1(Y1)1_2	0.75428	0.32029	2.35502	0.01852
FamS1(Y1)1_2/Intc1(Y1)1_2	-0.22916	0.11960	-1.91607	0.05536
FamS1(Y1)1_2/FamS1(Y1)1_2	0.09121	0.05878	1.55182	0.12071
Race1(Y1)1_2/Intc1(Y1)1_2	-0.06559	0.11484	-0.57118	0.56788
Race1(Y1)1_2/FamS1(Y1)1_2	0.06984	0.04859	1.43744	0.15059
Race1(Y1)1_2/Race1(Y1)1_2	0.10509	0.07346	1.43045	0.15259

LEVEL 1	TAU-HAT	STD.ERR.	Z-VALUE	PR > |Z|
ERROR(Y1)1_2/ERROR(Y1)1_2	9.96628	2.86223	3.48200	0.0050

LEVEL 2 COVARIANCE MATRIX

	Intc1(Y1)1_2	FamS(Y1)1_2	Race(Y1)1_2
Intc1(Y1)1_2	0.75428		
FamS1(Y1)1_2	-0.22916	0.09121	
Race1(Y1)1_2	-0.06559	0.06984	0.10509

LEVEL 2 CORRELATION MATRIX

	Intc1(Y1)1_2	FamS(Y1)1_2	Race(Y1)1_2
Intc1(Y1)1_2	1.0000		
FamS1(Y1)1_2	-0.8599	1.0000	
Race1(Y1)1_2	-0.2334	0.6904	1.0000

```
LEVEL 1 COVARIANCE MATRIX


              ERROR(Y1)1_2
ERROR(Y1)1_2     9.96628
```

The level-1 error variance is highly significant. From the level-2 part of the model, we see significant variation over occupations attributable to the *Intcept* and *Famstruc* predictors. In the final section of this example, empirical Bayes estimates and predicted probabilities for this model are discussed.

5.8.4 Empirical Bayes Estimates and Predicted Probabilities

In the previous example the following level-2 model was fitted to the data set **socmob1.psf**:

$$y_{ij} = \beta_0 + (\beta_1 + u_{1i})Famstruc_{ij} + (\beta_2 + u_{2i})Race_{ij} + e_{ij}.$$

It can be shown that the so-called empirical Bayes (or shrunken) estimator of y_{ij} is given by

$$y_{i\_s} = (\beta_0 + u_{0i}) + (\beta_1 + u_{1i})Famstruc_{ij} + (\beta_2 + u_{2i})Race_{ij}.$$

From the output obtained in the previous example, this simplifies to

$$y_{i\_s} = (2.20435 + u_{0i}) + (-0.2534 + u_{1i})Famstruc_{ij} + (-0.15686 + u_{2i})Race_{ij},$$

where *Famstruc* is 1 for an intact family background, and 2 otherwise, and *Race* = 1 for whites and 2 for blacks. Values for u_{0i}, u_{1i} and u_{2i} are contained in the file **socmob1.ba2**. A portion of this file is shown below.

Occupation	Residual	Estimate	Variance
1	u_{10}	-1.4972	0.0497
1	u_{11}	0.3788	0.0120
1	u_{12}	-0.1857	0.0206
...	...	...	...
17	$u_{17,0}$	0.6521	0.2994

17	$u_{17,1}$	-0.2019	-0.0645
17	$u_{17,2}$	-0.0651	0.0744

Empirical Bayes estimates for the occupation groups 1 and 17 are shown in Table 5.7. The y_{i_s}-values can be used to calculate the predicted probability that a son's current occupation will differ from that of the father. This probability is

$$P(different \mid occupation_i) = \frac{1}{1+\exp(-y_{i\_s})}.$$

Table 5.7: Empirical Bayes estimates for selected occupation groups

Occupation	Family Structure	Race	y_{i_s}
1	Intact	White	0.65710
	Intact	Black	0.48167
	Not intact	White	0.78248
	Not intact	Black	0.60705
2	Intact	White	2.17927
	Intact	Black	1.95733
	Not intact	White	1.72400
	Not intact	Black	1.50206

The probability that the son's occupation is the same occupation is 1-*P(different)*. In Figures 5.7 and 5.8, the probability of a son having the same occupation as the father is plotted by occupation. In Figure 5.7 the two white subpopulations are shown, and in Figure 5.8 the two black subpopulations. It is interesting to note that a son seemed more likely to have the same occupation as the father when coming from a non-intact family background, regardless of race.

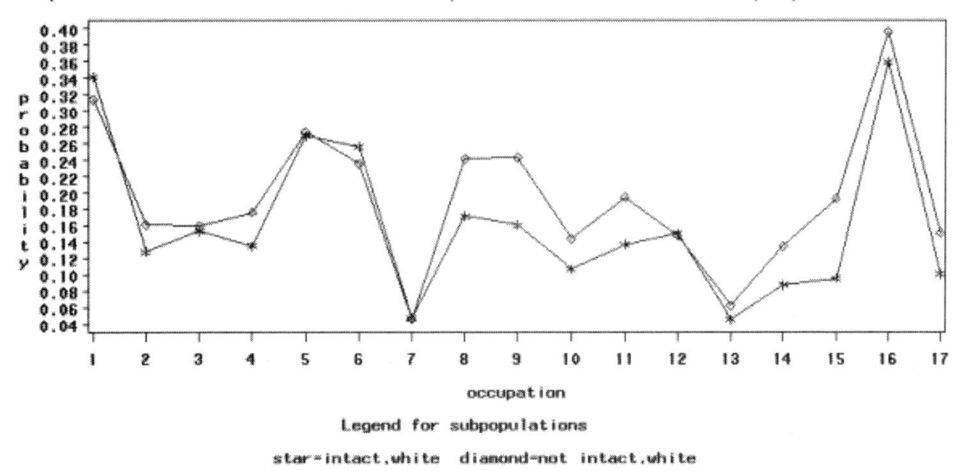

Figure 5.7: Plot of probabilities of son having the same occupation as the father-white subpopulations

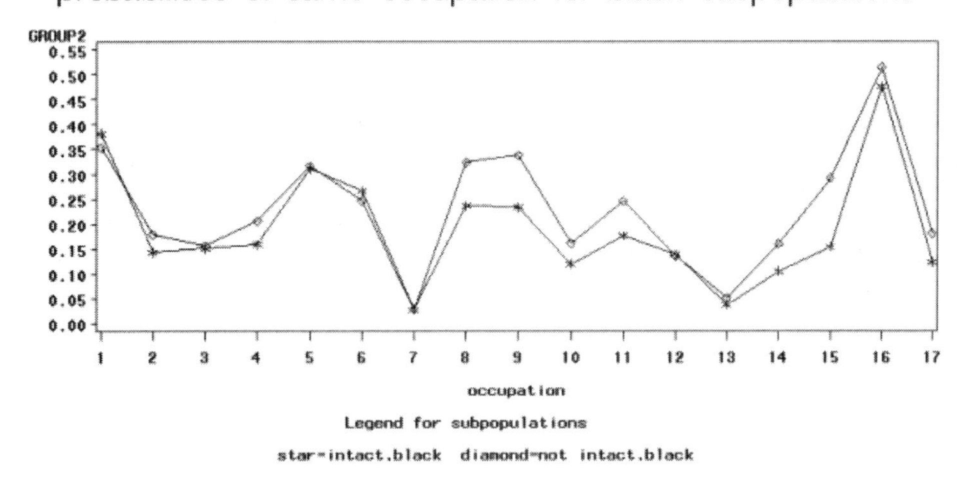

Figure 5.8: Plot of probabilities of son having the same occupation as the father-black subpopulations

5.9 Multilevel Nonlinear Regression Models

5.9.1 2-level Nonlinear Regression Model

The data set used contain repeated measurements on 82 striped mice and was obtained from the Department of Zoology at the University of Pretoria, South Africa (see du Toit, 1979). The complete description of this data is given in Section 5.2. For linear multilevel models fitted to this data, see Section 5.2. In this section, a logistic regression model is fitted to the data:

$$y_{ij} = b_{1i} / (1 + \exp(b_{2i} - b_{3i} x_{ij})) + e_{ij},$$

where

$$b_{1i} = \beta_1 + \gamma_1 gender + u_{1i}$$
$$b_{2i} = \beta_2 + \gamma_2 gender + u_{2i}$$
$$b_{3i} = \beta_3 + \gamma_1 gender + u_{3i}.$$

5.9.2 Creating the Syntax

Use the ***File, Open*** option to locate **mouse.psf** in the **tutorial** folder. A portion of the spreadsheet is shown below. The variable *iden2* (level 2 ID) is the number of a specific mouse while *iden1* (level 1 ID) corresponds to the different measurement occasions.

MOUSE.PSF

	iden2	iden1	weight	constant	time	timesq	gender
1	1.00	1.00	15.00	1.00	1.00	1.00	1.00
2	1.00	2.00	17.00	1.00	2.00	4.00	1.00
3	1.00	3.00	23.00	1.00	3.00	9.00	1.00
4	1.00	4.00	24.00	1.00	4.00	16.00	1.00
5	1.00	5.00	26.00	1.00	5.00	25.00	1.00
6	1.00	6.00	31.00	1.00	6.00	36.00	1.00
7	1.00	7.00	37.00	1.00	7.00	49.00	1.00
8	1.00	8.00	42.00	1.00	8.00	64.00	1.00
9	1.00	9.00	46.00	1.00	9.00	81.00	1.00
10	2.00	1.00	11.00	1.00	1.00	1.00	1.00

A preliminary inspection of the data reveals that the weight measurements increase *monotonically* over time, but reaches an asymptote value at maturity. It is apparent that a polynomial will describe the data quite accurately within the range of 1 to 9 weeks, but predictions will be unreliable outside the observed range of the data as is illustrated on page 275 of Pinheiro and Bates (2000).

From the main menu bar, select the ***Multilevel, Non-Linear Model*** option. To fit a non-linear model to repeated measurements data, click on the ***Title and Options*** option.

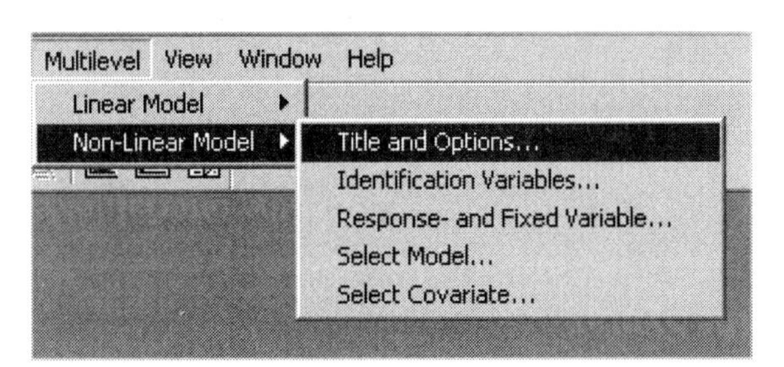

The ***Titles and Options*** dialog box enables the researcher to assign a title, and to specify the maximum number of iterations, the convergence criterion, the missing data code and the number of quadrature points. The quadrature points determine the accuracy of the numerical integration procedure. Since numerical integration turns out to be computationally intensive, one would generally choose a large number of points for 2 to 3 coefficient functions (30 – 80) but much fewer points (6 – 12) for non linear curves with more than 3 coefficients, *e.g.* the logistic + logistic (6 coefficients) function. Finally, we select the ***Use Full Maximum Likelihood*** option.

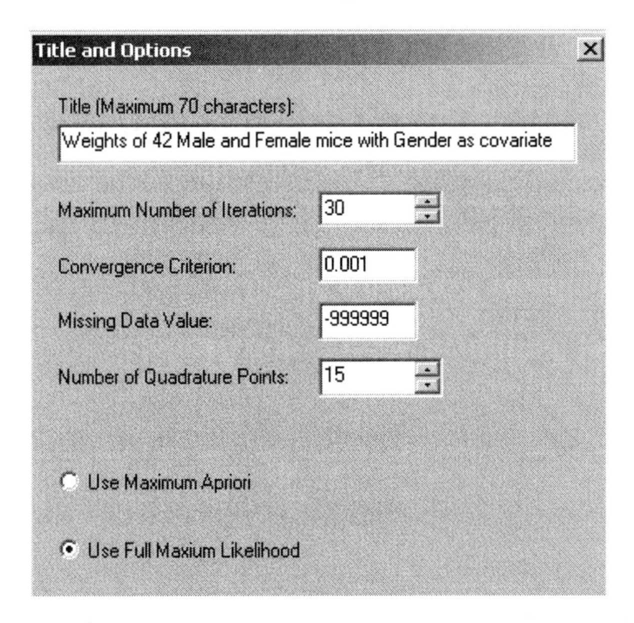

The ***Identification Variables*** dialog box is used to assign level-1 and level-2 identification variables as shown below.

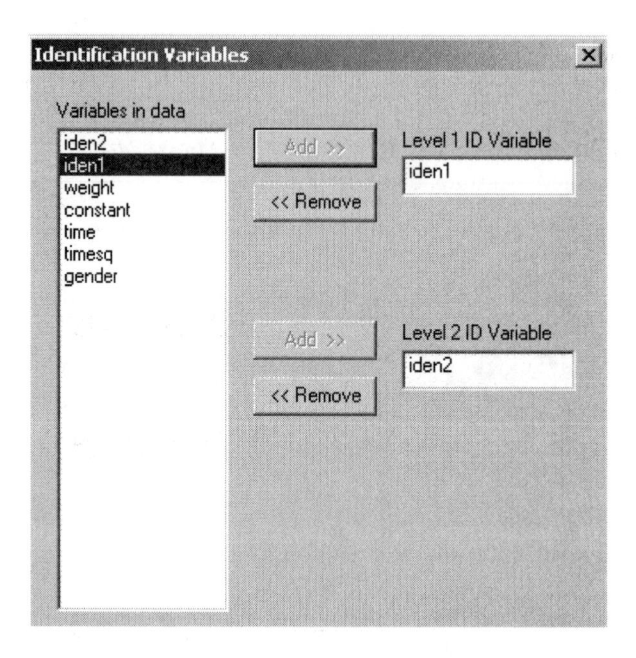

From the list of variables in the data set, *weight* is selected as response (y) variable and *time* as the fixed (x) variable. The next dialog box enables the researcher to specify an appropriate model for the data.

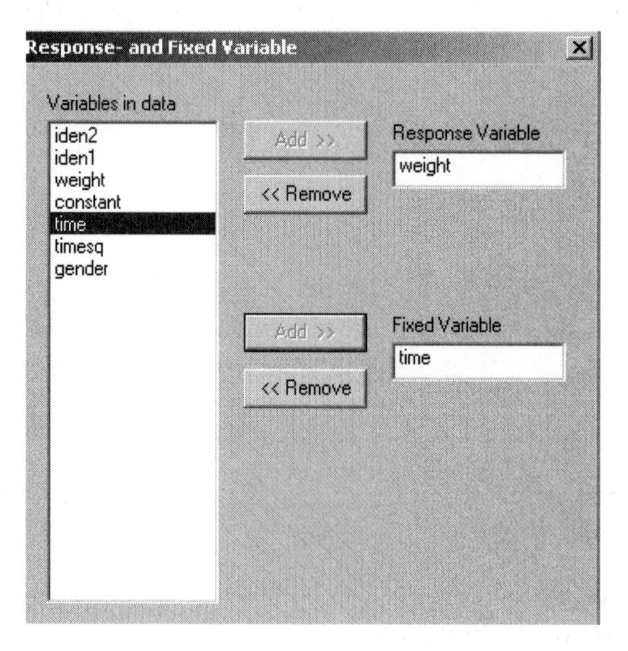

To obtain some idea of the pattern of weight measurements over time, we used the ***Data, Select Cases*** option and selected only those cases where *gender* = 1. Using the ***Output Options*** dialog box, the male data set is saved as **malmouse.psf**. From the ***Graphs*** menu, select ***Bivariate Plot*** and enter *weight* and *time* as the *Y* and *X* variables respectively.

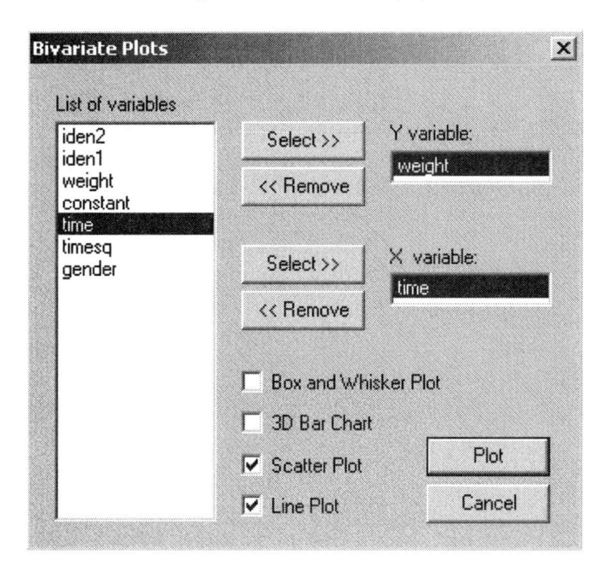

The figure below displays the gain in weight pattern over time. This curve follows a pattern more like a logistic than a polynomial model.

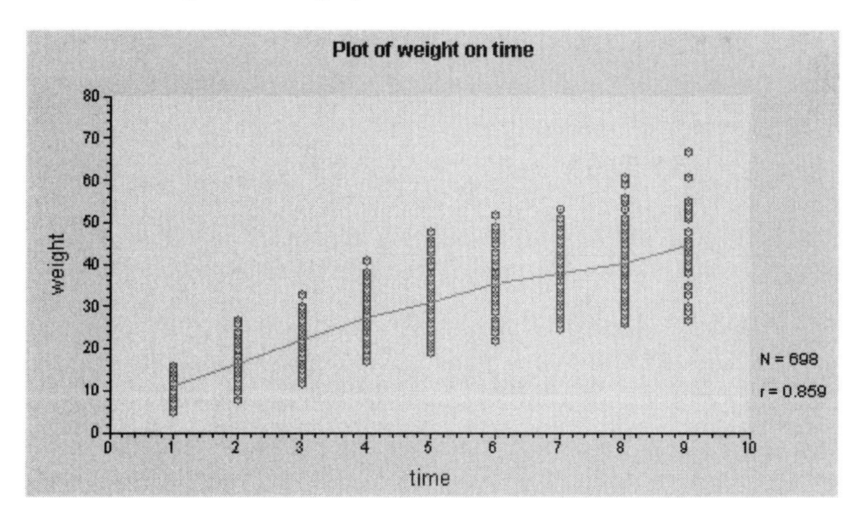

From the ***Select Model*** dialog box, select the single component logistic function and click ***Next*** to go to the ***Covariates*** dialog box.

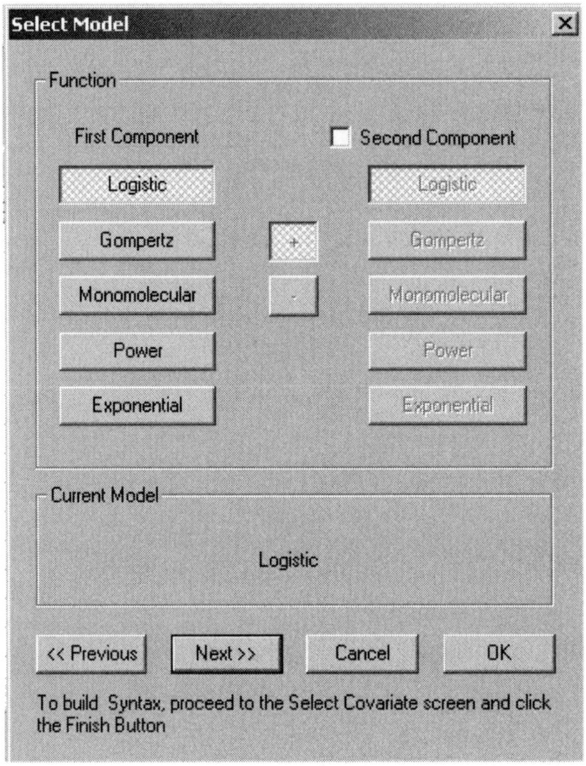

Select Model

Function

First Component

Second Component

Logistic

Gompertz

Monomolecular

Power

Exponential

Current Model

Logistic

<< Previous | Next >> | Cancel | OK

To build Syntax, proceed to the Select Covariate screen and click the Finish Button

It is realistic to assume that gender differences will result in different asymptote, potential growth and growth rate parameters. We therefore select *gender* as a level-2 covariate influencing all 3 coefficients of the logistic curve

$$\frac{b_1}{1+\exp(b_2 - b_3 t))}$$

that is

$$b_i = \beta_i + \gamma_i \, gender + u_i, \, i = 1,2,3.$$

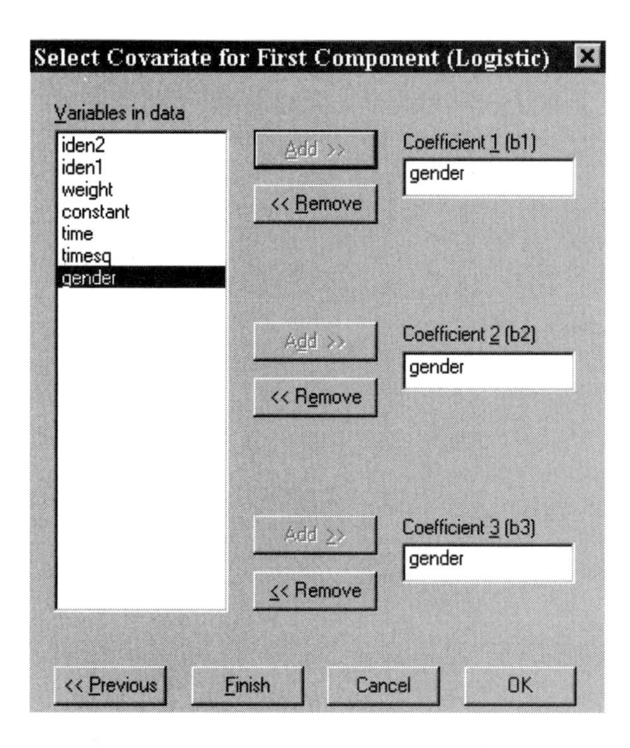

To generate the required syntax, click ***Finish*** in this dialog box. The syntax generated is shown below (see **mousel.pr2** in the **nonlinex** folder)

```
OPTIONS METHOD = ML CONVERGE = 0.001000 MAXITER = 30 QUADPTS = 15;
TITLE = Weights of 42 Male and Female mice with Gender as covariate;
SY=D:\lisrel850\TUTORIAL\mouse.psf;
ID1 = iden1;
ID2 = iden2;
RESPONSE = weight;
FIXED = time;
MODEL = Logistic;
COVARIATES b1 = gender
           b2 = gender
           b3 = gender;
```

A portion of the output is shown below.

malmouse.OUT

```
NUMBER OF LEVEL 2 UNITS :        82
NUMBER OF LEVEL 1 UNITS :       698

N2   :        1        2        3        4        5
N1   :        9        9        9        9        9

N2   :        9       10       11       12       13
N1   :        9        9        9        9        9
```

There are 82 level 2 units (42 male and 40 female mice). For each of the male mice there are 9 repeated measurements and for each of the female mice (numbers 43 – 82) there are 8 repeated measurements.

(i) Maximum A Priori (MAP) Estimates

Fixed parameters:

β_1	55.422
β_2	1.656
β_3	0.528

Covariance matrix of random coefficients:

98.032		
2.394	0.142	
-0.347	0.007	0.0172

Error Variance

2.804

(ii) Maximum Likelihood estimates

Fixed parameters	Estimate	Std. Error
β_1	45.155	0.878
β_2	1.573	0.0324
β_3	0.516	0.013

Covariates (gender)

	Estimate	Std. Error
γ_1	2.681	0.878
γ_2	0.103	0.0324
γ_3	0.040	0.013

Covariance matrix of the random coefficients

	Estimate	Std. Error
$Var(u_1)$	100.087	12.775
$Cov(u_2,u_1)$	2.106	0.382
$Var(u_2)$	0.135	0.018
$Cov(u_3,u_1)$	-0.600	0.152
$Cov(u_3,u_2)$	0.004	0.005
$Var(u_3)$	0.019	0.003

Level 1 error Variance

	Estimate	Std. Error
$Var(e)$	2.738	0.105

The random part of the output for this model shows a reasonable correspondence between the MAP and ML estimates. The MAP procedure, as implemented in LISREL, does not allow for covariates. This explains the differences in the estimated MAP and ML β-parameters.

6 FIRM Examples

In this chapter, two data sets are analyzed using both CATFIRM and CONFIRM. For more information on FIRM syntax, please see Chapter 8. An overview of this technique is given in Chapter 7.

The following examples are discussed in the following sections:

- The “auto repair” example
- CATFIRM analysis of the “head injuries” data set
- CONFIRM analysis of the “head injuries” data set
- CATFIRM analysis of the “mixed” data set
- CONFIRM analysis of the “mixed” data set.

6.1 Auto Repair Example

In the study of automobile costs, you may have in a database containing, in order of appearance, the variables *COUNTRY*, *MAKE*, *CYL*, *DISPL*, *TANK*, *PRICE*, *YEAR*, *REPAIR* which are respectively the car’s country of origin, its make, the number of cylinders it has, the engine displacement, fuel tank capacity, recommended retail price, year of manufracture and annual repair costs.

The following variables may be used to predict the repair costs (*REPAIR*), which is thus the dependent variable.

- *MAKE* : the make of the car, with 5 groupings (GM, Ford, Chrysler, Japanese and European). *MAKE* is clearly nominal, and would be used as a “free” predictor.
- *CYL* : the number of cylinders in the machine (4, 6, 8 or 12). *CYL* should be used as monotonic, in order to avoid the assumption that the repair costs difference between 4 and 8 cylinders is the same as between 8 and 12 cylinder cars.
- *PRICE* : manufacturer’s suggested retail purchase price when new. *PRICE* is a ratio-scale predictor that would be specified as monotonic—it seems generally accepted that cars that are more expensive to buy are more expensive to run, too.
- *YEAR* : year of manufacture. *YEAR* is an example of a predictor that looks to be on interval scale, but appearances may be deceptive. The best would be to use year as a “free” predictor, thus allowing the cost to have any type of response to *year*, including one in which isolated years are bad while their neighbors are good.

6.2 CATFIRM Analysis of the "head injuries" Data Set

6.2.1 Description of the Data

The "head injuries" data set of Titterington *et al* (1981) is an example of a data set in which FIRM is a potential method of analysis. The data set was gathered in an attempt to predict the final outcome of 500 hospital patients who had suffered head injuries. The *outcome* for each patient was that he or she was:

- *Dead or vegetative* (52% of the cases)
- had *severe* disabilities (10% of the cases) or
- had a *moderate or good* recovery (38% of the cases).

This outcome is predicted on the basis of 6 predictors assessed on the patients' admission to the hospital:

- *age* : The age of the patient. This is grouped into decades in the original data, and is grouped the same way here. It has eight classes.
- *EMV* : This is a composite score of three measures—of eye-opening in response to stimulation, motor response of best limb, and verbal response. This has seven classes, but is not measured in all cases, so that there are eight possible codes for this score — the seven measurements and an eighth "missing" category.
- *MRP* : This is a composite score of motor responses in all four limbs. This also has seven measurement classes with an eighth class for missing information.
- *Change* : The change in neurological function over the first 24 hours. This was graded 1, 2, or 3, with a fourth class for missing information.
- *Eye indicator* : A summary of diagnostics on the eyes. This too had three measurement classes, with a fourth for missing information.
- *Pupils* : Pupil reaction to light — present, absent, or missing.

In the CATFIRM analysis the dependent variable was treated as being a categorical variable.

6.2.2 FIRM Syntax File

The syntax for this analysis is contained in **headicat.pr2** and is shown and discussed below.

```
CATFIRM: den sum
    1 'Outcome'      3 'Dead/veg'  'Severe   '  'Mod/good'
    6
  'age'         2    1 'm'    0  8 '01234567'   4.9000  5.0000
  'EMV'         3    0 '1'    0  8 '?1234567'   4.9000  5.0000
  'MRP'         4    0 '1'    0  8 '?1234567'   4.9000  5.0000
  'Change'      5    0 '1'    0  4     '?123'   4.9000  5.0000
```

```
'Eye ind'     6    0 '1'   0  4     '?123'   4.9000  5.0000
 'Pupils'     7    0 'f'   0  3      '?12'   4.9000  5.0000
0 1000  25  0.50000  1.00000  25   .000      0  1  0  0  0  0  0  0
```

Note the following:

- The first line indicates that the outcome variable is categorical and both summary statistics (sum) and a dendrogram (den) are to be included in the output produced during analysis.
- The second line contains information on the dependent or outcome variable. It is the first variable in the data set, as indicated by 1 in the first field, and it is used here as a categorical variable. The outcome variable has three categories: 'Dead/veg', 'Severe', and 'Mod/good'.
- This is followed by six lines of data, one for each predictor variable. Predictors included range from 'age', a monotonic variable with 8 possible values to 'Pupils', a free variable with 3 possible values.
- The last two lines of the syntax file contain the output options for this analysis. The options specified are as follows:

  0: Table format. The option 0 gives the table as a percentage frequency breakdown of the dependent variable for all cases in the node, and for the cases broken down by the different categories of the predictor.
  1000: The detail output code is a single number reflecting the answers to all three questions about the printed information, and is calculated as follows

  Q1: Do you want details of splits not used?,
  Q2: Do you want cross tabs before grouping?, and
  Q3: Do you want cross tabs after grouping?.

  Score 1 for Yes and 2 for No; then compute this option as

$$1000 \times Q1 + Q2 + 2 \times Q3 - 1003.$$

  In this case, detailed analysis of splits is not requested (Q1 = 2), but cross-tabs are requested (Q2 = Q3 = 1).
  25: The minimum size a node must have to be considered for further splitting.
  0.50000: The raw significance level for a split to be made.
  1.0000: The conservative significance level that a predictor must attain for the split to be made.
  25: The maximum number of nodes to be analyzed.
  0.000: Constant to be added to χ^2.

0: The data for this analysis is in free format. If fixed format had been indicated, a format statement would have followed directly after this line of input.
1: Specifies use of FIRM 2.1 methodology rather than FIRM 2.0 methodology.
0: Not currently used, set to default value of 0.
0: Not currently used, set to default value of 0.
0: Not currently used, set to default value of 0.
0: Not currently used, set to default value of 0.
0: Not currently used, set to default value of 0.
0: Not currently used, set to default value of 0.

For more information on syntax, see the sections on the dependent and predictor variable specifications in Chapter 8.

6.2.3 CATFIRM Output

The syntax read from the syntax file is echoed to the output file. After a preliminary echo of the syntax from the syntax file, the file shows the analysis of each node. As can be seen from the output below, 14 nodes were obtained.

```
CATFIRM.  Formal Inference-based Recursive Modeling
 Program dimensions
 Maximum number of predictors          1500
 Maximum number of categories in
        Predictors                       16
        Dependent variable               16

     the data file (input):
                  D:\lisrel850\FIRMEX\HEADINJ.dat
     the detailed analysis of splits (output):
                  D:\lisrel850\FIRMEX\HEADINJ.spl
     the contingency tables made (output):
                  D:\lisrel850\FIRMEX\HEADINJ.tab
     the split rule table of splits made (output):
                  D:\lisrel850\FIRMEX\HEADINJ.spr
 Run now starting...
 All data in.          500 cases read with
                       500 retained.
 Starting    node             1    descended    from              0
 Split on    6 making descendants   2   3   2
 Starting node    2 descended from    1
 Split on    1 making descendants   4   4   5   5   6   6   7   7
 Starting node    3 descended from    1
 Starting node    4 descended from    2
 Split on    3 making descendants   8   8   8   8   8   8   9   9
 Starting node    5 descended from    2
 Split on    5 making descendants  11  10  11  12
```

```
Starting node    6 descended from    2
Split on   3 making descendants  15  13  13  13  14  14  15  15
Starting node    7 descended from    2
Split on   2 making descendants  16  16  16  16  17  17  17  18
Starting node    8 descended from    4
Starting node    9 descended from    4
Starting node   11 descended from    5
Starting node   12 descended from    5
Starting node   14 descended from    6
Starting node   15 descended from    6
Starting node   17 descended from    7

Mini-dendrogram of analysis

Legend:                     node number
                            splitting variable
                   ------------------------------------
                   horizontal line connects descendants

                                        1
                                        6
                             ---------------------------------------
                             2                                     3
                                1
          ---------------------------------------------------
           4              5                 6                 7
           3              5                 3                 2
       -------      -------------     ------------     -------------
       8     9    10     11     12    13    14    15    16    17    18
```

Summary information on the outcome and predictor variables is given next, followed by a list of the options in effect for the analysis.

```
 Outcome has   3 categories:

Dead/veg
Severe
Mod/good


There are    6 predictors as follows

Type        # cats Cat symbols    Use?  Split% Merge%
mono         8     01234567        May    4.90    5.00  age
float        8     ?1234567        May    4.90    5.00  EMV
float        8     ?1234567        May    4.90    5.00  MRP
float        4     ?123            May    4.90    5.00  Change
float        4     ?123            May    4.90    5.00  Eye ind
free         3     ?12             May    4.90    5.00  Pupils
```

```
Options in effect:
  Tables printed  as column percentages
  Detail output on file split
  Tables given before and after each step
  To be analysed, a group must:
          have at least   25 cases;
          be significant at the  0.500% level;
          be conservatively significant at the  1.000% level.
  The run will terminate when    25 groups have been formed.
  Standard Pearson X^2 statistic is used
  Tests use new (FIRM 2.1) procedures
```

The summary of results of the first split is given below. Node number 1 is the full data set. In node 1, all the predictors give highly significant splits on the dependent variable *Outcome*. The number of groups selected by CATFIRM varies from 2 to 4: under the heading `groups` we see how the categories break down. For example, using *EMV* would split the cases into 4 groups. Taking the Pearson χ^2 value for the 3×4 table and finding its p-value under the non-asymptotic distribution would give

$$P = 3.44 \times 10^{-20}\%.$$

Making the Bonferroni adjustment for the grouping that went into this final set of categories multiplies this p-value by 95, giving the Bonferroni significance as

$$< 3.27 \times 10^{-18}\%.$$

The multiple comparison p-value takes the original Pearson χ^2 for the 3×8 table before grouping the categories and enters it into the non-asymptotic distribution, giving a p-value of

$$< 5.00 \times 10^{-17}\%.$$

Since both the Bonferroni and the multiple comparison p-values are conservative, we take the smaller of them, $3.27 \times 10^{-18}\,\%$, to be the conservative significance level of the predictor.

```
Summary of results of node number     1 predecessor node number    0
Total group
no.  Name                   Signif %    Bonf sig % MC sig %    groups
 1 age                    6.40E-13    2.24E-11    1.62E-10 01 23 45 67
 2 EMV                    3.44E-20    3.27E-18    5.00E-17 12 345 ?6 7
 3 MRP                    2.04E-20    1.04E-18    2.32E-16 12 345 ?67
 4 Change                   0.0116      0.0579      0.0208 ?1 23
 5 Eye ind                3.88E-19    1.94E-18    1.40E-18 1 ?2 3
 6 Pupils                 5.08E-20    1.52E-19    3.47E-17 ?2 1
```

Detail on the characteristics of the best predictor is given next. The most significant split uses *Pupils*. It is a binary split, between the pooled class ? or 2 and class 1. This split has a Bonferroni significance level of $1.52 \times 10^{-19}\%$ and a multiple comparison significance level of $3.47 \times 10^{-17}\%$. The smaller of these (both being conservative) is taken as the significance level of the split.

The summary table below shows how the cases divide between these two—roughly three quarters go to node 2, where the recovery rate is quite variable, and the rest go to node 3, where 90% of the patients are dead or vegetative.

```
 Characteristics of the best predictor
   6 Pupils                        5.08E-20    1.52E-19    3.47E-17 ?2 1
***********************************************************************
 predictor    6    Pupils                *percent*  total number    500
                   ?,2      1    Total
 Dead/veg           39.4  90.2   51.8
 Severe             11.9   5.7   10.4
 Mod/good           48.7   4.1   37.8
   totals (100%)     378   122    500
          Raw significance of table is    5.08E-20

  The descendant nodes are numbered    2   3
```

The analysis continues with node number 2. CATFIRM lists the makeup of this node—it is all cases for which *Pupils* is ? or 2. As the analysis proceeds, this record grows to reflect the successive splits giving rise to the node. In this node, no significant split can be made on *Pupils* (not surprisingly, since the two classes of *Pupils* represented in this node were grouped because of their compatibility). All other predictors give significant splits, the conservative significance ranging from 0.1% for *Change* down to $8 \times 10^{-12}\%$ for *age*. Node 2 is split four ways on *age*, the cut points being at ages 20, 30 and 60. All four nodes are investigated again later in the analysis.

```
Summary of results of node number     2 predecessor node number    1

Makeup    Pupils     (?,2)
no.   Name                      Signif %    Bonf sig % MC sig %    groups
1 age                         2.30E-13    8.06E-12    4.37E-11 01 23 45 67
2 EMV                         9.44E-10    4.81E-08    6.21E-07 12 ?345 67
3 MRP                         8.34E-10    4.25E-08    1.24E-06 12345 ?6 7
4 Change                        0.0367      0.1836      0.1010 ?1 23
5 Eye ind                     2.64E-07    1.32E-06    5.84E-07 1 ?2 3
6 Pupils                        100.00      100.00      100.00 ?2
```

```
 Characteristics of the best predictor

1 age                        2.30E-13   8.06E-12   4.37E-11 01 23 45 67
***********************************************************************
predictor   1   age                  *percent*  total number    378
                    0,1      2,3      4,5      6,7    Total
Dead/veg           21.3     30.9     48.2     81.7   39.4
Severe              8.8     15.5     16.5      6.7   11.9
Mod/good           69.9     53.6     35.3     11.7   48.7
  totals (100%)     136       97       85       60    378

          Raw significance of table is    2.30E-13
  The descendant nodes are numbered    4    5    6    7
```

Node 3 is terminal. Its cases can not (at the significance levels selected) be split further. When there is no further significant split to be made within a node, the program adds a message to this effect at the end of the summary of results for that node. This was the case for node 3, where the best predictor, *Eye ind*, was not significant.

```
Summary of results of node number     3 predecessor node number    1
Makeup   Pupils    (1)
no.  Name                       Signif %    Bonf sig % MC sig %   groups
1 age                           100.00      100.00     100.00 01234567
2 EMV                           0.1354      6.9051     8.5868 12 345 ?67
3 MRP                           0.3558      4.6251     8.0063 123456 ?7
4 Change                        100.00      100.00     100.00 ?123
5 Eye ind                       0.1679      0.8394     0.3613 12 ?3
6 Pupils                        100.00      100.00     100.00 11

 Characteristics of the best predictor
5 Eye ind                       0.1679      0.8394     0.3613 12 ?3
 This predictor is not significant
```

The analysis continues in the same way with the successive nodes generated. The information on the splits actually made is duplicated in the dendrogram. The additional information in the summary file is on the predictors that did not give rise to splits. In the full data set, all the predictors were highly significant. In Node 2, *age* was as significant as it was in the full data set, so that the predictive power is *age* is different from that in *Pupils*. *EMV*, *MRP* and *Eye ind* had considerably less significance in Node 2 than in Node 1. While to some degree this is an inevitable consequence of the reduction in sample size going down the tree, in part it can also indicate overlap in the predictive information—that these predictors are correlated with *Pupils* and the common variability they have with *Pupils* comprises a lot of the information about *Outcome*. Scanning the summary file for this sort of information can often produce valuable insights.

The detailed information of splits is given in the output file **headicat.spl.** The first few lines of this file are given below. The split file details the tests that went into the final grouping of each predictor at each node. There are three slightly different layouts here, illustrated by the monotonic predictor *age*, the free predictor *Pupils* and the floating predictor *MRP*. For the monotonic predictor, only adjacent predictors may be merged.

The predictor *age* starts out with 8 groups, giving 7 pairwise χ^2 statistics that are listed after "Test statistics for grouping". The smallest of these is 1.3775 for merging categories 2 and 3. This value is well below the merge significance level for the run, so these two classes are joined into a composite, leaving 7 groups. The merge statistics for these revised groups are computed, the smallest of which is 1.6163 for merging classes 0 and 1, so these classes are merged. The analysis continues in the same way until 4 composite groups remain. At this stage, the smallest merge statistic—that for merging (23) with (45)—is significant at the 2% level, which is below the run's threshold for merging. Thus no further reduction of the classes of *age* occurs.

No more merging being possible, CATFIRM then attempts to split the composite categories. The line "Test stats for splitting" gives the details of this, showing a χ^2 statistic for each possible resplitting point of a composite category. The largest of these statistics is 5.0593, which is far from significant at the split significance level specified for the run, and so no resplitting takes place. Thus (01), (23), (45), (67) is the final grouping of the categories. The final 3 × 4 contingency table gives a Pearson χ^2 of 77.984. Entering this into χ^2 tables with 6 degrees of freedom gives a raw significance level of $6.4\times10^{-13}\,\%$. The Bonferroni multiplier, which allows for the grouping that went into reducing 8 groups into 4, is 35. The Bonferroni significance level is therefore

$$35\times6.4\times10^{-13}\,\% = 1.62\times10^{-10}\,\%.$$

```
    1Total group
****************************************************************************
Monotonic age
Table has chi-square   86.310, with df  14 and significance  1.62E-10
  8 groups:                       0     1     2     3     4     5     6     7
Test statistics for grouping:        1.6   9.8   1.4   4.0   1.7   7.7   5.1

Min stat is  1.3775, to merge (2) and (3).        d.f.   2, sig   51.1157
  7 groups:                       0     1    (2     3)    4     5     6     7
Test statistics for grouping:        1.6  10.7          7.9   1.7   7.7   5.1

Min stat is  1.6163, to merge (0) and (1).        d.f.   2, sig   45.3755
  6 groups:                      (0     1)   (2     3)    4     5     6     7
Test statistics for grouping:              9.5          7.9   1.7   7.7   5.1

Min stat is  1.7396, to merge (4) and (5).        d.f.   2, sig   43.0870
  5 groups:                      (0     1)   (2     3)   (4     5)    6     7
Test statistics for grouping:              9.5          7.8         7.1   5.1

Min stat is  5.0593, to merge (6) and (7).        d.f.   2, sig    7.6055
  4 groups:                      (0     1)   (2     3)   (4     5)   (6     7)
Test statistics for grouping:              9.5          7.8        14.4

Min stat is  7.7518, to merge (23) and (45).      d.f.   2, sig    2.0360
  4 groups:               (0     1) (2     3) (4     5) (6     7)
```

```
Test stats for splitting
     1.6        1.4        1.7        5.1
Max stat is  5.0593 to split group (67). d.f.   2 significance    7.6055
Best is  4 groups, with chi square   77.984 d.f.   6
Bonferroni multiplier, raw significance and MC significance
     35.                     6.40E-13              1.62E-10
```

The floating predictors, as exemplified by *MRP*, have an extra wrinkle to them. Not only can the monotonic portion of the scale be merged in the same way as with *age*, but the floating category can be merged with any of them. Thus the detail starts out with two lines—the test statistics for merging a successive pair on the scale 1-7 on the upper line, and those for merging the floating category with each monotonic class on the second line. CATFIRM checks the full list of (in this case 13) test statistics, and finds that the smallest is 0.7021, for merging classes `1` and `2`. These classes are merged and the analysis repeated. In the second stage, it happens that the smallest test statistic is for merging the category `?` with `6`, so these two categories are joined. Once this is done, there is no longer a floating category, just the monotonic categories `(12)`, `3`, `4`, `5`, `(?6)` and `7`, and the subsequent lines of the analysis look like those of a monotonic predictor. There is a slightly different twist at the last phase. Here the floating category is part of a three-class composite (?67), and so there are three possible binary splits—`?` vs `(67)`, `(?6)` vs `7` and `6` vs `(?7)`. The test statistics for these are shown on two lines of the printout—2.5 and 4.3 for the first two, and 1.4 for the third. In reading these two lines, ignore the rightmost `?` on the first line and the leftmost on the second. Again, the split statistic is not significant, and so the grouping `(12)` `(345)` `(?67)` is final.

```
Float MRP
Table has chi-square  117.765, with df  14 and significance  2.32E-16
  8 groups:                       1     2     3     4     5     6     7
Test statistics for grouping:       0.7   3.8   1.9   3.7   5.0   3.0
and for grouping ? with          17.9  16.6   6.1   8.8   1.4   0.9   4.5
Min stat is  0.7021, to merge (1) and (2).        d.f.   2, sig   72.5715
  7 groups:                      (1     2)    3     4     5     6     7
Test statistics for grouping:               5.9   1.9   3.7   5.0   3.0
and for grouping ? with          23.1   6.1   8.8   1.4   0.9   4.5

Min stat is  0.8839, to merge (?) and (6).        d.f.   2, sig   65.2141
  6 groups:                      (1     2)    3     4     5    (?     6)    7
Test statistics for grouping:               5.9   1.9   3.7   4.7         4.3

Min stat is  1.9242, to merge (3) and (4).        d.f.   2, sig   38.4126
  5 groups:                      (1     2)   (3     4)    5    (?     6)    7
Test statistics for grouping:               8.6         3.3   4.7         4.3

Min stat is  3.3063, to merge (34) and (5).       d.f.   2, sig   18.9915
  4 groups:                      (1     2)   (3     4     5)   (?     6)    7
Test statistics for grouping:             11.8               34.9         4.3

Min stat is  4.2724, to merge (?6) and (7).       d.f.   2, sig   11.8624
  3 groups:                      (1     2)   (3     4     5)   (?     6     7)
Test statistics for grouping:             11.8               56.1

Min stat is 11.8477, to merge (12) and (345).     d.f.   2, sig    0.2492
  3 groups:              (1     2) (3     4     5) (?     6     7     ?)
```

```
Test stats for splitting
                         0.7       1.5   3.3       2.5   4.3
                                                             1.4
Max stat is  4.2724 to split group (?67). d.f.   2 significance   11.8624
Best is  3 groups, with chi square  106.856 d.f.   4
Bonferroni multiplier, raw significance and MC significance
     51.                       2.04E-20             2.32E-16
```

Free predictors with more than three categories involve much more computation than monotonic or floating predictors with the same number of categories. Their analysis (though not the potential computational load) is illustrated by *Pupils*. Since any two categories may be merged, each step of the merge phase computes and lists the lower triangle of a matrix of pairwise χ^2 values—in this case of a 3×3 matrix. The first round of testing finds the categories ? and 2, and the single statistic FT 3.6 shows that this split is not significant.

If the splitting phase finds a split that is significant at the split significance level selected for this run, this composite is split, and CATFIRM returns to the first phase of looking for possible merges.

Since a c-category composite class of a free predictor can be resplit in 2^{c-1} ways, testing for resplitting of free predictors with many classes can become an enormous computational burden. It is partly for this reason and partly because of the method of implementation that CATFIRM has an immutable upper limit of 16 on the number of categories a free predictor may have.

In analyses having many categories, a substantial amount of computing time can sometimes be saved with a modest loss in the quality of the final grouping by skipping the resplitting phase. This happens automatically whenever the split significance level specified for the run is larger than the merge significance level.

```
Free Pupils
Table has chi-square  101.477 with d.f.   4 and signif  3.47E-17
    Merge phase     3 groups          ?     1
                          1         9.9
                          2         3.6  99.9
Min stat  3.62 to merge groups  1,  3 d.f.   2  signif   15.4512
    Merge phase     2 groups          ?,2
                          1          97.7
Min stat 97.71 to merge groups  1,  2 d.f.   2  signif  5.08E-20
Split phase    2           groups: (?    2 ) (1)
Test stats for splitting the group: ? 2
FT   3.6

Max stat is  3.6208 to split group numbered  1 as                 1
 d.f.   2 significance   15.4512
Best is  2 groups with chi squared   97.714 d.f.   2
 Bonferroni multiplier, raw significance and MC significance
     3.                       5.08E-20             3.47E-17
```

The other optional and potentially huge file produced on request by CATFIRM is the contingency tables before and/or after the grouping of categories. This is illustrated by the next section of printout from the output file **headicat.tab**. Scanning these tables is often helpful in gaining a better perspective to the grouping that went before, in particular regarding the number of cases in the groups that were merged.

The table for the total group before grouping with respect to the predictor *Pupils* (the most significant predictor) is given first. This is followed by the table for the total group after splitting on *Pupils*. Percentages are given for each category of the outcome separately by category of the predictor. The total number of cases in each group is also given, allowing the user to calculate the number of cases per combination of predictor and outcome categories should that be required.

```
    1Total group   before grouping
***********************************************************************
predictor    6    Pupils                    *percent*   total number     500
                      ?       1       2    Total
Dead/veg             61.5    90.2    38.6    51.8
Severe               15.4     5.7    11.8    10.4
Mod/good             23.1     4.1    49.6    37.8
  totals (100%)        13     122     365     500
           Raw significance of table is    3.47E-17

1Total group   after grouping
***********************************************************************
predictor    6    Pupils                    *percent*   total number     500
                    ?,2       1    Total
Dead/veg            39.4    90.2    51.8
Severe              11.9     5.7    10.4
Mod/good            48.7     4.1    37.8
  totals (100%)      378     122     500

     Raw significance of table is     5.08E-20
```

The final output file, **headicat.spr**, contains the split rule table of splits made during the analysis. The complete file is given below. There is nothing of interest to most starting users in this file. Its main use is for other programs that want to read and make use of the results of a FIRM analysis. For the contents, layout and format of the split rule table, please see the section on FIRM syntax in Chapter 7.

```
   1    6    2    3    2
   2    1    4    4    5    5    6    6    7    7
   4    3    8    8    8    8    8    8    9    9
   5    5   11   10   11   12
   6    3   15   13   13   13   14   14   15   15
   7    2   16   16   16   16   17   17   17   18
  -1   -1
 0
 0
```

The final file produced by CATFIRM is the full-scale dendrogram (see below). The dendrogram produced by CATFIRM is a very informative picture of the relationship between these predictors and the patients' final outcome. We see in it that the most significant separation was obtained by splitting the full sample on the basis of the predictor *Pupils*.

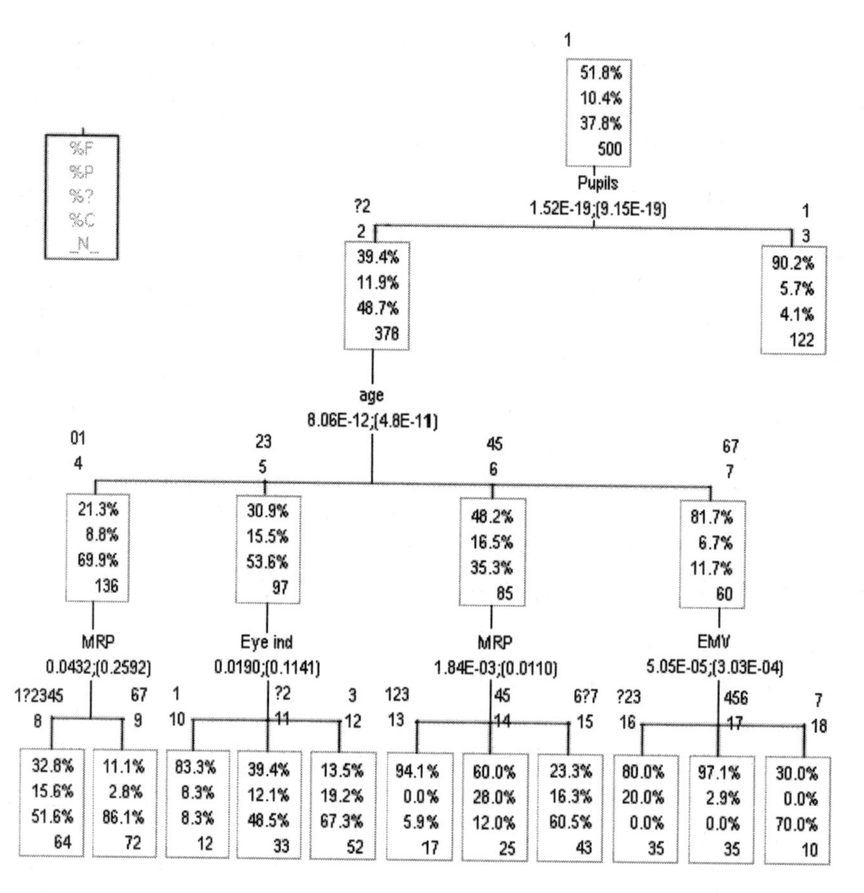

Figure 6.1: CATFIRM dendrogram of "head injuries" data set

The groups of 378 cases for which *Pupils* had the value 2 or the value ? (that is, missing) were statistically indistinguishable, and so are grouped together. They constitute one of the successor groups (node number 2), while those 122 for whom *Pupils* had the value 1 constitute the other (node number 3), a group with much worse outcomes—90% dead or vegetative compared with 39% of those in node 2.

Each of these nodes in turn is subjected to the same analysis. The cases in node number 2 can be split again into more homogeneous subgroups. The most significant such split is obtained by separating the cases into four groups on the basis of the predictor *age*. These are patients under 20 years old (node 4), patients 20 to 40 years old (node 5), those 40 to

60 years old (node 6) and those over 60 (node 7). The prognosis of these patients deteriorates with increasing *age*; 70% of those under 20 ended with moderate or good recoveries, while only 12% of those over 60 did.

Node 3 is terminal. Its cases can not (at the significance levels selected) be split further. Node 4 can be split using *MRP*. Cases with *MRP* = 6 or 7 constitute a group with a favorable prognosis (86% with moderate to good recovery), and the other MRP levels constitute a less-favorable group but still better than average.

These groups, and their descendants, are analyzed in turn in the same way. Ultimately no further splits can be made and the analysis stops. Altogether 17 nodes are formed, of which 12 are terminal and the remaining 5 intermediate. Each split in the dendrogram shows the variable used to make the split and the values of the splitting variable that define each descendant node. It also lists the statistical significance (p-value) of the split. Two p-values are given: that on the left is FIRM's conservative p-value for the split. The p-value on the right is a Bonferroni-corrected value, reflecting the fact that the split actually made had the smallest p-value of all the predictors available to split that node. So for example, on the first split, the actual split made has a p-value of $1.52 \times 10^{-19}\,\%$. But when we recognize that this was selected because it was the most significant of the 6 possible splits, we may want to scale up its p-value by the factor 6 to allow for the fact that it was the smallest of 6 p-values available (one for each predictor). This gives its conservative Bonferroni-adjusted p-value as $9.15 \times 10^{-19}\,\%$.

The dendrogram and the analysis giving rise to it can be used in two obvious ways—for making predictions, and to gain understanding of the importance of and interrelationships between the different predictors. Taking the prediction use first, the dendrogram provides a quick and convenient way of predicting the outcome for a patient. Following patients down the tree to see into which terminal node they fall yields 12 typical patient profiles ranging from 97.1% *dead/vegetative* to 86.1% with *moderate to good* recoveries. The dendrogram is often used in exactly this way to make predictions for individual cases. Unlike, for instance, predictions made using multiple regression or discriminant analysis, the prediction requires no arithmetic—merely the ability to take a future case down the tree. This allows for quick predictions with limited potential for arithmetic errors. A hospital could, for example, use the dendrogram to give an immediate estimate of a head-injured patient's prognosis.

6.3 CATFIRM Analysis of the "mixed" Data Set

6.3.1 Description of the Data

The "mixed" data set was provided by Gordon V. Kass. The original file contains the records of nearly 20,000 students at the University of the Witwatersrand. From this file, we randomly extracted 500 for use as a FIRM calibration data set and another 500 to illustrate validation. Most of the predictors are of type *real*, being percentage scores attained by the students for a variety of subjects in high school. In addition, the data set contains four character predictors:

- *mattype*: indicating which of several possible university entrance qualifying examinations the student wrote,
- *sex*: gender of the student,
- *faculty*: a subject-matter grouping like the colleges of many US universities, and
- *matmonth*: the month in which the student wrote the final school-leaving examination.

The dependent variable for the CATFIRM run was also of the type *character*—a promotion code. Possible values for this were *P* for a clear pass for the year; *F* for a clear fail for the year; and *C* for a partial credit on courses for the year. There was a number of other low-frequency codes that were changed as part of the initial work on the data file to make them all ? along with the genuinely missing values.

6.3.2 FIRM Syntax

The CATFIRM syntax for the analysis of the "mixed" data is given below:

```
CATFIRM: den sum
   15 'Promo code'    -1
   12
 'Afrikaans'      7     0 'r'   0  0 ''    0.9000  1.0000
   'Biology'     10     0 'r'   0  0 ''    0.9000  1.0000
   'English'      6     0 'r'   0  0 ''    0.9000  1.0000
   'Faculty'     13     0 'c'   0  0 ''    0.9000  1.0000
   'History'     12     0 'r'   0  0 ''    0.9000  1.0000
     'Latin'     11     0 'r'   0  0 ''    0.9000  1.0000
      'Math'      8     0 'r'   0  0 ''    0.9000  1.0000
   'Mattype'      3     0 'c'   0  0 ''    0.9000  1.0000
  'Matmonth'      4     0 'f'   0 13 '?123456789ABC'    0.9000   1.0000
   'Matyear'      5     0 'r'   0  0               ''   0.9000   1.0000
   'Science'      9     0 'r'   0  0               ''   0.9000   1.0000
       'Sex'      2     0 'c'   0  0               ''   0.9000   1.0000
 0 1000   25    .10000  1.00000  25    .500       0  1  0  0  0  0  0  0
```

- The first line of the syntax file specifies the inclusion of the summary statistics (keyword `sum`) and the dendrogram (keyword `den`) in the output file, rather than having this information written to various external files.
- The dependent variable, *Promo code*, is in the 15th position in the data set. The artificial code `-1` indicates that the dependent variable is a character variable and that FIRM should find the separate categories from the data.
- There are 12 predictors, as indicated by the number `12` in the third line of the syntax.

- The information for each predictor follows next, and is given in the following order: the name of the predictor, immediately followed by its position in the data. In the case of *Afrikaans*, for example, the predictor is in the 7th position in the data.
- `0` indicates that this predictor may be used for splitting. *Afrikaans* is a real predictor (type of predictor = `r`). The monotonic, free and float predictor types all require that the predictor's values be consecutive integers. The next two fields should contain information on this range, and are only required for these three predictor types. For a real predictor, `0`s are used instead.
- This is usually followed by the category codes. In this case, the absence of category codes is indicated by " ". Note that category codes and range of values are provided in the case of the predictor *Matmonth*, which is a free predictor.
- The last two entries in this line are the splitting and merging significance values. These levels are given in percent. In this example, a 0.9% significance level will be used for splitting and 1% for merging.
- The last line of the syntax file contains the output options for this CONFIRM analysis. The options specified are as follows:

```
0 1000  25   .10000  1.00000  25   .500      0  1  0  0  0  0  0  0
```

`0`: Table format. The option `0` gives the table as a percentage frequency breakdown of the dependent variable for all cases in the node, and for the cases broken down by the different categories of the predictor.
`1000`: The detail output code is a single number reflecting the answers to all three questions about the printed information, and is calculated as follows

Q1: Do you want details of splits not used?,
Q2: Do you want cross tabs before grouping?, and
Q3: Do you want cross tabs after grouping?.

Score `1` for `Yes` and `2` for `No`; then compute this option as

$$1000 \times Q1 + Q2 + 2 \times Q3 - 1003.$$

In this case, detailed analysis of splits is not requested (Q1 = `2`), but cross-tabs are requested (Q2 = Q3 = `1`).
`25`: The minimum size a node must have to be considered for further splitting.
`0.10000`: The raw significance level for a split to be made.
`1.0000`: The conservative significance level that a predictor must attain for the split to be made.
`25`: The maximum number of nodes to be analyzed.
`0.500`: Constant to be added to χ^2.

0: The data for this analysis is in free format. If fixed format had been indicated, a format statement would have followed directly after this line of input.
1: Specifies use of FIRM 2.1 methodology .
0: Not currently used, set to default value of 0.
0: Not currently used, set to default value of 0.
0: Not currently used, set to default value of 0.
0: Not currently used, set to default value of 0.
0: Not currently used, set to default value of 0.
0: Not currently used, set to default value of 0.

For more information on syntax, see the sections on the dependent and predictor variable specifications in Chapter 8.

6.3.3 CATFIRM Output

The first part of the output file for the CATFIRM analysis of the "mixed" data concerns the specification of syntax, input and output files.

```
The following lines were read from file H:\FIRM\DATA\MIXED.CAT


 CATFIRM: den sum
    15 'Promo code'    -1
    12
  'Afrikaans'     7    0 'r'   0  0 ''    0.9000  1.0000
    'Biology'    10    0 'r'   0  0 ''    0.9000  1.0000
    'English'     6    0 'r'   0  0 ''    0.9000  1.0000
    'Faculty'    13    0 'c'   0  0 ''    0.9000  1.0000
    'History'    12    0 'r'   0  0 ''    0.9000  1.0000
      'Latin'    11    0 'r'   0  0 ''    0.9000  1.0000
       'Math'     8    0 'r'   0  0 ''    0.9000  1.0000
    'Mattype'     3    0 'c'   0  0 ''    0.9000  1.0000
   'Matmonth'     4    0 'f'   0 13 '?123456789ABC'    0.9000  1.0000
    'Matyear'     5    0 'r'   0  0              ''    0.9000  1.0000
    'Science'     9    0 'r'   0  0              ''    0.9000  1.0000
        'Sex'     2    0 'c'   0  0              ''    0.9000  1.0000
  0 1000  25   .10000  1.00000  25   .500      0  1  0  0  0  0  0  0

CATFIRM.  Formal Inference-based Recursive Modeling
Program dimensions
Maximum number of predictors           1000
Maximum number of categories in
        Predictors                       16
        Dependent variable               16

    the data file (input):
                H:\FIRM\DATA\MIXED.dat
```

```
    the detailed analysis of splits (output):
                   H:\FIRM\DATA\MIXED.spl
    the contingency tables made (output):
                   H:\FIRM\DATA\MIXED.tab
    the split rule table of splits made (output):
                   H:\FIRM\DATA\MIXED.spr
Run now starting...
All data in.        500 cases read with
                    500 retained.
```

This is followed by basic splitting and merging information and the layout of the dendrogram.

```
Starting node    1 descended from    0
Split on   4 making descendants   2   3   2   4   5   3   4   5   5   4

Starting node    2 descended from    1
Split on 7 making descendants   6   6   6   6   7   7   7   7   7   7  7
 Starting node    3 descended from    1

Starting node    4 descended from    1
Split on  11 making descendants   9   8   8   8   8   9  10  10  10  10  10

Starting node    5 descended from    1
Starting node    6 descended from    2
Starting node    7 descended from    2
Starting node    8 descended from    4
Starting node    9 descended from    4
Starting node   10 descended from    4

Mini-dendrogram of analysis

 Legend:                    node number
                            splitting variable
                    -----------------------------------
                    horizontal line connects descendants

                         1
                         4
                    -------------------------------------------------
                    2               3          4                     5
                    7                         11
              ------------       ----------------------
              6          7          8          9          10
```

The output of summary and error information follows. This information is written to the output file if the keyword `sum` is used in the first line of the syntax file.

Immediately following the echo of the options, the different codes seen for the dependent variable are listed—these were F, P, ? and C. These values will be used in the run as the category labels. This is followed by the code list for the character predictors. The predictor *Mattype*, for example, took on 11 different values all of which (as it happens) were one character long (character predictors do not have to be single-character—they can have length up to 20). These values being necessarily different, they will be used as the category symbols for this predictor.

```
 ------------------------------------------
 |  Output of Summary and Error Information  |
 ------------------------------------------

              CATFIRM  Formal Inference-based Recursive Modeling.
              Categorical dependent variable
              Version 2.3  1999/09/04
              Copyright  1999  Douglas M Hawkins
              Applied Statistics
              University of Minnesota
 CATFIRM.  Formal Inference-based Recursive Modeling

 Program dimensions
 Maximum number of predictors          1003
 Maximum number of categories in
        Predictors                       16
        Dependent variable               16

 There are   12 predictors as follows
 Type      # cats Cat symbols    Use?  Split% Merge%
 real        0                    May    0.90   1.00  Afrikaans
 real        0                    May    0.90   1.00  Biology
 real        0                    May    0.90   1.00  English
 char        0                    May    0.90   1.00  Faculty
 real        0                    May    0.90   1.00  History
 real        0                    May    0.90   1.00  Latin
 real        0                    May    0.90   1.00  Math
 char        0                    May    0.90   1.00  Mattype
 free       13    ?123456789ABC   May    0.90   1.00  Matmonth
 real        0                    May    0.90   1.00  Matyear
 real        0                    May    0.90   1.00  Science
 char        0                    May    0.90   1.00  Sex
 Option  1 is       0.
 Option  2 is    1000.
 Option  3 is      25.
 Option  4 is       0.
 Option  5 is       1.
 Option  6 is      25.
 Option  7 is       1.
 Option  8 is       0.
 Option  9 is       1.
 All data in.          500 cases read with
                       500 retained.

 Character dependent var has values
    F
    P
    ?
    C

 Character predictors seen in the data and their values are:

Predictor     4          Faculty                 values seen
       A                    C                 M                    S
       D                    E                 F                    H
       B                    L
 These values will be abbreviated to: ACMSDEFHBL
 Predictor     8           Mattype                  values seen
       5                    1                 6                    J
       F                    7                 ?                    4
       3                    A                 2
```

```
These values will be abbreviated to: 516JF7?43A2

Predictor   12           Sex                         values seen
    M                    F
These values will be abbreviated to: MF
```

Next there is a listing of the cutpoints for the real predictors, which is needed when reading the rest of the summary output and the dendrogram. Just one real predictor–*Afrikaans*—from this list is shown below.

The cutpoints 45, 51, 54, 55, 57, 62, 64, 65 and 75 are the best we can find for dividing the data set up into 10 groups of about equal size, and are the starting cutpoints from which FIRM will do its subsequent merging to a final grouping.

```
Continuous predictors seen in the data and their cutpoints are:

Predictor    1   Afrikaans
Code    Max value in class
 ?         Invalid or missing
 0         <=       45.000
 1         <=       51.000
 2         <=       54.000
 3         <=       55.000
 4         <=       57.000
 5         <=       62.000
 6         <=       64.000
 7         <=       65.000
 8         <=       75.000
 9         >        75.000

Options in effect:
 Tables printed  as column percentages
 Detail output on file split
 Tables given before and after each step
 To be analysed, a group must:
        have at least   25 cases;
        be significant at the  0.100% level;
        be conservatively significant at the  1.000% level.
 The run will terminate when    25 groups have been formed.
 Modify chi sq statistic by 0.50000
 Tests use new (FIRM 2.1) procedures
```

This is followed by a summary output of the same format as used for the "head injuries" data set. The full data set can be split using almost any of the predictors. The most significant is a four-way split on *English* into the groupings {0,1,2,3,4,5}, {6}, {?,7} and {8,9}. The summary file output continues, showing the same information for the descendant nodes.

```
Summary of results of node number     1 predecessor node number    0

 Total group
 no.  Name                     Signif %    Bonf sig % MC sig %   groups
   1 Afrikaans                   0.0189      7.0146     0.9615 012 3 ?4 56789
   2 Biology                   1.49E-03      0.0284     9.1541 0123?456 789
   3 English                   1.18E-08    4.38E-06   1.75E-04 ? 012345 6 789
   4 Faculty                   7.45E-14    2.54E-09   7.39E-07 AM CE SFL DBH
```

```
   5 History                    8.75E-04      0.0805       4.3416 012 ?345 678
   6 Latin                        100.00      100.00       100.00 01?2345678
   7 Math                       4.75E-05    5.56E-03       0.1141 012 3?456 789
   8 Mattype                    7.57E-04      0.7744       6.5932 56421JAF37 ?
   9 Matmonth                   3.23E-06    4.10E-04       1.1351 ?347 1B6C
  10 Matyear                    1.65E-03      0.1518       0.2005 ? 01234 5678
  11 Science                    5.38E-05    6.29E-03       0.7419 012 ?34567 89
  12 Sex                          1.2017      1.2017       1.2017 M F

Characteristics of the best predictor

   4 Faculty                    7.45E-14    2.54E-09     7.39E-07 AM CE SFL DBH
*************************************************************************
predictor   4   Faculty                 *percent*  total number     500
                    A,M       C,E      S,F,L      D,B,H    Total
F                  17.9      34.9       30.3        9.1     25.0
P                  70.3      49.3       29.4       61.4     54.2
?                   9.7      11.8       19.3       29.5     14.2
C                   2.1       3.9       21.1        0.0      6.6
  totals (100%)     195       152        109         44      500
         Raw significance of table is   7.45E-14
```

The final output provided is the dendrogram, as shown below.

```
      |                                                            1
   split var                                                 +---------+
 P val (Bonf P)                                              |    25.0%|
        |                                                    |    54.2%|
 --------------                                              |    14.2%|
        |                                                    |     6.6%|
    levels                                                   |      500|
 Node  |                                                     +---------+
+---------+                                                       |
| % F     |                                                    Faculty
| % P     |                                              2.54E-09;(3.05E-08)
| % ?     |                                                       |
| % C     |                       -------------------------------------------------------
| _N_     |                 AM                 CE             SFL                    DBH
+---------+                 2 |                3 |            4 |                    5 |
                     +---------+        +---------+   +---------+          +---------+
                     |    17.9%|        |    34.9%|   |    30.3%|          |     9.1%|
                     |    70.3%|        |    49.3%|   |    29.4%|          |    61.4%|
                     |     9.7%|        |    11.8%|   |    19.3%|          |    29.5%|
                     |     2.1%|        |     3.9%|   |    21.1%|          |     0.0%|
                     |      195|        |      152|   |      109|          |       44|
                     +---------+        +---------+   +---------+          +---------+
                          |                                 |
                        Math                             Science
                 4.35E-03;(0.0522)                 3.10E-03;(0.0372)
                          |                                 |
                    -----------              ------------------------
                ?012        345789        0123          ?4          56789
                 6 |          7 |          8 |           9 |         10 |
          +---------+  +---------+  +---------+  +---------+  +---------+
          |    28.0%|  |     7.4%|  |    63.9%|  |    13.3%|  |    14.0%|
          |    54.0%|  |    87.4%|  |    13.9%|  |    16.7%|  |    51.2%|
          |    14.0%|  |     5.3%|  |    13.9%|  |    33.3%|  |    14.0%|
          |     4.0%|  |     0.0%|  |     8.3%|  |    36.7%|  |    20.9%|
          |      100|  |       95|  |       36|  |       30|  |       43|
          +---------+  +---------+  +---------+  +---------+  +---------+
```

In the complete group, 25% of the students failed, 54.2 % passed, and 6.6% of the students obtained some college credits for the year completed. In 14.2% of the cases information was incomplete or unavailable. The most important predictor was the

Faculty, which divided the group into four subsets: (AM), (CE), (SFL) and (DBH), representing subsets of the subject-matter groupings.

6.4 CONFIRM Analysis of the "head injuries" Data Set

6.4.1 Description of the data

The "head injuries" data set of Titterington *et al* (1981) discussed in Section 6.2 is analyzed here using CONFIRM. For the CATFIRM analysis of this data, see Section 6.2. The outcome is predicted on the basis of 6 predictors assessed on the patients' admission to the hospital:

- *age*. The age of the patient. This is grouped into decades in the original data, and is grouped the same way here. It has eight classes.
- *EMV*. This is a composite score of three measures—of eye-opening in response to stimulation, motor response of best limb, and verbal response. This has seven classes, but is not measured in all cases, so that there are eight possible codes for this score—the seven measurements and an eighth "missing" category.
- *MRP*. This is a composite score of motor responses in all four limbs. This also has seven measurement classes with an eighth class for missing information.
- *Change*. The change in neurological function over the first 24 hours. This was graded 1, 2, or 3, with a fourth class for missing information.
- *Eye indicator*. A summary of diagnostics on the eyes. This too had three measurement classes, with a fourth for missing information.
- *Pupils*. Pupil reaction to light—present, absent, or missing.

In the CONFIRM analysis the dependent variable is treated as being on the interval scale of measurement with values 1, 2, and 3. As the outcome is on the ordinal scale, this equally-spaced scale is not necessarily statistically appropriate in this data set and is used as a matter of convenience rather than with the implication that this is considered the best way to proceed.

6.4.2 FIRM Syntax

The CONFIRM syntax for the analysis of the "head injuries" data is given below:

```
CONFIRM: DEN SUM SPLIT RULE
    1 'Outcome'      0
    6
 'age'          2     1 'm'   0  8 '01234567'    0.9000  1.0000
 'EMV'          3     0 '1'   0  8 '?1234567'    0.9000  1.0000
 'MRP'          4     0 '1'   0  8 '?1234567'    0.9000  1.0000
 'Change'       5     0 '1'   0  4      '?123'    0.9000  1.0000
 'Eye ind'      6     0 '1'   0  4      '?123'    0.9000  1.0000
  'Pupils'      7     0 'f'   0  3       '?12'    0.9000  1.0000
 1  20   .00100   .50000  1.00000  50    0 0 1   .0000000       1  0  0  0  0
```

- The first line indicates that a CONFIRM analysis is required, and that a dendrogram (keyword `den`), summary statistics (keyword `sum`), the split file (keyword `split`) and the split rule file (keyword `rule`) are to be included in the output file.
- The second line contains information on the dependent or outcome variable. It is the first variable in the data set, as indicated by the `1` in the first field, and it is used here as a continuous variable, as indicated by the `0` in the third entry on this line. Because of this specification, no category names are required for the FIRM analysis. This is the dependent variable specification section of the syntax file.
- The third line has one entry—`6`—indicating the number of predictor variables to be used in the analysis. This is the start of the predictor specification section in the syntax file. The next six lines provide information on each of the predictors.
- The variable *age*, grouped into decades as described in the previous subsection, is the second variable in the data set (position of predictor = `2`). The next value, `1`, indicates that the variable is to be carried along but not used for splitting. *Age* is a monotonic predictor (type of predictor = `m`). The range of values of *age* is 0 to 8, as indicated in the next two fields. This is followed by the category codes, in this case `0` to `7`. The last two entries in this line are the splitting and merging significance values. These levels are given in percent. In this example, a 0.9% significance level will be used for splitting and 1% for merging.
- The information for the other predictors, *EMV, MRP*, *Change*, *Eye ind* and *Pupils*, are given in a similar way, concluding the predictor variable specification.
- The last line of the syntax file contains the output options for this CONFIRM analysis.

```
1  20   .00100   .50000  1.00000  50   0 0 1   .0000000      1  0  0  0  0
```

The options specified are as follows:

`1`: The detailed split file is requested.
`20`: Minimum number of cases in a group. In this case, no group smaller than 20 cases will be considered for splitting.
`0.00100`: The minimum proportion of SSD to analyze the group.
`0.50000`: The raw significance level.
`1.0000`: The conservative significance level that a predictor must attain for the split to be made.
`50`: The maximum number of nodes to be analyzed.
`0`: The external degrees of freedom on the variance estimate. In this example, no such information is available and the `0` indicates that.
`0`: The data for this analysis is in free format. If fixed format had been indicated, a format statement would have followed directly after this line of input.
`1`: The use of the pooled variance in *t*-statistic for the testing of pairs of categories for compatibility is requested.

.0000000: No external variance estimate is used.
1: Indicated use of FIRM 2.1 methodology rather than FIRM 2.0 methodology.
0: Not currently used, set to default value of 0.
0: Not currently used, set to default value of 0.
0: Not currently used, set to default value of 0.
0: Not currently used, set to default value of 0.

For more information on syntax, see the sections on the dependent and predictor variable specifications in Chapter 8.

6.4.3 CONFIRM Output

The first few lines of the output file contain information on the version of the FIRM analysis, copyright and contact information. This is followed by the syntax used in this analysis.

```
DATE: 04/21/2000
                              TIME: 09:59
                             C O N F I R M

                         Version 2.3  1999/09/04
                      Copyright 1999 Douglas M Hawkins
                           Applied Statistics
                         University of Minnesota
                    Formal Inference-based Recursive Modeling

                     Scientific Software International, Inc.
                        7383 North Lincoln Avenue, Suite 100
                         Lincolnwood, IL 60712-1704, U.S.A.
            Phone: (800)247-6113, (847)675-0720, Fax: (847)675-2140
                        Website: www.ssicentral.com
                    Techsupport: techsupport@ssicentral.com

 The following lines were read from file H:\FIRM\DATA\HEADINJ.CON

 CONFIRM: DEN SUM SPLIT RULE
     1 'Outcome'      0
     6
  'age'          2     1 'm'   0  8 '01234567'   0.9000  1.0000
  'EMV'          3     0 '1'   0  8 '?1234567'   0.9000  1.0000
  'MRP'          4     0 '1'   0  8 '?1234567'   0.9000  1.0000
  'Change'       5     0 '1'   0  4      '?123'   0.9000  1.0000
  'Eye ind'      6     0 '1'   0  4      '?123'   0.9000  1.0000
   'Pupils'      7     0 'f'   0  3       '?12'   0.9000  1.0000
  1  20   .00100   .50000  1.00000  50   0 0 1   .0000000     1  0  0  0  0
```

Initial information on the analysis is given next, followed by a mini-dendrogram. In both cases, it can be seen that the first split was on variable number 6, and two new nodes were formed. In the case of the first of these nodes, splitting next occurred using predictor 3, while the other node was split using predictor 1.

```
Run now starting....
 All data in.         500 cases read with
                      500 retained.
 Start FIRM processing

 Starting analysis of node   1
 Split on    6 making descendants   2   2   3

 Starting analysis of node   2
 Split on    3 making descendants   5   4   4   4   4   4   4   5

 Starting analysis of node   3
 Split on    1 making descendants   6   6   6   6   7   7   8   8

 Starting analysis of node   4

 Starting analysis of node   5

 Starting analysis of node   6
 Split on    5 making descendants  10   9  10  10

 Starting analysis of node   7
 Split on    3 making descendants  12  11  11  11  11  11  12  12

 Starting analysis of node   8
 Split on    2 making descendants  13  -1  13  13  13  13  13  14

 Starting analysis of node  10
 Split on    2 making descendants  16  15  15  16  16  16  16  16

 Starting analysis of node  11

 Starting analysis of node  12
 Split on    2 making descendants  17  -1  -1  17  17  18  19  19
 .
 .

 Mini-dendrogram of analysis

 Legend:                    node number
                            splitting variable
                  ------------------------------------
                  horizontal line connects descendants

                                              1
                                              6
                ------------------------------
               2                              3
               3                              1
           ---------        ---------------------------------------
           4       5        6            7                         8
                            5            3                         2
                   ----------        ----------              ---------
                   9       10       11        12             13       14
                            2                  2
                   ----------        -----------------
                  15       16       17        18       19
```

As the `sum`, `den`, `split` and `rule` keywords were included in the syntax file, the summary information is not written to an external file, but forms the next part of the output file.

```
     ---------------------------------------------
     |  Output of Summary and Error Information  |
     ---------------------------------------------

 Program dimensions:
 Maximum number of predictors          1003
 Maximum number of predictor categories   20

     Dependent variable no.    1 is named Outcome
     Predictor variables
  no  posn  name           no of cats    split  merge  may?     type
   1     2  age                     8    0.900  1.000  yes      mono
   2     3  EMV                     8    0.900  1.000  yes      flt
   3     4  MRP                     8    0.900  1.000  yes      flt
   4     5  Change                  4    0.900  1.000  yes      flt
   5     6  Eye ind                 4    0.900  1.000  yes      flt
   6     7  Pupils                  3    0.900  1.000  yes      free

 Run options in effect
 Full split/merge details of predictors
 For a group to be analyzed, it must: -
    contain at least    20 cases;
    have at least proportion 0.00100 of starting ssd.
 Minimum % raw significance to split   0.500
 Minimum % conservative significance to split   1.000
 Analysis will stop after   50 groups have been formed
 Error variance is pooled Anova MS
 New (FIRM 2.1 and newer) p-values used
 All data in.          500 cases read with
                       500 retained.
```

A summary of the options used in splitting and merging of the groups is followed by specific information on the splitting of each node. The primary use of this information is for exploration of possible splits that were not used.

When we look at the analysis of the first node, we see that all of the predictors were able to provide highly significant splits into 2, 3, or 4 category groupings. While all predictors are highly significant, *Pupils* gives the most significant split.

```
 **********************************************************************
     Analysis of group no.   1          previous group no.   0
  no.  name                      mc-sig(%)   bonf-sig(%), grouping
   1  age                        8.76E-13    4.70E-13     01 23 45 67
   2  EMV                        1.27E-21    1.31E-21     12 345 ?6 7
   3  MRP                        3.35E-21    1.93E-22     12 345 ?67
   4  Change                     5.99E-03     0.0202      ?1 23
   5  Eye ind                    1.30E-21    3.25E-21     1 ?2 3
   6  Pupils                     1.42E-22    2.02E-22     1? 2

     Best predictor
   6  Pupils                     1.42E-22    2.02E-22     1? 2
```

The one-way analysis of variance for the split actually made is given next, followed by the summary statistics of the descendant groups formed by the split. This information is duplicated in the dendrogram, and so it is not of particular interest here. What is interesting is the fact that in the whole data set, all predictors are highly significant, but their relative predictive powers seem to change as we go down the dendrogram. Look, for example, at the relative size of the p-values of *age* and of the clinical observations as one goes down the tree. A portion of the output is given below.

```
     Analysis of variance
                 Sum of squares     mean square        degrees of freedom
     Grouping          84.2132       84.2132                  1
    Error             353.9868        0.7108                 498

    F-value                           118.474
    significance                     6.75E-23
    Bonferroni P                     2.02E-22
    multiple comparison P            1.42E-22
    overall conservative P           1.42E-22
    adjusted for predictor count     8.53E-22

    Grouping is significant at the conservative  8.53E-22%   level

Statistics for grouping
Node          Mean         s.d.        size      s.e. (mean)
   2        1.185185     0.520980      135     0.0448389
   3        2.109589     0.934116      365     0.0488939

************************************************************************
    Analysis of group no.   2            previous group no.   1
 no.  name                          mc-sig(%)    bonf-sig(%),  grouping
  1  age                              100.00       100.00       01234567
  2  EMV                              0.1413       0.0493       123456 ?7
  3  MRP                            9.19E-03     1.12E-04       123456 ?7
  4  Change                           100.00       100.00       ?123
  5  Eye ind                        6.55E-03     5.14E-03       12 ?3
  6  Pupils                           0.1505       0.1505       1 ?

    Best predictor
  3  MRP                            9.19E-03     1.12E-04       123456 ?7

     Analysis of variance
                 Sum of squares     mean square       degrees of freedom
     Grouping          7.0729        7.0729                  1
    Error             29.2975        0.2203                 133

    F-value                            32.108
    significance                     8.65E-06
    Bonferroni P                     1.12E-04
    multiple comparison P            9.19E-03
    overall conservative P           1.12E-04
    adjusted for predictor count     6.74E-04

    Grouping is significant at the conservative  6.74E-04%   level

Statistics for grouping
Node          Mean         s.d.        size      s.e. (mean)
   4        1.073394     0.325076      109     0.0311366
   5        1.653846     0.845804       26     0.1658758
```

```
***********************************************************************
    Analysis of group no.   3          previous group no.   1
 no.  name                        mc-sig(%)    bonf-sig(%), grouping
  1  age                          1.22E-12     7.84E-13     0123 45 67
  2  EMV                          5.68E-09     3.68E-09     12 345 ?6 7
  3  MRP                          1.21E-08     4.03E-10     12345 ?67
  4  Change                         0.2076       0.2994     ?1 23
  5  Eye ind                      1.47E-06     3.30E-06     1 ?2 3
  6  Pupils                         100.00       100.00     2

    Best predictor
  1  age                          1.22E-12     7.84E-13     0123 45 67

     Analysis of variance
                 Sum of squares     mean square      degrees of freedom
     Grouping          56.6016        28.3008                  2
    Error             261.0148         0.7210                362

    F-value                          39.250
    significance                    3.73E-14
    Bonferroni P                    7.84E-13
    multiple comparison P           1.22E-12
    overall conservative P          7.84E-13
    adjusted for predictor count    3.92E-12

    Grouping is significant at the conservative  3.92E-12%   level

Statistics for grouping
Node         Mean         s.d.       size     s.e. (mean)
   6        2.385965    0.860374     228     0.0569796
   7        1.876543    0.913547      81     0.1015052
   8        1.321429    0.690379      56     0.0922558

***********************************************************************
    Analysis of group no.   4          previous group no.   2
 no.  name                        mc-sig(%)    bonf-sig(%), grouping
  1  age                            100.00       100.00     01234567
  2  EMV                            100.00       100.00     ?1234567
  3  MRP                            100.00       100.00     123456
  4  Change                         100.00       100.00     ?123
  5  Eye ind                        100.00       100.00     ?123
  6  Pupils                         100.00       100.00     ?1
    No prediction possible.

***********************************************************************
    Analysis of group no.   5          previous group no.   2
 no.  name                        mc-sig(%)    bonf-sig(%), grouping
  1  age                            100.00       100.00     01234567
  2  EMV                            100.00       100.00     ?1234567
  3  MRP                            100.00       100.00     ?7
  4  Change                         100.00       100.00     ?123
  5  Eye ind                        100.00       100.00     ?123
  6  Pupils                         100.00       100.00     1?
    No prediction possible.
```

As with the summary statistics, the use of the `den` keyword leads to the inclusion of the full dendrogram in the output file, rather than the writing of this information to an external file. The dendrogram thus forms the next part of the CONFIRM output file considered here.

```
                                 LEGEND:
                                           split var
                                          P val (Bonf P)
                                                 |
                   +------+                ------------
                   |   500|                      |
                   |1.8600|                   levels
                   |0.9371|                Node |
                   |1.0000|                  +------+
                   |3.0000|                  |N     |
                   +------+                  |Xbar  |
                       |                     |SD    |
                      Pupils                 |Min   |
                 1.42E-22;(8.53R-22)         |Max   |
                       |                     +------+
        --------------------------
       1?                        2
      2 |                      3 |
     +------+                 +------+
     |   135|                 |   365|
     |1.1852|                 |2.1096|
     |0.5210|                 |0.9431|
     |1.0000|                 |1.0000|
     |3.0000|                 |3.0000|
     +------+                 +------+
         |                         |
        MRP                       age
    1.12E-04;(6.74E-04)    7.84E-13;(3.92E-12)
         |                         |
    ----------     -----------------------------------
123456     ?7      0123      45                     67
 4 |      5 |      6 |      7 |                    8 |
+------+ +------+ +------+ +------+               +------+
|   109| |    26| |   228| |    81|               |    56|
|1.0734| |1.6538| |2.3860| |1.8765|               |1.3214|
|0.3251| |0.8458| |0.8604| |0.9135|               |0.6904|
|1.0000| |1.0000| |1.0000| |1.0000|               |1.0000|
|3.0000| |3.0000| |3.0000| |3.0000|               |3.0000|
+------+ +------+ +------+ +------+               +------+
                      |        |                     |
                    Eye ind   MRP                    EMV
         4.46-06;(2.23E-05)  3.75E-05;(1.88E-04)   1.23E-07;(6.16E-07)
                     |         |                     |
         ------------      ----------             ----------
        1       ?23      12345     ?67         ?23456      7
      9 |     10 |      11 |     12 |          13 |     14 |
     +------+ +------+  +------+ +------+      +------+ +------+
     |    18| |   210|  |    41| |    40|      |    46| |    10|
     |1.3333| |2.4762|  |1.3659| |2.4000|      |1.0870| |2.4000|
     |0.6860| |0.8137|  |0.6617| |0.8412|      |0.2849| |0.9661|
     |1.0000| |1.0000|  |1.0000| |1.0000|      |1.0000| |1.0000|
     |3.0000| |3.0000|  |3.0000| |3.0000|      |2.0000| |3.0000|
     +------+ +------+  +------+ +------+      +------+ +------+
                  |                  |
                 EMV                EMV
          1.99E-03;(9.94E-03) 0.0513;(0.2566)
                  |                  |
        ----------          -------------------
       12     ?34567      ?34        5        67
     15 |     16 |       17 |     18 |     19 |
     +------+ +------+   +------+ +------+ +------+
     |    17| |   193|   |     7| |    12| |    21|
     |1.5882| |2.5544|   |2.5714| |1.5833| |2.8095|
     |0.8703| |0.7627|   |0.7868| |0.7930| |0.5118|
     |1.0000| |1.0000|   |1.0000| |1.0000| |1.0000|
     |3.0000| |3.0000|   |3.0000| |3.0000| |3.0000|
     +------+ +------+   +------+ +------+ +------+
```

The full set of 500 cases has a mean outcome score of 1.8600. The first split is made on the basis of the predictor *Pupils*. Cases for which *Pupils* is `1` or `?` constitute the first descendant group (`Node 2`), while those for which it is 2 given `Node 3`. Note that this is the same predictor that CATFIRM elected to use, but the patients with missing values of *Pupils* are grouped with *Pupils* = `1` by the CONFIRM analysis, and with *Pupils* = `2` by CATFIRM. Going to the descendant groups, Node 2 is then split on *MRP* (unlike in CATFIRM, where the corresponding node was terminal), while Node 3 is again split four ways on *age*.

The overall tree is bigger than that produced by CATFIRM—11 terminal and 7 interior nodes—and produces groups with prognostic scores ranging from a grim 1.0734 up to 2.8095 on the 1 to 3 scale. As with CATFIRM, the means in these terminal nodes could be used for prediction of future cases, giving estimates of the score that patients in the terminal nodes would have.

The statistical model used in CONFIRM can be described as a *piecewise constant regression* model. For example, among the patients in Node 3, the outcome is predicted to be 2.48 for patients aged under 20, 2.25 for those aged 20 to 40, 1.88 for those aged 40 to 60, and 1.32 for those aged over 60. This approach contrasts sharply with, say, a linear regression in which the outcome would be predicted to rise by some constant amount with each additional year of *age*. While this piecewise constant regression model is seldom valid exactly, it is often a reasonable working approximation of reality.

The dendrogram is the most obviously and immediately useful output of a FIRM analysis. It is an extract of the much more detailed output, which may be written to file, or, by using the `split` keyword, included in the output file. If written to the file, this file will contain information on both split and table files, in contrast with CATFIRM where such information will be written to two separate files.

This output contains the following (often very valuable) information:

- An analysis of each predictor in each node, showing which categories of the predictor FIRM finds it best to group together, and what the conservative statistical significance level of the split by each predictor is;
- The number of cases flowing into each of the descendant groups;
- Summary statistics of the cases in the descendant nodes. In the case of CATFIRM, the summary statistics are a frequency breakdown of the cases between the different classes of the dependent variable. With CONFIRM, the summary statistics given are the arithmetic mean and standard deviation of the cases in the node. This summary information is echoed in the dendrogram, as discussed previously.

The information on the splitting of the first group is shown below. The actual output file contains similar information on all the groups.

The listing starts with the summary statistics of the grouping of the cases in that node by the different classes of the predictor. The categories of a free or character predictor are arranged in descending order of the dependent variable mean—all the other predictors are listed in their original order.

```
        -----------------------------------------
        | Output of Detailed Analysis of Splits |
        -----------------------------------------

*******************************************************************
Analysis of group no. 1   previous group no. 0  mean = 1.860 size = 500

      predictor no.  1   age
      statistics before merging
  Cate        0      1      2      3      4      5      6      7
  mean     2.11   2.30   1.99   1.95   1.61   1.71   1.31   1.00
  size       55    111     77     55     61     58     61     22
  s.d.     0.96   0.92   0.91   0.95   0.88   0.88   0.67   0.00
```

The summary table is followed by the one-way analysis of variance for the ungrouped cross-classification.

```
  Anova  SSE     DFE     SSH     DFH  R-squared      F          Sign(%)
      372.0      492    66.2      7     0.1510   12.5046  0.8757E-12
```

Next are the details of merging the categories. For a monotonic predictor like *age*, the only possible merges are between adjacent categories. CONFIRM lists a Student's *t*-statistic between each pair of categories on the line below. For example, the *t*-value for merging classes 0 and 1 is 1.3, that for merging 1 and 2 is -2.4; that for merging 6 and 7 is -1.4. The smallest *t*-value is -0.3 for merging classes 2 and 3. This is done, and the number of classes for merging becomes 7, with 6 possible mergings and their associated *t*-values. These *t*-values are listed on the second "merge stats" line, which shows that the least significantly different mergeable pair is 4 and 5, with a *t*-value of 0.6. This merge in turn takes place, as this *t*-value is not significant at the merge significance level selected in this run. This reduction continues until at the final line, both the Student's *t*-values (for merging the composite categories (01), (23), (45), and (67)) are significant, and the merge testing stops.

```
  Group         0    1    2    3    4    5    6    7
  merge stats  1.3 -2.4 -0.3 -2.1  0.6 -2.5 -1.4
  Group         0    1    23    4    5    6    7
  merge stats  1.3 -2.9   -2.7  0.6 -2.5 -1.4
  Group         0    1    23    45    6    7
  merge stats  1.3 -2.9   -2.9   -2.5 -1.4
  Group         01    23    45    6    7
  merge stats   -2.6   -2.9   -2.5 -1.4
  Group         01    23    45    67
  merge stats   -2.6   -2.9   -3.4

      statistics after splitting / merging
  Cate        01      23      45      67
  mean      2.23    1.97    1.66    1.23
  size       166     132     119      83
  s.d.      0.93    0.92    0.88    0.59
```

```
Anova  SSE       DFE      SSH      DFH   R-squared      F         Sign(%)
   375.2         496    63.0        3    0.1437    27.7403   0.1344E-13

    predictor no.  2    EMV
    statistics before merging
Cate       ?       1       2       3       4       5       6       7
mean     2.07    1.00    1.19    1.58    1.81    1.84    2.25    2.63
size       28      19      64      52     111      96      65      65
s.d.     0.94    0.00    0.53    0.82    0.93    0.91    0.94    0.74
Anova  SSE       DFE      SSH      DFH   R-squared      F         Sign(%)
   341.2         492    97.0        7    0.2214    19.9861   0.1274E-20
Group        ?    1    2    3    4    5    6    7
merge stats -4.3  0.9  2.5  1.7  0.3  3.0  2.6
            -4.3 -4.7 -2.5 -1.5 -1.3  0.9  3.0
Group        ?    1    2    3    45    6    7
merge stats -4.3  0.9  2.5  1.9   3.6  2.6
            -4.3 -4.7 -2.5 -1.5  0.9  3.0
Group        ?    12    3    45    6    7
merge stats -5.1   2.9  1.9   3.6  2.6
            -5.1 -2.5 -1.5  0.9  3.0
Group        12    3    45    ?6    7
merge stats    2.9  1.9   3.5   3.3
Group        12    345    ?6    7
merge stats    6.0    4.1   3.2

    statistics after splitting / merging
Cate       12     345     ?6      7
mean     1.14    1.78    2.19    2.63
size       83     259      93      65
s.d.     0.47    0.90    0.94    0.74
Anova  SSE       DFE      SSH      DFH   R-squared      F         Sign(%)
   344.9         496    93.3        3    0.2128    44.7058   0.1382E-22
```

A slightly different format is used for a floating predictor, as illustrated by the predictor *MRP*. Here, while the floating category is on its own, there are two lines of statistics for each stage—the first for merging ? with 1, ? with 2, ? with 3... At the first merge phase, these lines show that the least significant is for categories 3 with 4, and so these categories are merged. The second stage merges 1 with 2 and the third 6 with 7. At the fourth stage, the smallest t is for merging ? with the composite category (67), and after this is done, ? no longer floats, and so for the last two stages there is only a single line of merge outputs.

```
    predictor no.  3    MRP
    statistics before merging
Cate       ?       1       2       3       4       5       6       7
mean     2.14    1.21    1.31    1.61    1.54    1.87    2.30    2.37
size       21      38      61      33     114      30      91     112
s.d.     0.91    0.58    0.67    0.90    0.81    0.94    0.89    0.90
Anova  SSE       DFE      SSH      DFH   R-squared      F         Sign(%)
   342.6         492    95.6        7    0.2182    19.6192   0.3351E-20
Group        ?    1    2    3    4    5    6    7
merge stats -4.1  0.6  1.6 -0.4  1.9  2.4  0.6
            -4.1 -3.9 -2.3 -3.0 -1.2  0.8  1.1
Group        ?    1    2    34    5    6    7
merge stats -4.1  0.6  1.9   1.8  2.5  0.6
            -4.1 -3.9 -3.0 -1.2  0.8  1.1
```

```
Group      ?    12    34    5    6    7
merge stats -4.3   2.6   1.9  2.5  0.6
            -4.3 -3.0 -1.2  0.8  1.1
Group      ?    12    34    5    67
merge stats -4.3   2.6   1.9  2.9
            -4.3 -3.0 -1.2  1.0

Group      12    34    5    ?67
merge stats   2.6   1.9  2.8
Group      12    345    ?67
merge stats   3.2    8.4

    statistics after splitting / merging
Cate      12     345     ?67
mean     1.27   1.61   2.32
size       99    177    224
s.d.     0.64   0.85   0.89
Anova   SSE      DFE      SSH      DFH   R-squared     F         Sign(%)
   346.2        497   92.0        2    0.2099    66.0065  0.3781E-23

    predictor no.  4   Change
    statistics before merging
Cate      ?      1      2      3
mean     1.82   1.60   1.97   2.13
size      132    143    115    110
s.d.     0.94   0.87   0.92   0.95
Anova   SSE      DFE      SSH      DFH   R-squared     F         Sign(%)
   419.1        496   19.1        3    0.0437     7.5530  0.5987E-02
Group      ?    1    2    3
merge stats -2.0  3.2  1.3
            -2.0  1.3  2.6
Group      ?    1    23
merge stats -2.0  4.5
            -2.0  2.3
Group      ?1    23
merge stats   4.1

    statistics after splitting / merging
Cate      ?1     23
mean     1.71   2.05
size      275    225
s.d.     0.91   0.94
Anova   SSE      DFE      SSH      DFH   R-squared     F         Sign(%)
   423.6        498   14.6        1    0.0333    17.1594  0.4038E-02

    predictor no.  5   Eye ind
    statistics before merging
Cate      ?      1      2      3
mean     1.92   1.09   1.70   2.22
size      110     96     73    221
s.d.     0.97   0.36   0.89   0.90
Anova   SSE      DFE      SSH      DFH   R-squared     F         Sign(%)
   351.4        496   86.8        3    0.1982    40.8602  0.1305E-20
Group      ?    1    2    3
merge stats -7.0  4.6  4.6
            -7.0 -1.7  3.0
Group      1    ?2    3
merge stats  6.9   4.6
```

```
     statistics after splitting / merging
Cate       1       ?2      3
mean     1.09    1.83    2.22
size       96     183     221
s.d.     0.36    0.94    0.90
Anova   SSE      DFE       SSH       DFH    R-squared      F           Sign(%)
   353.5         497    84.7          2     0.1933     59.5594   0.6500E-21
```

The final type of output is that for a free predictor (like *Pupils*). This is much simpler than in the corresponding CATFIRM case. Here the first stage of the analysis is to sort the categories of the predictor into ascending order of their mean values of the dependent variable. Thereafter, the analysis proceeds just like that of a monotonic predictor in these re-ordered categories. This implies that if any two categories of a free predictor are to end up joined together, the composite class will necessarily include any other categories whose mean score lay between the mean scores of the joined groups. This implication is not a consequence of the definition of a free predictor, but accords with common-sense expectations.

```
    predictor no.   6    Pupils
     statistics before merging
Cate       1       ?       2
mean     1.14    1.62    2.11
size      122      13     365
s.d.     0.45    0.87    0.93
Anova   SSE      DFE       SSH       DFH    R-squared      F           Sign(%)
   351.3         497    86.9          2     0.1983     61.4491   0.1422E-21
Group         1     ?     2
merge stats   1.9   2.1
Group        1?     2
merge stats   10.9

     statistics after splitting / merging
Cate       1?      2
mean     1.19    2.11
size      135     365
s.d.     0.52    0.93
Anova   SSE      DFE       SSH       DFH    R-squared      F           Sign(%)
   354.0         498    84.2          1     0.1922    118.4738   0.6746E-22
```

Inclusion of the `rule` keyword leads to the inclusion of the split rule table shown below table in the output file. Apart from the first section summarizing the splits made, it has the additional information on the categories of the character predictors and the cutpoints of the real predictors.

```
   1   6   2   2   3
   2   3   5   4   4   4   4   4   4   5
   3   1   6   6   6   6   7   7   8   8
   6   5  10   9  10  10
   7   3  12  11  11  11  11  11  12  12
   8   2  13  -1  13  13  13  13  13  14
  10   2  16  15  15  16  16  16  16  16
  12   2  17  -1  -1  17  17  18  19  19
  -1  -1
 0
 0
```

6.5 CONFIRM Analysis of the "mixed" Data Set

6.5.1 Description of the Data

The "mixed" data set was provided by Gordon V. Kass and is described in detail in Section 6.3, where it was analyzed using CATFIRM.

The dependent variable for the CATFIRM run was of the type *character*—a promotion code. A second measure of the student's overall achievement in relation to the requirements for going on to the second-year curriculum was the aggregate score. This is the average score received for all courses taken in the first year at University, and is suitable for analysis using CONFIRM.

6.5.2 FIRM Syntax File

The syntax file for the analysis of the "mixed" data using CONFIRM is given below:

```
confirm:   sum den ! add dendrogram and summary information to output
file
     17 'U aggr'      0    ! Number of dependent variable
     12                    ! Number of predictors
   'Afrikaans'      7     0 'r'    0  0 ''    0.9000  1.0000
     'Biology'     10     0 'r'    0  0 ''    0.9000  1.0000
     'English'      6     0 'r'    0  0 ''    0.9000  1.0000
     'Faculty'     13     0 'c'    0  0 ''    0.9000  1.0000
     'History'     12     0 'r'    0  0 ''    0.9000  1.0000
       'Latin'     11     0 'r'    0  0 ''    0.9000  1.0000
        'Math'      8     0 'r'    0  0 ''    0.9000  1.0000
     'Mattype'      3     0 'c'    0  0 ''    0.9000  1.0000
    'Matmonth'      4     0 'f'    0 13 '?123456789ABC'    0.9000   1.0000
     'Matyear'      5     0 'r'    0  0                ''    0.9000   1.0000
     'Science'      9     0 'r'    0  0                ''    0.9000   1.0000
         'Sex'      2     0 'c'    0  0                ''    0.9000   1.0000
1  20    .00100  1.00000  1.00000  20    0 0 1    .0000000   1  0  0  0  0
```

- The first line indicates that a CONFIRM analysis is required, and that a dendrogram (keyword `den`), summary statistics (keyword `sum`), the split file (keyword `split`) and the split rule file (keyword `rule`) are to be included in the output file. Note the use of the `!` to start a comment after the end of the input on the first line.
- The second line contains information on the dependent or outcome variable. It is the 17th variable in the data set, as indicated by `17` in the first field, and it is used here as a continuous variable, as indicated by the `0` in the third entry on this line. Because of this specification, no category names are required for the FIRM analysis. This is the dependent variable specification section of the syntax file.

- The third line has one entry—12—indicating the number of predictor variables to be used in the analysis. This is the start of the predictor specification section on the syntax file. The next twelve lines provide information on each of the predictors.
- The variable *Afrikaans* is the seventh variable in the data set (position of predictor = 7), and is a real predictor (type of predictor = r). The 1 indicates that this variable is to be carried over but not used for splitting.
- The monotonic, free, and float predictor types all require that the predictor's values be consecutive integers. The next two fields should contain information on this range, and are only required for these three predictor types. For a real predictor, 0s are used instead.
- This is usually followed by the category codes. In this case, the absence of category codes is indicated by " ". The last two entries in this line are the splitting and merging significance values. These levels are given in percent. In this example, a 0.9% significance level will be used for splitting and 1% for merging.
- The predictor *Matmonth* is a free predictor, with values ranging from 0 to 13. Codes for the values are assigned as ?, 1, 2, ..., 9, A, B, C. Splitting and merging significance levels are the same as for all other predictors.
- The last line of the syntax file contain the output options for this CONFIRM analysis.

```
1  20   .00100  1.00000  1.00000  20   0 0 1   .0000000   1  0  0  0  0
```

- The options specified are as follows:

1: The detailed split file is requested.
20: Minimum number of cases in a group. In this case, no group smaller than 20 cases will be considered for splitting.
0.00100: The minimum proportion of SSD to analyze the group.
0.5000: The raw significance level.
1.0000: The conservative significance level that a predictor must attain for the split to be made.
20: The maximum number of nodes to be analyzed.
0: The external degrees of freedom on the variance estimate. In this example, no such information is available and the 0 indicates that.
0: The data for this analysis is in free format. If fixed format had been indicated, a format statement would have followed directly after this line of input.
1: The use of the pooled variance in *t*-statistic for the testing of pairs of categories for compatibility is requested.
.0000000: No external variance estimate is used.
1: Indicated use of FIRM 2.1 methodology rather than FIRM 2.0 methodology.
0: Not currently used, set to default value of 0.

0: Not currently used, set to default value of 0.
0: Not currently used, set to default value of 0.
0: Not currently used, set to default value of 0.

For more information on syntax, see the sections on the dependent and predictor variable specifications in Chapter 8.

6.5.3 CONFIRM Output

The first section of the output file echoes the syntax file and also contains the data set name and number of observations used in the analysis.

```
confirm:  sum den split rule! add dendrogram and summary information to output
file
    17 'U aggr'      0    ! Number of dependent variable
    12                    ! Number of predictors
  'Afrikaans'     7     0 'r'   0  0 ''   0.9000  1.0000
    'Biology'    10     0 'r'   0  0 ''   0.9000  1.0000
    'English'     6     0 'r'   0  0 ''   0.9000  1.0000
    'Faculty'    13     0 'c'   0  0 ''   0.9000  1.0000
    'History'    12     0 'r'   0  0 ''   0.9000  1.0000
      'Latin'    11     0 'r'   0  0 ''   0.9000  1.0000
       'Math'     8     0 'r'   0  0 ''   0.9000  1.0000
    'Mattype'     3     0 'c'   0  0 ''   0.9000  1.0000
   'Matmonth'     4     0 'f'   0 13 '?123456789ABC'   0.9000  1.0000
    'Matyear'     5     0 'r'   0  0               ''  0.9000  1.0000
    'Science'     9     0 'r'   0  0               ''  0.9000  1.0000
        'Sex'     2     0 'c'   0  0               ''  0.9000  1.0000
 1  20    .00100  1.00000  1.00000  20   0 0 1   .0000000      1  0  0  0  0

 CONFIRM.  Formal Inference - based Recursive Modeling
 Program dimensions
 Maximum number of predictors                           1000
 Maximum number of categories in predictors               20
     the data file (input):
                    H:\FIRM\DATA\MIXED.dat
     Run now starting....
 All data in.           500 cases read with
                        500 retained.
 Start FIRM processing
```

Initial information on the analysis is given next, followed by a mini-dendrogram. In both cases, it can be seen that the first split was on variable number 7, and three new nodes were formed. In the case of the last of these nodes, splitting next occurred using predictor 11.

```
 Starting analysis of node    1
 Split on      7 making descendants    3    2    2    2    2    2    3    3    3
4    4

 Starting analysis of node    2
 Starting analysis of node    3
 Starting analysis of node    4
```

```
Starting to draw full-size dendrogram
Full-size dendrogram complete
Starting to draw mini-dendrogram

Mini-dendrogram of analysis

Legend:                     node number
                            splitting variable
                  ------------------------------------
                  horizontal line connects descendants

                    1
                    7
                    ----------------------------------------
                    2                    3                  4
```

As the sum, den, split and rule keywords were included in the syntax file, the summary information is not written to an external file, but forms the next part of the output file.

```
     -----------------------------------------
     |  Output of Summary and Error Information  |
     -----------------------------------------

                 CONFIRM  Formal Inference - based Recursive Modeling
                 Continuous dependent variable

                 Version 2.3  1999/09/04
                 Copyright 1999 Douglas M Hawkins
                 Applied Statistics
                 University of Minnesota

 Program dimensions:
 Maximum number of predictors           1003
 Maximum number of predictor categories   20

    Dependent variable no.   17 is named U aggr
    Predictor variables
  no  posn  name           no of cats   split  merge  may?     type
   1    7  Afrikaans              0    0.900  1.000  yes      real
   2   10  Biology                0    0.900  1.000  yes      real
   3    6  English                0    0.900  1.000  yes      real
   4   13  Faculty                0    0.900  1.000  yes      char
   5   12  History                0    0.900  1.000  yes      real
   6   11  Latin                  0    0.900  1.000  yes      real
   7    8  Math                   0    0.900  1.000  yes      real
   8    3  Mattype                0    0.900  1.000  yes      char
   9    4  Matmonth              13    0.900  1.000  yes      free
  10    5  Matyear                0    0.900  1.000  yes      real
  11    9  Science                0    0.900  1.000  yes      real
  12    2  Sex                    0    0.900  1.000  yes      char

 Run options in effect
 Full split/merge details of predictors
 For a group to be analyzed, it must: -
    contain at least    20 cases;
    have at least proportion 0.00100 of starting ssd.
 Minimum % raw significance to split   1.000
```

```
Minimum % conservative significance to split   1.000
Analysis will stop after   20 groups have been formed
Error variance is pooled Anova MS
New (FIRM 2.1 and newer) p-values used
All data in.           500 cases read with
                       500 retained.
```

Next there is a listing of the cutpoints for the real predictors, which is needed when reading the rest of the summary output and the dendrogram. Just one real predictor–*Afrikaans*—from this list is shown below.

The cutpoints 45, 51, 54, 55, 57, 62, 64, 65 and 75 are the best we can find for dividing the data set up into 10 groups of about equal size, and are the starting cutpoints from which FIRM will do its subsequent merging to a final grouping.

```
Character predictors seen in the data and their values are:

Predictor    4  Faculty                                      values seen
      A                    C                    M                     S
      D                    E                    F                     H
      B                    L
These values will be abbreviated to: ACMSDEFHBL

Predictor    8  Mattype                                      values seen
       5                   1                    6                     J
       F                   7                    ?                     4
       3                   A                    2
These values will be abbreviated to: 516JF7?43A2

Predictor   12  Sex                                          values seen
       M                   F
These values will be abbreviated to: MF

Continuous predictors seen in the data and their cutpoints are:

Predictor    1   Afrikaans

Code     Max value in class
 ?          Invalid or missing
 0          <=       45.000
 1          <=       51.000
 2          <=       54.000
 3          <=       55.000
 4          <=       57.000
 5          <=       62.000
 6          <=       64.000
 7          <=       65.000
 8          <=       75.000
 9          >        75.000
```

A summary of the options used in splitting and merging of the groups is followed by specific information on the splitting of each node. The primary use of this information is for exploration of possible splits that were not used.

When we look at the analysis of the first node, we see that four of the predictors (*Biology*, *English*, *Math* and *Science*) are able to provide splits significant at better than 1% conservative. Of these, the predictor *Math* is the most significant. It divides the students into three subsets—{01234}, {?567} and {89}—effectively distinguishing between the bottom 50% of the students, the next 30% plus those with missing Math grades, and the top 20%. Alternatively, we can use the table of cutpoints (given in the summary file, but now shown here) to turn these into ranges of Math scores. The first group is those at or below 57%, the second those above 57% but not exceeding 75% (together with the missing class), and those with scores above 75%. We will need to use the actual cutpoints given in the summary file (or, in this case, the complete output file) if we want to apply the model to the prediction of future students' performance, but the interpretation in terms of approximate percentages of the pool is probably more informative when it comes to a better understanding of the connection of the Math score to the aggregate first-year grade.

```
*************************************************************************
    Analysis of group no.   1          previous group no.   0
 no.  name                        mc-sig(%)    bonf-sig(%), grouping
  1  Afrikaans                     27.5146       9.6823     ?01234567 89
  2  Biology                      2.46E-03     7.22E-05     01234 ?56 789
  3  English                        0.0108     2.51E-03     ?012345 6789
  4  Faculty                        1.5561      15.1396     BLSC FEHAMD
  5  History                        7.5641       1.5349     ?01234567 8
  6  Latin                          100.00       100.00     ?012345678
  7  Math                         9.37E-04     4.67E-05     01234 ?567 89
  8  Mattype                        100.00       100.00     7FJ51A3?642
  9  Matmonth                       4.6522      13.1064     74 3B1C?6
 10  Matyear                       20.4593       3.2037     0123 ?45678
 11  Science                      2.14E-03     4.60E-04     012 ?34567 89
 12  Sex                            100.00       100.00     MF

    Best predictor
  7  Math                         9.37E-04     4.67E-05     01234 ?567 89
```

The one-way analysis of variance for the split actually made is given next, followed by the summary statistics of the descendant groups formed by the split. This information is duplicated in the dendrogram, and so it is not of particular interest here.

```
Analysis of variance

                  Sum of squares    mean square      degrees of freedom
     Grouping       15069.3285      7534.6643               2
     Error         196253.2795       394.8758             497

     F-value                          19.081
     significance                   1.04E-06
     Bonferroni P                   4.67E-05
     multiple comparison P          9.37E-04
     overall conservative P         4.67E-05
     adjusted for predictor count   5.60E-04

     Grouping is significant at the conservative  5.60E-04%   level

  Statistics for grouping
  Node         Mean          s.d.        size     s.e. (mean)
     2      42.308411    20.600566       214     1.4082257
     3      49.346154    19.753955       234     1.2913564
     4      60.288462    17.114764        52     2.3733908
```

```
*************************************************************************
     Analysis of group no.   2          previous group no.   1
  no.  name                       mc-sig(%)    bonf-sig(%), grouping
   1  Afrikaans                    100.00       100.00      ?0123456789
   2  Biology                      100.00       100.00      ?0123456789
   3  English                      3.7527       5.1131      012345 6789
   4  Faculty                     10.8388      66.2082      S MCLAHFED
   5  History                      100.00       100.00      ?012345678
   6  Latin                        100.00       100.00      ?024568
   7  Math                         100.00       100.00      01234
   8  Mattype                      100.00       100.00      7J1A24653
   9  Matmonth                     100.00       100.00      4BC13
  10  Matyear                      100.00       100.00      012345678
  11  Science                      100.00       100.00      ?0123457
  12  Sex                          100.00       100.00      MF

     Best predictor
   3  English                      3.7527       5.1131      012345 6789

      Analysis of variance
                  Sum of squares     mean square      degrees of freedom
      Grouping         3210.5718     3210.5718                1
     Error            87183.0731      411.2409               212

     F-value                           7.807
     significance                      0.5681
     Bonferroni P                      5.1131
     multiple comparison P             3.7527
     overall conservative P            3.7527
     adjusted for predictor count     45.0319
     No prediction possible.
```

As with the summary statistics, the use of the den keyword leads to the inclusion of the full dendrogram in the output file, rather than the writing of this information to an external file. The dendrogram thus forms the next part of the CONFIRM output file. The dendrogram is the most obviously and immediately useful output of a FIRM analysis. It is an extract of the much more detailed output, which may be written to a file, or, by using the split keyword, included in the output file. If written to a file, this file will contain information on both split and table files, in contrast with CATFIRM where such information will be written to two separate files.

This output contains the following (often very valuable) information:

- An analysis of each predictor in each node, showing which categories of the predictor FIRM finds it best to group together, and what the conservative statistical significance level of the split by each predictor is.
- The number of cases flowing into each of the descendant groups.
- Summary statistics of the cases in the descendant nodes. In the case of CATFIRM, the summary statistics are a frequency breakdown of the cases between the different classes of the dependent variable. With CONFIRM, the summary statistics given are the arithmetic mean and standard deviation of the cases in the node. This summary information is echoed in the dendrogram, as discussed previously.

The split file contains information on the splitting of all groups. The listing starts with the summary statistics of the grouping of the cases in that node by the different classes of the predictor. The categories of a free or character predictor are arranged in descending order of the dependent variable mean—all the other predictors are listed in their original order.

Inclusion of the `rule` keyword leads to the inclusion of the table in the output file. Apart from the first section summarizing the splits made, it has the additional information on the categories of the character predictors and the cutpoints of the real predictors.

7 New Statistical Features

7.1 System Files

PRELIS and LISREL generate several system files through which they can communicate with each other. These system files are binary files. Some of these system files can be used directly by users. Here we present these system files and their uses.

7.1.1 The PRELIS System File

The majority of software packages make use of a data file format that is unique to that package. These files are usually stored in a binary format so that reading and writing data to the hard drive is as fast as possible. Examples of file formats are

- Microsoft Excel (*.**xls**)
- SPSS for Windows (*.**sav**)
- Minitab (*.**mtw**)
- SAS for Windows 6.12 (*.**sd2**)
- STATA (*.**dta**)
- STATISTICA (*.**sta**)
- SYSTAT (*.**sys**)

These data system files usually contain all the known information about a specific data set.

LISREL for Windows can import any of the above and many more file formats and convert them to a *.**psf** file. The *.**psf** file is analogous to, for example, a *.**sav** file in the sense that one can retrieve 3 types of information from a PRELIS system file:

- **General information:**

  This part of the *.**psf** file contains the number of cases in the data set, the number of variables, global missing value codes and the position (number) of the weight variable.

- **Variable information:**

  The variable name (8 characters maximum), and variable type (for example, continuous or ordinal), and the variable missing value code. For an ordinal variable the following information is also stored: the number of categories, and category labels (presently 4 characters maximum) or numeric values assigned to each category.

- **Data information:**

  Data values are stored in the form of a rectangular matrix, where the columns denote the variables and the rows the cases. Data are stored in double precision to ensure accuracy of all numerical calculations.

  By opening a *.**psf** file, the main menu bar expands to include the ***Data***, ***Transformation***, ***Statistics***, ***Graphs*** and ***Multilevel*** menus. Selecting any of the options from these menus will activate the corresponding dialog box. The ***File*** and ***View*** menus are also expanded. See Chapter 2 for more details.

  An important feature of LISREL 8.50 is the use of *.**psf** files as part of the LISREL or SIMPLIS syntax. In doing so, one does not have to specify variable names, missing value codes, or the number of cases.

  The folders **msemex** and **missingex** contain many examples to illustrate this new feature.

  When building LISREL or SIMPLIS syntax from a path diagram, it is sufficient to select an appropriate *.**psf** file from the ***Add/Read*** option in the ***Setup***, ***Variables*** dialog box. For more information and examples, see Chapter 4.

7.1.2 The Data System File

A data system file, or a *.**dsf** for short, is created each time PRELIS is run. Its name is the same as the PRELIS syntax file but with the suffix *.**dsf**. The *.**dsf** contains all the information about the variables that LISREL needs in order to analyze the data, *i.e.*, variable labels, sample size, means, standard deviations, covariance or correlation matrix, and the location of the asymptotic covariance matrix, if any. The *.**dsf** file can be read by LISREL directly instead of specifying variable names, sample size, means, standard deviations, covariance or correlation matrix, and the location of the asymptotic covariance matrix using separate commands in a LISREL or SIMPLIS syntax file.

To specify a *.**dsf** in the SIMPLIS command language, write:

```
System file from File filename.DSF
```

This line replaces the following typical lines in a SIMPLIS syntax file (other variations are possible):

```
Observed Variables: A B C D E F
Means from File filename
Covariance Matrix from File filename
Asymptotic Covariance Matrix from File filename
Sample Size: 678
```

To specify a ***.dsf** in the LISREL command language, write:

```
SY=filename.DSF
```

This replaces the following typical lines in a LISREL syntax file (other variations are possible):

```
DA NI=k NO=n
ME=filename
CM=filename
AC=filename
```

As the ***.dsf** is a binary file, it can be read much faster than the syntax file. To make optimal use of this, consider the following strategy, assuming the data consists of many variables, possibly several hundreds, and a very large sample.

Use PRELIS to deal with all problems in the data, *i.e.*, missing data, variable transformation, recoding, definition of new variables, etc., and compute means and the covariance matrix or correlation matrix, say, and the asymptotic covariance matrix, if needed. Specify the ***.dsf** file, select the variables for analysis and specify the model in a LISREL or SIMPLIS syntax file. Several different sets of variables may be analyzed in this way, each one being based on a small subset of the variables in the ***.dsf**. The point is that there is no need to go back to PRELIS to compute new summary statistics of each LISREL model. With the SIMPLIS command language, selection of variables is automatic in the sense that only the variables included in the model will be used.

The use of the ***.dsf** is especially important in simulations, as these will go much faster. The ***.dsf** also facilitates the building of SIMPLIS or LISREL syntax by drawing a path diagram.

7.1.3 The Model System File

A model system file, or ***.msf** for short, is created each time a LISREL syntax file containing a path diagram request is run. Its name is the same as the LISREL syntax file but with the suffix ***.msf**. The ***.msf** contains all the information about the model that LISREL needs to produce a path diagram, *i.e.*, type and form of each parameter,

parameter estimates, standard errors, *t*-values, modification indices, fit statistics, etc. Usually, users do not have a direct need for the ***.msf**.

For a complete discussion of this topic, please see the *LISREL 8: New Statistical Features Guide*.

7.2 Multiple Imputation

Multivariate data sets, where missing values occur on more than one variable, are often encountered in practice. Listwise deletion may result in discarding a large proportion of the data, which in turn, tends to introduce bias.

Researchers frequently use *ad hoc* methods of imputation to obtain a complete data set. The multiple imputation procedure implemented in LISREL 8.50 is described in detail in Schafer (1997) and uses the EM algorithm and the method of generating random draws from probability distributions via Markov chains.

In what follows, it is assumed that data are missing at random and that the observed data have an underlying multivariate normal distribution.

7.2.1 Technical Details

EM algorithm:

Suppose $\mathbf{y} = (y_1, y_2, \ldots, y_p)'$ is a vector of random variables with mean $\boldsymbol{\mu}$ and covariance matrix $\boldsymbol{\Sigma}$ and that $y_1, y_2, \ldots, y_n$ is a sample from **y**.

Step 1: (M-Step)

Start with an estimate of $\boldsymbol{\mu}$ and $\boldsymbol{\Sigma}$, for example the sample means and covariances $\bar{\mathbf{y}}$ and **S** based on a subset of the data, which have no missing values. If each row of the data set contains a missing value, start with $\boldsymbol{\mu} = 0$ and $\boldsymbol{\Sigma} = \mathbf{I}$.

Step 2: (E-Step)

Calculate $E(\mathbf{y}_{imiss} \mid \mathbf{y}_{iobs}; \hat{\boldsymbol{\mu}}, \hat{\boldsymbol{\Sigma}})$ and $Cov(\mathbf{y}_{imiss} \mid \mathbf{y}_{iobs}; \hat{\boldsymbol{\mu}}, \hat{\boldsymbol{\Sigma}})$, $i = 1, 2, \ldots, N$.

Use these values to obtain an update of $\boldsymbol{\mu}$ and $\boldsymbol{\Sigma}$ (M-step) and repeat steps 1 and 2 until $(\hat{\boldsymbol{\mu}}_{k+1}, \hat{\boldsymbol{\Sigma}}_{k+1})$ are essentially the same as $(\hat{\boldsymbol{\mu}}_k, \hat{\boldsymbol{\Sigma}}_k)$.

Markov chain Monte Carlo (MCMC):

In LISREL 8.50, the estimates of $\boldsymbol{\mu}$ and $\boldsymbol{\Sigma}$ obtained from the EM-algorithm are used as initial parameters of the distributions used in Step 1 of the MCMC procedure.

Step 1: (P-Step)

Simulate an estimate $\boldsymbol{\mu}_k$ of $\boldsymbol{\mu}$ and an estimate $\boldsymbol{\Sigma}_k$ of $\boldsymbol{\Sigma}$ from a multivariate normal and an inverted Wishart distribution respectively.

Step 2: (I-Step)

Simulate $\mathbf{y}_{imiss} \mid \mathbf{y}_{iobs}$, $i = 1, 2, ..., N$ from conditional normal distributions with parameters based on $\boldsymbol{\mu}_k$ and $\boldsymbol{\Sigma}_k$.

Replace the missing values with simulated values and calculate $\boldsymbol{\mu}_{k+1} = \bar{\mathbf{y}}$ and $\boldsymbol{\Sigma}_{k+1} = \mathbf{S}$ where $\bar{\mathbf{y}}$ and $\mathbf{S}$ are the sample means and covariances of the completed data set respectively. Repeat Steps 1 and 2 m times. In LISREL, missing values in row i are replaced by the average of the simulated values over the m draws, after an initial burn-in period. See Chapter 3 for a numerical example.

7.3 Full Information Maximum Likelihood (FIML) for Continuous Variables

Suppose that $\mathbf{y} = (y_1, y_2, \cdots, y_p)'$ has a multivariate normal distribution with mean $\boldsymbol{\mu}$ and covariance matrix $\boldsymbol{\Sigma}$ and that $\mathbf{y}_1, \mathbf{y}_2, ..., \mathbf{y}_n$ is a random sample of the vector $\mathbf{y}$.

Specific elements of the vectors $\mathbf{y}_k$, $k = 1, 2, \ldots, n$ may be unobserved so that the data set comprising of n rows (the different cases) and p columns (variables 1, 2, . . . , p) have missing values.

Let $\mathbf{y}_k$ denote a vector with incomplete observations, then this vector can be replaced by $\mathbf{y}_k^* = \mathbf{X}_k \mathbf{y}_k$ where $\mathbf{X}_k$ is a selection matrix, and $\mathbf{y}_k$ has typical elements $(y_{k1}, y_{k2}, \ldots, y_{kp})$ with one or more of the y_{kj}s missing, $j = 1, 2, \ldots, p$.

Example:

Suppose $p = 3$, and that variable 2 is unobserved, then

$$\mathbf{y}_k^* = \begin{bmatrix} y_{k1} \\ y_{k3} \end{bmatrix} = \begin{bmatrix} 1 & 0 & 0 \\ 0 & 0 & 1 \end{bmatrix} \begin{bmatrix} y_{k1} \\ y_{k2} \\ y_{k3} \end{bmatrix}.$$

From the above example it can easily be seen that $\mathbf{X}_k$ is based on an identity matrix with rows deleted according to missing elements of $\mathbf{y}_k$.

If an observed vector $\mathbf{y}_k$ contains no unobserved values, then $\mathbf{X}_k$ is equal to the identity matrix and hence $\mathbf{y}_k^* = \mathbf{y}_k$.

Without loss in generality, $(\mathbf{y}_1, \mathbf{y}_2, \ldots \mathbf{y}_n)$ can be replaced with $(\mathbf{y}_1^*, \mathbf{y}_2^*, \ldots \mathbf{y}_n^*)$ where $\mathbf{y}_k^*$, $k = 1, 2, \ldots, n$ has a normal distribution with mean $\mathbf{X}_k \boldsymbol{\mu}$ and covariance matrix $\mathbf{X}_k \boldsymbol{\Sigma} \mathbf{X}_k'$. The log-likelihood for the non-missing data is $\sum_{k=1}^{n} \log f(\mathbf{y}_k^*, \boldsymbol{\mu}_k, \boldsymbol{\Sigma}_k)$, where $f(\mathbf{y}_k^*, \boldsymbol{\mu}_k, \boldsymbol{\Sigma}_k)$ is the pdf of $\mathbf{X}_k \mathbf{y}_k$ given the parameters $\boldsymbol{\mu}_k = \mathbf{X}_k \boldsymbol{\mu}$ and $\boldsymbol{\Sigma}_k = \mathbf{X}_k \boldsymbol{\Sigma} \mathbf{X}_k'$.

In practice, when data are missing at random, there are usually M patterns of missingness, where $M < n$. When this is the case, the computational burden of evaluating n likelihood functions is considerably decreased.

It is customary to define the chi-square statistic as $\chi^2 = F_0 - F_1$, where $F_0 = -2 \ln L_0$, $F_1 = -2 \ln L_1$, and where $\ln L_1$ denotes the log-likelihood (at convergence) when no restrictions are imposed on the parameters ($\boldsymbol{\mu}$ and $\boldsymbol{\Sigma}$). The quantity $\ln L_0$ denotes the log-likelihood value (at convergence) when parameters are restricted according to a postulated model. The degrees of freedom equals ν, where $\nu = p + p(p+1)/2 - k$ and k is the number of parameters in the model.

See the examples in Section 4.8. Additional examples are given in the **missingex** folder.

7.4 Multilevel Structural equation Modeling

7.4.1 Multilevel Structural equation Models

Social science research often entails the analysis of data with a hierarchical structure. A frequently cited example of multilevel data is a dataset containing measurements on children nested within schools, with schools nested within education departments.

The need for statistical models that take account of the sampling scheme is well recognized and it has been shown that the analysis of survey data under the assumption of a simple random sampling scheme may give rise to misleading results.

Iterative numerical procedures for the estimation of variance and covariance components for unbalanced designs were developed in the 1980s and were implemented in software packages such as MLWIN, SAS PROC MIXED and HLM. At the same time, interest in latent variables, that is, variables that cannot be directly observed or can only imperfectly be observed, led to the theory providing for the definition, fitting and testing of general models for linear structural relations for data from simple random samples.

A more general model for multilevel structural relations, accommodating latent variables and the possibility of missing data at any level of the hierarchy and providing the combination of developments in these two fields, was a logical next step. In papers by Goldstein and MacDonald (1988), MacDonald and Goldstein (1989) and McDonald (1993), such a model was proposed. Muthén (1990, 1991) proposed a partial maximum likelihood solution as simplification in the case of an unbalanced design. An overview of the latter can be found in Hox (1993).

General two-level structural equation modeling is available in LISREL 8.50. Full information maximum likelihood estimation is used, and a test for goodness of fit is given. An example, illustrating the implementation of the results for unbalanced designs with missing data at both levels of the hierarchy, is also given.

7.4.2 A General Two-level Structural equation Model

Consider a data set consisting of 3 measurements, *math 1*, *math 2*, and *math 3*, made on each of 1000 children who are nested within N = 100 schools. This data set can be schematically represented for school i as follows

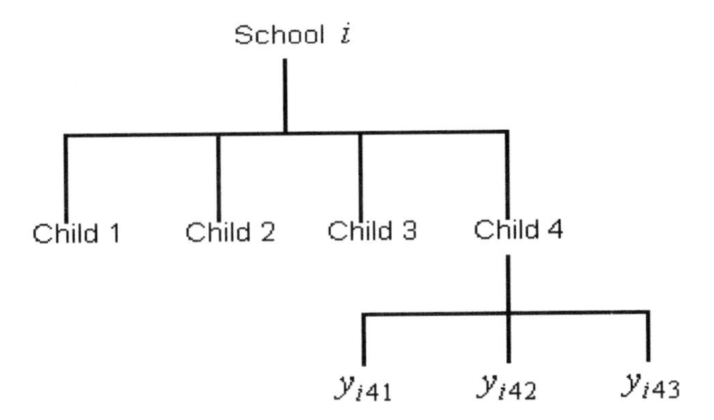

For the i-th level-2 unit (school), we can write

$$\mathbf{y}_i = \begin{pmatrix} \mathbf{y}_{i1} \\ \mathbf{y}_{i2} \\ \mathbf{y}_{i3} \\ \mathbf{y}_{i4} \end{pmatrix},$$

where for child 4 within school i

$$\mathbf{y}_{i4}' = \begin{pmatrix} y_{i41} & y_{i42} & y_{i43} \\ math1 & math2 & math3 \end{pmatrix}.$$

A model which allows for between- and within-schools variation in math scores is the following simple variance component model

$$\begin{aligned} \mathbf{y}_{i1} &= \mathbf{v}_i + \mathbf{u}_{i1} \\ \mathbf{y}_{i2} &= \mathbf{v}_i + \mathbf{u}_{i2} \\ \mathbf{y}_{i3} &= \mathbf{v}_i + \mathbf{u}_{i3} \\ \mathbf{y}_{i4} &= \mathbf{v}_i + \mathbf{u}_{i4}, \end{aligned}$$

or $\mathbf{y}_{ij} = \mathbf{v}_i + \mathbf{u}_{ij}, \quad i = 1, 2, \ldots N$, where it is assumed that $\mathbf{v}_1, \mathbf{v}_2, \ldots \mathbf{v}_N$ are i.i.d. $N(0, \mathbf{\Sigma}_B)$ and that $\mathbf{u}_{i1}, \ldots, \mathbf{u}_{iN}$ are i.i.d $N(0, \mathbf{\Sigma}_W)$. It is additionally assumed that

$$Cov(\mathbf{v}_i, \mathbf{u}_{ij}) = 0, i = 1, \ldots, N;\ j = 1, 2, \ldots n_i.$$

From the distributional assumptions it follows that

$$Cov(\mathbf{y}_i, \mathbf{y}_i') = \begin{pmatrix} \boldsymbol{\Sigma}_B + \boldsymbol{\Sigma}_W & \boldsymbol{\Sigma}_B & \boldsymbol{\Sigma}_B & \boldsymbol{\Sigma}_B \\ \boldsymbol{\Sigma}_B & \boldsymbol{\Sigma}_B + \boldsymbol{\Sigma}_W & \boldsymbol{\Sigma}_B & \boldsymbol{\Sigma}_B \\ \boldsymbol{\Sigma}_B & \boldsymbol{\Sigma}_B & \boldsymbol{\Sigma}_B + \boldsymbol{\Sigma}_W & \boldsymbol{\Sigma}_B \\ \boldsymbol{\Sigma}_B & \boldsymbol{\Sigma}_B & \boldsymbol{\Sigma}_B & \boldsymbol{\Sigma}_B + \boldsymbol{\Sigma}_W \end{pmatrix}$$

or $Cov(\mathbf{y}_i, \mathbf{y}_i') = \mathbf{11}' \otimes \boldsymbol{\Sigma}_B + \mathbf{I} \otimes \boldsymbol{\Sigma}_W$.

It also follows that

$$E(\mathbf{y}_i) = \mathbf{0}.$$

In practice, the latter assumption $E(\mathbf{y}_i) = \mathbf{0}$ is seldom realistic, since measurements such as math scores do not have zero means. One approach to this problem is to use grand mean centering. Alternatively, one can add a fixed component to the model $\mathbf{y}_{ij} = \mathbf{v}_i + \mathbf{u}_{ij}$, so that

$$\mathbf{y}_{ij} = \mathbf{X}_{ij}\boldsymbol{\beta} + \mathbf{v}_i + \mathbf{u}_{ij}, \tag{7.1}$$

where $\mathbf{X}_{ij}$ denotes a design matrix and $\boldsymbol{\beta}$ a vector of regression coefficients.

Suppose that for the example above, the only measurements available for child 1 are *math 1* and *math 3* and for child 2 *math 2* and *math 3.*

Let $\mathbf{S}_{i1}$ and $\mathbf{S}_{i2}$ be selection matrices defined as follows

$$\mathbf{S}_{i1} = \begin{bmatrix} 1 & 0 & 0 \\ 0 & 0 & 1 \end{bmatrix}, \text{ therefore } \mathbf{S}_{i1}\mathbf{v}_i = \begin{bmatrix} v_{i1} \\ v_{i3} \end{bmatrix}$$

and

$$\mathbf{S}_{i2} = \begin{bmatrix} 0 & 1 & 0 \\ 0 & 0 & 1 \end{bmatrix}, \text{ therefore } \mathbf{S}_{i2}\mathbf{v}_i = \begin{bmatrix} v_{i2} \\ v_{i3} \end{bmatrix}.$$

In general, if p measurements were made, $\mathbf{S}_{ij}$ (see, for example, du Toit, 1995) consists of a subset of the rows of the $p \times p$ identity matrix $\mathbf{I}_p$, where the rows of $\mathbf{S}_{ij}$ correspond to the response measurements available for the (i, j)-th unit.

The above model can be generalized to accommodate incomplete data by the inclusion of these selection matrices. Hence

$$\mathbf{y}_{ij} = \mathbf{X}_{(y)ij}\boldsymbol{\beta} + \mathbf{S}_{ij}\mathbf{v}_i + \mathbf{S}_{ij}\mathbf{u}_{ij} \tag{7.2}$$

where $\mathbf{X}_{(y)}$ is a design matrix of the appropriate dimensions.

If we further suppose that we have a $q \times 1$ vector of variables $\mathbf{x}_i$ characterizing the level-2 units (schools), then we can write the observed data for the i-th level-2 unit as

$$\mathbf{y}_i^{'} = [\mathbf{y}_{i1}^{'}, \mathbf{y}_{i2}^{'}, \ldots, \mathbf{y}_{in_i}^{'}, \mathbf{x}_i^{'}],$$

where

$$\mathbf{y}_{ij}^{'} = [y_{ij1}, y_{ij2}, \ldots, y_{ijp}]$$

and

$$\mathbf{x}_i^{'} = [x_{i1}, x_{i2}, \ldots, x_{iq}]. \tag{7.3}$$

We assume that $\mathbf{y}_{ij}$ and $\mathbf{x}_i$ can be written as

$$\mathbf{y}_{ij} = \mathbf{X}_{(y)ij}\boldsymbol{\beta}_y + \mathbf{S}_{ij}\mathbf{v}_i + \mathbf{S}_{ij}\mathbf{u}_{ij},\ j = 1, 2, \ldots n_i \tag{7.4}$$

$$\mathbf{x}_i = \mathbf{X}_{(x)i}\beta_x + \mathbf{R}_i\mathbf{w}_i,\ i = 1, 2, \ldots N \tag{7.5}$$

where $\mathbf{X}_{(y)}$ and $\mathbf{X}_{(x)}$ are design matrices for fixed effects, and $\mathbf{S}_{ij}$ and $\mathbf{R}_i$ are selection matrices for random effects of order $p_{ij} \times p$ and $q_i \times q$ respectively. Note that (7.4) defines two types of random effects, where $\mathbf{v}_i$ is common to level-3 units and $\mathbf{u}_{ij}$ is common to level-1 units nested within a specific level-2 unit.

Additional distributional assumptions are

$$\begin{aligned} &Cov(\mathbf{w}_i) = \boldsymbol{\Sigma}_{xx}, \quad i = 1, 2, \cdots, N \\ &Cov(\mathbf{y}_{ij}, \mathbf{w}_i) = \boldsymbol{\Sigma}_{xy}, \quad i = 1, 2, \cdots, N;\ j = 1, 2, \ldots, n_i \\ &Cov(\mathbf{u}_{ij}, \mathbf{w}_i) = \mathbf{0}. \end{aligned} \tag{7.6}$$

From (7.4) and (7.5), it follows that

$$\mathbf{y}_i = \begin{bmatrix} \mathbf{X}_{(y)i}\boldsymbol{\beta}_y + \mathbf{S}_i\mathbf{v}_i + \sum_{j=1}^{n_i}\mathbf{Z}_{ij}\mathbf{u}_{ij} \\ \mathbf{X}_{(x)i}\boldsymbol{\beta}_x + \mathbf{R}_i\mathbf{r}_i \end{bmatrix} \tag{7.7}$$

where

$$\mathbf{X}_{(y)i} = \begin{bmatrix} \mathbf{X}_{(y)i1} \\ \vdots \\ \mathbf{X}_{(y)in_i} \end{bmatrix}, \mathbf{S}_i = \begin{bmatrix} \mathbf{S}_{i1} \\ \vdots \\ \mathbf{S}_{in_i} \end{bmatrix}, \mathbf{R}_i = \begin{bmatrix} \mathbf{R}_{i1} \\ \vdots \\ \mathbf{R}_{in_i} \end{bmatrix},$$

and

$$\mathbf{Z}_{ij} = \begin{bmatrix} \mathbf{0} \\ \vdots \\ \mathbf{0} \\ \mathbf{S}_{ij} \\ \mathbf{0} \\ \vdots \\ \mathbf{0} \end{bmatrix}.$$

From the distributional assumptions given above, it follows that

$$\mathbf{y}_i \sim N(\boldsymbol{\mu}_i, \boldsymbol{\Sigma}_i),$$

where

$$\boldsymbol{\mu}_i = \begin{bmatrix} \mathbf{X}_{(y)i} & \mathbf{0} \\ \mathbf{0} & \mathbf{X}_{(x)i} \end{bmatrix} \begin{bmatrix} \boldsymbol{\beta}_y \\ \boldsymbol{\beta}_x \end{bmatrix} = \mathbf{X}_i\boldsymbol{\beta}, \tag{7.8}$$

and

$$\boldsymbol{\Sigma}_i = \begin{bmatrix} \mathbf{V}_i & \mathbf{S}_i\boldsymbol{\Sigma}_{yx}\mathbf{R}_i' \\ \mathbf{R}_i\boldsymbol{\Sigma}_{xy}\mathbf{S}_i' & \mathbf{R}_i\boldsymbol{\Sigma}_{xx}\mathbf{R}_i' \end{bmatrix} \tag{7.9}$$

where

$$\mathbf{V}_i = Cov\begin{pmatrix} \mathbf{y}_{i1} \\ \vdots \\ \mathbf{y}_{in_i} \end{pmatrix} = \mathbf{S}_i \mathbf{\Sigma}_B \mathbf{S}_i^{'} + \sum_{j=1}^{n_i} \mathbf{Z}_{ij} \mathbf{\Sigma}_W \mathbf{Z}_{ij}^{'}.$$

Remark

If $\mathbf{R}_i = \mathbf{I}_q$ and $\mathbf{S}_{ij} = \mathbf{I}_p$, corresponding to the case of no missing y or x variables, then $\mathbf{S}_i \mathbf{\Sigma}_{yx} \mathbf{R}_i^{'} = \mathbf{1} \otimes \mathbf{\Sigma}_{yx}$ where $\mathbf{1}^{'}$ is a $n_i \times 1$ row vector (1,1, ...,1).

Furthermore, for $\mathbf{S}_{ij} = \mathbf{I}_p$, $j = 1, 2, \ldots, n_i$

$$\mathbf{V}_i = \mathbf{I}_{n_i} \otimes \mathbf{\Sigma}_W + \mathbf{1}\mathbf{1}^{'} \otimes \mathbf{\Sigma}_B$$

(see, for example, MacDonald and Goldstein, 1989). The unknown parameters in (7.8) and (7.9) are $\boldsymbol{\beta}$, $vecs\mathbf{\Sigma}_B$, $vecs\mathbf{\Sigma}_W$, $vecs\mathbf{\Sigma}_{xy}$ and $vecs\mathbf{\Sigma}_{xx}$.

Structural models for the type of data described above may be defined by restricting the elements of $\boldsymbol{\beta}$, $\mathbf{\Sigma}_B$, $\mathbf{\Sigma}_W$, $\mathbf{\Sigma}_{xy}$, and $\mathbf{\Sigma}_{xx}$ to be some basic set of parameters $\boldsymbol{\gamma}^{'} = (\gamma_1, \gamma_2, \ldots, \gamma_k)$.

For example, assume the following pattern for the matrices $\mathbf{\Sigma}_W$ and $\mathbf{\Sigma}_B$, where $\mathbf{\Sigma}_W$ refers to the within (level-1) covariance matrix and $\mathbf{\Sigma}_B$ to the between (level-2) covariance matrix:

$$\begin{aligned} \mathbf{\Sigma}_W &= \mathbf{\Lambda}_W \mathbf{\Psi}_W \mathbf{\Lambda}_W^{'} + \mathbf{D}_W \\ \mathbf{\Sigma}_B &= \mathbf{\Lambda}_B \mathbf{\Psi}_B \mathbf{\Lambda}_B^{'} + \mathbf{D}_B. \end{aligned} \tag{7.10}$$

Factor analysis models typically have the covariance structures defined by (7.10). Consider a confirmatory factor analysis model with 2 factors and assume $p = 6$.

$$\mathbf{\Lambda}_W = \begin{bmatrix} \lambda_{11} & 0 \\ \lambda_{21} & 0 \\ \lambda_{31} & 0 \\ 0 & \lambda_{42} \\ 0 & \lambda_{52} \\ 0 & \lambda_{62} \end{bmatrix}, \quad \mathbf{\Psi}_W = \begin{bmatrix} \psi_{11} & \psi_{12} \\ \psi_{21} & \psi_{22} \end{bmatrix},$$

and

$$D_W = \begin{bmatrix} \theta_{11} & & \\ & \ddots & \\ & & \theta_{66} \end{bmatrix}.$$

If we restrict all the parameters across the level-1 and level-2 units to be equal, then

$$\gamma' = [\lambda_{11}, \lambda_{21}, \ldots, \lambda_{44}, \psi_{11}, \psi_{21}, \psi_{22}, \theta_{11}, \ldots, \theta_{66}]$$

is the vector of unknown parameters.

7.4.3 Maximum Likelihood for General Means and Covariance Structures

In this section, we give a general framework for normal maximum likelihood estimation of the unknown parameters. In practice, the number of variables ($p + q$) and the number of level-1 units within a specific level-2 unit may be quite large, which leads to $\mathbf{\Sigma}_i$ matrices of very high order. It is therefore apparent that further simplification of the likelihood function derivatives and Hessian is required if the goal is to implement the theoretical results in a computer program. These aspects are addressed in du Toit and du Toit (forthcoming).

Denote the expected value and covariance matrix of $\mathbf{y}_i$ by $\boldsymbol{\mu}_i$ and $\mathbf{\Sigma}_i$ respectively (see (7.8) and (7.9)). The log-likelihood function of $\mathbf{y}_1, \mathbf{y}_2, \ldots, \mathbf{y}_N$ may then be expressed as

$$\ln L = -\frac{1}{2}\sum_{i=1}^{N}\{n_i \ln 2\pi + \ln|\mathbf{\Sigma}| + tr\mathbf{\Sigma}_i^{-1}(\mathbf{y}_i - \boldsymbol{\mu}_i)(\mathbf{y}_i - \boldsymbol{\mu}_i)'\} \tag{7.11}$$

Instead of maximizing $\ln L$, maximum likelihood estimates of the unknown parameters are obtained by minimizing $-\ln L$ with the constant term omitted, *i.e.*, by minimizing the following function

$$F(\gamma) = \frac{1}{2}\sum_{i=1}^{N}\{\ln|\mathbf{\Sigma}_i| + tr\mathbf{\Sigma}_i^{-1}\mathbf{G}_{y_i}\}, \tag{7.12}$$

where

$$\mathbf{G}_{y_i} = (\mathbf{y}_i - \boldsymbol{\mu}_i)(\mathbf{y}_i - \boldsymbol{\mu}_i)'. \tag{7.13}$$

Its minimum $\frac{\partial F(\boldsymbol{\gamma})}{\partial \boldsymbol{\gamma}} = \mathbf{0}$ yields the normal maximum likelihood estimator $\hat{\boldsymbol{\gamma}}$ of the unknown vector of parameters $\boldsymbol{\gamma}$.

Unless the model yields maximum likelihood estimators in closed form, it will be necessary to make use of an iterative procedure to minimize the discrepancy function. The optimization procedure (Browne and du Toit, 1992) is based on the so-called Fisher scoring algorithm, which in the case of structured means and covariances may be regarded as a sequence of Gauss-Newton steps with quantities to be fitted as well as the weight matrix changing at each step. Fisher scoring algorithms require the gradient vector and an approximation to the Hessian matrix.

7.4.4 Fit Statistics and Hypothesis Testing

The multilevel structural equation model, $M(\boldsymbol{\gamma})$, and its assumptions imply a covariance structure $\boldsymbol{\Sigma}_B(\boldsymbol{\gamma})$, $\boldsymbol{\Sigma}_W(\boldsymbol{\gamma})$, $\boldsymbol{\Sigma}_{xy}(\boldsymbol{\gamma})$, $\boldsymbol{\Sigma}_{xx}(\boldsymbol{\gamma})$ and mean structure $\boldsymbol{\mu}(\boldsymbol{\gamma})$ for the observable random variables where $\boldsymbol{\gamma}$ is a $k \times 1$ vector of parameters in the statistical model. It is assumed that the empirical data are a random sample of N level-2 units and $\sum_{i=1}^{N} n_i$ level-1 units, where n_i denotes the number of level-1 units within the i-th level-2 unit. From this data, we can compute estimates of $\boldsymbol{\mu}$, $\boldsymbol{\Sigma}_B$, …, $\boldsymbol{\Sigma}_{xx}$ if no restrictions are imposed on their elements. The number of parameters for the unrestricted model is

$$k^* = m + 2\left[\frac{1}{2}p(p+1)\right] + pq + \frac{1}{2}q(q+1)$$

and is summarized in the $k^* \times 1$ vector $\boldsymbol{\pi}$. The unrestricted model $M(\boldsymbol{\pi})$ can be regarded as the "baseline" model.

To test the model $M(\boldsymbol{\gamma})$, we use the likelihood ratio test statistic

$$c = -2\ln L(\hat{\boldsymbol{\gamma}}) - 2\ln L(\hat{\boldsymbol{\pi}}) \tag{7.14}$$

If the model $M(\boldsymbol{\gamma})$ holds, c has a χ^2-distribution with $d = k^* - k$ degrees of freedom. If the model does not hold, c has a non-central χ^2-distribution with d degrees of freedom and non-centrality parameter λ that may be estimated as (see Browne and Cudeck, 1993):

$$\hat{\lambda} = \max\{(c-d), 0\} \quad (7.15)$$

These authors also show how to set up a confidence interval for λ.

It is possible that the researcher has specified a number of competing models $M_1(\gamma_1), M_2(\gamma_2), \ldots, M_k(\gamma_k)$. If the models are nested in the sense that $\gamma_j : k_j \times 1$ is a subset of $\gamma_i : k_i \times 1$, then one may use the likelihood ratio test with degrees of freedom $k_i - k_j$ to test $M(\gamma_j)$ against $M(\gamma_i)$.

Another approach is to compare models on the basis of some criteria that take parsimony as well as fit into account. This approach can be used regardless of whether or not the models can be ordered in a nested sequence. Two strongly related criteria are the AIC measure of Akaike (1987) and the CAIC of Bozdogan (1987).

$$\text{AIC} = c + 2d \tag{7.16}$$

$$\text{CAIC} = c + (1 + \ln \sum_{i=1}^{N} n_i) d \tag{7.17}$$

The use of c as a central χ^2-statistic is based on the assumption that the model holds exactly in the population. A consequence of this assumption is that models that hold approximately in the population will be rejected in large samples.

Steiger (1990) proposed the root mean square error of approximation (RMSEA) statistic that takes particular account of the error of approximation in the population

$$RMSEA = \sqrt{\frac{\hat{F}_0}{d}}, \tag{7.18}$$

where $\hat{F}_0$ is a function of the sample size, degrees of freedom and the fit function. To use the RMSEA as a fit measure in multilevel SEM, we propose

$$\hat{F}_0 = \max\left\{\left[\frac{c-d}{N}\right], 0\right\} \tag{7.19}$$

Browne and Cudeck (1993) suggest that an RMSEA value of 0.05 indicates a close fit and that values of up to 0.08 represent reasonable errors of approximation in the population.

7.4.5 Starting Values and Convergence Issues

In fitting a structural equation model to a hierarchical data set, one may encounter convergence problems unless good starting values are provided. A procedure that appears to work well in practice is to start the estimation procedure by fitting the unrestricted model to the data. The first step is therefore to obtain estimates of the fixed components ($\boldsymbol{\beta}$) and the variance components ($\boldsymbol{\Sigma}_B, \boldsymbol{\Sigma}_{xy}, \boldsymbol{\Sigma}_{xx}$ and $\boldsymbol{\Sigma}_W$). Our experience with the Gauss-Newton algorithm (see, for example, Browne and du Toit, 1992) is that convergence is usually obtained within less than 15 iterations, using initial estimates $\boldsymbol{\beta} = \mathbf{0}, \boldsymbol{\Sigma}_B = \mathbf{I}_p, \boldsymbol{\Sigma}_{xy} = \mathbf{0}, \boldsymbol{\Sigma}_{xx} = \mathbf{I}_q$ and $\boldsymbol{\Sigma}_W = \mathbf{I}_p$. At convergence, the value of $-2\ln L$ is computed.

Next, we treat

$$\mathbf{S}_B = \begin{bmatrix} \hat{\boldsymbol{\Sigma}}_B & \hat{\boldsymbol{\Sigma}}_{yx} \\ \hat{\boldsymbol{\Sigma}}_{xy} & \hat{\boldsymbol{\Sigma}}_{xx} \end{bmatrix} \text{ and } \mathbf{S}_W = \begin{bmatrix} \hat{\boldsymbol{\Sigma}}_W & \mathbf{0} \\ \mathbf{0} & \mathbf{0} \end{bmatrix}$$

as sample covariance matrices and fit a two-group structural equation model to the between- and within-groups. Parameter estimates obtained in this manner are used as the elements of the initial parameter vector $\boldsymbol{\gamma}_0$.

In the third step, the iterative procedure is restarted and γ_k updated from γ_{k-1}, $k = 1,2,\ldots$ until convergence is reached.

The following example illustrates the steps outlined above. The data set used in this section forms part of the data library of the Multilevel Project at the University of London, and comes from the Junior School Project (Mortimore *et al*, 1988). Mathematics and language tests were administered in three consecutive years to more than 1000 students from 49 primary schools, which were randomly selected from primary schools maintained by the Inner London Education Authority.

The following variables were selected from the data file:

- *School* School code (1 to 49)
- *Math1* Score on mathematics test in year 1 (score 1 - 40)
- *Math2* Score on mathematics test in year 2 (score 1 - 40)
- *Math3* Score on mathematics test in year 3 (score 1 - 40)

The school number (*School*) is used as the level-2 identification.

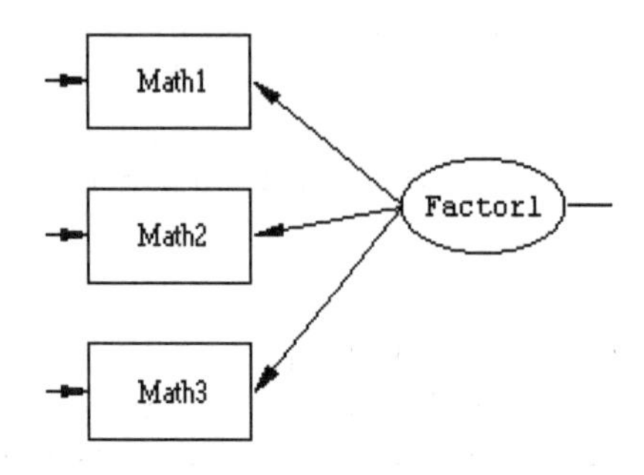

Figure 7.1: Confirmatory factor analysis model

A simple confirmatory factor analysis model (see Figure 7.1) is fitted to the data:

$$\mathbf{\Sigma}_B = \boldsymbol{\lambda}\boldsymbol{\Psi}\boldsymbol{\lambda}' + \mathbf{D}_B,$$
$$\mathbf{\Sigma}_W = \boldsymbol{\lambda}\boldsymbol{\Psi}\boldsymbol{\lambda}' + \mathbf{D}_W,$$

where

$$\boldsymbol{\lambda}' = (1, \lambda_{21}, \lambda_{31})$$

and $\mathbf{D}_B$ and $\mathbf{D}_W$ are diagonal matrices with diagonal elements equal to the unique (error) variances of *Math1*, *Math2* and *Math3.* The variance of the factor is denoted by $\boldsymbol{\Psi}$. Note that we assume equal factor loadings and factor variances across the between- and within-groups, leading to a model with 3 degrees of freedom. The SIMPLIS (see Jöreskog and Sörbom, 1993) syntax file to fit the factor analysis model is shown below. Note that the between- and within-groups covariance matrices are the estimated $\mathbf{\Sigma}_B$ and $\mathbf{\Sigma}_W$ obtained in the first step by fitting the unrestricted model. These estimates may also be obtained by deleting the variable names *eng1*, *eng2*, and *eng3* in the `RESPONSE` command of the syntax file **jsp1**.**pr2** in the **mlevelex** folder.

```
Group 1: Between Schools JSP data (Level 2)
Observed Variables: Math1 Math2 Math3
Covariance matrix
3.38885
2.29824    5.19791
2.31881    3.00273    4.69663
Sample Size=24    ! Taken as (n1+n2+...nN)/N rounded to
                  ! nearest integer
Latent Variables: Factor1
```

```
Relationships
Math1=1*Factor1
Math2-Math3=Factor1

Group 2: Within Schools JSP data (Level 1)
Covariance matrix
47.04658
38.56798    55.37006
30.81049    36.04099    40.71862
Sample Size=1192  ! Total number of pupils
! Set the Variance of Factor1  Free ! Remove comment to
                                    !free parameter
Set the Error Variance of Math1  Free
Set the Error Variance of Math2  Free
Set the Error Variance of Math3  Free
Path Diagram
LISREL OUTPUT ND=3
End of Problem
```

Table 7.1: Parameter estimates and standard errors for factor analysis model

	SIMPLIS		Multilevel SEM	
	Estimate	**Standard error**	**Estimate**	**Standard error**
Factor loadings				
λ_{11}	1.000	-	1.000	-
λ_{21}	1.173	0.031	1.177	0.032
λ_{31}	0.939	0.026	0.947	0.028
Factor variance				
Ψ	32.109	1.821	31.235	1.808
Error variances (between)				
Math1	1.640	0.787	1.656	0.741
Math2	2.123	1.059	2.035	0.942
Math3	1.868	0.779	1.840	0.734
Error variances (within)				
Math1	14.114	0.810	14.209	0.890
Math2	10.274	0.884	10.256	0.993
Math3	11.910	0.699	11.837	0.806
Chi-square	36.233		46.56	
Degrees of freedom	3		3	

Table 7.1 shows the parameter estimates, estimated standard errors and χ^2-statistic values obtained from the SIMPLIS output and from the multilevel SEM output respectively.

Remarks:

1. The between-groups sample size of 26 used in the SIMPLIS syntax file was computed as $\frac{1}{N}\sum_{i=1}^{N} n_i$, where N is the number of schools and n_i the number of children within school *i*. Since this value is only used to obtain starting values, it is not really crucial how the between-group sample size is computed. See, for example, Muthén (1990,1991) for an alternative formula.
2. The within-group sample size of 1192 used in the SIMPLIS file syntax is equal to the total number of school children.
3. The number of missing values per variable is as follows:

   Math1: 38
   Math2: 63
   Math3: 239

   The large percentage missing for the *Math3* variable may partially explain the relatively large difference in χ^2-values from the SIMPLIS and multilevel SEM outputs.
4. If one allows for the factor variance parameter to be free over groups, the χ^2 fit statistic becomes 1.087 at 2 degrees of freedom. The total number of multilevel SEM iterations required to obtain convergence equals eight.

In conclusion, a small number of variables and a single factor SEM model were used to illustrate the starting values procedure that we adopted. The next section contains additional examples, also based on a schools data set. Another example can be found in du Toit and du Toit (forthcoming). Also see the **msemex** folder for additional examples.

7.4.6 Practical Applications

The example discussed in this section is based on school data that were collected during a 1994 survey in South Africa.

A brief description of the **SA_Schools.psf** data set in the **msemex** folder is as follows:

N = 136 schools were selected and the total number of children within schools $\sum_{i=1}^{N} n_i = 6047$, where n_i varies from 20 to 60. A description of the variables is given in Table 7.2.

Table 7.2: Description of variables in SA_School.psf

Name	Description	Number missing
Pupil	Level-1 identification	0
School	Level-2 identification	0
Intcept	All values equal to 1	0
Grade	0 = Grade 2 1 = Grade 3 2 = Grade 4	0
Language	0 = Black 1 = White	0
Gender	1 = Male 2 = Female	1
Mothedu	Mother's level of education on a scale from 1 to 7	783
Fathedu	Father's level of education on a scale from 1 to 7	851
Read Speech Write Arithm	Teacher's evaluation on a scale from 1 to 5: 1 = Poor 5 = Excellent	482 470 467 451
Socio	Socio-economic status indicator, scale 0 to 5 on school level	0
Classif	Classification: total correct out of 30 items	23
Compar	Comparison: total correct out of 23 items	27
Verbal	Verbal Instructions: total correct out of 50 items	20
Figure	Figure Series: total correct out of 24 items	118
Pattcomp	Pattern Completion: total correct out of 24 items	109
Knowled	Knowledge: total correct out of 32 items	112
Numserie	Number Series: total correct out of 15 items	2305

The variables *Language* and *Socio* are school-level variables and their values do not vary within schools. Listwise deletion of missing cases results in a data set containing only 2691 of the original 6047 cases.

For this example, we use the variables *Classif*, *Compar*, *Verbal*, *Figure*, *Pattcomp* and *Numserie* from the schools data set discussed in the previous section. Two common factors are hypothesized: word knowledge and spatial ability. The first three variables are assumed to measure *wordknow* and the last three to measure *spatial.* A path diagram of the hypothesized factor model is shown in Figure 7.2.

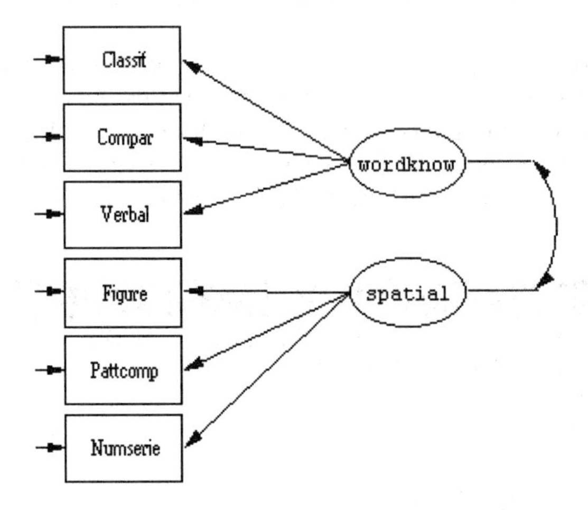

Figure 7.2: Confirmatory factor analysis model for 6 variables

The between- and within-school structural equation models are

$$\begin{aligned} \mathbf{\Sigma}_W &= \mathbf{\Lambda}_W \mathbf{\Psi}_W \mathbf{\Lambda}_W^{'} + \mathbf{D}_W \\ \mathbf{\Sigma}_B &= \mathbf{\Lambda}_B \mathbf{\Psi}_B \mathbf{\Lambda}_B^{'} + \mathbf{D}_B. \end{aligned} \tag{7.20}$$

where

$$\mathbf{\Lambda}_W = \mathbf{\Lambda}_B = \begin{bmatrix} 1 & 0 \\ \lambda_{21} & 0 \\ \lambda_{31} & 0 \\ 0 & 1 \\ 0 & \lambda_{52} \\ 0 & \lambda_{62} \end{bmatrix},$$

and where factor loadings are assumed to be equal on the between (schools) and within (children) levels. The 2 x 2 matrices $\mathbf{\Psi}_B$ and $\mathbf{\Psi}_W$ denote unconstrained factor covariance matrices. Diagonal elements of $\mathbf{D}_B$ and $\mathbf{D}_W$ are the unique (error) variances.

Gender and *Grade* differences were accounted for in the means part of the model,

$$E(y_{ijk}) = \beta_{k0} + \beta_{k0} Gender_{ijk} + \beta_{k2} Grade_{ijk},$$

where the subscripts i, j and k denote schools, students and variables k, respectively.

From the description of the school data set, we note that the variable *Numserie* has 2505 missing values. An inspection of the data set reveals that the pattern of missingness can hardly be described as missing at random. To establish how well the proposed algorithm perform in terms of the handling of missing cases, we have decided to retain this variable in this example. The appropriate LISREL 8.50 syntax file for this example is given below.

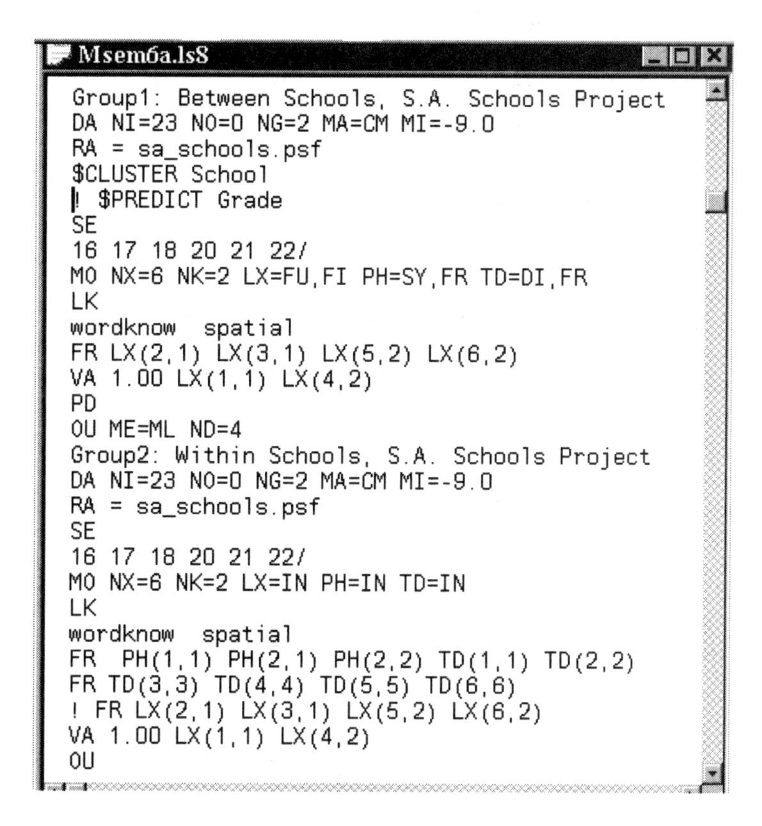

```
Msem6a.ls8
Group1: Between Schools, S.A. Schools Project
DA NI=23 NO=0 NG=2 MA=CM MI=-9.0
RA = sa_schools.psf
$CLUSTER School
! $PREDICT Grade
SE
16 17 18 20 21 22/
MO NX=6 NK=2 LX=FU,FI PH=SY,FR TD=DI,FR
LK
wordknow  spatial
FR LX(2,1) LX(3,1) LX(5,2) LX(6,2)
VA 1.00 LX(1,1) LX(4,2)
PD
OU ME=ML ND=4
Group2: Within Schools, S.A. Schools Project
DA NI=23 NO=0 NG=2 MA=CM MI=-9.0
RA = sa_schools.psf
SE
16 17 18 20 21 22/
MO NX=6 NK=2 LX=IN PH=IN TD=IN
LK
wordknow  spatial
FR  PH(1,1) PH(2,1) PH(2,2) TD(1,1) TD(2,2)
FR TD(3,3) TD(4,4) TD(5,5) TD(6,6)
! FR LX(2,1) LX(3,1) LX(5,2) LX(6,2)
VA 1.00 LX(1,1) LX(4,2)
OU
```

Table 7.3 shows the estimated between-schools covariance matrix $\hat{\Sigma}_B$ when no restrictions are imposed on its elements, and the fitted covariance matrix $\Sigma_B(\hat{\gamma})$ where γ is the vector of parameters of the CFA models given in (7.20).

Table 7.3: Estimated between-schools covariance matrix, Σ_B

(i) $\hat{\Sigma}_B$ unrestricted

	Classif	Compar	Verbal	Figure	Pattcomp	Numserie
Classif	1.32					
Compar	1.32	2.75				
Verbal	2.98	3.76	10.93			
Figure	2.17	2.88	7.29	5.85		
Pattcomp	2.26	2.75	6.94	5.37	5.57	
Numserie	1.56	2.03	5.25	4.12	4.01	3.32

(ii) $\Sigma_B(\hat{\gamma})$ for the CFA model

	Classif	Compar	Verbal	Figure	Pattcomp	Numserie
Classif	1.61					
Compar	1.79	3.97				
Verbal	2.26	3.35	7.03			
Figure	2.64	3.91	4.93	6.24		
Pattcomp	2.41	3.57	4.50	5.42	5.28	
Numserie	1.83	2.71	3.42	4.12	3.75	3.17

Likewise, Table 7.4 shows $\hat{\Sigma}_W$ for the unrestricted model and $\Sigma_W(\hat{\gamma})$ for the CFA model.

Table 7.4: Estimated within-schools covariance matrix, Σ_W

(i) $\hat{\Sigma}_W$ unrestricted

	Classif	Compar	Verbal	Figure	Pattcomp	Numserie
Classif	8.47					
Compar	4.56	18.72				
Verbal	5.45	7.54	17.01			
Figure	4.39	7.12	8.28	16.11		
Pattcomp	4.26	7.14	8.28	9.41	16.08	
Numserie	3.03	4.47	5.64	7.55	6.04	7.63

(ii) $\Sigma_W(\hat{\gamma})$ for the CFA model

	Classif	Compar	Verbal	Figure	Pattcomp	Numserie
Classif	8.40					
Compar	4.21	18.50				
Verbal	5.31	7.85	17.23			
Figure	4.61	6.83	8.61	15.99		
Pattcomp	4.21	6.23	7.85	9.52	16.18	
Numserie	3.20	4.73	5.97	7.23	6.60	7.77

The goodness of fit statistics for the CFA model are shown in Table 7.5.

Table 7.5: Goodness-of-fit statistics

(6047 students, 136 schools)

$\chi^2 = 243.408$	degrees of freedom = 20
RMSEA=0.061	

Parameter estimates and estimated standard errors are given in Table 7.6.

It is typical of SEM models to produce large χ^2-values when sample sizes are large, as in the present case. The RMSEA may be a more meaningful measure of goodness of fit and the value of 0.061 indicates that the assumption of equal factor loadings between and within schools is reasonable.

Table 7.6: Parameter estimates and standard errors

	Estimate	**Standard error**
Factor loadings		
λ_{11}	1.0	
λ_{21}	1.480	0.042
λ_{31}	1.866	0.047
λ_{42}	1.0	
λ_{52}	0.912	0.015
λ_{62}	0.693	0.012
Factor covariances (between schools)		
Φ_{11}	1.212	0.186
Φ_{21}	2.644	0.356
Φ_{22}	5.940	0.777

Table 7.6: Parameter estimates and standard errors (continued)

	Estimate	Standard error
Error variances (between schools)		
Classif	0.399	0.078
Compar	1.321	0.221
Verbal	2.811	0.400
Figure	0.304	0.091
Pattcomp	0.333	0.089
Numserie	0.314	0.068
Factor covariances (within schools)		
Φ_{11}	2.844	0.127
Φ_{21}	4.615	0.146
Φ_{22}	10.437	0.298
Error variances (within schools)		
Classif	5.554	0.119
Compar	12.276	0.263
Verbal	7.328	0.225
Figure	5.553	0.168
Pattcomp	7.496	0.183
Numserie	2.760	0.098

7.5 Multilevel Non-Linear Models

7.5.1 Introduction to Multilevel Modeling

The analysis of data with a hierarchical structure has been described in the literature under various names. It is known as hierarchical modeling, random coefficient modeling, latent curve modeling, growth curve modeling or multilevel modeling. The basic underlying structure of measurements nested within units at a higher level of the hierarchy is, however, common to all. In a repeated measurements growth model, for example, the measurements or outcomes are nested within the experimental units (second level units) of the hierarchy.

Ignoring the hierarchical structure of data can have serious implications, as the use of alternatives such as aggregation and disaggregation of information to another level can induce high collinearity among predictors and large or biased standard errors for the estimates. Standard fixed parameter regression models do not allow for the exploration of variation between groups, which may be of interest in its own right. For a discussion of the effects of these alternatives, see Bryk and Raudenbush (1992), Longford (1987) and Rasbash (1993).

Multilevel or hierarchical modeling provides the opportunity to study variation at different levels of the hierarchy. Such a model can also include separate regression coefficients at different levels of the hierarchy that have no meaning without recognition of the hierarchical structure of the population. The dependence of repeated measurements belonging to one experimental unit in a typical growth curve analysis, for example, is taken into account with this approach. In addition, the data to be analyzed need not be balanced in nature. This has the advantage that estimates can also be units for which a very limited amount of information is available.

7.5.2 Multilevel Non-Linear Regression Models

It was pointed out by Pinheiro and Bates (2000) that one would want to use nonlinear latent coefficient models for reasons of *interpretability*, *parsimony*, and more importantly, *validity* beyond the observed range of the data.

By increasing the order of a polynomial model, one can get increasingly accurate approximations to the true, usually nonlinear, regression function, within the range of the observed data. High order polynomial models often result in multicollinearity problems and provide no theoretical considerations about the underlying mechanism producing the data.

There are many possible nonlinear regression models to select from. Examples are given by Gallant (1987) and Pinheiro and Bates (2000). The Richards function (Richards, 1959) is a generalization of a family of non-linear functions and is used to describe growth curves (Koops, 1988). Three special cases of the Richards function are the logistic, Gompertz and Monomolecular functions, respectively. Presently, one can select curves of the form

$$y = f_1(x) + f_2(x) + e$$

where the first component, $f_1(x)$ may have the form:

- logistic: $\dfrac{b_1}{(1 + s\exp(b_2 - b_3 x)}$
- Gompertz: $b_1 \exp(-b_2 \exp(-b_3 x))$
- Monomolecular: $b_1(1 + s\exp(b_2 - b_3 x))$
- power: $b_1 x^{b_2}$
- exponential: $b_1 \exp(-b_2 x)$

The second component, $f_2(x)$, may have the form:

- logistic: $\dfrac{c_1}{(1+s\exp(c_2-c_3x)}$
- Gompertz: $c_1\exp(-c_2\exp(-c_3x))$
- Monomolecular: $c_1(1+s\exp(c_2-c_3x))$
- power: $c_1x^{c_2}$
- exponential: $c_1\exp(-c_2x)$

In the curves above s denotes the sign of the term $\exp(b_2-b_3x)$ and is equal to 1 or -1.

Since the parameters in the first three functions above have definite physical meanings, a curve from this family is preferred to a polynomial curve, which may often be fitted to a set of responses with the same degree of accuracy. The parameter b_1 represents the time asymptotic value of the characteristic that has been measured, the parameter b_2 represents the potential increase (or decrease) in the value of the function during the course of time t_1 to t_p, and the parameter b_3 characterizes the rate of growth.

The coefficients $b_1, b_2, \ldots, c_3$ are assumed to be random, and, as in linear hierarchical models, can be written as level-2 outcome variables where

$$\begin{aligned}
b_1 &= \beta_1 + u_1 \\
b_2 &= \beta_2 + u_2 \\
b_3 &= \beta_3 + u_3 \\
c_1 &= \beta_4 + u_4 \\
c_2 &= \beta_5 + u_5 \\
c_3 &= \beta_6 + u_6.
\end{aligned}$$

It is assumed that the level-2 residuals $u_1, u_2, \ldots, u_6$ have a normal distribution with zero means and covariance matrix $\mathbf{\Phi}$. In LISREL, it may further be assumed that the values of any of the random coefficients are affected by some level-2 covariate so that, in general,

$$\begin{aligned}
b_1 &= \beta_1 + \gamma_1 z_1 + u_1 \\
b_2 &= \beta_2 + \gamma_2 z_2 + u_2 \\
b_3 &= \beta_3 + \gamma_3 z_3 + u_3 \\
c_1 &= \beta_4 + \gamma_4 z_4 + u_4 \\
c_2 &= \beta_5 + \gamma_5 z_5 + u_5 \\
c_3 &= \beta_6 + \gamma_6 z_6 + u_6.
\end{aligned}$$

where z_i denotes the value of a covariate and γ_i the corresponding coefficient.

It is usually sufficient to select a single component $(f_1(x))$ to describe a large number of monotone increasing (or decreasing) growth patterns.

To describe more complex patterns, use can be made of two-component regression models. LISREL allows the user to select any of the 5 curve types as component 1 and to combine it with any one of the 5 curve types for the second component.

Valid choices are, for example,

- Monomolecular + Gompertz
- logistic
- exponential + logistic
- logistic + logistic

The unknown model parameters are the vector of fixed coefficients $(\boldsymbol{\beta})$, the vector of covariate coefficients $(\boldsymbol{\gamma})$, the covariance matrix $(\boldsymbol{\Phi})$ of the level-2 residuals and the variance (σ^2) of the level-1 measurement errors. See Chapter 5 for an example of fitting of a multilevel nonlinear model. Additional examples are given in the **nonlinex** folder.

7.5.3 Estimation Procedure for Multilevel Non-Linear Regression Models

In linear multilevel models, y has a normal distribution, since y is a linear combination of the random coefficients. For example, the intercept-and-slopes-as-outcomes model is

$$y = b_1 + b_2 x + e$$

where $b_1 = u_1$, $b_2 = u_2$ and (u_1, u_2) is assumed to be normally distributed.

A multilevel nonlinear model is a regression model which cannot be expressed as a linear combination of its coefficients and therefore y is no longer normally distributed. The probability density function of y can be evaluated as the multiple integral

$$f(y) = \int_{b_1} \ldots \int_{c_3} f(y, b_1, \ldots, c_3)\, db_1 \ldots dc_3$$

that, in general, cannot be solved in closed form.

To evaluate the likelihood function

$$L = \prod_{i=1}^{n} f(y_i),$$

one has to use a numerical integration technique. We assume that e has a $N(0,\sigma^2)$ distribution and that $(b_1, b_2, b_3, c_1, c_2, c_3)$ has a $N(\mathbf{0}, \mathbf{\Phi}^2)$ distribution.

In the multilevels procedure, use is made of a Gauss quadrature procedure to evaluate the integrals numerically. The ML method requires good starting values for the unknown parameters. The estimation procedure is described by Cudeck and du Toit (in press).

Starting values

Once a model is selected to describe the nonlinear pattern in the data, for example as revealed by a plot of y on x, a curve is fitted to each individual using ordinary non-linear least squares.

In step 1 of the fitting procedure, these OLS parameter estimates are written to a file and estimates of $\boldsymbol{\beta}$ and $\mathbf{\Phi}$ are obtained by using the sample means and covariances of the set of fitted parameters. Since observed values from some individual cases may not be adequately described by the selected model, these cases can have excessively large residuals, and it may not be advisable to include them in the calculation of the $\boldsymbol{\beta}$ and $\mathbf{\Phi}$ estimates.

In step 2 of the model fitting procedure, use is made of the MAP (Maximum Aposterior) estimator of the unknown parameters.

Suppose that a single component $f_1(b, x)$ is fitted to the data. Since

$$f(b \mid y) = f(b) f(y \mid b) / f(y),$$

where $f(b \mid y)$ is the conditional probability density function of the random coefficients b given the observations, it follows that

$$\ln f(b \mid y) = \ln f(b) + \ln f(y \mid b) + k,$$

where $k = -\ln f(y)$.

The MAP procedure can briefly be described as follows:

Step a:

Given starting values of $\boldsymbol{\beta}$, $\boldsymbol{\Phi}$ and σ^2, obtain estimates of the random coefficients $\hat{b}_i$ from

$$\frac{\partial}{\partial b_i} \ln f(b_i \mid y_i) = 0, \quad i = 1, 2, \ldots, n.$$

Step b:

Use the estimates $\hat{b}_1, \hat{b}_2, \ldots, \hat{b}_n$ and $Cov(\hat{b}_1), \ldots, Cov(\hat{b}_n)$ to obtain new estimates of $\boldsymbol{\beta}$, $\boldsymbol{\Phi}$ and σ^2 (see Herbst, A. (1993) for a detailed discussion).

Repeat steps a and b until convergence is attained.

For many practical purposes, results obtained from the MAP procedure may be sufficient. However, if covariates are included in the model, parameter estimates are only available via the ML option, which uses the MAP estimates of $\boldsymbol{\beta}$, $\boldsymbol{\Phi}$ and σ^2 as starting values.

7.6 FIRM: Formal Inference-based Recursive Modeling

7.6.1 Overview

Recursive modeling is an attractive data-analytic tool for studying the relationship between a dependent variable and a collection of predictor variables. In this, it complements methods such as multiple regression, analysis of variance, neural nets, generalized additive models, discriminant analysis and log linear modeling. It is particularly helpful when simple models like regression do not work, as happens when the relationship connecting the variables is not straightforward—for example if there are synergies between the predictors. It is attractive when there is missing information in the data. It can also be valuable when used in conjunction with a more traditional statistical data analysis, as it may provide evidence that the model assumptions underlying the traditional analysis are valid, something that is otherwise not easy to do.

The output of a recursive model analysis is a "dendrogram", or "tree diagram". This shows how the predictors can be used to partition the data into successively smaller and more homogeneous subgroups. This dendrogram can be used as an immediate pictorial way of making predictions for future cases, or "interpreted" in terms of the effects on the dependent variable of changes in the predictor. The maximum problem size the codes distributed in this package can handle is 1000 predictors; the number of cases is limited only by the available disk space.

Recursive partitioning (RP) is not attractive for all data sets. Going down a tree leads to rapidly diminishing sample sizes and the analysis comes to an end quite quickly if the initial sample size was small. As a rough guide, the sample size needs to be in three digits before recursive partitioning is likely to be worth trying. Like all feature selection methods, it is also highly non-robust in the sense that small changes in the data can lead to large changes in the output. The results of a RP analysis therefore always need to be thought of as *a* model for the data, rather than as *the* model for the data. Within these limitations though, the recursive partitioning analysis gives an easy, quick way of getting a picture of the relationship that may be quite different than that given by traditional methods, and that is very informative when it is different.

FIRM is made available through LISREL by the kind permission of the author, Professor Douglas M. Hawkins, Department of Applied Statistics, University of Minnesota.

7.6.2 Recursive Partitioning Concepts

The most common methods of investigating the relationship between a dependent variable Y and a set of predictors $X_1, X_2, \ldots, X_p$ are the linear model (multiple regression, analysis of variance and analysis of covariance) for a continuous dependent variable, and the log linear model for a categorical dependent. Both methods involve strong model assumptions. For example, multiple regression assumes that the relationship between Y and the X_i is linear, that there are no interaction terms other than those explicitly included in the model; and that the residuals satisfy strong statistical assumptions of a normal distribution and constant variance. The log linear model for categorical data assumes that any interaction terms not explicitly included are not needed; while it provides an overall χ^2 test-of-fit of the model, when this test fails it does not help to identify what interaction terms are needed. Discriminant analysis is widely used to find rules to classify observations into the different populations from which they came.

It is another example of a methodology with strong model assumptions that are not easy to check.

A different and in some ways diametrically opposite approach to all these problems is modeling by recursive partitioning. In this approach, the calibration data set is successively split into ever-smaller subsets based on the values of the predictor variables. Each split is designed to separate the cases in the node being split into a set of successor groups, which are in some sense maximally internally homogeneous.

An example of a data set in which FIRM is a potential method of analysis is the "head injuries" data set of Titterington *et al* (1981) (data kindly supplied by Professor Titterington). As we will be using this data set to illustrate the operation of the FIRM codes, and since the data set is included in the FIRM distribution for testing purposes, it may be appropriate to say something about it. The data set was gathered in an attempt to predict the final outcome of 500 hospital patients who had suffered head injuries. The outcome for each patient was that he or she was:

- *Dead or vegetative* (52% of the cases);
- had *severe* disabilities (10% of the cases);
- or had a *moderate or good* recovery (38% of the cases).

This outcome is to be predicted on the basis of 6 predictors assessed on the patients' admission to the hospital:

- *Age* : The age of the patient. This was grouped into decades in the original data, and is grouped the same way here. It has eight classes.
- *EMV* : This is a composite score of three measures—of eye-opening in response to stimulation; motor response of best limb; and verbal response. This has seven classes, but is not measured in all cases, so that there are eight possible codes for this score, these being the seven measurements and an eighth "missing" category.
- *MRP* : This is a composite score of motor responses in all four limbs. This also has seven measurement classes with an eighth class for missing information.
- *Change* : The change in neurological function over the first 24 hours. This was graded 1, 2 or 3, with a fourth class for missing information.
- *Eye indicator* : A summary of diagnostics on the eyes. This too had three measurement classes, with a fourth for missing information.
- *Pupils* : Eye pupil reaction to light, namely present, absent, or missing.

7.6.3 CATFIRM

Figure 7.3 is a dendrogram showing the end result of analyzing the data set using the CATFIRM syntax, which is used for a categorical dependent variable like this one. Syntax can be found in the file **headicat.pr2**.

The dendrogram produced by CATFIRM is a very informative picture of the relationship between these predictors and the patients' final outcome. We see in it that the most significant separation was obtained by splitting the full sample on the basis of the predictor *Pupils*.

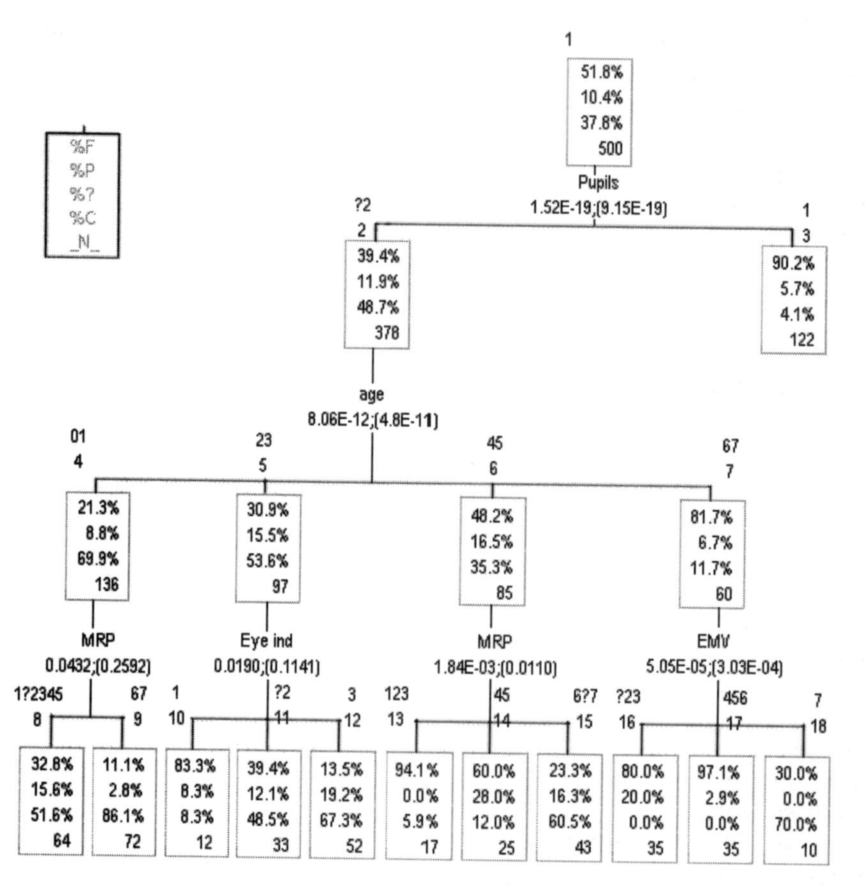

Figure 7.3: CATFIRM dendrogram of "head injuries" data set

The groups of 378 cases for which *Pupils* had the value 2 or the value ? (that is, missing) were statistically indistinguishable, and so are grouped together. They constitute one of the successor groups (node number 2), while those 122 for whom *Pupils* had the value 1 constitute the other (node number 3), a group with much worse outcomes—90% dead or vegetative compared with 39% of those in node 2.

Each of these nodes in turn is subjected to the same analysis. The cases in node number 2 can be split again into more homogeneous subgroups. The most significant such split is obtained by separating the cases into four groups on the basis of the predictor *Age*. These are patients under 20 years old, (node 4), patients 20 to 40 years old (node 5), those 40 to

60 years old (node 6) and those over 60 (node 7). The prognosis of these patients deteriorates with increasing age; 70% of those under 20 ended with moderate or good recoveries, while only 12% of those over 60 did.

Node 3 is terminal. Its cases cannot (at the significance levels selected) be split further. Node 4 can be split using *MRP*. Cases with *MRP* = 6 or 7 constitute a group with a favorable prognosis (86% with moderate to good recovery); and the other MRP levels constitute a less-favorable group but still better than average.

These groups, and their descendants, are analyzed in turn in the same way. Ultimately no further splits can be made and the analysis stops. Altogether 18 nodes are formed, of which 12 are terminal and the remaining 6 intermediate. Each split in the dendrogram shows the variable used to make the split and the values of the splitting variable that define each descendant node. It also lists the statistical significance (p-value) of the split. Two p-values are given: that on the left is FIRM's conservative p-value for the split. The p-value on the right is a Bonferroni-corrected value, reflecting the fact that the split actually made had the smallest p-value of all the predictors available to split that node. So for example, on the first split, the actual split made has a p-value of 1.52×10^{-19}. But when we recognize that this was selected because it was the most significant of the 6 possible splits, we may want to scale up its p-value by the factor 6 to allow for the fact that it was the smallest of 6 p-values available (one for each predictor). This gives its conservative Bonferroni-adjusted p-value as 9.15×10^{-19}.

The dendrogram and the analysis giving rise to it can be used in two obvious ways: for making predictions, and to gain understanding of the importance of and interrelationships between the different predictors. Taking the prediction use first, the dendrogram provides a quick and convenient way of predicting the outcome for a patient. Following patients down the tree to see into which terminal node they fall yields 12 typical patient profiles ranging from 96% *dead/vegetative* to 100% with *moderate to good* recoveries. The dendrogram is often used in exactly this way to make predictions for individual cases. Unlike, say, predictions made using multiple regression or discriminant analysis, the prediction requires no arithmetic—merely the ability to take a future case "down the tree". This allows for quick predictions with limited potential for arithmetic errors. A hospital could, for example, use Figure 7.3 to give an immediate estimate of a head-injured patient's prognosis.

The second use of the FIRM analysis is to gain some understanding of the predictive power of the different variables used and how they interrelate. There are many aspects to this analysis. An obvious one is to look at the dendrogram, seeing which predictors are used at the different levels. A rather deeper analysis is to look also at the predictors that were not used to make the split in each node, and see whether they could have discriminated between different subsets of cases. This analysis is done using the "summary" file part of the output, which can be requested by adding the keyword `sum` on the first line of the syntax file. In this head injuries data set, the summary file shows that all 6 predictors are very discriminating in the full data set, but are much less so as soon as the initial split has been performed. This means that there is a lot of overlap in their predictive information: that different predictors are tapping into the same or related basic physiological indicators of the patient's state.

Another feature often seen is an *interaction* in which one predictor is predictive in one node, and a different one is predictive in another. This may be seen in this data set, where *Age* is used to split Node 4, but *Eye ind* is used to split Node 5. Using different predictors at different nodes at the same level of the dendrogram is one example of the interactions that motivated the ancestor of current recursive partitioning codes—the Automatic Interaction Detector (AID). Examples of diagnosing interactions from the dendrogram are given in Hawkins and Kass (1982).

7.6.4 CONFIRM

Figure 7.3 illustrates the use of CATFIRM, which is appropriate for a categorical dependent variable. The other analysis covered by the FIRM package is CONFIRM, which is used to study a dependent variable on the interval scale of measurement. Figure 7.4 shows an analysis of the same data using CONFIRM. The dependent variable was on a three-point ordinal scale, and for the CONFIRM analysis we treated it as being on the interval scale of measurement with values 1, 2 and 3 scoring 1 for the outcome "*dead or vegetative*"; 2 for the outcome "*severe disability*" and 3 for the outcome "*moderate to good recovery*". As the outcome is on the ordinal scale, this equal-spaced scale is not necessarily statistically appropriate in this data set and is used as a matter of convenience rather than with the implication that it may be the best way to proceed.

The full data set of 500 cases has a mean outcome score of 1.8600. The first split is made on the basis of the predictor *Pupils*. Cases for which *Pupils* is 1 or ? constitute the first descendant group (Node 2), while those for which it is 2 give Node 3. Note that this is the same predictor that CATFIRM elected to use, but the patients with missing values of *Pupils* are grouped with *Pupils* = 1 by the CONFIRM analysis, and with *Pupils* = 2 by CATFIRM. Going to the descendant groups, Node 2 is then split on *MRP* (unlike in CATFIRM, where the corresponding node was terminal), while Node 3 is again split four ways on *Age*. The overall tree is bigger than that produced by CATFIRM—13 terminal and 8 interior nodes—and produces groups with prognostic scores ranging from a grim 1.0734 up to 2.8095 on the 1 to 3 scale. As with CATFIRM, the means in these terminal nodes could be used for prediction of future cases, giving estimates of the score that patients in the terminal nodes would have.

The statistical model used in CONFIRM can be described as a *piecewise constant regression* model. For example, among the patients in Node 3, the outcome is predicted to be 2.48 for patients aged under 20, 2.25 for those aged 20 to 40, 1.88 for those aged 40 to 60, and 1.32 for those aged over 60. This approach contrasts sharply with, say, a linear regression in which the outcome would be predicted to rise by some constant amount with each additional year of *Age*. While this piecewise constant regression model is seldom valid exactly, it is often a reasonable working approximation of reality.

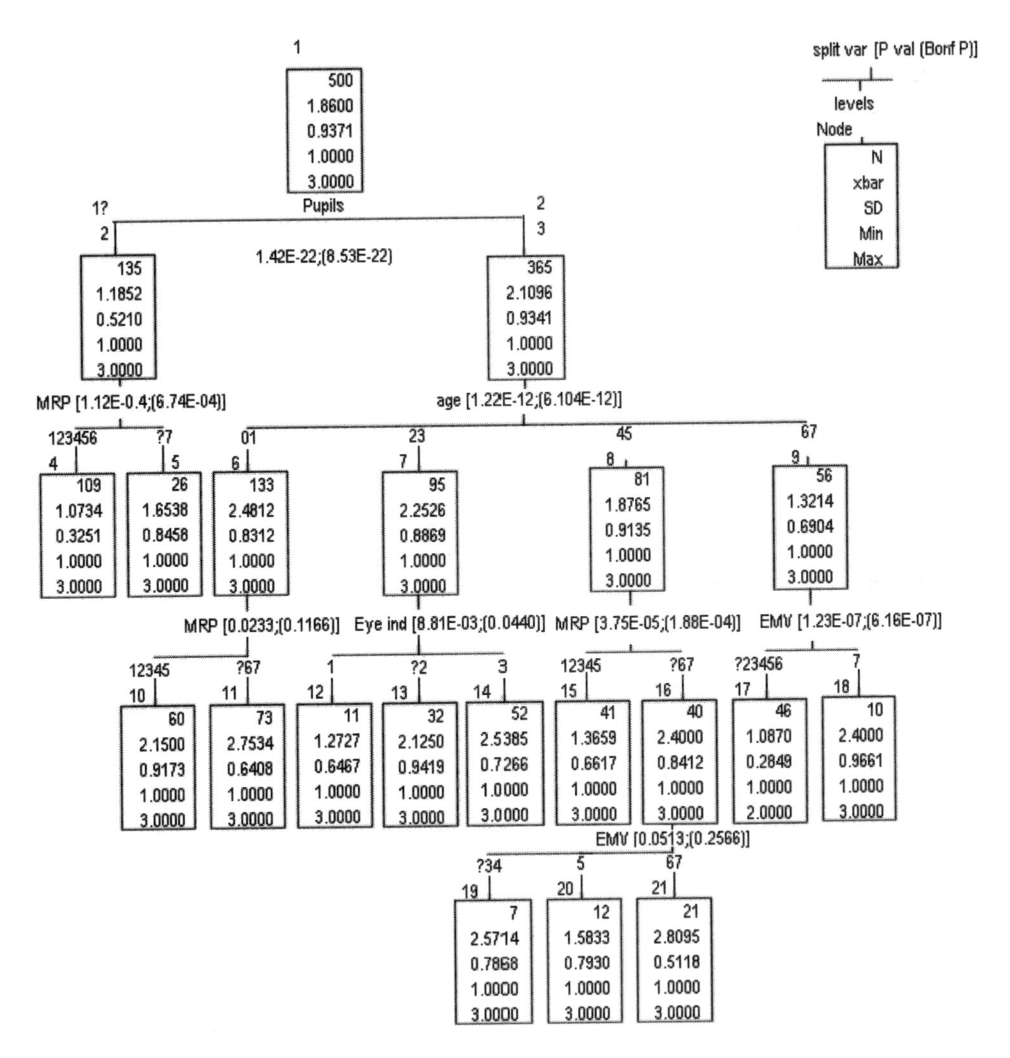

Figure 7.4: CONFIRM dendrogram of "head injuries" data set

7.6.5 Other Detailed Output Files

The dendrograms of Figures 7.3 and 7.4 produced by the FIRM codes are the most obviously and immediately useful output of a FIRM analysis. They are an extract of the much more detailed output, a sample of which is given in Chapter 6. This output contains the following (often very valuable) additional information:

- An analysis of each predictor in each node, showing which categories of the predictor FIRM finds it best to group together, and what the conservative statistical significance level of the split by each predictor is;
- The number of cases owing into each of the descendant groups;
- Summary statistics of the cases in the descendant nodes. In the case of CATFIRM, the summary statistics are a frequency breakdown of the cases between the different classes of the dependent variable. With CONFIRM, the summary statistics given are the arithmetic mean and standard deviation of the cases in the node. This summary information is echoed in the dendrogram.

7.6.6 FIRM Building Blocks

There are three elements to analysis by a recursive partitioning method such as FIRM:

- deciding which predictor to use to define the split;
- deciding which categories of the predictor should be grouped together so that the data set is not split more ways than are really necessary (for example, is a binary, three-way, four-way etc. split on this variable required); and
- deciding when to stop growing the tree.

Different implementations of recursive partitioning handle these issues in different ways. The theoretical basis for the procedures implemented in FIRM and used to produce these dendrograms is set out in detail in Hawkins and Kass (1982) and Kass (1980). Readers interested in the details of the methodology should refer to these two sources and to the papers they reference. The Ph.D. thesis by Kass (1972) included a PL/1 program for CHAID ("chi-squared automatic interaction detector")—CATFIRM is based on a translation of this code into FORTRAN. The methodology (but no code) is set out in a perhaps more widely accessible medium in the paper by Kass (1975). The 1981 technical report by Heymann discusses a code XAID ("extended automatic interaction detector") which extended the framework underlying CHAID to an interval-scale dependent variable. CONFIRM is a development of this program. The FIRM methodology has been applied to a wide variety of problems—for example, Hooton *et al* (1981) was an early application in medicine, and Hawkins, Young and Rusinko (1997) illustrates an application to drug discovery.

After this brief illustration of two FIRM runs, we will now go into some of the underlying ideas of the methodology. Readers who have not come across recursive partitioning before would be well advised to experiment with the CONFIRM and CATFIRM before going too far into this detail. See Chapter 6 for FIRM examples and also the **firmex** folder.

7.6.7 Predictor and Dependent Variable Types

Writings on measurement types often distinguish four scales of measurement:

- *Nominal*, in which the measurement divides the individuals studied into different classes. Eye color is an example of a nominal measure on a person.
- *Ordinal*, in which the different categories have some natural ordering. Social class is an example of an ordinal measure. So is a rating by the adjectives "Excellent", "Very good", "Good", "Fair", "Poor".
- *Interval*, in which a given difference between two values has the same meaning wherever these values are in the scale. Temperature is an example of a measurement on the interval scale—the temperature difference between 10 and 15 degrees is the same as that between 25 and 30 degrees.
- *Ratio*, which goes beyond the interval scale in having a natural zero point.

Each of these measurement scales adds something to the one above it, so an interval measure is also ordinal, but an ordinal measure is generally not interval.

FIRM can handle two types of dependent variable: those on the nominal scale (analyzed by CATFIRM), and those on the interval scale (analyzed by CONFIRM). There is no explicit provision for a dependent variable on the ordinal scale such as, for example, a qualitative rating "Excellent", "Good", "Fair" or "Poor". Dependents like this have to be handled either by ignoring the ordering information and treating them using CATFIRM, or by trying to find an appropriate score for the different categories. These scores could range from the naive (for example the 1-2-3 scoring we used to analyze the head injuries data with CONFIRM) to conceptually sounder scorings obtained by scaling methods.

Both FIRM approaches use the same predictor types and share a common underlying philosophy. While FIRM recognizes five predictor types, two going back to its Automatic Interaction Detection roots are fundamental:

- Nominal predictors (which are called "*free*" predictors in the AID terminology),
- Ordinal predictors (which AID labels "*monotonic*" predictors).

In practical use, some ordinal predictors are just categorical predictors with ordering among their scales, while others are interval-scaled. Suppose for example that you are studying automobile repair costs, and think the following predictors may be relevant:

- *make*: The make of the car. Let's suppose that in the study it is sensible to break this down into 5 groupings: GM, Ford, Chrysler, Japanese and European.
- *cyl*: The number of cylinders in the engine: 4, 6, 8 or 12.
- *price*: The manufacturer's suggested retail purchase price when new.
- *year*: The year of manufacture.

make is clearly nominal, and so would be used as a free predictor. In a linear model analysis, you would probably use analysis of variance, with *make* a factor. Looking at these predictors, *cyl* seems to be on the interval scale, and so might be used as an ordered (monotonic) predictor. If we were to use *cyl* in a linear regression study, we would be making a strong assumption that the repair costs difference between a 4- and an 8-cylinder car is the same as that between an 8- and a 12-cylinder car. FIRM's monotonic predictor type involves no such global assumption.

price is a ratio-scale predictor that would be specified as monotonic—it seems generally accepted that cars that are more expensive to buy are also more expensive to maintain. In fact, though, even if this were false, so long as the relationship between price and repair costs was reasonably smooth (and not necessarily even monotonic) FIRM's piecewise constant model fitting would be able to handle it.

Finally, *year* is an example of a predictor that looks to be on an interval scale (since a one-year gap is a one-year gap whether it is the gap between 1968 and 1969 or that between 1998 and 1999), but appearances may be deceptive. While some years' cars do certainly seem to be more expensive to maintain than other years' cars, it is not obvious that this relationship is a smooth one. A "lemon" year can be preceded and followed by years in which a manufacturer makes good, reliable cars. For this reason it would be preferable to make *year* a free predictor rather than monotonic. This will allow the cost to have any type of response to *year*, including one in which certain isolated years are bad while their neighbors are good. If the repair cost should turn out to vary smoothly with *year* then little would have been lost and that little could be recovered by repeating the run, respecifying *year* as monotonic.

7.6.8 Internal Grouping to get Starting Groups

Internally, the FIRM code requires all predictors to be reduced to a modest number (no more than 16 or 20) of distinct categories, and from this starting set it will then merge categories until it gets to the final grouping. Thus in the car repair cost example, *make*, *cyl* and possibly *year* are ready to run. The predictor *price* is continuous—it likely takes on many different values. For FIRM analysis it would therefore have to be categorized into a set of price ranges. There are two ways of doing this: you may decide ahead of time what sensible price ranges are and use some other software to replace the dollar value of the price by a category number. This option is illustrated by the predictor *Age* in the "head injuries" data. A person's age is continuous, but in the original data set it had already been coded into decades before we got the data, and this coded value was used in the FIRM.

You can also have FIRM do the grouping for you. It will attempt to find nine initial working cutpoints, breaking the data range into 10 initial classes. The downside of this is that you may not like its selection of cutpoints: if we had had the subjects' actual ages in the head injuries data set, FIRM might have chosen strange-looking cutpoints like 14, 21, 27, 33, 43, 51, 59, 62, 68 and 81. These are not as neat or explainable as the ages by decades used by Titterington, *et al.*

Either way of breaking a continuous variable down into classes introduces some approximation. Breaking age into decades implies that only multiples of 10 years can be used as split points in the FIRM analysis. This leaves the possibility that, when age was used to split node 3 in CATFIRM with cut points at ages 20, 40 and 60, better fits might have been obtained had ages between the decade anniversaries been allowed as possible cut points: perhaps it would have been better to split at say age 18 rather than 20, 42 rather than 40 and 61 rather than 60. Maybe a completely different set of cut points could have been chosen. All this, whether likely or not, is certainly possible, and should be borne in mind when using continuous predictors in FIRM analysis.

Internally, all predictors in a FIRM analysis will have values that are consecutive integers, either because they started out that way or because you allowed FIRM to translate them to consecutive integers. This is illustrated by the "head injuries" data set. All the predictors had integer values like 1,2,3. Coding the missing information as 0 then gave a data set with predictors that were integer valued.

7.6.9 Grouping the Categories

The basic operation in modeling the effect of some predictor is the reduction of its values to different groups of categories. FIRM does this by testing whether the data corresponding to different categories of the dependent variable differ significantly, and if not, it puts these categories together. It does this differently for different types of predictors:

- *free* predictors are on the nominal scale. When categories are tested for possibly being together, any classes of a free predictor may be grouped. Thus in the car repairs, we could legitimately group any of the categories GM, Ford, Chrysler, Japanese and European together.
- *monotonic* predictors are on the ordinal scale. One example might be a rating with options "Poor", "Fair", "Good", "Very good" and "Excellent". Internally, these values need to be coded as consecutive integers. Thus in the "head injuries" data the different age groups were coded as 1, 2, 3, 4, 5, 6, 7 and 8. When the classes of a predictor are considered for grouping, only groupings of contiguous classes are allowed. For example, it would be permissible to pool by age into the groupings {1,2}, {3}, {4,5,6,7,8}, but not into {1,2}, {5,6}, {3,4,7,8}, which groups ages 3 and 4 with 7 and 8 while excluding the intermediate ages 4 and 5.

The term "monotonic" may be misleading as it suggests that the dependent variable has to either increase or decrease monotonically as the predictor increases. Though this is usual when you specify a predictor as monotonic, it is not necessarily so. FIRM fits "step functions", so it could be sensible to specify a predictor as monotonic if the dependent variable first increased with the predictor and then decreased, so long as the response was fairly smooth.

Consider, for example, using a person's age to predict physical strength. If all subjects were aged under, say, 18, then strength would be monotonically increasing with age. If all subjects were aged over 25, then strength would be monotonically decreasing with age. If the subjects span both age ranges, then the relationship would be initially increasing, then decreasing. Using age as a monotonic predictor will still work correctly in that only adjacent age ranges will be grouped together. The fitted means, though, would show that there were groups of matching average strength at opposite ends of the age spectrum.

If you have a predictor that is on the ordinal scale, but suspect that the response may not be smooth, then it would be better, at least initially, to specify it as free and then see if the grouping that came back looked reasonably monotonic. This would likely be done in the car repair costs example with the predictor *year*.

7.6.10 Additional Predictor Types

These are the two basic predictor types inherited from FIRM's roots in AID. In addition to these, FIRM has another three predictor types:

1. The *floating* predictor is a variant of the monotonic predictor. A floating predictor is one whose scale is monotonic except for a single "floating" class whose position in the monotonic scale is unknown. This type is commonly used to handle a predictor which is monotonic but which has missing values on some cases. The rules of grouping are that the monotonic portion of the scale can be grouped only monotonically, but that the floating class may be grouped with any other class or classes on the scale. Like the free and monotonic predictors, this predictor type must also be coded in the data set as consecutive integers, and in these implementations of FIRM, the missing class is required to be the first one. So for example you might code a math grade of A, B, C, D, F or missing as 0 for missing, with A, B, C, D and F coded as 1, 2, 3, 4 and 5, for a total of 6 categories.
2. A *real* predictor is continuous: for example, the purchase price of the automobile; a patient's serum cholesterol level; a student's score in a pretest. FIRM handles real predictors by finding 9 cutpoints that will divide the range of values of the predictor into 10 classes of approximately equal frequency. Putting the predictor into one of these classes then gives a monotonic predictor. The real predictor may contain missing values. FIRM interprets any value in the data set that is not a valid number to be missing. For example, if a real predictor is being read from a data set and the value recorded is ? or . or *M*, then the predictor is taken to be missing on that case. Any real

predictor that has missing values will be handled as a floating predictor—the missing category will be isolated as one category and the non-missing values will be broken down into the 10 approximately equal-frequency classes.

3. A *character* predictor is a free predictor that has not been coded into successive digits. For example, if gender is a predictor, it may be represented by M, F and ? in the original data base. To use it in FIRM, you could use some other software to recode the data base into, say, 1 for M, 2 for F, and 0 for ?, and treat it as a free predictor. Or you could just leave the values as is, and specify to FIRM that the predictor is of type character. FIRM will then find the distinct gender codes automatically.

   The character predictor type can also sometimes be useful when the predictor is integer- valued, but not as successive integers. For example, the number of cylinders of our hypothetical car, 4, 6, 8 or 12, would best be recoded as 1, 2, 3 and 4, but we could declare it character and let FIRM find out for itself what different values are present in the data. Releases of FIRM prior to 2.0 supported only free, monotonic and floating predictors. FIRM 2.0 and up implement the real and character types by internally recoding real predictors as either monotonic or floating predictors, and character as free predictors. Having a variable as type character does not remove the restriction on the maximum number of distinct values it can have—this remains at 16 for CATFIRM and (in the current release) 20 for CONFIRM. As of FIRM 2.0, the dependent variable in CATFIRM may also be of type character. With this option, CATFIRM can find out for itself what are the different values taken on by the dependent variable.

7.6.11 Missing Information and Floating Predictors

The monotonic and free predictor types date right back to the early 1960s implementation of AID; the real and character predictor types are essentially just an extra code layer on top of the basic types to make them more accessible. As the floating predictor type is newer, and also is not supported by all recursive modeling procedures, some comments on its potential uses may be helpful. The floating predictor type is mainly used for handling missing information in an otherwise ordinal predictor. Most procedures for handling missing information have at some level an underlying "missing at random" assumption. This assumption is not always realistic: there are occasions in which the fact that a predictor is missing in a case may itself convey strong information about the likely value of the dependent variable.

There are many examples of potentially informative missingness; we mention two. In the "head injuries" data, observations on the patient's eye function could not be made if the eye had been destroyed in the head injury. This fact may itself be an indicator of the severity of the injury. Thus, it does not seem reasonable to assume that the eye measurements are in any sense missing at random; rather it is a distinct possibility that missingness could predict a poor outcome. More generally in clinical contexts, clinicians tend to call for only tests that they think are needed. Thus, patients who are missing a particular diagnostic test do not have a random outcome on the missing test, but are biased towards those in whom the test would have indicated normal function.

Another example we have seen is in educational data, where Kass and Hawkins attempted to predict college statistics grades on the basis of high school scores. In their student pool, high school math was not required for college entry, and students who had not studied math in high school therefore had missing values for their high school math grade. But it was often the academically weaker students who chose not to study math in high school, and so students missing this grade had below average success rates in college. For the math grade to be missing was therefore in and of itself predictive of below-average performance.

FIRM's floating predictors provide a way to accommodate this predictive missingness. If a predictor really is missing at random, then the floating category will tend to be grouped with other categories near the middle of the ordinal part of the scale. But if the missingness is predictive, then the floating category will be either segregated out on its own or grouped with categories near one end of the ordinal part of the scale. In other words, predictors may be missing at random or in an informative way; FIRM can handle both situations, and the grouping of the floating category will provide a diagnosis of whether the missing information is random or informative. Inspection of the groupings produced by FIRM for the "head injuries" data does indeed suggest informative missingness on several predictors. For example *EMV* and *MRP*, with their 7-class monotonic plus floating class show this by the missing class being grouped with classes 6 and 7 on the scale.

No special predictor type is needed for missing information on a free predictor; all that is necessary is to have an extra class for the missing values. For example, if in a survey of adolescents you measured family type in four classes:

- 1: living with both birth parents;
- 2: with single parent;
- 3: with blended family;
- 4: with adoptive family;
- then respondents for whom the family type was not observed would define a fifth class, and the five classes would define a free (nominal) scale.

In the "head injuries" data set, *age* was always observed, and is clearly monotonic. *EMV*, *MRP*, *Change* and *Eye indicator* all have a monotonic scale but with missing information. Therefore, they are used as floating predictors. With *Pupils*, we have a choice. The outcome was binary, with missing information making a third category. Some thought shows that this case of three categories may be handled by making the predictor either free or floating—with either specification any of the categories may be grouped together and so we will get the same analysis.

We already mentioned that it is informative to see where the "missing information" category ends up in the grouping. Apart from the use mentioned of diagnosing whether missingness is predictive, this grouping may provide a useful fill-in for missing information on a predictor. For example, suppose the FIRM analysis groups the missing information category with categories 4, 5 and 6 of some floating predictor. If you are planning to do some other analyses, for example multiple regression, that cannot handle missing information, then it might be sensible to replace missing values on this predictor with one of the values 4, 5 or 6 for purposes of this other analysis. The justification for this substitution would be that, at least insofar as the dependent variable is concerned, cases missing this predictor behave like those with the predictor having values 4-6. The middle of this range, 5, is then a plausible fill-in value.

A second data set included in the **firmex** folder illustrates the use of character and real predictors. This data set (called **mixed.dat**) was randomly extracted from a much larger file supplied by Gordon V. Kass. It contains information on the high school records of students—their final scores in several subjects; when each wrote the University entrance examination and which of several possible examinations was taken; and the outcome of their first year at the university. This outcome is measured in two ways: the promotion code is a categorical measure of the student's overall achievement in relation to the requirements for going on to the second-year curriculum and can be analyzed using CATFIRM. The aggregate score is the average score received for all courses taken in the first University year, and is suitable for analysis using CONFIRM. Apart from having real and character predictors, this file also contains much missing information—for example, the high school Latin scores are missing for the large number of students who did not study Latin in high school. Some of this missing information could be predictively missing.

Figure 7.5 shows the results of analyzing the promotion code using CATFIRM, while Figure 7.6 shows the results of analyzing the University aggregate score using CONFIRM.

Both dendrograms are smaller than those of the "head injuries" data set, reflecting the fact that there is much more random variability in the student's outcomes than there is in the head injuries data. The FIRM models can isolate a number of significantly different groups, but there is substantial overlap in the range of performance of the different subgroups.

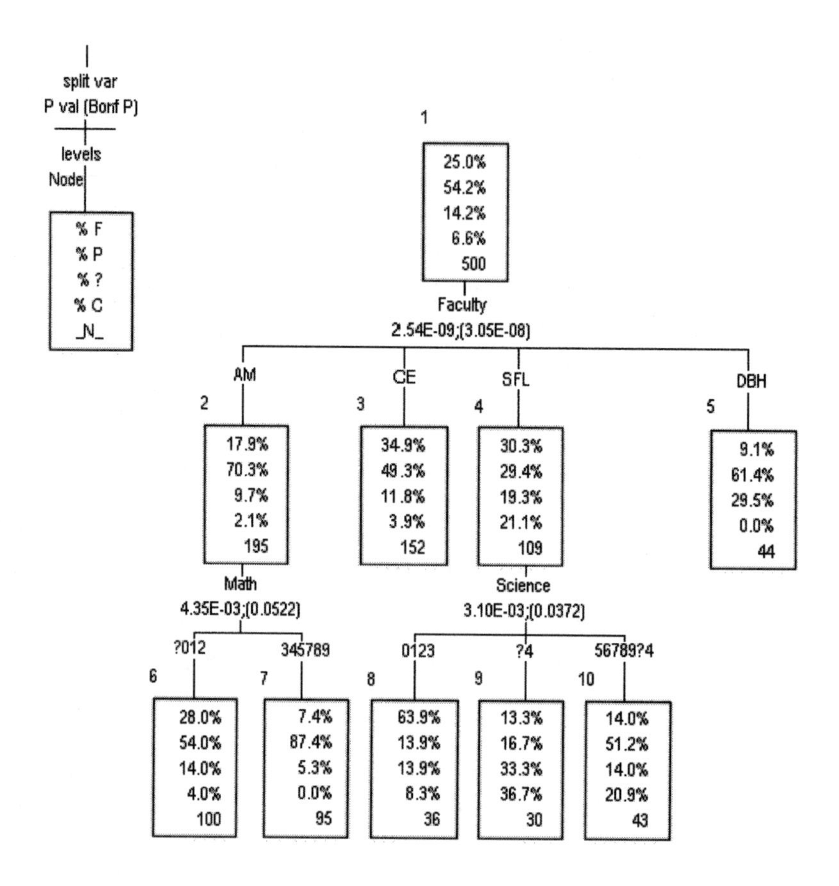

Figure 7.5: CATFIRM dendrogram of education data set

The CONFIRM analysis has terminal nodes ranging from a mean score of 42.2% up to 66.6%. This range is wide enough to be important academically (it corresponds to two letter grades), even though it is comparable with the within-node standard deviation of scores.

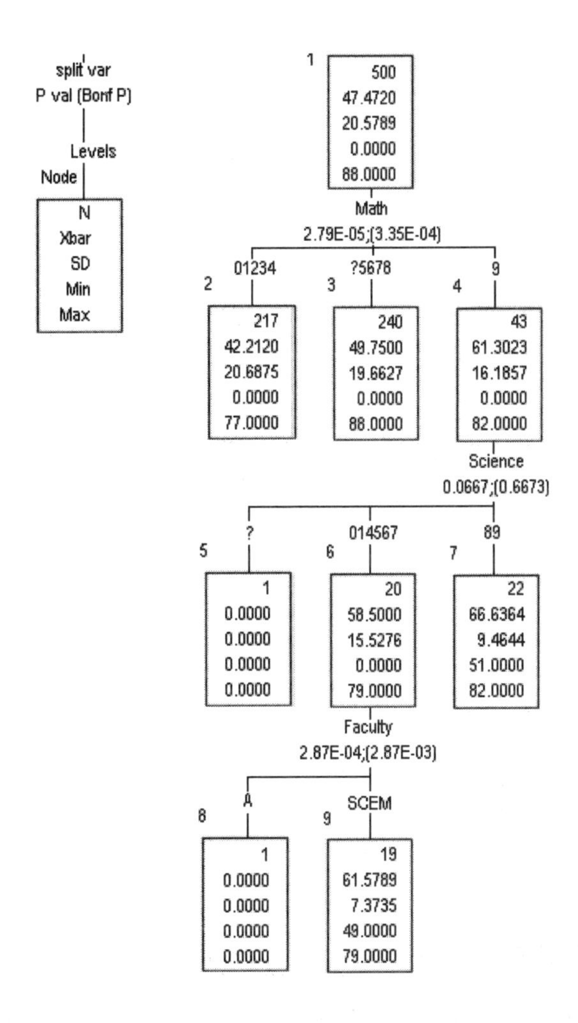

Figure 7.6: CONFIRM dendrogram of education data set

7.6.12 Outliers

The CONFIRM output also illustrates how FIRM reacts to outliers in the data. Nodes 5 and 8 in Figure 7.6 each contain a single student who dropped out and whose overall final score is recorded as zero—a value all but impossible for someone who was physically present at all final exams. FIRM discovers that it is able to make a statistically highly significant split on a predictor variable that happens to "fingerprint" this case, and it does so.

7.6.13 Overall Method of Operation

CONFIRM with a categorical (free) predictor is like a one-way analysis of variance on that predictor, except that CONFIRM groups together different predictor classes that seem not to differ significantly. CONFIRM for an ordinal predictor is a possible alternative to linear regression, with the difference that while linear regression assumes a constant rate of increase as you go from one level of the ordinal predictor to another, CONFIRM uses a "staircase" model where the dependent variable either remains constant as you go from one level to the next, or jumps by some (not necessarily constant) amount.

Most other recursive partitioning procedures fix the number of ways a node splits, commonly considering only binary splits. FIRM's method of reducing the classes of the predictors differs from these approaches. A c-category predictor may group into any of 1 through c categories, the software deciding what level of grouping the data appear to require. Both CONFIRM and CATFIRM follow the same overall approach. If a predictor has c classes, the cases are first split into c separate groups corresponding to these classes. Then tests are carried out to see whether these classes can be reduced to fewer classes by pooling classes pairwise. This is done by finding two-sample test statistics (Student's t for CONFIRM, chi-squared for CATFIRM) between each pair of classes that could legally be grouped together. The groups that can "legally" be joined depend on the predictor type:

- For a free predictor and a character predictor, any two groups can be joined.
- For a monotonic predictor, only adjacent groups may be joined.
- For a float predictor, any pair of adjacent groups can be joined. In addition, the floating category can join any other group.
- A real predictor works like a monotonic predictor if it has no missing information. If it is missing on some cases, then it works like a float predictor.

If the most similar pair fails to be significantly different at the user-selected significance level, then the two classes are merged into one composite class, reducing the number of groups by 1. The pairwise tests are then repeated for the reduced set of c - 1 classes. This process continues until no legally poolable pair of simple or composite classes is separated by a non-significant test statistic, ending the "merge" phase of the analysis. Let's illustrate this process with CONFIRM's handling of the floating predictor *EMV* in the "head injuries" data. The process starts with the following overall summary statistics for the grouping of the cases by *EMV*:

```
predictor no. 2 EMV
statistics before merging
Cate    ?    1    2    3    4    5    6    7
mean 2.07 1.00 1.19 1.58 1.81 1.84 2.25 2.63
size   28   19   64   52  111   96   65   65
s.d.  .94  .00  .53  .82  .93  .91  .94  .74
```

Any of the categories can merge with its immediate left or right neighbors, and in addition the float category can merge with any group. The "split" output file lists the following summary statistics. The first line of "merge stats" shows the Student's *t*-value for merging the two groups between which it lies; the second shows the Student's *t*-value distinguishing the float category from the category to the right. So, for example, the *t*-value testing whether we can merge categories 3 and 4 is 1.7; and the *t*-value for testing whether we can merge the float category with category 4 is -1.5.

```
Group          ?     1      2     3     4     5     6     7
merge stats       -4.3     .9   2.5   1.7    .3   3.0   2.6
                   4.3   -4.7  -2.5  -1.5  -1.3    .9   3.0


Group          ?     1      2     3     45    6     7
merge stats       -4.3     .9   2.5   1.9   3.6   2.6
                   4.3   -4.7  -2.5  -1.5    .9   3.0


Group          ?    12      3    45     6     7
merge stats       -5.1    2.9   1.9   3.6   2.6
                   5.1   -2.5  -1.5    .9   3.0


Group               12      3    45    ?6     7
merge stats        2.9    1.9   3.5   3.3


Group               12    345    ?6     7
merge stats        6.0    4.1   3.2
```

The closest groups are 4 and 5, so these are merged to form a composite group. Once this is done, the closest groups are 1 and 2, so these are merged. Next ? and 6 are merged. Following this, there no longer is a floating category, so the second line of "merge stats" stops being produced. Finally, group 3 is merged with the composite 45. At this stage, the nearest groups (?6 and 7) are separated by a significant Student's *t*-value of 3.2, so the merging stops. The final grouping on this predictor is {12}, {345}, {?6}, {7}.

To protect against occasional bad groupings formed by this incremental approach, FIRM can test each composite class of three or more categories to see whether it can be resplit into two that are significantly different at the split significance level set for the run. If this occurs, then the composite group is split and FIRM repeats the merging tests for the new set of classes. This split test is not needed very often, and can be switched off to save execution time by specifying a split significance level that is smaller than the merge significance level.

The end result of this repeated testing is a grouping of cases by the predictor. All classes may be pooled into one, indicating that that predictor has no descriptive power in that node. Otherwise, the analysis ends with from 2 to c composite groups of classes, with no further merging or splitting possible without violating the significance levels set.

The last part of the FIRM analysis of a particular predictor is to find a formal significance level for it. In doing this, it is essential to use significance levels that reflect the grouping of categories that has occurred between the original c-way split and the final (say k-way) split. Hawkins and Kass (1982) mention two ways of measuring the overall significance of a predictor conservatively—the "Bonferroni" approach, and the "Multiple comparison" approach. Both compute an overall test statistic of the k-way classification: for CATFIRM a Pearson χ^2-statistic, and for CONFIRM a one-way analysis of variance F-ratio. The Bonferroni approach takes the p-value of the resulting test statistic and multiplies it by the number of implicit tests in the grouping from c categories to k. There are two possible multiple comparison p-values. The default in (versions of FIRM through FIRM 2.0) computes the p-value of the final grouping as if its test statistic had been based on a c-way classification of the cases. The default introduced in FIRM 2.1 is the p-value of the original c-way classification, but the earlier multiple comparison test can be selected if desired.

Since both the Bonferroni and multiple comparison approaches yield conservative bounds on the p-value, the smaller of the two values is taken to be the overall significance of the predictor.

The final stage is the decision of whether to split the node further and, if so, on which predictor. FIRM does this by finding the most significant predictor on the conservative test, and making the split if its conservative significance level multiplied by the number of active predictors in the node is below the user-selected cutoff. The FIRM analysis stops when none of the nodes has a significant split.

Some users prefer not to split further any nodes that contain only a few, or very homogeneous, cases. Some wish to limit the analysis to a set maximum number of nodes. The codes contain options for these preferences, allowing the user to specify threshold sizes (both codes) and a threshold total sum of squares (CONFIRM) that a node must have to be considered for further splitting, and to set a maximum number of nodes to be analyzed.

7.6.14 Comparison with Other Codes

Before discussing the details of using the codes, we mention some historical development and other related approaches. The original source on recursive partitioning was the work of Morgan and Sonquist (1963) and their "Automatic Interaction Detector" (AID). This covered monotonic and free predictors, and made only binary splits. At each node, the split was made where the explained sum of squares was greatest, and the analysis terminated when all remaining nodes had sums of squared deviations below some user-set threshold.

It was soon found that the lack of a formal basis for stopping made the procedure prone to overfitting. In addition, the use of explained sum of squares without any modification for the number of starting classes or the freedom to pool them also made AID tend to prefer predictors with many categories to those with fewer, and to prefer free predictors to monotonic.

Breiman *et al* (1984) built on the AID foundation with their Classification and Regression Trees approach. This has two classes of predictors—categorical (corresponding to AID's "free" predictors) and continuous; unlike FIRM, their method does not first reduce continuous (in FIRM terminology "real") predictors down to a smaller number of classes, nor does it draw any distinction between these and monotonic predictors with just a few values. The Breiman *et al* methodology does not have an equivalent to our "floating" category. Instead, missing values of predictors are handled by "surrogate splits", by which when a predictor is missing on some case, other predictors are used in its stead. This approach depends on the predictive power of the missingness being captured in other non-missing predictors.

Breiman *et al* use only binary splits. Where FIRM chooses between different predictors on the basis of a formal test statistic for identity, they use a measure of "node purity", with the predictor giving the purest descendant nodes being the one considered for use in splitting. This leads to a tendency to prefer categorical predictors to continuous, and to prefer predictors with many distinct values to predictors with few. FIRM's use of conservative testing tends to lead to the opposite biases, since having more categories, or being free rather than monotonic, increases the Bonferroni multipliers applied to the predictor's raw significance level.

With a continuous dependent variable, their measure of node purity comes down to essentially a two-sample t (which FIRM would also use for a binary split), but their purity measure for a categorical dependent variable is very different than FIRM's. FIRM looks for statistically significant differences between groups while they look for splits that will classify to different classes. So the Breiman *et al* approach seeks two-column cross classifications with different modal classes—that is, in which one category of the dependent variable has the largest frequency in one class while a different category has the largest frequency in the other class. FIRM looks instead for groupings that give large chi-squared values, which will not necessarily have different modal classes. This important difference has implications for the use of the two procedures in minimal-model classification problems. As it is looking for strong relationships rather than good classification rules, a FIRM analysis may produce a highly significant split, but into a set of nodes all of which are modal for the same class of the dependent variable.

An example of this is the split of Node 4 of the head injuries data set (Figure 7.3). This creates two descendant nodes with 52% and 86% moderate to good recoveries. If all you care about is predicting whether a patient will make a moderate to good recovery, this split is pointless—in both descendant nodes you would predict a moderate to good recovery. But if you want to explore the relationship between the indicator and the outcome, this split would be valuable, as the proportions of favorable outcomes are substantially different.

It might also be that a clinician might treat two patients differently if it is known one of them has a 86% chance of recovery while the other has a 52% chance, even though both are expected to recover. In this case, too, the FIRM split of the node could provide useful information.

Two major differences between FIRM and the Classification and Regression Trees method relate to the rules that are used to decide on the final size of tree. FIRM creates its trees by "forward selection". This means that as soon as the algorithm is unable to find a node that can be split in a statistically significant way, the analysis stops. Since it is possible for a non-explanatory split to be needed before a highly explanatory one can be found at a lower level, it is possible for the FIRM analysis to stop too soon and fail to find all the explanatory power in the predictors. It is easy to construct artificial data sets in which this happens.

You can protect yourself against this eventuality by initially running FIRM with a lax splitting criterion to see if any highly significant splits are hidden below non-significant ones; if not, then just prune away the unwanted non-significant branches of the dendrogram. Particularly when using the printed dendrogram, this is a very easy operation as all the splits are labeled with their conservative significance level. We do not therefore see this theoretical possibility of stopping too soon as a major practical weakness of FIRM's forward testing approach.

Breiman *et al* use the opposite strategy of "backward elimination". First, a deliberately very oversized tree is created by continuing to split nodes, whether the splits produce some immediate benefit or not. Then the oversized tree is pruned from the bottom, undoing splits that appear not to correspond to distinct patterns in the data.

The second, more profound difference is in the procedure used to decide whether a particular split is real, or simply a result of random sampling fluctuations giving the appearance of an explanatory split. FIRM addresses this issue using the Neyman Pearson approach to hypothesis testing. At each node, there is a null hypothesis that the node is homogeneous, and this is tested using a controlled, conservative significance level. In the most widely-used option, Breiman *et al* decide on the value of a split using cross validation. We will do no more than sketch the implications, as the method is complicated (see for example Breiman, *et al*, 1984 or McLachlan, 1992). An original very oversized tree is formed. The sample is then repeatedly (typically 10 times) split into a "calibration" subsample and a "holdback" sample. The same tree analysis is repeated on the calibration subsample as was applied to the full data and this tree is then applied to the holdback sample, to get an "honest" measure of its error rate. The picture obtained from these 10 measures is then used to decide how much of the detail of the original tree was real and how much due just to random sampling effects. This leads to a pruning of the original tree, cutting away all detail that fails to survive the cross-validation test.

As this sketch should make clear, an analysis using the Breiman *et al* methodology generally involves far more computing than a FIRM analysis of the same data set. As regards stopping rules, practical experience with the two approaches used for a continuous dependent variable seems to be that in data sets with strong relationships, they tend to produce trees of similar size and form. In data sets with weak, albeit highly significant relationships, FIRM's trees appear to be generally larger and more detailed. See Hawkins and McKenzie (1995) for some further examples and discussion of relative tree sizes of the two methods.

In view of the very different objectives of the two approaches to analysis of categorical data (getting classification rules and uncovering relationships respectively), there are probably few useful general truths about the relative performance of the two approaches for a categorical dependent variable.

Another recursive partitioning method is the Fast Algorithm for Classification Trees (FACT), of Loh and Vanichsetakul (1988). This method was designed for a categorical dependent variable, and treats the splitting by using a discriminant analysis paradigm.

The KnowledgeSEEKER procedure is also based on the CHAID algorithm and produces results similar to those of CATFIRM. The main technical difference from FIRM (other than the more user-friendly interface) lies in the conservative bounds used to assess a predictor, where KnowledgeSEEKER uses a smaller Bonferroni multiplier than does FIRM.

The public domain SAS procedure TREEDISC (available on request by Warren Sarle of SAS Institute) also implements an inference-based approach to recursive partitioning. The SAS Data Mining module also contains a variety of RP approaches including all three widely-used approaches.

This list is far from exhaustive—there are many other RP approaches in the artificial intelligence arena—but perhaps covers some of the major distinct approaches.

7.7 Graphical Display of Data

Graphics are often a useful data-exploring technique through which the researcher may familiarize her or himself with the data. Relationships and trends may be conveyed in an informal and simplified visual form via graphical displays.

In addition to the display of a bar chart, histogram, 3-D bar chart, line and bivariate scatter plot, LISREL 8.50 for Windows now includes options to display a pie chart, box-and-whisker plot and a scatter plot matrix. By opening a PRELIS system file (see Section 7.1 and Chapter 2), the PSF toolbar appears and one may select either of the ***Univariate***, ***Bivariate*** or ***Multivariate*** options from the ***Graphs*** menu.

7.7.1 Pie Chart

A pie chart display of the percentage distribution of a variable is obtained provided that

- The variable is defined as ordinal. Variables types are defined via the ***Data, Define Variables*** option.
- The variable does not have more than 15 distinct values.

Pie charts may be customized by using the graph editing dialog boxes discussed in Chapter 2. The pie chart below is obtained by opening the dataset **income.psf** from the **mlevelex** folder. Select the ***Graphs, Univariate*** option to go to the ***Univariate Plots*** dialog box. Select the variable *group* and click the ***Pie Chart*** radio button.

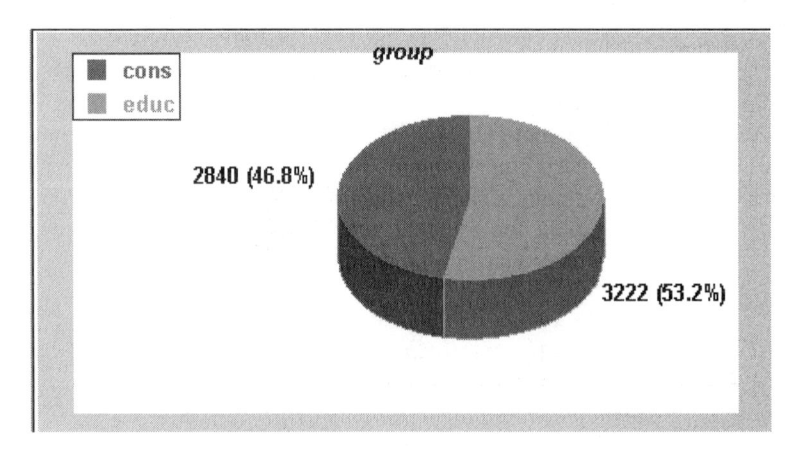

The pie chart above shows an almost even distribution of respondents in the construction (cons) and education (educ) sectors.

7.7.2 Box-and-Whisker Plot

A box-and-whisper plot is useful for depicting the locality, spread and skewness of a data set. It also offers a useful way of comparing two or more data sets with each other with regard to locality, spread and skewness. Open **income.psf** from the **mlevelex** folder and select the ***Graphs, Bivariate*** option to obtain the ***Bivariate Plots*** dialog box shown below. Create a plot by clicking ***Plot*** once the variables have been selected and the ***Box and Whisker plot*** check box have been checked.

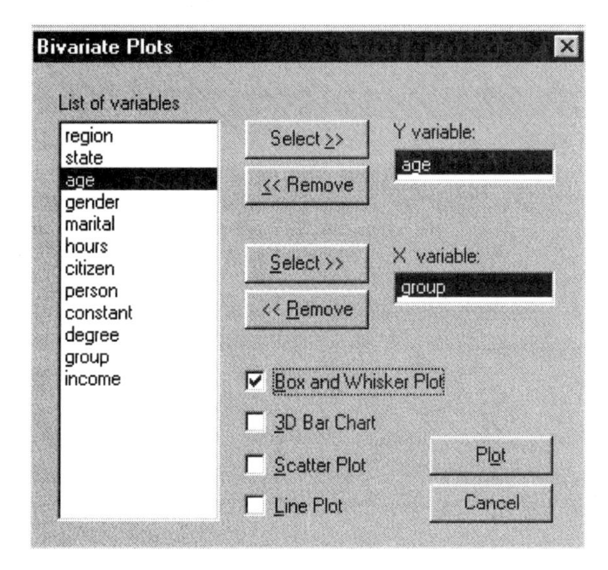

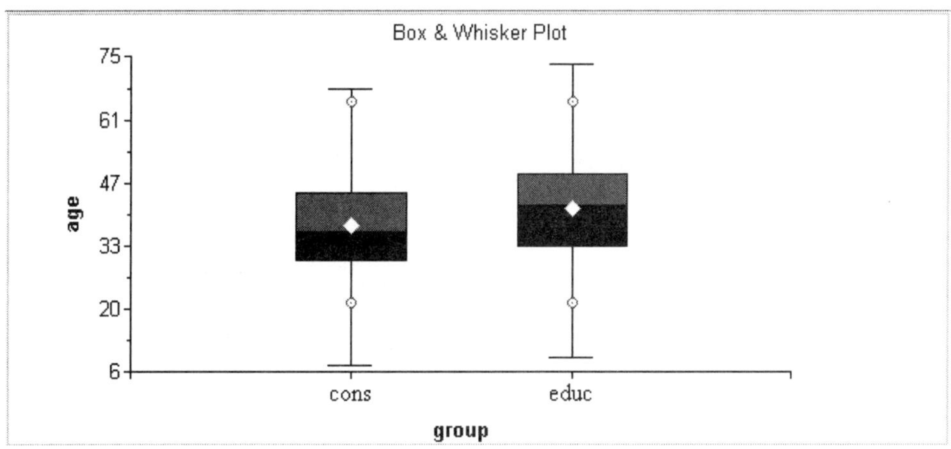

The bottom line of a box represents the first quartile (q_1), the top line the third quartile (q_3), and the in-between line the median (me). The arithmetic mean is represented by a diamond.

The whiskers of these boxes extend to $1.5(q_3 - q_1)$ on both sides of the median. The length of the whiskers is based on a criterion of Tukey (1977) for determining whether outliers are present. Any data point beyond the end of the whiskers is then considered an outlier. Two red circles are used to indicate the minimum and maximum values.

Graph parameters that may be changed include the axes and description thereof, the symbols used and the colors assigned to the symbols/text. To change any of these, simply double-click on the symbol/text to be changed to activate a dialog box in which changes can be requested.

For symmetric distributions the mean equals the median and the in-between line divides the box in two equal parts.

From the graphics display above we conclude

- The age distribution of respondents from the construction sector are positively skewed while the opposite holds for the education sector.
- The mean age of respondents from the education sector is greater than those from the construction sector.

For a large sample from a normal distribution, the plotted minimum and maximum values should be close to the whiskers of the plot. This is not the case for either of the groups and we conclude that the distribution of age values in this data is not bell-shaped. This observation may be further confirmed by making histogram displays of *Age* using the datasets **cons.psf** and **educ.psf**, these being subsets of the **income.psf** dataset.

7.7.3 Scatter Plot Matrix

The scatter plot matrix, as implemented in LISREL, provides an organized way of simultaneously looking at bivariate plots of up to ten variables. Except for the diagonal, each frame contains a scatter plot. The diagonals contain the variable names and the minimum and maximum values of the corresponding variables. Each scatterplot is a visual summary of linearity, nonlinearity, and separated points (see, for example, Cook and Weisberg (1994)).

Select ***Open*** from the ***File*** menu and select the PRELIS system file **fitchol.psf** from the **tutorial** folder.

As a first step, assign category labels to the four categories of the variable *Group*:

> Weightlifters ($n = 17$)
> Students (control group, $n = 20$)
> Marathon athletes ($n = 20$)
> Coronary patients ($n = 9$).

To do this, select the ***Define Variables*** option from the ***Data*** menu. In the ***Define Variables*** dialog box, click on the variable *Group* and then click ***Category Labels***. On the ***Category Labels*** dialog box, assign the labels as shown below and click ***OK*** when done.

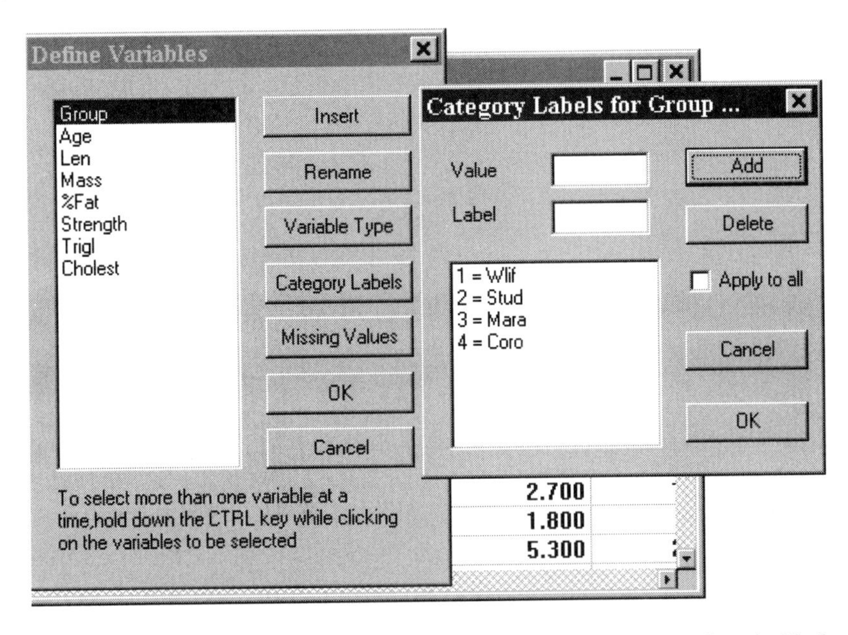

In order to create a scatter plot matrix of the variables *Age*, *Trigl* and *Cholest*, the ***Multivariate*** option on the ***Graphs*** menu is used.

Statistics Graphs Multilevel View Window
Univariate ...
Bivariate...
Mutlivariate...
Fitchol

	Group	Age
1	1.000	22.000
2	1.000	30.000
3	1.000	26.000
4	1.000	23.000
5	1.000	29.000

As a result, the ***Scatter Plot Matrix*** dialog box appears. The variables *Age*, *Trigl* and *Cholest* are selected and added to the ***Selected*** window. The variable *Group*, representing the four groups to which we have just assigned category labels, is added to the ***Mark by*** window. Click ***Plot*** to create the 3 x 3 scatter plot matrix.

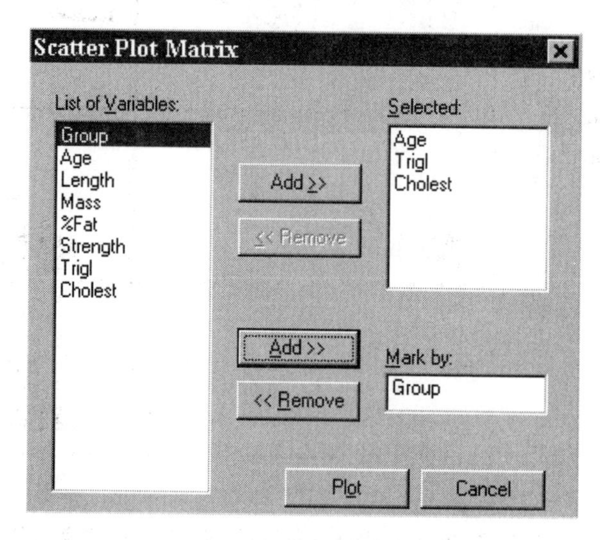

The scatter plot matrix is shown below. In the first row of plots at the top of the matrix, *Cholest* is plotted against *Age* and *Trigl* respectively. The third plot in this row contains the variable name and the observed minimum and maximum values of *Cholest.*

The first plot in the second row shows the scatter plot of *Trigl* against *Age.* The other scatter plots are essentially mirror images of the three discussed this far and are the plots for (*Cholest*, *Trigl*), (*Trigl*, *Age*) and (*Cholest*, *Age*) respectively. The diagonal elements of the matrix contain the names, minimum and maximum values of each variable.

From the display, it is apparent that the (*Age*, *Trigl*) and (*Age*, *Cholest*) scatter plots of coronary patients, denoted by a "+" symbol, are clustered together, away from the main cluster of points formed by the other three groups. In the (*Age*, *Age*) segment the minimum and maximum values of the *Age* variable are given.

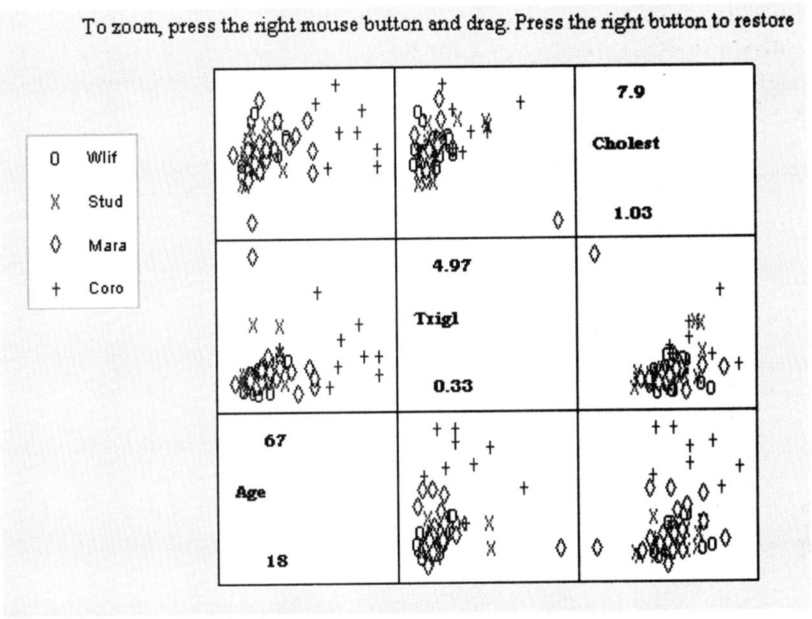

To zoom, press the right mouse button and drag. Press the right button to restore

0 Wlif
X Stud
◊ Mara
+ Coro

7.9
Cholest
1.03
4.97
Trigl
0.33
67
Age
18

In order to take a closer look at the (*Trigl*, *Cholest*) scatter plot, the right mouse button is held down and the plotting area of this scatter plot is selected as shown below.

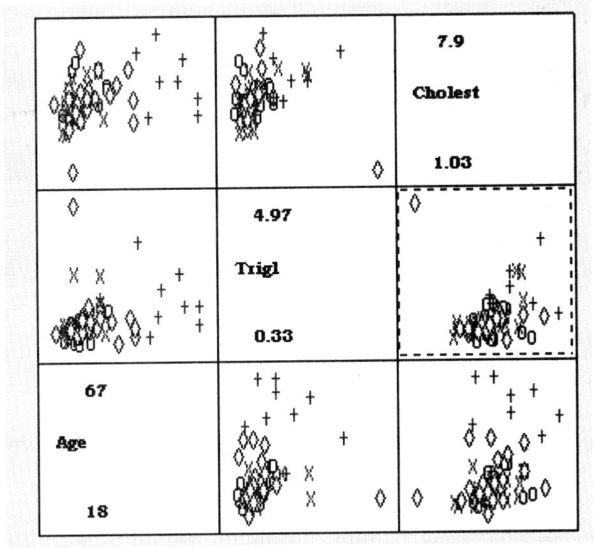

7.9
Cholest
1.03
4.97
Trigl
0.33
67
Age
18

Releasing the mouse button produces a closer look at the selected area as shown below.

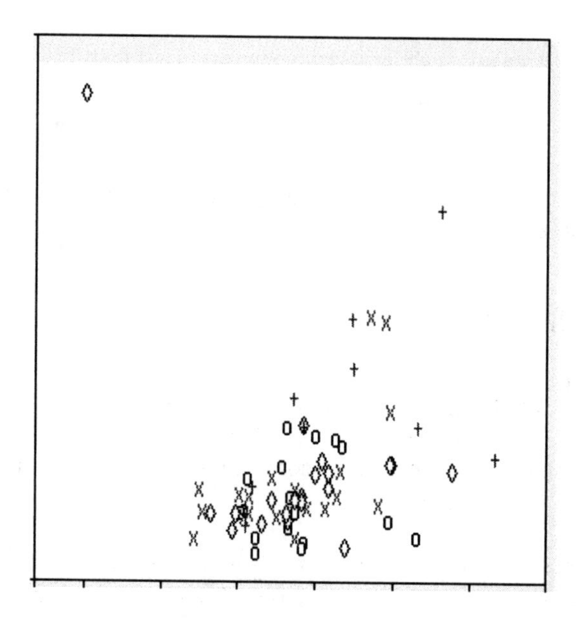

From the graph, we see that the (*Trigl*, *Cholest*) measurements for the weightlifters (plot symbol "0") are closer to each other than is the case for any of the other three groups. The most dispersed points are for the coronary patients (plot symbol "+"). To return to the scatter plot matrix, press the right mouse button.

8 Syntax

This chapter contains new syntax introduced in LISREL 8.50 as well as previously undocumented syntax.

8.1 LISREL Syntax

8.1.1 RA Command

The RA command is used to read records containing values of the variables for analysis (raw data, as opposed to summary data).

The raw data are read one case after another. There must be NI data values for each case. Preferably, data are read from an external file. Then, the program will determine the sample size if NO on the DA line is default or set to zero. Or, if NO is erroneously set too large, the program will terminate input when an end-of-file is read and will use the correct data count in the computations. If NO is set too small, only NO cases will be read, regardless of how many cases exist. When the data are included in the command file, NO should indicate the number of cases correctly, otherwise the program will stop with an error message.

Examples:

```
RA = c:\lisrel850\tutorial\growth.dat
RA FI=c:\lisrel\tutorial\growth.dat
RA = growth.psf
```

From the above it follows that a ***.psf** file (See Section 7.1) can be used in version 8.50 as part of the LISREL syntax.

Note that if a ***.psf** file is used, the LA statement may be omitted. the DA line is still required, but on this line it is sufficient to specify the number of variables only. For example:

```
DA NI = 8
RA = growth.psf
```

8.1.2 FIML and Missing Data

To fit a SEM model to data with missing values, add the MI keyword to the DA line, e.g.;

```
DA NI = 6 NO = 840 MI = -99
```

indicates that a missing value code of -99 is to be used.

Note:

The MI = <value> keyword only has an effect when raw data are used as input to the LISREL analysis, for example

```
RA = cholesterol.dat
```

If a *.**psf** file is used and a missing value code is defined via the ***Define Variables*** option from the ***Data*** menu, the MI command may be omitted.

See the **missingex** and **msemex** folders for examples.

8.2 SIMPLIS Syntax

8.2.1 FIML and Missing Data

The Missing Value Code command is required to fit a SEM model to data with missing values.

```
Missing Value Code <value>
```

Note:

This command should be followed by the following 2 commands:

```
Sample Size: <nsize>
Raw data from file <filename>
```

where filename can have the *.**psf** file extension in LISREL 8.50. See the **missingex** and **msemex** folders for examples.

8.3 PRELIS Syntax

8.3.1 Multiple Imputation and the EM Command

This command is used to estimate the population means and covariance matrix of a dataset with values missing at random, using the Expected Maximization algorithm. In addition, missing values are estimated.

```
EM CC = ccvalue  IT = itvalue,
```

where `ccvalue` is the convergence criterion (default `CC=0.00001`) and `itvalue` the number of iterations (default `IT=200`). These default values are created when ***Multiple Imputation*** is selected from the ***Statistics*** menu. Note that the `CC` and `IT` keywords have to be specified. If either one or both of these keywords is omitted, the EM procedure will not be invoked.

Also see the multiple imputation example in Chapter 3.

8.3.2 Multiple Imputation and the MC command

This command is used to estimate missing values using a Monte Carlo Markov Chain procedure (MCMC).

```
MC CC=ccvalue IT=itvalue,
```

where `ccvalue` is the convergence criterion (default `CC=0.00001`) and `itvalue` the number of iterations (default `IT=200`).

Also see the multiple imputation example in Chapter 3.

8.4 Multilevel Modeling Syntax

8.4.1 Introduction

A new feature of LISREL 8.50 is the ability to fit non-linear multilevel regression models to repeated measurement data.

A typical example is a data set consisting of height measurements made on different occasions (for example, 6, 7, ..., 15 years) on each of 50 individuals. In this context, individuals are the level-2 units and the height measurements are the level-1 units.

Suppose that the following models provide reasonable descriptions of the change in height over time.

Model 1 (quadratic):

$$y_t = b_0 + b_1 t + b_2 t^2 + e_t, t = 6, 7, \ldots, 15$$

Model 2 (logistic):

$$y_t = b_1 / [1 + \exp(-b_2 b_3 t)] + e_t$$

A plot of both the quadratic and logistic functions exhibit a non-linear trend over time. Model 1, however, falls in the class of linear hierarchical models since the response variable (y_t) can be expressed as a linear combination of the regression coefficients b_0, b_1 and b_2. Since this is not true for model 2, the latter model falls under the class of non-linear multilevel models.

To create syntax for these models by using dialog boxes, the ***.psf** file containing the relevant data is first opened. As an example, open the file **japan.psf** located in the **tutorial** folder. By clicking on the ***Multilevel*** menu, one can select either ***Linear Model*** or ***Non-Linear Model***.

8.4.2 Linear Multilevel Models

The drop-down menu associated with the ***Linear Model*** has four options, as shown below:

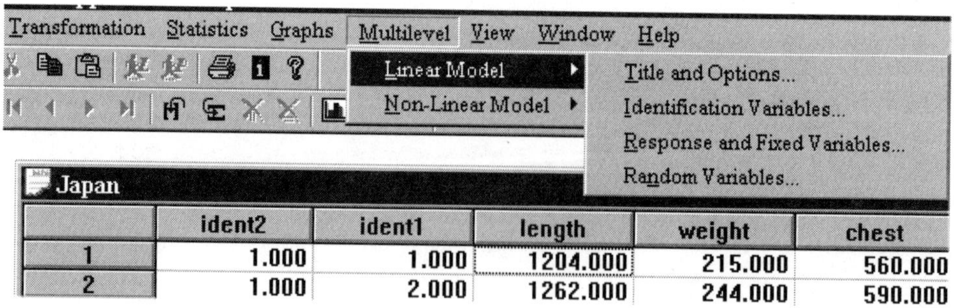

	ident2	ident1	length	weight	chest
1	1.000	1.000	1204.000	215.000	560.000
2	1.000	2.000	1262.000	244.000	590.000

The ***Title and Options***, ***Identification Variables***, ***Response and Fixed Variables*** and ***Random Variables*** dialog boxes can be used to create a multilevel syntax file (see the examples in Chapter 5). See *LISREL 8: New Statistical Features* for a description of these commands.

The basic structure of the multilevel syntax file is as given below, and the required commands are indicated.

Command	
`OPTIONS` <list of options>;	Required
`IDn` = name of variable identifying level *n* units;	Required
`RANDOMn` = names of variables random on level *n* of the model;	Required
`RESPONSE` = name(s) of response variables(s);	Required
`FIXED` = names of variables included as fixed effects in the model;	Required
`DUMMY` = name of variable to create dummy variables from;	Optional
`MISSING_DAT` = integer value;	Optional
`MISSING_DEP` = integer value;	Optional
`TITLE` = job title;	Optional
`COVnVAL` = starting values for level *n* random coefficient covariance matrix;	Optional*
`COVnPAT` = pattern for level *n* random coefficient covariance matrix;	Optional*
`FIXVAL` = starting values for fixed effect parameters;	Optional*
`CONTRAST` = name of contrast file;	Optional*
`SUBPOP` = names of variables to be used to construct subpopulations;	Optional*
`WEIGHT1`=name of level-1 weight variable;	Optional*

The last two optional commands are used in conjunction with categorical response variables (see Section 4.8 for an example). None of the optional commands marked with an asterisk can be generated by using dialog boxes. Once a syntax file is created, the user can add any of these commands with the LISREL text editor or any other text editor (such as Notepad).

When the multilevel syntax file is constructed or edited outside the interface, the following guidelines should be kept in mind:

- All commands start with a keyword and conclude with a semi-colon.
- The maximum length of a command in the syntax file is 127 characters.
- There is no specific required order in which commands have to be given, with the exception of the `OPTIONS` command, which must always be the first line.
- Blank lines are allowed between commands.
- The commands may be given in either upper or lower case letters.
- A keyword can be a mixture of upper and lower case letters.

Commands which are generated by using the dialog boxes automatically conform to the above rules. *LISREL 8: New Statistical Features* and the Help file contain a complete description of the syntax for linear multilevel models, with the exception of the `DUMMY` command, which is new to LISREL 8.50 for Windows.

DUMMY Command

The optional DUMMY command is used to create dummy variables for a selected variable. Names for the dummy variables are denoted by $dummy_1, dummy_2, \ldots, dummy_k$, where k equals the number of distinct values of the selected variable. For example, if the variable *TIME* has 4 distinct values, 1, 2, 3, and 4, then the command

```
DUMMY = TIME;
```

will result in the creation of 4 dummy variables:

TIME	$dummy_1$	$dummy_2$	$dummy_3$	$dummy_4$
1	1	0	0	0
2	0	1	0	0
3	0	0	1	0
4	0	0	0	1

When the syntax file is created by opening a ***.psf** file and selecting the ***Multilevel, Linear Model, Response and Fixed Variables*** option, the DUMMY command is automatically generated. This is accomplished by selecting an appropriate variable using the ***Response and Fixed Variables*** dialog box.

The user can change the default names of the dummy variables by the inclusion of the PREFIX keyword in the DUMMY command.

DUMMY Command: PREFIX Keyword

The syntax for the use of this keyword is

```
PREFIX = <prefix to be used for dummy variables>
```

For example

```
DUMMY = TIME PREFIX = TIM;
```

A prefix name of longer than 5 characters will be truncated to the first 5 characters.

Example:

```
DUMMY = time PREFIX = OCCASION;
```

will result in dummy variables named *OCCAS1*, *OCCAS2*,...

Remark:

The `DUMMY` command is useful when fitting a model to the data when no restrictions are imposed on the means or covariances. A typical syntax file in this case will have the form

```
OPTIONS MAXITER=30;
TITLE Demonstration of dummies;
SY=JAPAN.PSF;
ID2=ident2;
RESPONSE=chest;
DUMMY=time PREFIX=TIM;
FIXED=TIM1:TIM9;
RANDOM2=TIM1:TIM9;
```

Note that, for this 2-level model, one does not include the `ID1` or `RANDOM1` commands.

8.4.3 Non-linear Multilevel Models

The drop-down menu associated with the ***Non-Linear Model*** menu has five options, as shown below:

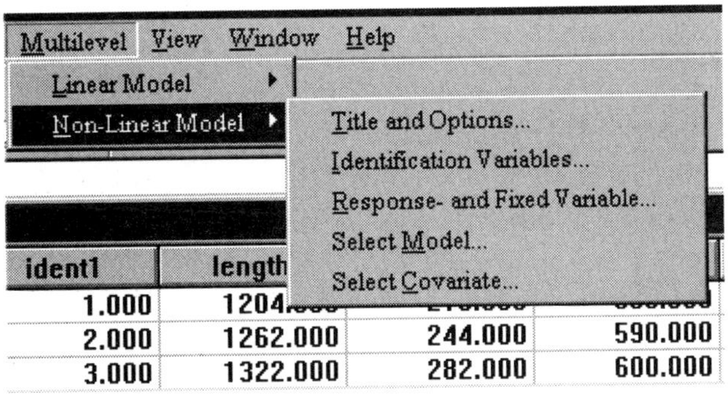

The ***Title and Options***, etc. dialog boxes can be used to create a syntax file (see the examples in Chapter 5 for more details).

The basic structure of the syntax file is shown below.

`OPTIONS` <list of options>;	Required
`ID1` = name of variable identifying level-1 units;	Required
`ID2` = name of variable identifying level-2 units;	Required

RESPONSE = name of response variable;	Required
FIXED = name of predictor;	Required
MODEL = model selected;	Required
MISSING_DAT = integer value;	Optional
COVARIATES b1 = varname1	Optional

```
COVARIATES b1 = varname1
           b2 = varname2
                    .
           c6 = varname6;
```

The basic guidelines for constructing or changing the syntax file are the same as those given for linear multilevel models. Each of the new commands is described below.

OPTIONS Command

The options are used to control the estimation procedure and are listed below:

```
METHOD = MAP/ML
CONVERGE = ccvalue
MAXITER = maxitvalue
QUADPTS = number of quadrature points.
```

Default:

```
METHOD = MAP
CONVERGE = 0.001
MAXITER = 30
QUADPTS = 10
```

QUADPTS refers to the number of quadrature points to be used in the numerical integration procedure.

Examples:

```
OPTIONS METHOD=MAP CONVERGE=0.0001 MAXITER=50;
```

The method of estimation is maximum a priori (see Chapter 7) and the number of iterations equals 50. Note that if METHOD = MAP, then 10 times the specified value of MAXITER is used.

```
OPTIONS METHOD=ML CONVERGE=0.001 MAXITER=25 QUADPTS=20;
```

The method of estimation is maximum likelihood (see Chapter 7 for details). The number of quadrature points equals 20. Note that it is almost always advisable to start with a smaller number of quadrature points (*e.g.* `QUADPOINTS = 6`) for models with more than 3 coefficients to get an indication of the amount of computation time required for the analyses.

As a rough guideline, suppose a model with m ($m > 3$) random coefficients is selected and that one iteration requires a CPU time of 5 seconds. One may then expect the CPU time to increase by a factor $\left(\frac{n}{6}\right)^{m-2}$ if the number of quadrature points are increased from 6 to n.

ID*n* Commands

Level-2 subjects are generally people or objects such as machines, trees or animals. The measurements made on these level-2 units are the level-1 units.

Examples:

```
ID2 = CaseNo;
ID1 = Occasion;
```

```
ID2 = Tree;
ID1 = Time;
```

RESPONSE Command

The class of models considered in LISREL allows for a monotone increase (or decrease) in the values of the dependent (response) variable over values of the predictor (fixed) variable.

Examples:

```
RESPONSE = weight;
RESPONSE = yield;
RESPONSE = cholesterol;
```

FIXED Command

Presently LISREL allows for a single predictor of the change in values of the response variable at the different measurement occasions. The most commonly used predictor is time, but predictors may also be variables such as concentration levels, pressure (in pounds per square inch), distance, etc.

Examples:

```
FIXED = time;
FIXED = pressure;
```

MISSING_DAT Command

Use of the `MISSING_DAT` command results in a listwise deletion of any row containing a missing value code.

Example:

```
MISSING_DAT = -99;
```

The above command will result in a listwise deletion of any row in the data set that contains a value of `-99`. Note that deletion is based on only that subset of the data used for a given analysis.

MODEL Command

The type of model to be fitted to the data is specified by using the `MODEL` command.

```
MODEL = function1 + function2 ;
```

Presently, there are five functions available, these being the logistic, Gompertz, Monomolecular, power and exponential curves. For the definition of these curves, see Chapter 7. A number of examples are given in the **nonlinex** folder.

Examples:

```
MODEL = Exponential;
MODEL = Logistic + Logistic;
```

As a general guideline, one should only use single-component models when the number of repeated measurements is relatively small (for example, less than 15).

COVARIATES Command

This keyword allows the user to model random coefficients as level-2 outcome variables.

The syntax for this command is

```
COVARIATES b1=varname1
           b2=varname2
           .
           c6=varname6;
```

Remarks:

- `varnames` may be different or the same.
- a subset of coefficients can be selected, e.g.

```
COVARIATES b1=gender
           c1=gender;
```

- A restriction in LISREL 8.50 is that each coefficient can be associated with only one covariate.

8.5 FIRM Syntax

8.5.1 Introduction

FIRM works with a number of files other than those users look at directly. Some are scratch files that exist only for the duration of the run and then are erased. These are of no concern at all (except to the limited extent that an aborted run causes them not to be deleted at the end of the run and they then waste unnecessary disk space).

The FIRM syntax file forms the basis of any CATFIRM or CONFIRM analysis. The CATFIRM and CONFIRM syntax files each have four sections:

- one to specify CATFIRM and CONFIRM analysis and output options,
- one for the dependent variable,
- one for the predictors,
- and one for the analysis options.

The input is in free format. Character arguments must be put in quotes; all other arguments must be entirely numeric. Spacing between the entries is largely irrelevant. It is generally safe to put entries other than character strings on multiple lines, but you can get into trouble by putting two or more things on the same line when the programs expect them on different lines. See the examples of syntax files in Section 7.6 for more information.

8.5.2 Specifying Analyses and Output Options

The first line of the syntax file specifies whether a CATFIRM or CONFIRM analysis should be carried out. Examples are

```
CATFIRM: den sum
CONFIRM: den sum split rule
CONFIRM: den
```

The options `den` and `sum` specify that a dendrogram and summary information should be added to the output file. One can also add the options `split` and `rule` to add details of the split procedure and split rule table to the output file.

8.5.3 Dependent Variable Specification

The first section of the syntax file is the dependent variable section. This section specifies the position, name, number, and names of categories of the dependent variable in a CATFIRM or CONFIRM analysis.

Examples:

(i) `8 'Repair costs' 0`
(ii) `1 'Outcome' 3 'Dead/veg' 'severe' 'mod/good'`
(iii) `'promotion code' -1`

The options defining aspects of the dependent variable to be used in the analysis are now discussed in the order in which they should be entered.

Position of the dependent variable in the data set

The first information entered is the position of the dependent variable in the data. For each case, a number of data values are to be read. Any of these may be the dependent variable, and this first question is which of them it is. In the automobile repair cost example (see Chapter 7), it is repair costs that we want to predict, so this is the dependent variable. If it is, for example, the eighth item in the list, its position is `8`.

Name of the dependent variable

In the auto repair data discussed in Chapter 7, for example,

```
'Repair costs'
```

is a valid dependent variable name. Up to 50 characters may be used for the name of a variable.

Number of categories for the dependent variable

If the dependent variable is categorical (so that you will be analyzing it using CATFIRM), then type the number of different categories it has (see (ii) above).

If the dependent variable is continuous (so that you will be analyzing it using CONFIRM), then type zero to this question (see (i) above).

If the dependent variable is a character variable so that you will have FIRM find the separate categories, enter the artificial code `-1` which signals a character dependent variable (see (iii) above).

Warning:

If you specify that the dependent variable is categorical, CATFIRM expects that its categories will have integer values 1, 2, 3, ... If your coding starts at `0` (so that the categories are 0, 1, 2, ...), then the `0`s will be regarded as data errors and the FIRM run will fail. The way around this is to specify the dependent as being of type `character` and let FIRM discover the different codes.

Names of the categories of the dependent variable

CATFIRM looks for as many category names for the dependent variable as you said there were categories, so make sure you have that many names. The names can be on the same line or on different lines. If you gave the number of categories as `0` or `-1`, then do not put any category names in the syntax file (see (i) and (ii) above).

If you indicated that the dependent variable had 1 or more categories (which implies that you will be running it using CATFIRM), a name for each category must be provided. Give the categories names of no more than 20 characters (see (ii) above).

For the auto repair costs example, where we clearly want a CONFIRM analysis, the dependent variable is specified as having 0 categories. FIRM will infer from this information that you want a CONFIRM analysis, and will not look for category names.

8.5.4 Predictor Variable Specification

Number of predictors

In the first line of the predictor variable specification section, you need a line specifying how many predictors there are. In the auto repair example, the effects of *MAKE*, *CYL*, *PRICE* and *YEAR* are to be studied, while the effects of *COUNTRY*, *TANK* and *DISPL* are not of interest. In this case, the number of predictors would be specified as 4.

This is followed by the information on the predictors. The information on each predictor must start on a new line. If you want to, you can split the information on a predictor across several lines. The following are examples of predictor variable specifications. Note that there are nine pieces of information, given in the order discussed below.

```
(i)       'History' 12 0 'r' 0 0 ' ' 0.9 1.0
(ii)      'Eye pupils' 7 0 'f' 0 3 '?12' 0.9 1.0
(iii)     'Eye ind' 6 0 '1' 0 4 '?pfg' 0.9 1.0
(iv)      'Age' 2 1 'm' 0 4 '0123' 0.9 1.0
(v)       'Year' 15 'm' 1984 5 '01234' 0.95 1.0
```

The predictor's name

Each line should start with the name of the predictor. Up to 50 characters may be used for the name of a variable.

The position of the predictor in the data set

For the auto repair example, the positions of the predictors we want to use will be 2, 3, 6 and 7. These values should follow the name of the predictor.

Whether or not the variable may be used for splitting

Enter a `0` if it may be used for splitting, or a `1` if it is to be carried along for analysis but not used for splitting.

This may look like a rather dumb question, but it is not. Commonly, all predictors are available for splitting, so that whichever predictor gives the most significant split, the data will be split on that predictor. On occasion though, there will be potential predictors whose effects you would like to study but that you do not want to use for splitting. You then include these predictors in the analysis, but tag them as unusable. FIRM will carry out the same analysis for unusable as for usable predictors, showing how the categories of the predictor are grouped and computing the significance of the split on the unusable predictor, but will not use them to make splits.

The type of the predictor

Enter `m` if the predictor is monotonic ((iv) above); `f` if it is free ((ii) above); and `1` (the digit 'one') if it is a floating predictor (see (iii) above), the smallest value of whose scale is used to indicate that the predictor is missing on that case. The code `r` is used for a real predictor (see (i) above) and the code `c` for a character predictor.

In the auto repair example, *MAKE* and *YEAR* would be specified as free or character and *CYL* as monotonic. *PRICE* would be specified as real (if it was measured in dollars), or monotonic (if it had been coded into ranges already) or floating (if it had been coded and some cars had missing price information).

The monotonic, free and float predictor types all require that the predictor's values be consecutive integers. The next two fields should contain information on this range, and are only required for these three predictor types.

The smallest value a monotonic, float or free predictor has in the data

The first field in the range is the smallest value of the predictor. Most commonly, the values will run from 1 up, or from 0 up, but any other smallest value can be handled. For example, if a data set contained *YEAR* with values going from 1980 to 1994, then you would give 1980 as the smallest value, and the data from the consecutive years would be the consecutive integers required for these predictor types.

The range of a monotonic, float or free predictors

The range of monotonic, float and free predictors is then specified by listing the number of categories. A predictor taking values 0, 1, 2, 3 for example has 4 categories, so the response for this question is `4`.

Predictor category symbol list

The category symbol list is only applicable to monotonic, free and float predictors. This option calls for a one-character label to be associated with each category of a predictor. For example, if a predictor has categories corresponded to missing, poor, fair, good and excellent, then it will make reading the output easier if you label these categories '?', 'p', 'f', 'g', and 'e' respectively. List the symbols with no intervening spaces—so type in '?pfge'. For real and character predictors, FIRM will specify its own symbols. For a real predictor, '?' is used for "missing or invalid" and '0,1,2,3,4,5,6,7,8,9' used for the remaining 10 categories. For a character predictor, FIRM will use the first character of each value as its symbol provided these are all different. If they are not all different, then the symbols used are a; b; c; d; e; ...

Percentage significance for splitting and percentage significance for merging

FIRM operates by carrying out sequences of two-sample tests to group the categories of each predictor. This field provides information on the significance levels to be used for the test to split a composite category and to merge a pair of (simple or composite) categories respectively.

Normally, the split significance level should be a smaller value than the merge; if it is not, then the splitting test of merged categories will be bypassed. If the split significance level is set higher than the merge significance level, then only merging (no splitting) will be carried out. This can lead to some savings in execution time in CATFIRM for free predictors with many categories, but is not otherwise useful.

The significance levels are given in percent. For example, if you wish to test at the 0.9% significance level for splitting and 1% for merging, then enter the values `0.9 1.` Usually, you will make these significances quite small—like 1%. If the significance levels are set to smaller values, then there will tend to be fewer final categories on any predictor since a more stringent test will be applied for keeping categories separate.

For real and character predictors, you do not need to put in the actual values for items 3, 6 or 7 as the FIRM codes will figure these out for themselves. You DO, however, need to put "placeholders" in—for example a zero for 3 and 6, and " " for 7 (see example (i) above).

8.5.5 Output Options Specification

The final input in the syntax file deals with output options. All of these have defaults, which (for numeric queries) are obtained by entering the value 0 in the field concerned. The dependent variable and predictor variable sections of the syntax file are almost identical for CATFIRM and CONFIRM, but the output options are different. These options for CATFIRM and CONFIRM are discussed separately.

Examples of run option specifications are:

CONFIRM:

```
Option no: 1  2    3     4  5    6 7 8 9 10   11 12 13 14 15
---------------------------------------------------------
           1 20 0.001 .5 1.0 50 0 0 1 0.0  1  0  0  0  0
```

CATFIRM:

```
Option no: 1   2    3  4    5  6    7 8 9 10 11 12 13 14 15
---------------------------------------------------------
           3 1000 25 0.1 1.0 50 0.5 0 1  0  0  0  0  0  0
```

CATFIRM output options

An example of a CATFIRM output options line is

```
Option no: 1   2    3  4    5  6    7 8 9 10 11 12 13 14 15
---------------------------------------------------------
           3 1000 25 0.1 1.0 50 0.5 0 1  0  0  0  0  0  0
```

The syntax file should contain fifteen numbers including the nine selected options described below. The last six numbers are 0 0 0 0 0 0 and are not presently used. Note that there are nine pieces of information, given in the order discussed below.

Table format

The value 0, 1, 2, or 3 is used to control the printing of the contingency tables generated in the solution. You may print these in 4 possible formats—as percentages of the column totals; as overall percentages of the grand total; as percentages of the row totals; or as raw counts. CATFIRM gives contingency tables with the dependent variable forming the rows and the predictor the columns.

Thus, the generally most useful (and default) option of `0` gives the table as a percentage frequency breakdown of the dependent variable for all cases in the node, and for the cases broken down by the different categories of the predictor. The last line of each contingency table is the total number of cases at that value of the predictor.

The second option (enter `1`) gives the individual cell frequencies, and the row and column totals, as percentages of the grand total frequency in that table.

The third option (enter `2`) produces a percentage frequency breakdown of the different categories of the predictor for each level of the dependent variable.

The fourth option (enter `3`) gives the actual frequencies in each cell, and the marginal totals, of the contingency table (see the example given above).

Detail output code

Both FIRM programs will, on request, produce details of the analysis of each predictor in each node, showing the successive steps in the splitting and merging that produces the final grouping of each predictor, and giving the test statistics at each stage of the grouping. This output can be very useful, but is voluminous, particularly where many nodes are created, where there are many predictors, and/or where the predictors have many levels. This option allows you to request or to suppress the gory detail. If the detail is requested, it will be put on a separate file—the detailed split file. Whether you request this detailed output or not, FIRM will give an overall summary of each predictor in each node, showing how its categories group and what the final significance of the split is.

The detail output code is a single number reflecting the answers to all three questions about the printed information, and is calculated as follows

Q1: Do you want details of splits not used?
Q2: Do you want cross tabs before grouping?, and
Q3: Do you want cross tabs after grouping?.

Score `1` for `Yes` and `2` for `No`; then compute this option as

$$1000 \times Q1 + Q2 + 2 \times Q3 - 1003.$$

CATFIRM will optionally produce a separate file of the contingency tables associated with each predictor in each node. Like the detailed split output, this file, while often useful, can be very big, and so CATFIRM provides the option of limiting or suppressing it. Two questions are asked about cross tabulations—whether you want to see the cross tabs before grouping and whether you want to see them after grouping. If you select both, then the tables file will contain the full cross tabulation before the grouping of the predictor's categories takes place, as well as again for the finally grouped categories for each predictor in each node.

Minimum size of group to be analyzed

The next field specifies the minimum size a node must have to be considered for further splitting. There are two reasons for using this option. One is that the chi-squared approximation to Pearson's χ^2 statistic deteriorates when frequencies in the contingency table get small so that you can get unremarkable splits with huge χ^2-values.

The other reason is that in practical terms you may not be interested in finding out about the splits possible in nodes containing only a few cases. If no value is set for this option, the default value of 50 is used. Give the actual value. Entering a zero will cause CATFIRM to use its default.

Raw significance level for a split to be made

This option is usually superseded by the conservative significance level for splitting (see below), but if you want to use the raw rather than the conservative significance level, you could set this option to the raw significance level you want to use, and specify a large *p*-value for the conservative splitting significance level option.

Conservative significance level for group splitting

There are two fields relating to splitting—the minimum raw significance and the minimum conservative significance level that a split must attain to be used. When the analysis of a predictor is complete, the result is an R x C contingency table, the C representing the number of composite categories after grouping. Associated with this table is a Pearson χ^2-value. The raw significance level is obtained by entering χ^2 in a χ^2-table with $(R \times 1)(C \times 1)$ degrees of freedom (old FIRM *p*-values) or a non-asymptotic approximation (new FIRM *p*-values). Setting the first of these significance levels to, for example, α means that the raw significance level must attain at least α for the split to be made.

The conservative significance level is a more realistic value that takes into account the effect on χ^2 of the grouping that occurs in analyzing the predictor. Setting the second significance level to α means that the conservative significance must attain at least $\alpha\%$

for the split to be made. Usually, the testing will be driven by the conservative and not the raw significance level, since the conservative *p*-value is always at least as large as the raw; setting the two values equal ensures this. The default for both significance levels is 1%.

The criterion for splitting a node is the conservative *p*-value for the predictor, multiplied by the number of predictors that were active in that node. So, if for example you have a node in which 7 predictors are active and the smallest has a conservative *p*-value of 0.6%, the multiplicity-adjusted *p*-value for splitting will be 4.2%. This is small enough to justify a split if you specified a conservative 5% cutoff for splitting, but not if you specified a conservative 1%.

Larger values of the *p*-value may be used to get a deliberately oversized tree that can then be inspected and pruned. The *p*-value should otherwise be small, particularly in very large samples.

Enter the value of the conservative significance level for a group to be split, expressed as a percentage. Zero will lead to the default. This significance level is applied to the Bonferroni-adjusted conservative *p*-value computed for the splitting variable (the node's conservative *p*-value multiplied by the number of active predictors at that node).

Maximum number of groups to analyze

Give the number. Analysis will terminate when this many nodes have been investigated for splitting. This limit applies to the number of nodes analyzed (not created); FIRM may form more nodes than this, and the excess nodes will be left unanalyzed when the run ends.

Constant to be added to Chi-square

Normally, you will enter a zero, but entering some other value will lead to the heuristic modification sketched below.

Choosing zero gives the standard Pearson χ^2. Entering a nonzero constant A instead, causes CATFIRM to compute, instead of χ^2, a statistic whose summand is

$$\frac{(\text{observed-expected})^2}{\text{expected-A}}.$$

Giving the value zero for A gives the usual Pearson χ^2. Choosing an A value bigger than zero reduces the apparent significance of splits that produce a tiny group on one side—

the conventional Pearson χ^2 overstates the significance of such splits and the adjustment reduces this tendency. This means that CATFIRM will be less inclined to make splits that strip off a handful of cases.

If there is a theoretical justification for this modification to Pearson's χ^2, we are unaware of it, but as an heuristic it does seem to have some value in data sets that were or became sparse since it discourages the formation of groupings in which the expected counts are small. If this option is used, A should be small (for example around 0.5 or 1, see example given above) or there could be serious distortion of the statistical significance of the reported χ^2-values.

Format of the data

Enter 0 for free format and 1 for fixed. If you do specify fixed format, then the option line must be followed by a FORTRAN format specification for reading the data.

Free format is the more common and convenient, and will be the option selected for the majority of FIRM users. Free format can be used whenever the data values in the file are separated from each other by any of the following

- One or more blanks
- A single tab between each pair of values
- A single comma between each pair of values.

While data files are mainly of this type nowadays, some data files (particularly older ones) may contain values that are not separated by any spaces and that need to be read by specifying their format. (These remarks assume that you know how to set up the required format command—this is probably true of most people who have fixed format data.)

If you said that the data set has to be read in fixed format, a FORTRAN format specification should follow. CATFIRM's data have to be read with INTEGER specifications. If it is necessary to use fixed format and the data file contains variables that are neither the dependent variable nor predictors to be analyzed, then it will be slightly more efficient computationally to give a format specification that skips over these values. When this is done, the "position in the data" needed for both dependent and predictor variables refers to the position amongst those variables actually read.

If you have any real or character predictors, then you cannot use fixed format but must use free format, and so need to have the data values separated by spaces, commas or tabs. If you have a data file containing unseparated values with character or real predictors, then you will have to replace it with either a file in which the values are separated by blanks, tabs or commas, or one in which the real and character predictors have been recoded into consecutive integers. You will need to use some other software to make either of these conversions.

Old or new *p*-value technology

The user has to specify whether the new or old *p-value* technology should be used. Specify a `1` for FIRM 2.1 methodology, and a `0` for FIRM 2.0 methodology.

FIRM 2.1 introduced two changes to the way *p*-values are calculated and made them the default (though you can still select the pre-FIRM 2.1 methods by answering `n` to this question).

- One change relates to the "multiple comparison" *p*-value reported for a predictor. Unless the predictor is collapsed all the way down to a single category, the multiple comparison *p*-value used will be that of the predictor before the grouping. (The pre-FIRM 2.1 *p*-value takes the χ^2-value from the collapsed table but enters it using degrees of freedom from the original uncollapsed table).
- The second change is that, instead of using the χ^2-distribution to get *p*-values for the Pearson χ^2-statistics, FIRM will use a non-asymptotic approach due to Mielke and Berry (1985). This may give somewhat more reliable *p*-values in tables with small frequencies than does the conventional asymptotic χ^2. The last stage of Bonferroni correction—that of multiplying the conservative *p*-value of a node by the number of active predictors—dates from FIRM 2.2, and is also bypassed by selecting the "old *p*-value" option.

Finally, these values should be followed by 6 zeros, for a total of 15 output options. The additional 6 zeros are not currently used by CATFIRM, but could be used in future versions for additional output options.

This completes the syntax specification for a CATFIRM run. The analysis is initiated by using *the **Run PRELIS*** icon button.

CONFIRM Output Options

The dependent variable and predictor input sections for a CONFIRM run are exactly as described for CATFIRM. FIRM figures out whether you are planning to use CONFIRM or CATFIRM on the basis of whether your dependent variable has zero categories (implying CONFIRM), a positive value (implying CATFIRM) or a negative value (implying CATFIRM with a character dependent variable). The predictor options are also identical. CONFIRM uses many of the same output options as CATFIRM, but some work a bit differently or come in a different order as described below. An example of a CONFIRM output options line is

```
Option no: 1  2    3     4   5    6 7 8 9 10   11 12 13 14 15
-------------------------------------------------------------
           1 20 0.001 .5 1.0 50 0 0 1 0.0  1  0  0  0  0
```

CONFIRM currently uses 11 options. As with CATFIRM, the syntax file should include a total of 15 output options, leaving 4 free for additional options that may be added to future FIRM versions. The 11 current options are given below in the order in which they should be entered.

Detailed split file

Like CATFIRM, CONFIRM is able to produce the details of the analysis of each predictor at each node, giving the values of the two-sample *t*-statistics used to decide whether to merge and split the groups. While not as large as the corresponding CATFIRM output, this file can still be large, and users frequently suppress it.

Enter `0` to suppress the detail and `1` to request it.

Minimum number of cases to analyze a group

CONFIRM allows you to set thresholds on both the size and variability a node must have to be considered for splitting. Using these thresholds prevents the study of very small or very homogeneous nodes. This option, in combination with the next, is used to control these thresholds.

Give the number. If, for example, you enter a value of `10`, then no group smaller than 10 cases will be considered for splitting (see example above).

Minimum proportion of SSD to analyze the group

Give a proportion of the sum of squared deviations of the full sample that a group must have to be considered for splitting. By setting this option, you can suppress the analysis of groups that are small and/or homogeneous, even if they are capable of being split in highly significant ways.

The "minimum proportion of SSD" option can be described as follows. Suppose you give a value of `0.001`. Then any node whose sum of squared deviations from the mean fell below 0.001 times that of the original full sample would not be considered for splitting. This is the criterion that was used in the original AID procedure. It is not of much interest today, so you should probably set its value very small (for example 0.0001) unless you know of some good reason not to do so.

Raw significance level to split a group

As with CATFIRM, normally you will leave this at its default setting, letting the following run parameter guide the splitting. As with CATFIRM, you can set this parameter and thereby override normal conservative testing. The default value for this option is 1%. In the example above, a value of 0.5% is specified.

Conservative significance level required to split a group

Set the actual conservative significance level expressed as a percentage. The default value for this option is 1%.

Maximum number of groups to analyze

Set the actual number here. When this many nodes have been investigated for splitting, the analysis terminates. The number of nodes formed may exceed this maximum. If this happens, the excess nodes will not be un-analyzed at the end of the run.

The external degrees of freedom for variance

Normally, CONFIRM uses conventional *t*- and *F*-tests, estimating the background variance from the variance of the cases within nodes.

Sometimes, though, you may have some information, independent of the data, about the true within-group variance of cases. This information will typically take the form of some variance estimate with associated degrees of freedom. For example, if 25 measurements were made of the dependent variable under constant conditions, and it has been found that these 25 measurements have a variance of 3.02, this external information is incorporated if the value 24 (25 - 1) external degrees of freedom should be used. This option is used in conjunction with the external variance estimate (see below). This external information on variance will be included in all the *t*- and *F*-tests CONFIRM does on the data set.

If you do not have any such information, enter `0` for this option.

Is data file in free or fixed format?

Enter `0` for free format and `1` for fixed. If you do specify fixed format, then the option line must be followed by a FORTRAN format specification for reading the data.

Free format is the more common and convenient, and will be the option selected for the majority of FIRM users. Free format can be used whenever the data values in the file are separated from each other by any of the following

- One or more blanks
- A single tab between each pair of values

- A single comma between each pair of values.

While data files are mainly of this type nowadays, some data files (particularly older ones) may contain values that are not separated by any spaces and that need to be read by specifying their format. (These remarks assume that you know how to set up the required format command—this is probably true of most people who have fixed format data.)

If you said that the data set has to be read in fixed format, a FORTRAN format specification should follow. While CATFIRM's data have to be read with INTEGER specifications, CONFIRM (one of whose variables is continuous) requires REAL specifications for all variables—dependent and predictor. If it is necessary to use fixed format and the data file contains variables that are neither the dependent variable nor predictors to be analyzed, then it will be slightly more efficient computationally to give a format specification that skips over these values. When this is done, the "position in the data" needed for both dependent and predictor variables refers to the position amongst those variables actually read.

If you have any real or character predictors, then you cannot use fixed format but must use free format, and so need to have the data values separated by spaces, commas or tabs. If you have a data file containing unseparated values with character or real predictors, then you will have to replace it with either a file in which the values are separated by blanks, tabs or commas, or one in which the real and character predictors have been recoded into consecutive integers. You will need to use some other software to make either of these conversions.

Error variance option

The first stage of the analysis of each predictor in each node is to form the count, mean values, and sum of squared deviations from the mean of the cases with each value of the predictor variable. These numbers form a one-way analysis of variance layout. When subsequently testing pairs of categories for compatibility, there are two natural candidates for the variance term in the denominator of the t-statistic—the pooled variance of just those two groups being tested, and the pooled error variance of the one-way analysis of variance. In this option, the user selects which of these to use.

Neither is uniformly superior to the other—the pooled variance brings mode information to bear on the test (which is good), but if the data contain outliers or heteroscedasticity, then pooling may contaminate the good information for a particular pair of categories with bad information from other categories (which is bad). On balance, the pooled option is the more generally attractive.

Specify `1` for the two-group variance, and `0` for pooled.

External variance estimate

If you have some external estimate of the within-group variance, give this estimate as the tenth option (its degrees of freedom was given as the "external degrees of freedom for variance" option (see earlier in this section). This external information on variance will be included in all the *t*- and *F*-tests CONFIRM does on the data set.

If you do not have any such information, enter 0 for this option.

P-value technology

As of FIRM 2.1, the default multiple comparison *p*-value of a predictor is the *p*-value of the *F* ratio obtained before any collapsing unless the predictor is merged down to a single class. In earlier implementations, the multiple comparison *p*-value was found by testing the explained and residual sums of squares of the final collapsed one-way layout as if they had been computed using the full number of levels of the predictor. There is a second, more obscure change. In the earlier FIRM implementations, the user-supplied *p*-values for splitting and merging were not applied directly, but were modified to allow for the multiple testing in merging the original categories. The current default undoes this modification and uses the user-supplied split and merge *p*-values as they stand in the *t*-tests for splitting and merging categories.

Give a 1 to use FIRM 2.1 methodology, and a 0 to use FIRM 2.0 methodology.

Note:

FIRM makes provision for 15 output options. While the remaining options are not currently used, you should specify 15 values for the output options, setting the excess values to zero.

8.5.6 Optional Output

Summary

The summary statistics of the grouping and the statistical significance of the splits are reported in this section of the output, if requested.

For a detailed discussion of the summary statistics produced by FIRM, please see the examples in Chapter 6.

Dendrogram

The dendrogram shows all splits made; listing which predictor was used for the split, the conservative statistical significance of the split, and which categories of the predictor define which descendant node. Sometimes, this dendrogram is all that is used from the output, though ignoring the summary file does leave a lot of potentially valuable information unexplored.

For a detailed discussion of the dendrograms produced by FIRM, please see the examples in Chapter 6.

Split rule table

This FIRM output file contains a terse history of the splitting process which gave rise to the final dendrogram.

The first part of the split rule table contains one line for each node that is split in the analysis. The first number in the line is the node number. The second is the number of the predictor that was used to split that node. The remaining numbers show to which descendant node the cases with each value of that predictor go, starting from the lowest category of the predictor and going up to the highest. All entries are, and must be, integer fields exactly four columns wide. The format of the file is perhaps best illustrated by an example. Imagine that node 10 were split on the third predictor, and that the range of values of the third predictor was 3, 4, 5, 6, 7 and 8, and that the cases where this predictor was 3, 5 or 7 defined descendant node 18, those with the predictor 4 defined descendant node 18, and those with the predictor 6 or 8 defined descendant node 19. (Clearly this predictor is free). Then the split rule table would have a line

```
10 3 17 18 17 19 17 19
```

This first part of the split rule table is terminated with a line containing the two values `-1 -1`. This line is a flag that the end of the dendrogram has been reached.

On occasion, some categories of a predictor are absent in a node—the category may have been represented in the original full sample but is empty in the current node. FIRM can resolve some such situations by logic—in a monotonic predictor if the empty class's neighbors both classify to the same descendant group then so should the empty class—but logic will not always give a good rule for empty categories. When FIRM has no basis for resolving an empty category, it signals by giving the category the destination node—`1`, a catchall destination node for dead values of a predictor used for splitting. Such cases would be left unpredicted in future use of the model, or would be predicted by the summary statistics of the immediate ancestor node.

Following this section of the split rule table, there may be two more sections: one for any real variables in the run, and another for any character variables. The section for the real variables starts out with the number of real predictors in the data. For example, in the MIXED data set, there were 8 real predictors. There is a record for each of these predictors. Each record starts with the position of the predictor in the syntax file—for example the code `4` would indicate that the predictor was the fourth predictor listed in the syntax file. The second entry is a `0` if there were any cases with missing information, and a 1 if there were none, and the third entry is the number of categories used (this currently defaults to 10). Following this is a list of the cutpoints between the categories. There should be one fewer cutpoint than there are categories. The final entry for each real predictor is the category symbol list, aligned in the first columns of its line in the file. By default, these category symbols are set to ?0123456789.

Following this is a corresponding section for any character variables. The first item is the number of character predictors. Then for each character predictor, the split rule file lists its position in the syntax file, a 0 if there are any cases with missing information on the predictor, and a 1 otherwise, then the number of the distinct nodes seen in the data. The distinct codes are listed four to a line, exactly 20 columns per code. The final entry for the predictor is the category symbol list.

If the dependent variable is of type character in a CATFIRM run, it too will have an entry in the character variable list. This precedes the details for the character predictors and differs from them only in that there is no list of category symbols. By convention, the dependent variable is assigned a predictor number in the syntax file that is one higher than the number of predictors.

Split file

This output file from FIRM contains the full details of the test statistics used in the reduction of each predictor at each node to its final grouping. It lists the values of the test statistics (Pearson χ^2 for CATFIRM, Student's t for CONFIRM) used in deciding how to merge and split the different categories of each predictor at each node. This output makes up a large file in most problems, and so while you may quite routinely produce it in case you may want to look at it later, you probably will not use much of it very often. Its main value is to help you understand the route by which a particular predictor arrived at its final grouping—you may wonder, for example, why two categories you expected to differ were merged, and you may find the answer in a small sample size in one or the other of the categories, or in unexpectedly similar means.

For a detailed discussion of the split file produced by FIRM, please see the examples in Chapter 6.

Tables file

This file contains the cross tabulations of the dependent variable against each predictor at each node that you may request or suppress using a run time option. Depending on the detail option selected, the tables are shown before the grouping, after the grouping, or both.

8.5.7 Versions and Computer Resources Needed

In addition to main storage, the programs require scratch space on disk. In a problem with N cases and M predictors, CATFIRM requires disk space for a scratch file of MN bytes, while CONFIRM requires $3N(M + 12)$ bytes.

Most of the execution time on typical data sets is consumed by data manipulation—reading the data in from disk and creating the initial working files. This is an order MN activity. The smaller remainder of time is taken up by the reduction of categories, which is an order M activity.

Apart from that implied by available disk space, there is no limit on the number of cases. There are, however, limits on the number of predictors, and on the number of categories the variables may have. At the time of writing, the PC versions have a limit of 1000 predictors and 20 variable categories in CONFIRM, 1000 predictors and 16 categories in CATFIRM.

9 References

Agresti, A. (1990). *Categorical Data Analysis*. Wiley: New York.

Aish, A.M., and Jöreskog, K.G. (1990). A panel model for political efficacy and responsiveness: An application of LISREL 7 with weighted least squares. *Quality and Quantity*, **24**, 405-426.

Akaike, H. (1987). Factor analysis and AIC. *Psychometrika*, **52**, 317-332.

Anderson, T.W., and Rubin, H. (1956). *Statistical inference in factor analysis*. In: Proceedings of the Third Berkeley Symposium, Volume V. University of California Press: Berkeley.

Biblarz, T.J., and Raftery, A.E. (1993). The effects of Family Disruption on Social Mobility. *American Sociological Review*, 97-109.

Bollen, K.A. (1995). Structural equation models that are nonlinear in latent variables: A least squares estimator. In: P.M. Marsden (Ed.) *Sociological Methodology*. Blackwell: Cambridge, MA.

Bollen, K.A. (1996). An alternative two stage least squares (2SLS) estimator for latent variable equations. *Psychometrika*, **61**, 109-121.

Bozdogan, H. (1987). Model selection and Akaike's information criteria (AIC). *Psychometrika*, **52**, 345-370.

Breiman, L., Friedman, J.H., Olshen, R.A., and Stone, C.J. (1984). *Classification and Regression Trees*. Wadsworth: Belmont.

Browne, M.W., and Cudeck, R. (1993). Alternative ways of assessing model fit. In K.A. Bollen and J.S. Long (Eds.) *Testing Structural Equation Models*. Sage: Newbury Park, CA.

Browne, M.W., and du Toit, S.H.C. (1992). Automated fitting of nonstandard models for mean vectors and covariance matrices. *Multivariate Behavioral Research*, **27**(2), 269-300.

Bryk, A.S., and Raudenbush, S.W. (1992). *Hierarchical Linear Models*. Sage: Newbury Park, CA.

Cook, R.D., and Weissberg, S. (1994). *An Introduction to Regression Graphics*. Wiley: New York.

Crowder, M.J., and Hand, D.J. (1990). *Analysis of Repeated Measures*. Monographs on Statistics and Applied Probability 41, Chapman and Hall/CRC.

Cudeck, R. and du Toit, S.H.C. (2001). Nonlinear multilevel models for repeated measures data. (in press). In: Duan, N. and Reise, S.P. (Eds.) *Multilevel modeling: Methodological advances, issues and applications*. Lawrence Erlbaum Associates, Publishers: Mahwah, NJ.

Du Toit, M. (1995). *The Analysis of Hierarchical and Unbalanced Complex Survey Data using Multilevel Models*. Unpublished Ph.D. dissertation, University of Pretoria: Pretoria.

Du Toit, S.H.C. (1979). *The analysis of growth curves*. Unpublished Ph.D. dissertation, University of South Africa: Pretoria.

Du Toit, S.H.C., Steyn, A.G.W. and Stumpf, R.H. (1986). *Graphical Exploratory Data Analysis*. Springer-Verlag: New York.

French, J.V. (1951). *The description of aptitude and achievement tests in terms of rotated factors*. Psychometric Monographs, **5**.

Gallant, A.R. (1987). *Nonlinear Statistical Models*. Wiley: New York.

Goldberger, A.S. (1964). *Econometric theory*. Wiley: New York.

Goldstein, H., and McDonald, R.P. (1988). A general model for the analysis of multilevel data. *Psychometrika*, **53**, 455-467.

Guilford, J.P. (1956). The structure of intellect. *Psychological Bulletin*, **53**, 267-293.

Hawkins, D.M., and Kass, G.V. (1982). Automatic Interaction Detection. In: Hawkins, D.M. (Ed.) *Topics in Applied Multivariate Analysis*, Cambridge University Press: Cambridge.

Hawkins, D.M., and McKenzie, D.P. (1995). *A data-based comparison of some recursive partitioning procedures*. In: Proceedings, Statistical Computing Section, American Statistical Association, 245-252.

Hawkins, D.M., Young, S.S., and Rusinko, A. (1997). Analysis of a large structure-activity data set using recursive partitioning. To appear in *Quantitative Structure Activity Relationships*.

Herbst, A. (1993). *The Statistical Modelling of Growth.* Unpublished Ph.D. dissertation, University of Pretoria: Pretoria.

Heymann, C. (1981). *XAID—an Extended Automatic Interaction Detector.* Internal Report SWISK 28, Council for Scientific and Industrial Research: Pretoria.

Holzinger, K., and Swineford, F. (1939). *A study in factor analysis: The stability of a bifactor solution.* Supplementary Educational Monograph no. 48. University of Chicago Press: Chicago.

Hooton, T.M., Haley, R.W., Culver, D.H., White, J.W., Morgan, W.M., and Carroll, R.J. (1981). The joint associations of multiple risk factors with the occurrence of nosocomial infections. *American Journal of Medicine*, **70**, 960-970.

Hox, J.J. (1993). Factor analysis of multilevel data: Gauging the Muthén model. In: Oud, J.H.L. and van Blokland-Vogelesang, R.A.W. (Eds.). *Advances in longitudinal and multivariate analysis in the behavioral sciences.* Nijmegen, NL, 141-156.

Jöreskog, K.G. (1979). Basic ideas of factor and component analysis. In: Jöreskog. K.G. and Sörbom, D. (Eds.): *Advances in factor analysis and structural equation models.* Abt Books: Cambridge, MA., 5-20.

Jöreskog, K.G. (1990). New developments in LISREL: Analysis of ordinal variables using polychoric correlations and weighted least squares. *Quality and Quantity*, **24**, 387-404.

Jöreskog, K.G. (1994). On the estimation of polychoric correlations and their asymptotic covariance matrix. *Psychometrika*, **59**, 381-389.

Jöreskog, K.G., and Sörbom, D. (1989). *LISREL 7-A guide to the program and applications.* SPSS Publications: Chicago.

Jöreskog, K.G., and Sörbom, D. (1993a). *New features in PRELIS 2.* Scientific Software International: Chicago.

Jöreskog, K.G., and Sörbom, D. (1993b). *New features in LISREL 8.* Scientific Software International: Chicago.

Jöreskog, K.G., and Sörbom, D. (1996a). *PRELIS 2: User's Reference Guide.* Scientific Software International: Chicago.

Jöreskog, K.G., and Sörbom, D. (1996b). *LISREL 8: User's Reference Guide.* Scientific Software International: Chicago.

Jöreskog, K.G., and Sörbom, D. (1996c). *LISREL 8: Structural Equation Modeling with the SIMPLIS Command Language*. Scientific Software International: Chicago.

Jöreskog, K.G., and Yang, F. (1996). Nonlinear structural equation models: The Kenny-Judd model with interaction effects. In: Marcoulides, G.A. and Schumacker, R.E. (Eds.) *Advanced structural equation modeling: Issues and techniques*. Lawrence Erlbaum Associates, Publishers: Mahwah, NJ.

Kanfer, R., and Ackerman, P.L. (1989). Motivation and cognitive abilities: an integrative/aptitude-treatment interaction approach to skill acquisition. *Journal of Applied Psychology*, Monograph, **74**, 657-690.

Kass, G.V. (1980). An exploratory technique for investigating large quantities of categorical data. *Applied Statistics*, **29**, 119-127.

Kass, G.V. (1975). *Significance testing in, and an extension to Automatic Interaction Detection*. Ph.D. Thesis, University of the Witwatersrand: Johannesburg.

Klein, L.R. (1950). *Economic fluctuations in the United States 1921-1941*. Cowles Commission Monograph No. 11.

Koops, W.J. (1988). Multiphasic Analysis of Growth Curves in Chickens. *Poultry Science*, **67**, 33-42.

Kreft, I., and de Leeuw, J. (1988). *Introducing multilevel modeling*. Sage: Newbury Park, CA.

Loh, W-Y., and Vanichsetakul, N. (1988). Tree structured classification via generalized discriminant analysis. *Journal of the American Statistical Association*, **83**, 715-725.

Longford, N.T. (1987). A fast scoring algorithm for maximum likelihood estimation in unbalanced mixed models with nested effects. *Biometrika*, **741(4)**, 817-827.

McDonald, R.P. (1993). A general model for two-level data with responses missing at random. *Psychometrika*, **58(4)**, 575-585.

McDonald, R.P., and Goldstein, H. (1989). Balanced versus unbalanced designs for linear structural relations in two-level data. *Br. J. Math. Stat. Psych.*, **42**, 215-232.

McLachlan, G.J. (1992). *Discriminant Analysis and Statistical Pattern Recognition*. Wiley: New York.

Magidson, J. (1977). Toward a causal model approach for adjusting for pre-existing differences in the non-equivalent control group situation. *Evaluation Quarterly*, **1**, 399-420.

Mardia, K.V., Kent, J.T., and Bibby, J.M. (1980). *Multivariate analysis*. Academic Press: New York.

Mielke, P.W., and Berry, K.J. (1985). Non-asymptotic inferences based on the chi-square statistic for $r \times c$ contingency tables. *Journal of Statistical Planning and Inference*, **12**, 41-45.

Morgan, J. A., and Sonquist, J. N. (1963). Problems in the analysis of survey data and a proposal. *Journal of the American Statistical Association*, **58**, 415-434.

Mortimore, P., Sammons, P., Stoll, L., Lewis, D., and Ecob, R. (1988). *School Matters, the Junior Years*. Wells, Open Books.

Muthén, B. (1990). *Means and Covariance Structure Analysis of Hierarchical Data.* UCLA Statistics series, no 62: Los Angeles.

Muthén, B. (1991). Multilevel Factor Analysis of Class and Student Achievement Components. *Journal of Educational Measurement.* **28**, 338-354.

Pinheiro, J.C., and Bates, D.M. (2000). *Mixed-Effects Models in S and S-Plus*. Springer-Verlag: New York.

Quinn-Curtis, Inc. (1998). *Charting Tools for Windows*, WIN-BMC-100: Needham, MA.

Rasbash, J. (1993). *ML3E Version 2.3 Manual Supplement*. University of London: Institute of Education.

Reyment, R., and Jöreskog, K.G. (1993). *Applied factor analysis in the natural sciences.* Cambridge University Press: Cambridge.

Richards, F.J. (1959). A flexible growth function for empirical use. *Journal of Experimental Botany*, **10**, 29(290-300).

Ryan, B.F., and Joiner, B.L. (1994). *Minitab Handbook* (third Edition). Wadsworth: Belmont, CA.

Schafer, J.L. (1997). *Analysis of Incomplete Multivariate Data.* Monographs on Statistics and Applied Probability 72, Chapman and Hall/CRC.

Schumacker, R.E., and Marcoulides, G.A. (Eds.) (1998). *Interaction and nonlinear effects in structural equation models.* Lawrence Erlbaum Associates, Publishers: Mahwah, N.J.

Sörbom, D. (1981). Structural equation models with structured means. In: Jöreskog. K.G. and Wold, H. (Eds.) *Systems under indirect observation: Causality, structure and prediction*. North-Holland Publishing Co: Amsterdam.

Steiger, J.H. (1990). Structural model evaluation and modification: An interval estimation approach. *Multivariate Behavioral Research*, **25**, 173-180.

Thurstone, L.L. (1938). *Primary mental abilities*. Psychometric Monographs, **1**.

Titterington, D.M., Murray, G.D., Murray, L.S., Spiegelhalter, D.J., Skene, A.M., Habbema, J.D.F., and Gelpke, G.J. (1981). Comparison of discrimination techniques applied to a complex data set of head injured patients. *Journal of the Royal Statistical Society*, **A**, 144, 145-161.

Tukey, J.W. (1977). *Exploratory Data Analysis*. Addison-Wesley Publishing Company: Reading, MA.

Wheaton, B., Muthén, B., Alwin, D., and Summers, G. (1977). Assessing reliability and stability in panel models. In: Heise, D.R. (Ed.) *Sociological Methodology 1977*. Jossey-Bass: San Francisco.

Young, S.S., and Hawkins, D.M. (1995). Analysis of a 2^9 full factorial chemical library. *Journal of Medicinal Chemistry*, **38**, 2784-2788.

Subject Index

C

D

E

G

H

I

K

L

M

O

P

Q

R

T